D0622716

Work Accommodation and Retention in Mental Health

Izabela Z. Schultz · E. Sally Rogers
Editors

Work Accommodation and Retention in Mental Health

 Springer

Editors
Izabela Z. Schultz
University of British Columbia
Vancouver, BC
Canada
ischultz@telus.net

E. Sally Rogers
Boston University
Boston, MA
USA
erogers@bu.edu

ISBN 978-1-4419-0427-0   e-ISBN 978-1-4419-0428-7
DOI 10.1007/978-1-4419-0428-7
Springer New York Dordrecht Heidelberg London

Library of Congress Control Number: 2010937649

© Springer Science+Business Media, LLC 2011
All rights reserved. This work may not be translated or copied in whole or in part without the written permission of the publisher (Springer Science+Business Media, LLC, 233 Spring Street, New York, NY 10013, USA), except for brief excerpts in connection with reviews or scholarly analysis. Use in connection with any form of information storage and retrieval, electronic adaptation, computer software, or by similar or dissimilar methodology now known or hereafter developed is forbidden.
The use in this publication of trade names, trademarks, service marks, and similar terms, even if they are not identified as such, is not to be taken as an expression of opinion as to whether or not they are subject to proprietary rights.

Printed on acid-free paper

Springer is part of Springer Science+Business Media (www.springer.com)

Preface

Mental health disabilities are burgeoning and are repeatedly implicated when describing the challenges our society faces in global health and the societal burden of disability, now and in the future. Central to these issues is the work impairment associated with mental health problems. Research suggests that the numbers of individuals with mental health disabilities both in and out of the workforce are large and growing. Relative to physical disabilities, mental health disabilities have been the focus of vocational and occupational interventions and related research much more recently. Therefore our understanding of the most effective ways to intervene to improve work outcomes among individuals with mental health disabilities is more tentative, more fragmented, and less evidence-informed than in the physical rehabilitation world.

Mental health disabilities are repeatedly cited as reasons not only for unemployment but also for underemployment, lost productivity in the workplace (presenteeism and absenteeism), reduced quality of life, exploding medical costs, and public and private insurance programs. Historically, those involved in the provision of rehabilitation services have relied on the medical model of care for mental health disabilities with a correspondingly heavy emphasis on establishing a diagnosis and treating psychiatric symptoms. Such approaches to care may result in a stabilization of mental health symptoms, but have not been demonstrated to be sufficient for improving work function. Recent research has enhanced our understanding of the relationship between the functional impairments that individuals with mental health disabilities face in various life domains, in particular the role of worker. To intervene effectively and improve work outcomes, we must go beyond the medical model, with its reliance on diagnosis, symptoms, and stabilization, to incorporate an emphasis on work function, work environment, psychosocial factors, and employment outcomes. Our laws, norms, and cultural values also direct us to strive for full inclusion of individuals with all disabilities, including mental health disabilities, in the workplace. Not only is such inclusion mandated by existing laws and societal expectations but it also makes sense economically, from the employers', compensation systems', health care, and societal perspectives.

While our understanding of work outcomes for individuals with mental health disabilities is growing, there is, at the same time, no overarching, multisystem framework for conceptualizing and implementing interventions that promote the engagement and retention of individuals with mental health disabilities in the workforce. Over the past few decades, there has been a proliferation of model development, research, and evaluation of vocational rehabilitation approaches for mental health disabilities. These efforts have moved the field to a somewhat more integrated and functional approach, but have not been guided by a conceptual framework. An integrated, biopsychosocial approach to vocational rehabilitation in mental health that combines clinical and occupational interventions is needed to address the complexities of the problems faced by persons with mental health disabilities and the equally complex work environments in which they are expected to function.

Despite the growing knowledge generated by recent studies, few research findings have been integrated and translated into policy, clinical or vocational rehabilitation practice, or new paradigms of service delivery. There is no accepted framework for assessing the work function of individuals with mental health disabilities, for prevention or early intervention of those working who become disabled, or for return to work services. Vocational and clinical rehabilitation professionals, employers, case and disability managers, and those involved with the work life of individuals with mental health disabilities need information on effective, evidence-based, integrated clinical and vocational interventions, as well as disability management and prevention approaches specifically tailored for individuals with mental health disabilities. Such information is needed because, unlike many medical conditions, mental health disabilities affect multiple areas of functioning in complex ways, including the domains of social, interpersonal, cognitive, and family, in addition to the vocational domain. Further, studies repeatedly demonstrate that individuals with mental health disabilities fare worse than individuals with other disabilities when they do receive vocational rehabilitation interventions. Such complexities in terms of the disability itself, the effectiveness of existing vocational rehabilitation approaches, and the ecology of the workplace call for an integrated, multisystem approach to maximize positive work outcomes. In this book, we attempt to systematically integrate existing knowledge, frameworks, and interventions that can assist clinical and vocational rehabilitation and related professionals to provide best or evidence-informed interventions for a range of mental health disabilities.

Complicating the complexities of mental health disability are the negative attitudes that prevail in society at large and in the workplace. Studies suggest that stigma remains a significant factor in the rehabilitation process and in the promotion of full inclusion of individuals with mental health disabilities in the workplace. Social stigma associated with mental disorders may be more disabling in the workplace than a primary mental health condition.

Further, the role of social and support factors in the maintenance of work disability is frequently referenced but still poorly understood. Many individuals with significant work disabilities can and do perform well in the world of work, if they are provided carefully constructed work accommodations that take into account the social, organizational, and interpersonal issues in developing and implementing such supports. For individuals in the workplace who acquire a mental health disability, there are few systematic or legally defensible ways of identifying those at risk for significant problems and intervening with them early, before chronic disability and unemployment occur. This lack of a coordinated approach continues to be the norm, despite consistently promising outcome data on the use of an interdisciplinary, clinically, and vocationally integrated approach to services. The aim of this interdisciplinary book is to bridge this gap, and to discuss new developments in approaches to the clinical and vocational rehabilitation of individuals with mental health disabilities.

When we refer to individuals with mental health disabilities and work accommodations, what do we mean and what is the scope of our definitions? Throughout the book, authors of the chapters use the term "people with mental health disabilities" to refer to individuals with significant mental health problems, psychiatric disabilities or conditions, acquired and traumatic brain injuries, or cognitive impairments. All of these terms are meant to describe individuals who have significant impairments due to mental health conditions (e.g., anxiety and mood disorders, severe mental illness, such as schizophrenia and bipolar disorder, and brain trauma) that interfere with functioning and employment and, in particular, where reasonable accommodation or protection under the Americans with Disabilities Act or other legislation may be indicated. We intend this book to cover an array of mental health impairments broadly and, whenever possible, from a cross-diagnostic perspective, thus bridging the traditional separation between emotional and neuropsychological conditions that are often associated with different rehabilitation models and approaches. We also intend for the reader to obtain a broad view of accommodations in the workplace. We describe many interventions that go beyond the legal definition of reasonable accommodations (requiring disclosure of disability) to a wide array of system- and individual-oriented interventions that can assist persons with mental health disabilities to obtain, retain, or advance in employment despite their disability.

The book focuses on the current science related to work disability among individuals with a variety of mental health impairments, including new research and clinical and occupational models. Our book relies on the concepts of "knowledge exchange" that strive to integrate the recent advances in knowledge about the prediction of occupational disability and intervention in mental health. Diversity among persons with mental health disabilities has been recognized from diagnostic, functional, sociodemographic, and psychological perspectives. Clearly, individuals with mental health disabilities vary significantly in their functioning, ranging from those who acquire a moderately disabling condition while at work to individuals who because of a mental health disability have not entered the workforce.

Our book constitutes a state-of-the-art, integrated, research- and evidence-based resource to facilitate the transfer of knowledge and the development of new, effective clinical and occupational practices and policies for individuals with mental health disabilities. It synthesizes and critically reviews current research on mental health disabilities and provides broad epidemiological and economic information with implications for the identification of those at risk, intervention, case management, and disability prevention.

The book focuses on the functional and occupational impact of mental health disabilities and the most effective intervention approaches to help individuals become engaged in and retain employment or return to work. The conceptual and methodological issues and controversies, together with directions for future research and practice, are also highlighted. The book is further complemented by "how to," practice-oriented information, with illustrations of best organizational and rehabilitation practices, solutions, and case scenarios.

The chapters in this multidisciplinary book were written by distinguished researchers and providers of services from the USA, Canada, the UK, and Australia, all recognized experts in the fields of occupational and vocational rehabilitation, medicine, psychology, and neuropsychology, to provide the best "state-of-the-science" information about mental health disabilities. Implications for best and evidence-informed practices in clinical and vocational rehabilitation are drawn from the body of knowledge of each one of the biopsychosocial conditions and from integrative themes cutting across these seemingly disparate conditions.

The intended readers of our book include vocational and occupational providers of all kinds and at all levels, including occupational and rehabilitation physicians, vocational rehabilitation providers, occupational therapists, psychiatrists, psychologists, neuropsychologists, and social workers who assist individuals with mental health disabilities to choose, get, or keep work. Nurses, particularly occupational health nurses and nurse case managers, are also the intended audience, as are occupational therapists, physiotherapists, as well as vocational rehabilitation and disability management professionals. Human resources, labor relations, and management professionals who deal with the growing challenge of mental health disabilities in the workplace, and the management consultants and disability insurance and compensation professionals who work in the disability management and disability determination fields will find this book beneficial as well.

Moreover, the book provides vital background information for policy makers and mental health advocates in healthcare, Social Security, human resources, employment fairness, and disability rehabilitation entitlement across different levels of government, compensation, disability insurance, and healthcare systems. Importantly, we provide the reader with the knowledge of concepts and empirical evidence to guide their practice, as well as, in some chapters, a "toolkit" for serving individuals with mental health disabilities.

The book consists of the following major sections. Part I covers *Conceptual Issues in Job Accommodations in Mental Health*. This section provides a comprehensive overview of the legal, epidemiological, and economic considerations and ramifications in the area of mental health disability. Part II, *Mental Health Disabilities and Work Functioning*, is designed to provide important information about the predictors of work capacity and work outcomes for individuals with mental health disabilities and how diagnostic categories such as anxiety, depression, psychosis, brain injury, and personality disorders affect occupational functioning. Part III, *Employment Interventions for Persons with Mental Health Disabilities*, leads with a critical discussion of disclosure in the workplace, including the why and how of effective disclosure of one's mental health disability in a work context.

The rest of this section is devoted to descriptions of best practice interventions for individuals with anxiety and mood disorders, serious mental illness, mild cognitive disorders, and traumatic brain injuries. We conclude this section with two important approaches that focus on the system-level issues related to improving work outcomes for individuals with mental health disabilities.

Part IV focuses on *Barriers and Facilitators to Job Accommodations in the Workplace*. We examine employer attitudes toward accommodations, barriers, and facilitators in providing accommodations, and the role of stigma in work and job accommodations. In Part V, *Evidence-Informed Practice in Job Accommodation*, we examine the role of social processes, organizational culture, and known best practices in providing accommodations. Finally, in Part VI, *Future Directions*, we tie these preceding chapters together in an integrative article that weaves together what we know and do not yet know about improving employment outcomes for individuals with mental health disabilities. Current and future research, policy, and practice directions emerge from this overview.

It is likely that the reader of this book will find that some of the chapters offer disparate explanations or conflicting perspectives. Given the state of our knowledge about work functioning and vocational rehabilitation for individuals with mental health disabilities, this is to be expected. To some extent, it is a function of the multidisciplinary perspectives of our experts. That is, some experts focus on the legal aspects of disability, some on the biomedical, neuropsychological, or psychological aspects, while others focus on the ecological and workplace aspects of disability and accommodation. Our book attempts to provide best practice information that improves the work outcomes of individuals with mental health disabilities without regard to theoretical underpinnings.

We would like to thank all of the authors of the book chapters for their unique and valuable contributions, and for helping to realize this book on work accommodations and retention for individuals with mental health disabilities. We would also like to acknowledge the support and assistance of many colleagues on both the research and clinical sides of occupational disability through discussion, exchange of information, and suggestions.

In addition, we would like to thank Ms. Alanna Winter and Ms. Alison Stewart, Research Coordinators from the University of British Columbia, for their technical and research contributions to the development of this book. Ms. Mihiko Maru and Ms. Emily Green from the Center for Psychiatric Rehabilitation provided invaluable research and editorial help as well. We also appreciate ongoing support from our publishers at Springer.

Vancouver, BC   Izabela Z. Schultz
Boston, MA   E. Sally Rogers

Contents

Contributors

David Adler, M.D.
Professor of Psychiatry, Department of Psychiatry, Institute for Clinical Research and Health Policy Studies, Tufts Medical Center, 800 Washington Street, Boston, MA, 02111, USA
Dadler@tuftsmedical center.org

Sheila H. Akabas, Ph.D.
Professor of Social Work, Center for Social Policy and Practice in the Workplace, Columbia University School of Social Work, 1255 Amsterdam Avenue, New York, NY, 10027, USA
sa12@columbia.edu

Marjorie L. Baldwin, Ph.D.
Department of Economics, W.P. Carey School of Business,
Arizona State University, PO BOX 873806, Tempe, AZ, 85287-3806, USA
Marjorie.Baldwin@asu.edu

Jane K. Burke-Miller, Ph.D.
Department of Psychiatry, Center on Mental Health Services Research and Policy, University of Illinois at Chicago, 1601 W. Taylor Street MC – 912, Chicago, IL, 60612, USA
jburke@psych.uic.edu

Claudia Center, J.D.
The Legal Aid Society, Employment Law Center, 600 Harrison Street, Suite 120, San Francisco, CA 94107, USA
ccenter@las-elc.org

Hong Chang, Ph.D.
Statistician, Assistant Professor of Medicine Institute for Clinical Research and Health Policy Studies, Tufts Medical Center, 800 Washington Street, Box 345, Boston, MA, 02111, USA
hchang@tuftsmedicalcenter.org

Judith A. Cook, Ph.D.
Professor and Director, Department of Psychiatry, Center on Mental Health
Services Research and Policy, University of Illinois at Chicago, 1601 W. Taylor
Street, MC 912, Chicago, IL, 60612, USA
cook@ripco.com

Carolyn S. Dewa, Ph.D.
Work & Well-being Research & Evaluation Program,
Centre for Addiction and Mental Health, Toronto, ON, Canada M5S 2S1
Carolyn_Dewa@camh.net

Shengli Dong, M.S., M.Ed.
Graduate Student, Department of Counseling and Personnel Services, University
of Maryland, 3214 Benjamin Building, College Park, MD, 20742, USA
yerliang@umd.edu

Danielle Duplassie, Ph.D.
Registered Clinical Counsellor, Director, The Shanti Counselling Centre,
#104 – 3999 Henning Drive, Burnaby, BC, Canada, V5C 6P9

Susan L. Ettner, Ph.D.
Professor, Division of General Internal Medicine and Health Services Research,
David Geffen School of Medicine at UCLA, 911 Broxton Plaza, Room 106,
Box 951736, Los Angeles, CA, 90095-1736, USA
settner@mednet.ucla.edu

Terri Ferguson, M.A.
Graduate Student, Department of Counseling and Personnel Services, University
of Maryland, 3214 Benjamin Building, College Park, MD, 20742, USA
Terri8@umd.edu

Robert T. Fraser, Ph.D.
Harborview Medical Center, Box 359745325 Ninth Avenue, Seattle, WA, 98104, USA
rfraser@u.washington.edu

Karen A. Gallie, Ph.D.
Director/CQU Academic and Past-President, OHS Educator Chapter,
Safety Institute of Australia
abcsolaris@gmail.com

Lauren B. Gates, Ph.D.
Senior Research Scientist and Director, Center for Social Policy and Practice in
the Workplace, Columbia University School of Social Work, 1255 Amsterdam
Avenue, New York, NY, 10027, USA
lbg13@columbia.edu

Annabel Greenhill, M.A.
Project Manager, Institute for Clinical Research and Health Policy Studies, Tufts
Medical Center, 800 Washington Street, Box 345, Boston, MA, 02111, USA
Agreenhill@tuftsmedicalcenter.org

Rebecca Gewurtz, Ph.D.
Assistant Professor, School of Rehabilitation Science Rm. 447, IAHS, McMaster
University, 1280 Main Street West, Hamilton, ON, Canada, L8S 4L8
gewurtz@mcmaster.ca

Thomas J. Guilmette, Ph.D.
Department of Psychology, Providence College, 118 Albertus Magnus Hall, One
Cunningham Square, Providence, RI 02918, USA
tguilmet@providence.edu

Anthony J. Giuliano, Ph.D.
Clinical Instructor, Department of Psychiatry, Beth Israel Deaconess Medical
Center, Harvard Medical School, 330 Brookline Ave. Boston, MA, 02215, USA
ajgiulia@bidmc.harvard.edu

Douglas B. Hanson, M.A.
Registered Clinical Counsellor, Certified Rehabilitation Counsellor
#12-1599 Dufferin Crescent, Nanaimo, BC, Canada, V9S 5L5
douglas.hanson@waterman-associates.com

Henry G. Harder, Ph.D.
Professor, Disability Management & Psychology, University of Northern British
Columbia, 3333 University Way, Prince George, BC, V2N 42N, Canada
harderh@unbc.ca

Jodi Hawley, Ph.D.
Vocational Rehabilitation Consultant, Visions for Rehabilitation, 12593 Ocean
Cliff Drive, White Rock, BC, Canada, V4A 5Z6
jhawley@telus.net

Richard C. Hermann, M.S., M.D.
Scientist, Institute for Clinical Research and Health Policy Studies, Tufts Medical
Center, 800 Washington Street, Boston, MA, 02111, USA
Rhermann@tuftsmedicalcenter.org

Curt Johnson, M.S.
Rehabilitation Counselor, II Harborview Medical Center, 325 Ninth Avenue,
Box 359745, Seattle, WA, 98104, USA

Bonnie Kirsh, Ph.D.
Associate Professor, Department of Occupational Science and Occupational
Therapy, University of Toronto, 500 University Avenue, Toronto, ON, Canada,
M5G 1V7
Graduate Department of Rehabilitation Sciences, University of Toronto, 500
University Avenue, Toronto, ON, Canada, M5G 1V7
bonnie.kirsh@utoronto.ca

Terry Krupa, Ph.D.
Associate Professor, School of Rehabilitation Therapy, Queen's University,
Louise D. Acton Building, Room 200, 31 George Street, Kingston, ON, Canada,
K7L 3N6
terry.krupa@queensu.ca

Debra Lerner, M.S., Ph.D.
Senior Scientist, Institute for Clinical Research and Health Policy Studies, Tufts
Medical Center, 800 Washington Street, Box 345, Boston, MA, 02111, USA
Dlerner@tuftsmedicalcenter.org

Chia Huei Lin, M.Ed.
Graduate Student, Department of Counseling and Personnel Services, University
of Maryland, 3214 Benjamin Building, College Park, MD, 20742, USA
emarslin@umd.edu

Megan Kash MacDonald, M.S.
Research Worker, Institute of Psychiatry, King's College, Box P086,
De Crespigny Park, London, SE5 8AD, UK
megankash@gmail.com

Kim L. MacDonald-Wilson, Sc.D.
Assistant Professor, Department of Counseling and Personnel Services,
University of Maryland, 3214 Benjamin Building, College Park, MD, 20742, USA
kmacdona@umd.edu

Steven C. Marcus, Ph.D.
Research Associate Professor, School of Social Policy and Practice,
University of Pennsylvania, 3701 Locust Walk, Caster Building, Room C16,
Philadelphia, PA, 19104-6214, USA
and
Center for Health Equity Research and Promotion,
Philadelphia VA Medical Center, Philadelphia, PA, USA
marcuss@sp2.upenn.edu

David McDaid, M.Sc.
Research Fellow, Health Policy and Health Economics, European Observator on
Health Systems and Policies, London School of Economics and Political Science,
Houghton Street, London, WC2A 2AE, UK
D.Mcdaid@Ise.ac.uk

Ruth A. Milner, M.Sc.
Senior Consultant, Clinical Research Support, Department of Biostatistics,
BC Children's Hospital, University of British Columbia, 2329 West Mall,
Vancouver, BC, Canada, V6T 1Z4
ramilner@interchange.ubc.ca

Jason Peer, Ph.D.
Postdoctoral Fellow, VA Capitol Healthcare Network Mental Illness Research
Education and Clinical Center, 10 North Greene Street, Suite 6A, Baltimore, MD,
21201, USA
Jason.Peer@va.gov

Katherine Perch, B.A.
Graduate Student, Institute for Clinical Research and Health Policy Studies, Tufts
Medical Center, 800 Washington StreetBox 345, Boston, MA, 02111, USA
Katherine.Perch@tufts.edu

Lisa A. Razzano, Ph.D.
Research Director, Department of Psychiatry, Center on Mental Health Services
Research and Policy, University of Illinois at Chicago, 1601 W. Taylor Street,
MC 912, Chicago, IL, 60612, USA
Razzano@psych.uic.edu

E. Sally Rogers, Sc.D.
Sargent College of Health & Rehabilitation, Center for Psychiatric Rehabilitation,
Boston University, 940 Commonwealth Avenue West, Boston, MA 02215, USA
erogers@bu.edu

William H. Rogers, Ph.D.
Senior Statistician Institute for Clinical Research and Health Policy Studies, Tufts
Medical Center, 800 Washington Street, Box 345, Boston, MA, 02111, USA
Wrogers@tuftsmedicalcenter.org

Zlatka Russinova, Ph.D.
Senior Research Associate and Research Assistant Professor, Center for
Psychiatric Rehabilitation, Boston University, 940 Commonwealth Avenue, West,
Boston, MA, 02215, USA
zlatka@bu.edu

Izabela Z. Schultz, Ph.D.
Department of Educational and Counselling Psychology and Special Education,
University of British Columbia, Scarfe Library 297, 2125 Main Mall, Vancouver,
BC, V6T 1Z4, Canada
ischultz@telus.net

Alison Stewart, M.A.
Research Coordinator, Vocational Rehabilitation Counselling Program, University
of British Columbia, Scarfe Library 297, 2125 Main Mall, Vancouver, BC,
Canada, V6T 1Z4
GF Strong Rehabilitation Centre, 4255 Laurel St, Vancouver, BC, Canada,
V5Z 2G9
alison.stewart@telus.net

David Strand, B.A.
Harborview Medical Center, 325 Ninth Avenue, Box 359745, Seattle, WA, 98104,
USA

Pamela Sherron Targett, M.Ed.
Director of Special Projects, Virginia Commonwealth University, VCU RRTC
1314 W. Main Street, Richmond, VA, 23284, USA
psherron@vcu.edu

Wendy Tenhula, Ph.D.
VA Capitol Health Care Network Mental Illness Research Education and Clinical
Center, 10 North Greene Street, Suite 6A, Baltimore, MD, 21201, USA;
and
Department of Psychiatry, University of Maryland and School of Medicine,
Baltimore, MD, USA
Wendy.Tenhula@va.gov

Pamella Thomas, M.D.
Medical Director of Wellness and Health Promotion, Lockheed Martin
Aeronautics Company, Tufts Medical Center, 86 South Cobb Drive, Marietta,
GA, 30063-0454, USA
pamella.thomas@lmco.com

Jaye Wald, Ph.D.
Department of Psychiatry, University of British Columbia,
2255 Wesbrook Mall, Vancouver,
BC, Canada V6T 2A1
jwald@interchange.ubc.ca

Jian Li Wang
Departments of Psychiatry and of Community Health Sciences, Faculty of
Medicine, University of Calgary, Room 4D69, Teaching, Research & Wellness
Building. 3280 Hospital Dr. NW, Calgary, T2N 4Z6, Canada
jlwang@ucalgary.ca

Paul Wehman, Ph.D.
Professor and Chairman of Department of Physical Medicine and Rehabilitation/
Medical College of Virginia, Joint Appointment in Department of Special
Education and Disability Policy, School of Education and Department of
Rehabilitation Counseling, School of Allied Health Professions, Virginia
Commonwealth University, 1314 W. Main Street, Box 980677, Richmond,
VA, 23284, USA
pwehman@mcvh-vcu.edu

Alanna Winter, M.A.
Research Coordinator, University of British Columbia, Scarfe Library 297, 2125
Main Mall, Vancouver, BC, Canada, V6T 1Z4
alanna.winter@worksafebc.com

Part I
Conceptual Issues in Job Accommodation in Mental Health

Chapter 1
Law and Job Accommodation in Mental Health Disability

Claudia Center

For many years I suffered from a severe and continuous nervous breakdown tending to melancholia – and beyond. During about the third year of this trouble I went, in devout faith and some faint stir of hope, to a noted specialist in nervous diseases, the best known in the country. This wise man put me to bed and applied the rest cure, to which a still-good physique responded so promptly that he concluded there was nothing much the matter with me, and sent me home with solemn advice to "live as domestic a life as far as possible," to "have but two hours' intellectual life a day," and "never to touch pen, brush, or pencil again" as long as I lived. This was in 1887.

I went home and obeyed those directions for some three months, and came so near the borderline of utter mental ruin that I could see over. Then, using the remnants of intelligence that remained, and helped by a wise friend, I cast the noted specialist's advice to the winds and went to work again – work, the normal life of every human being; work, in which is joy and growth and service, without which one is a pauper and a parasite – ultimately recovering some measure of power.

> Charlotte Perkins Gilman, "*Why I Wrote the Yellow Wallpaper,*"
> THE FORERUNNER (Oct. 1913).

Introduction

Approximately 20% of the population – more than 40 million people in the U.S. – live with psychiatric disorders (United States Department of Health and Human Services 1999; National Institute of Mental Health 2003), and face significant barriers to employment, including discrimination, harassment, job loss, and unemployment (President's New Freedom Commission on Mental Health 2003).[1] Among people with severe

---

[1] "According to surveys conducted over the past five decades, employers have expressed more negative attitudes about hiring workers with psychiatric disabilities than any other group. Economists have found unexplained wage gaps that are evidence of discrimination against those with psychiatric disabilities" (President's New Freedom Commission on Mental Health 2003, p. 34).

C. Center (✉)
The Legal Aid Society, Employment Law Center, 600 Harrison Street, Suite 120,
San Francisco, CA 94107, USA
e-mail: ccenter@las-elc.org

I.Z. Schultz and E.S. Rogers (eds.), *Work Accommodation and Retention in Mental Health,*
DOI 10.1007/978-1-4419-0428-7_1, © Springer Science+Business Media, LLC 2011

psychiatric illness, the unemployment rate approaches 90% (New Freedom Commission on Mental Health 2003). Individuals with less severe disabilities – while more likely to be employed than severely disabled persons – still experience a 26% unemployment rate, and are far less likely than persons without disabilities to have a job (National Organization of Disability 2001; New Freedom Commission on Mental Health 2003).[2] At the same time, the vast majority of persons with disabilities want to work (Stodard et al. 1998).[3] In the United States, the loss of productivity that can be attributed to the lack of integration of persons with mental health disabilities into the workplace is $63 billion per year (New Freedom Commission on Mental Health 2003).

Despite recent advances in understanding, discrimination on the basis of mental health conditions is rampant. Individuals who have mental health disabilities or who receive psychiatric care often experience outright discrimination when seeking employment. Even when persons with disabilities manage to obtain employment, they report extraordinary levels of on-the-job discrimination:

> [M]ore than 3 out of 10 employed people with disabilities (36%) say they have encountered some form of discrimination in the workplace due to their disabilities, the most prevalent of which is not being offered a job for which they are qualified. More than half (51%) of those who have experienced discrimination say they have been refused a job due to their disabilities.
>
> Other forms of discrimination include: being denied a workplace accommodation (40%), being given less responsibility than coworkers (32%), being paid less than other workers with similar skills in similar jobs (29%), being refused a job promotion (28%), and being refused a job interview (22%) (Stodard et al. 1998, p. 3).

Additionally, persons with mental health disabilities routinely encounter discriminatory provisions in employer-provided benefits (Mental Health Liaison Group 2000).

While disclosing information about these often hidden conditions may improve work conditions, disclosure also creates the risk of stigma and misunderstandings. Numerous surveys have demonstrated that psychiatric conditions carry great social stigma, and are linked with the fear of unpredictable and violent behavior (New Freedom Commission on Mental Health 2003; Wahl 1998, 2001; Fink and Tasman 1992; Edwards and Allen 1996).[4] An overwhelming 90% of adults surveyed in 1997 agreed that mental illness continued to be stigmatized (Kong 1997). Unsurprisingly, many choose not to reveal their status, pursuing their work lives with the stress of a hidden disability.

Between 1996 and 1998, approximately 1,400 persons with mental health conditions were surveyed. The vast majority of these individuals – nearly 80% – reported overhearing hurtful or offensive comments about mental illness, and seeing offensive portrayals of persons with psychiatric disabilities in the mass media. As a result, 79% worried about being viewed unfavorably by others, 36% reported being often treated

---

[2] People with psychiatric disabilities "earn a median wage of only about $6 per hour versus $9 per hour for the general population" (New Freedom Commission on Mental Health 2003, p. 34).

[3] 67% of unemployed persons with disabilities want to work (NOD, p. 2). 79% of working-age people with disabilities want to work (Stoddard et al. 1998).

[4] "61% of Americans think that people with schizophrenia are likely to be dangerous to others" (New Freedom Commission on Mental Health 2003, p. 20).

as less competent by others, and 26% reported being shunned or avoided because of their disorder (Globe 1997).

The impact in the workplace is particularly severe: 71% of employees avoided disclosures of their mental health histories in the employment setting, choosing instead to hide their disabilities (Wahl 1999). One in three (31%) reported that they had been turned down for a job for which they were qualified after revealing that they were receiving mental health services (Wahl 1999). Fear of disclosing a mental health disability in the workplace is not limited to concerns about job loss; persons with mental conditions fear that disclosure will also cause coworkers not to want to socialize, work, or eat lunch with them (Huff 2000).

In 1999, the United States Surgeon General issued a groundbreaking report on mental health, noting: "Stigmatization of people with mental disorders has persisted throughout history. It is manifested by bias, distrust, stereotyping, fear, embarrassment, anger, and/or avoidance. … It reduces patients' access to resources and opportunities (e.g., housing, jobs) and leads to low self-esteem, isolation, and hopelessness. … In its most overt and egregious form, stigma results in outright discrimination and abuse. More tragically, it deprives people of their dignity and interferes with their full participation in society" (DHHS 1999, p. 6).

At the same time, on-the-job discrimination, segregation, and harassment are significant risk factors for depression and other mental health conditions (World Health Organization 2001;[5] McManamy[6]). Thus, for those who already suffer multiple injustices due to disability, race, gender, sexual orientation, or socioeconomic status, exclusion from the workplace only worsens mental health.

Conversely, successful employment and economic self-sufficiency promote recovery and wellness. In addition to providing a livelihood, employment offers the support and skills people need to become engaged, independent members of the community. More concretely, employment is often the primary avenue to health benefits, including mental health care.

Studies show that modest reasonable accommodations can be implemented successfully for persons with mental health disabilities (Blanck 1996).[7] At the same time, research demonstrates that most persons with mental disabilities – and many employers – are unaware of the rights and remedies available under civil rights statutes (Granger 1996). Granger (1996) surveyed employees and employers regarding

---

[5]Noting that societal discrimination such as racism is associated with psychological distress and depression.

[6]"Women in low-skill, high-demand jobs are more likely to be depressed … with job discrimination, sex discrimination, and sexual harassment possible causes" (McManamy).

[7]See also Peter Blanck, *The Economics of the Employment Provisions of the Americans with Disabilities Act: Part I – Workplace Accommodations*, 46 DePaul L. Rev. 877, 902 (1997). In a series of studies conducted at Sears, Roebuck and Co. from 1978 to 1996, a time period before and after [the ADA's] July 26, 1992 effective date, nearly all of the 500 accommodations sampled required little or no cost. During the years 1993 to 1996, the average direct cost for accommodations was $45, and from 1978 to 1992, the average direct cost was $121. The Sears studies also show that the direct costs of accommodating employees with hidden disabilities (for example, emotional and neurological impairments comprising roughly 15 percent of the cases studied) are even lower than the overall average of $45).

their understanding of the employment provisions of the Americans with Disabilities Act and their application to mental health disabilities, and found that employees and employers lacked sufficient awareness of accommodations enabling individuals with disabilities to obtain and maintain employment. The Institute concluded that education and training regarding non-discrimination and accommodations must be made available to individuals with mental health disabilities, employers, and others (1996).

With vigorous enforcement and widespread education and dissemination of information, persons with mental health disabilities can achieve greater access to employment opportunities. Such an endeavor, however, requires sustained effort and continuing resources.

Disability Nondiscrimination Laws

The Americans with Disabilities Act of 1990 (ADA) envisions equal opportunity and full integration in the workplace for persons with all disabilities, mental and physical. Title I of the act prohibits employers with 15 or more employees from discriminating against qualified workers with disabilities – including mental health disabilities – in regard to all aspects of employment, including job applications, hiring, advancement, firing, promotions, job training, compensation, and any other "terms, conditions, and privileges" of employment (42 U.S.C. §§ 12111(5), 12112(a)).

Importantly, the law requires employers to provide "reasonable accommodations" to employees with disabilities, unless the accommodations impose an "undue hardship" on the employer. Reasonable accommodations are workplace or job modifications that enable an employee with disability to successfully perform the job, or to achieve equal employment opportunity. There are a number of established and well recognized reasonable accommodations for persons with mental health disabilities, including job restructuring, leaves of absence, flexible scheduling, and part-time work.

Federal employers are covered by Section 501 of the Rehabilitation Act of 1973 (29 U.S.C. § 701, *et seq.*), which imposes requirements comparable to those of the ADA. Indian tribes are not covered by any of these laws; however, their contracts with federal or state governments may contain nondiscrimination provisions.

Note on the Definition of "Disability" and the ADA Amendments Act of 2008

What constitutes a disability emerged as one of the most disputed issues in employment law in both the federal and the state courts. Following the enactment of the ADA, federal courts chipped away at the law's protected class by adopting very narrow rules for the analysis of who meets the statutory definition of "disability." In *Sutton v. United*

Airlines and its companion cases (527 U.S. 471 (1999)), the Supreme Court denied protection under the ADA to individuals who are currently functioning well in spite of their disabilities due to mitigating measures (*Sutton*, 527 U.S. at 471;[8] *Murphy v. United Parcel Service*, 527 U.S. 516 (1991);[9] *Albertson's, Inc. v. Kirkingburg*, 527 U.S. 555 (1999)).[10] In *Toyota Motor Mfg., Ky. v. Williams*, the Court construed the statutory terms "strictly to create a demanding standard for qualifying as disabled," and held that to be substantially limited in performing a major life activity under the ADA "an individual must have an impairment that prevents or severely restricts the individual from doing activities that are of central importance to most people's daily lives" (534 U.S. 184 at 197, 198 (2002)). As a result of these and other cases, individuals with a variety of serious impairments, including mental health conditions, have been precluded from having their cases decided on the merits because the courts find at the outset that they do not have a disability protected by the ADA (American Bar Association 1998;[11] Allbright 2003;[12] Parry 2000).[13]

The ADA Amendments Act of 2008 (ADAAA) significantly changed the meaning of disability in an effort to reject these narrow judicial constructions, directing courts to construe disability "in favor of broad coverage ... to maximum extent permitted by the terms of this Act" (42 U.S.C. § 12102(4)(A), as amended). Perhaps the most prominent feature of the ADA Amendments Act is its overturning of the Supreme Court's rule on mitigating measures established in *Sutton*. Disability must now be assessed without considering the ameliorative effects of mitigating measures (42 U.S.C. § 12102(4)(E)(i), as amended). In addition, the Act now explicitly covers impairments that are episodic or in remission, but that would be substantially limiting if active (42 U.S.C. § 12102(4)(D), as amended). The amendments reject the strict and demanding interpretation of the terms "substantially" and "major" set forth by the Supreme Court in *Williams* (42 U.S.C. § 12102(4)(B), as amended; S. 3406 (2008) at Section 2 (Findings and Purposes)), and expand the concept of "major life activities" to include the functioning of bodily systems (42 U.S.C. § 12102(2)(B), as amended). Finally, the Act establishes limited protections for persons who can show that a prohibited action occurred because of an

---

[8] Finding that persons must be considered in the "mitigated" state to determine whether they are "disabled" under the ADA, in context of rejected applicants with severe myopia who sought pilot jobs.

[9] Finding that a person with high blood pressure that was treated with medication was not disabled.

[10] Finding that an individual with monocular vision who was able to subconsciously correct his visual impairment was not disabled.

[11] Reflecting the federal court's increasingly restrictive application of the law, particularly with regard to the definition of disability, a comprehensive survey of 1,200 ADA cases decided by federal appellate courts since 1992 found that employers won over 90% of litigated cases.

[12] Reviewing the 514 case decisions from the year 2000 appearing in the Federal Reporting, and determining that "96.4 percent resulted in employer wins."

[13] Updating the ABA's survey to include all cases decided in 1999 and finding employers winning 95.7% of the time in federal appellate court.

actual or perceived impairment, even where such impairment is not substantially limiting (42 U.S.C. §§ 12102(3)(A), 12201(h), as amended).

No Discrimination or Harassment

Under the ADA, employers may not treat qualified workers with disabilities more poorly than workers without disabilities. To be protected, the employee must be "disabled" as defined by the statute, and must be "qualified" to perform the basic, or essential, functions of the job. Additionally, the disability must be known to the employer and the poor treatment must be significant enough to constitute an adverse employment action (e.g., termination, demotion, discipline, failure to promote, or lower pay for the same work).

The rule against discrimination prohibits disability-based harassment in the workplace (*Flowers v. South Regional Physician Services*, 247 F.3d 229, 233–35 (5th Cir. 2001); *Fox v. Gen. Motors Corp.*, 247 F.3d 169, 176 (4th Cir. 2001); *Haysman v. Food Lion*, 893 F. Supp. 1092, 1107 (S.D. Ga. 1995)). As with sexual harassment, the employee must show that he has been subjected to hostile, offensive, or intimidating comments or conduct that are "sufficiently severe or pervasive" to alter the conditions of the person's employment and to create a hostile working environment (*Flowers*, 247 F.3d at 236; *Fox*, 247 F.3d at 177). While the harasser need not explicitly reference disability, the offensive conduct must be based on disability.

Unlike in the context of sex or race, it is often entirely appropriate and even required for an employer to discuss disability-related matters with an employee who is receiving accommodations, or who has requested accommodation. There may be a stark difference of opinion regarding the nature of an employer's comments or inquiries. They may be viewed as harassment. Alternatively, they may be legitimate statements or questions made as part of the "reasonable accommodation process" or to address the ordinary logistics of reasonable accommodation.

For example, an employer cannot make disability-related inquiries or issue a "warning" to an employee simply because she has heard that he has a mental health impairment. This could be evidence of harassment. However, if the employee has requested an accommodation, but the nature of the accommodation is unclear, or the basis for needing accommodation is not explained, an employer can lawfully seek reasonable medical documentation or schedule a meeting to discuss the accommodation.

No Unnecessary Policies that Screen Out Persons with Disabilities

An employer may not adopt a policy or practice that "screens out or tends to screen out" an individual on the basis of disability, unless the employer can demonstrate that the policy is job-related and consistent with business necessity (42 U.S.C. §

12112(b)(6)). Employers are usually successful in arguing that requiring compliance with established safety rules, such as Department of Transportation (DOT) and Occupational Safety and Health Act (OSHA) standards, is job-related and consistent with business necessity (*Albertson's v. Kirkingburg*, 527 U.S. 555, 567 (1999); *Morton v. United Parcel Service*, 272 F.3d 1249, 1260 (9th Cir. 2001); *Tate v. Farmland Industries, Inc.*, 268 F.3d 989, 993 (10th Cir. 2001); *Chevron v. Echazabal*, 536 U.S. 73, 84 (2002)). By contrast, employer-created policies found to screen out a person with a disability are less likely to be upheld (*Cripe v. City of San Jose*, 261 F.3d 877, 890–94 (9th Cir. 2001); *Belk v. Southwestern Bell Tel. Co.*, 194 F.3d 946, 951–53 (8th Cir. 1999); *Hendricks-Robinson v. Excel Corp.*, 154 F.3d 685 (7th Cir. 1998); *Prewitt v. United States Postal Service*, 662 F.2d 292 (5th Cir. 1981); *Bates v. UPS*, 511 F.3d. 974 (9th Cir. 2007) (en banc)).

Reasonable Accommodation

The ADA requires employers to provide "reasonable accommodations" to applicants and employees with disabilities. A reasonable accommodation is a modification to the job or the workplace that enables the worker with disability to successfully perform the essential functions of the job, or to enjoy equal benefits and privileges of employment (29 C.F.R. § 1630.2(o)(1)).

To require an employer to accommodate, it must be "plausible" that the accommodation will be successful (*Kimbro v. Atlantic Richfield Co.*, 889 F.2d 869, 878 (9th Cir. 1989); *Prewitt*, 662 F.2d at 310). The requested modification must be facially "reasonable" in the run of cases or "reasonable" given the special circumstances of the situation (*U.S. Airways v. Barnett*, 535 U.S. 391, 401–02 (2002)).

An employer is not required to provide accommodation if it can demonstrate that it would impose an "undue hardship," which means a significant difficulty or expense, considering an employer's resources and circumstances (42 U.S.C. §§ 12112(b)(5), 12111(10)). An employer cannot claim undue hardship due to the fears or prejudices of coworkers, or low employee morale about a proposed accommodation (29 C.F.R. App. § 1630.15(d)).

An accommodation should be effective in meeting the needs of the individual (*Barnett*, 535 U.S. at 400). If more than one accommodation would be effective, the employer may choose which one to implement (29 C.F.R. App. § 1630.9; United States Equal Employment Opportunity Commission [EEOC], *Enforcement Guidance on Reasonable Accommodation and Undue Hardship* (1999)).

The duty to accommodate is a continuing one, and is not satisfied by a single attempt if the employee needs additional modification (*Humphrey v. Memorial Hosps. Ass'n*, 239 F.3d 1128, 1138 (9th Cir. 2001); *Kimbro*, 889 F.2d at 869; EEOC, *Reasonable Accommodation and Undue Hardship*). An unreasonable delay in providing accommodation becomes unlawful when it amounts to a denial of accommodation (*Taylor v. Phoenixville Sch. Dist.*, 184 F.3d 296 (3d Cir (1999)); *Smith v. Midland Brake, Inc.*, 180 F.3d 1154, 1173 (10th Cir. 1999); *Krocka v. Riegler*, 958 F. Supp. 1333, 1342 (N.D. Ill. 1997)).

An employee is not required to accept a reasonable accommodation offered by an employer. However, the employer may satisfy its obligations by making such an offer, and an employee who rejects accommodation may no longer be considered "qualified" to perform the job (29 C.F.R. § 1630.9(d)).

Requesting Accommodation

Individuals with mental health disabilities face a difficult decision when it comes to disclosing or not disclosing their condition to an employer. Disclosure may be necessary when an accommodation is needed to enable an employee to perform her job, and the change cannot be obtained without alerting an employer to the situation. It may make for a more supportive work environment, or result in a loss of privacy and the risk of stigma and harassment. It is a personal decision, with both benefits and costs.

Requests for accommodations are not required to be made at the beginning of employment, but can be made at any time (EEOC, *Reasonable Accommodation and Undue Hardship*). However, it may be in the employee's best interest to make the request *before* performance begins to suffer or conduct problems arise. An employer does not have a legal obligation to provide an accommodation if they are not on notice that the employee has a disability and needs a modification (42 U.S.C. § 12112(b)(5)(A); 2 C.C.R. § 7293.9; *see also* 29 C.F.R. § 1630.9).

To request accommodation, the individual with disability does not need to use the word "accommodation," cite particular federal or state laws, or fill out a special form. An employee may use plain English to convey to the employer that some form of modification is needed due to disability. She may email the employer or make an oral request (EEOC, *Enforcement Guidance on Psychiatric Disabilities* (1997); EEOC, *Reasonable Accommodation and Undue Hardship; Barnett v. U.S. Airways*, 228 F.3d 1105, 1112 (9th Cir. 2000)). The employee must provide information specific enough for the employer to understand that the employee may have a mental health disability as defined by law (EEOC, Compliance Manual, Definition of the Term "Disability" (1995); EEOC, *Psychiatric Disabilities*).

Simply disclosing "stress" or an "emotional" problem may be insufficient (EEOC, *Psychiatric Disabilities*). At the same time, revealing a specific diagnosis or detailing every diagnostic feature may not be necessary. In fact, disclosing a diagnosis without more – without referencing or explaining the limitations it causes – may be insufficient, particularly where the need for accommodation is not obvious.

An employee is wise to make or confirm a request for accommodation in writing and retain a copy. A simple email, reiterating a conversation about accommodation, is persuasive evidence. (The employee should print out a copy of the email and keep it at home – should there be a dispute in the future, the employee may not have access to the work computer.) Further, to be safe, an employee seeking accommodation should use words such as "disability," "limiting," "major life activities," and "accommodation."

An employee must disclose to someone who represents the employer, such as a supervisor or human resources person. Medical information received by an employer must be maintained as a confidential medical record, and medical information may only be disclosed to persons with a need to know (42 U.S.C. § 12112(d)(3)(B), (d)(4)(C); EEOC, *Psychiatric Disabilities*; EEOC, *Reasonable Accommodation and Undue Hardship*). An employee is not required to disclose medical information to coworkers (EEOC, *Psychiatric Disabilities*; EEOC, *Reasonable Accommodation and Undue Hardship*).

A person other than the employee may request a reasonable accommodation on behalf of the employee. For example, a family member, friend, health care professional, or other representative may make the request (EEOC, *Psychiatric Disabilities*; EEOC, *Reasonable Accommodation and Undue Hardship*).

In some cases, where it is obvious that some form of accommodation is needed, the employer may be deemed to be on notice of a need for accommodation even without a specific request, either where the employee's disability interferes with communication or where the need for accommodation is obvious (EEOC, *Reasonable Accommodation and Undue Hardship*; 29 C.F.R. § 1630.2(o)(1)(iii), (o)(3); *Snead v. Metro. Prop. & Cas. Ins. Co.*, 237 F.3d 1080, 1087; *Barnett*, 228 F.3d at 1112; *Schmidt v. Safeway Inc.*, 864 F. Supp. 991, 997 (D. Or. 1994)). For example, in a case where an employee has a mental health disability that makes communication difficult, the employer may have a heightened duty to begin and engage in the process (*Taylor v. Phoenixville*, 184 F.3d at 296; *Bultemeyer v. Fort Wayne Cmty. Sch.*, 100 F.3d 1281, 1285 (7th Cir. 1996)). As a practical matter, however, the employee (or the employee's advocate) should affirmatively request an accommodation to make certain that the employer is on notice.

Pros and Cons of Disclosing a Mental Health Condition

The decision to disclose a mental health disability to an employer is an extremely personal one. Employees considering disclosing a mental health disability should weigh both the costs and benefits of the move. These include:

- need for accommodation to perform the job;
- need for accommodation to avoid discipline or termination;
- need for accommodation to protect health or mental health;
- whether the modification may be obtained without disclosing disability;
- risk of stigma and harassment;
- risk of loss of privacy; and
- potential for more successful and supportive employment experience.

Reasonable Medical Documentation

If an employee requests an accommodation and the disability or the need for accommodation is not obvious, the employer may request "reasonable documentation" showing the employee's right to accommodation (EEOC, *Psychiatric*

Disabilities; EEOC, *Reasonable Accommodation and Undue Hardship*). The documentation should be limited to a doctor's note or other medical documents showing that the employee has a disability and needs accommodation (EEOC, *Reasonable Accommodation and Undue Hardship*). Absent unusual circumstances, a request for a complete release of medical records exceeds the employer's legitimate need for information, and is not permitted (EEOC, *Reasonable Accommodation and Undue Hardship*).

A supportive health care provider should be able to draft a simple letter that confirms the existence of a disability and the need for accommodation without revealing intimate or potentially embarrassing psychiatric traits or symptoms.

If an employee provides insufficient information to support the request for reasonable accommodation, the employer may request that the employee be examined by a health professional of the employer's choice (EEOC, *Reasonable Accommodation and Undue Hardship*). Documentation may be insufficient if it does not verify the existence of a "disability" or explain the need for accommodation, the health care professional lacks relevant expertise, or circumstances indicate fraud or lack of credibility (EEOC, *Enforcement Guidance on Inquiries and Examinations of Employees* (2000)). Prior to seeking an exam, the employer should explain why the documentation is insufficient and allow the individual an opportunity to provide the missing information in a timely manner (EEOC, *Inquiries and Examinations of Employees*).

Any such examination or inquiry must be limited to verifying a disability, the need for reasonable accommodation, and related limitations (EEOC, *Reasonable Accommodation and Undue Hardship*). Employers should also consider alternatives like having their health professional consult with the individual's health professional, with the employee's consent (EEOC, *Inquiries and Examinations of Employees*).

Interactive Process

Once an employee requests accommodation, the ADA imposes an obligation upon both sides to engage where necessary in an interactive process to identify and implement an effective accommodation (29 C.F.R. § 1630.2(o)(3); *Humphrey*, 239 F.3d at 1137; *Hansen v. Henderson*, 233 F.3d 521, 523 (7th Cir. 2000); *Barnett*, 228 F.3d at 1111; *Fjellestad v. Pizza Hut of Am.*, 188 F.3d 944, 952 (8th Cir. 1999); *Smith*, 180 F.3d at 1172; *Loulseged v. Akzo Nobel Inc.*, 178 F.3d 731, 735–36 (5th Cir. 1999); *Cehrs v. Northeast Ohio Alzheimer's Research Ctr.*, 155 F.3d 775, 783–784 (6th Cir. 1998); *Criado v. IBM Corp.*, 145 F.3d 437, 444 (1st Cir. 1998); *Mengine v. Runyon*, 114 F.3d 415, 419 (3d Cir. 1997); *Taylor v. Phoenixville*, 184 F.3d at 317; *Hendricks-Robinson*, 154 F.3d at 693; *Bultemeyer*, 100 F.3d at 1285–86; *Feliberty v. Kember Corp.*, 98 F.3d 274, 280 (7th Cir. 1996); *Taylor v. Principal Fin. Group, Inc.*, 93 F.3d 155, 165 (5th Cir. 1996); *Beck v. Univ. of Wis.*, 75 F.3d 1130, 1135–36 (7th Cir. 1996); *Grenier v. Cyanamid Plastics, Inc.*, 70 F.3d 667, 677 (1st Cir. 1995); *see also Moses v. Am. Nonwovens, Inc.*, 97 F.3d 446, 448 (11th Cir. 1996)).

The purpose of the interactive process is to ensure that the employee and employer have a dialogue resulting in effective accommodation – a job modification that enables the employee to perform her essential job functions (*Barnett*, 228 F.3d at 1114.). The process is designed to be informal and flexible so as to respond to the unique needs and abilities of individual employees. One accommodation may not effectively accommodate all employees with the same or similar disabilities.

Where necessary, the interactive process includes the following steps:

- analyzing the particular job involved and determining its purpose and essential functions;
- consulting with the individual with a disability to ascertain the precise job-related limitations imposed by the individual's disability and how those limitations could be overcome with a reasonable accommodation;
- in consultation with the individual to be accommodated, identifying potential accommodations and assessing the effectiveness each would have in enabling the individual to perform the essential functions of the position; and
- considering the preference of the individual to be accommodated and selecting the accommodation that is most appropriate for both the employee and the employer (29 C.F.R. App. §1630.9).

Additionally, in some instances, it may be necessary for the employer to consult qualified experts to gather the information needed to identify an appropriate reasonable accommodation, including the employee's physician, rehabilitation specialists, and others with expert knowledge about dealing with the particular disability (EEOC, *Psychiatric Disabilities Prilliman v. United Air Lines, Inc.*, 53 Cal. App. 4th 935, 950 1997).

The interactive process is a two-way street, and requires both parties to communicate directly, exchange necessary information, and act in good faith (*Barnett*, 228 F.2d at 1114–15; *Bultemeyer*, 100 F.3d at 1285). The employer's participation is critical, as the employee usually possesses incomplete, and often inaccurate, information about the options available, including the existence of appropriate vacancies (*Taylor v. Phoenixville*, 184 F.3d at 316; *Smith*, 180 F.3d at 1173; *Aka. v. Wash. Hosp. Ctr.*, 156 F.3d 1284, 1304 n.27 (D.C. Cir. 1998); *Mengine*, 114 F.3d at 420; *Miller v. Ill. Dep't of Corr.*, 107 F.3d 483, 486–87 (7th Cir. 1997); *Bultemeyer*, 100 F.3d at 1285–86). The employee must also participate, and may be required to submit medical documentation and attend scheduled meetings. Persons with mental health disabilities may want to enlist help from third-party advocates.

An employer that fails to engage in the good faith, interactive process violates the ADA when a reasonable accommodation would have been possible but for the employer's failure (*Humphrey*, 239 F.3d at 1137–38; *Barnett*, 228 F.3d at 1116). Similarly, employees who fail to participate fully in the interactive process, or who unreasonably reject proposed effective accommodations, may lose their rights under the ADA (29 C.F.R. pt. 1630 app. § 1630.9(d) (1997); *Loulseged*, 178 F.3d at 740; *Stewart v. Happy Herman's Cheshire Bridge, Inc.*, 117 F.3d 1278, 1286–87 (11th Cir. 1997); *Beck*, 75 F.3d at 1136).

Accommodations for Persons with Mental Health Disabilities

There is no finite list of the types of workplace changes that can constitute a reasonable accommodation. The nature of an accommodation is limited only by the creativity of the employer and employee (*Higgins v. New Balance Athletic Shoe, Inc.*, 194 F.3d 252 (1st Cir. 1999) (relocating loudspeaker); *Stewart v. Brown County*, 86 F.3d 107, 112 (7th Cir. 1996) (window blinds); *Lyons v. Legal Aid Soc'y*, 68 F.3d 1512, 1517 (2d Cir. 1995) (paid parking); *EEOC v. Newport News Shipbuilding & Drydock Co.*, 949 F. Supp. 403, 408 (E.D. Va. 1996) (new office with separate air conditioner)). However, the form of an effective accommodation for employees with mental disabilities may not be immediately apparent. The EEOC encourages mental health professionals, including psychiatric rehabilitation counselors, to make suggestions about particular accommodations and to help employers and employees communicate effectively about reasonable accommodation (EEOC, *Psychiatric Disabilities; see also Cehrs*, 155 F.3d 775; *Haschmann v. Time Warner Entm't*, 151 F.3d 591 (7th Cir. 1998); *Criado*, 145 F.3d 437; *Rascon v. U.S. West Communications*, 143 F.3d 1324 (10th Cir. 1998); *Ralph v. Lucent Techs., Inc.*, 135 F.3d 166 (1st Cir. 1998)). Ideally, an employer and employee who engage sincerely in the interactive process will arrive at a solution that allows the employee to function effectively in the workplace without imposing an undue hardship on the employer.

For persons with mental health conditions, the accommodation may be a shift in schedule or duties rather than a new computer program or an assistive technology device. Often, the cost to the employer lies not in specific dollars expended, but in the management time spent on increased training or supervision (Blanck 1998).[14]

Leave of Absence

A leave of absence is one of the most common types of accommodations requested by employees with a mental health disability (29 C.F.R. App. § 1630.2(o); EEOC, *Technical Assistance Manual* (1992); *Humphrey*, 239 F.3d 1128, 1135–36; *Nunes v. Wal-Mart Stores*, 164 F.3d 1243, 1247 (9th Cir. 1999); *Cehrs*, 155 F.3d at 782–83; *Haschmann*, 151 F.3d at 601–02; *Criado*, 145 F.3d at 444; *Rascon*, 143 F.3d at 1333–34). As with other accommodations, there must be some basis for believing that the leave will be effective – that it is plausible that the employee will return to work upon recovery (*Humphrey*, 239 F.3d at 1135–36; *Kimbro*, 889 F.2d at 879).

When requesting a leave of absence under the ADA, the employee should include a return to work date, even if it needs to be modified later on. This is because an indefinite leave is more likely to be considered "unreasonable" or to impose an undue hardship

---

[14]The Job Accommodation Network (JAN), a consultation service of the President's Committee on Employment of People with Disabilities, reports that 80 percent of suggested accommodations cost less than $500.

(*Hudson v. MCI Telecomm. Corp.*, 87 F.3d 1167, 1169 (10th Cir. 1996); *Myers v. Hose*, 50 F.3d 278, 283 (4th Cir. 1995)).[15] Courts have tended to look more favorably on employees who have requested a leave of fixed duration that enables them to complete a well-defined course of treatment with reasonable prospects for recovery (*Rascon*, 143 F.3d at 1334). By contrast, a leave that is "erratic" or unexplained may not be required (*Waggoner v. Olin Corp.*, 169 F.3d 481, 484–85 (7th Cir. 1999)).

Additionally, before requesting a reasonable accommodation for a leave of absence under the ADA, employees should consider whether they qualify for a leave under the Family Medical Leave Act (FMLA), a law that gives employees with serious health conditions up to 12 weeks of job-protected leave, and does not include an "undue hardship" defense for employers.

An employer may be required to provide as a reasonable accommodation a leave that is longer than the maximum leave permitted under its internal policy (ADA, 42 U.S.C. § 12111(9)(B); 29 C.F.R. § 1630.2(o)(2)(ii); *Barnett*, 535 U.S. at 397–98). EEOC guidelines prohibit penalizing an employee for missing work during a reasonable accommodation leave, and applying policies automatically terminating employees after they have been on leave for a certain period of time, when an employee with a disability requires a longer leave (EEOC, *Reasonable Accommodation and Undue Hardship*).

Once an employee's leave imposes an undue hardship on the operation of the employer's business, the employer may end the leave and replace the employee (EEOC, *Psychiatric Disabilities; see also Watkins v. J & S Oil Co.*, 164 F.3d 55, 62 (1st Cir. 1998); *Walton v. Mental Health Ass'n of Southeastern Pa.*, 168 F.3d 661, 671 (3d Cir. 1999); *Schmidt*, 864 F. Supp. at 996). An undue hardship may occur even with a shorter leave where the employee was hired to complete a specific, time-sensitive task (*Micari v. Trans World Airlines, Inc.*, 43 F. Supp. 2d 275, 281 (E.D.N.Y. 1999); *Stubbs v. Marc Ctr.*, 950 F. Supp. 889, 893–94 (C.D. Ill. 1997)). Before terminating the employee, however, the employer must first consider whether there is an available vacancy to which the employee with disability may be transferred as a further accommodation (EEOC, *Reasonable Accommodation and Undue Hardship*). If such a transfer is possible without undue hardship, then the employee would continue her leave and, at its conclusion, be placed in the new position.

A central question in the context of a reasonable accommodation leave of absence is "how long is too long." Under the statutory scheme, each case should be evaluated individually to determine whether the length of the leave is imposing an undue hardship on the employer (ADA, 42 U.S.C. § 12112(b)(5)(A); EEOC, *Psychiatric Disabilities; Humphrey*, 239 F.3d at 1136, n.14; *Nunes*, 164 F.3d at 1246–47; *Rascon*, 143 F.3d at 1333–36; *Nowak*, 142 F.3d at 1004). Relevant factors include the individual employer's policies or the collective bargaining agree-

---

[15]Relatedly, some courts have held that regular attendance is an "essential job function." *See, e.g.*, Nesser v. Trans World Airlines, Inc., 160 F.3d 442, 445-46 (8th Cir. 1998); Nowak v. St. Rita High Sch., 142 F.3d 999, 1003 (7th Cir. 1998); Rogers v. Int'l Marine Terminals, Inc., 87 F.3d 755, 759 (5th Cir. 1996); Tyndall v. Nat'l Educ. Ctrs., 31 F.3d 209, 213 (4th Cir. 1994); Jackson v. Veterans Admin., 22 F.3d 277, 279-80 (11th Cir. 1994). *But see Humphrey*, 239 F.3d at 1135 n.11.

ment (*Nunes*, 164 F.3d at 1247; *Rascon*, 143 F.3d at 1335). Courts tend to draw the line at about 12 months or slightly longer (*Garcia-Ayala v. Lederle Parenterals, Inc.*, 212 F.3d 638 (1st Cir. 2000); *Treanor v. MCI Telecomm. Corp.*, 200 F.3d 570 (8th Cir. 2000); *Nunes*, 164 F.3d at 1247–48; *Cehrs*, 155 F.3d at 782; *Myers*, 50 F.3d at 283).

A leave of absence provided under the ADA is generally unpaid, although the employee may use sick days and other paid time off (EEOC, *Technical Assistance Manual*).

Modified or Part-Time Work Schedule

Modification of a regular work schedule may be considered a reasonable accommodation, unless it would cause the employer an undue hardship (EEOC, *Reasonable Accommodation and Undue Hardship*). Schedule changes may include flexibility in work hours or the work week, or part-time work. For example, some psychiatric medications cause grogginess and lack of concentration in the morning, and a later schedule might enable the employee to successfully perform his job duties (EEOC, *Psychiatric Disabilities*). An individual who needs to visit a therapist regularly could be given time off for the appointment, and work later to make up for the time.

Some employers have successfully argued that they are not required to provide an employee with a part-time schedule as a reasonable accommodation, because this eliminates an essential job function or creates a new job (*Rodal v. Anesthesia Group of Onondaga*, 369 F.3d 113, 120 (2d Cir. 2004); *see also Lamb v. Qualex, Inc.*, 33 Fed. Appx. 49, 56–57, 2002 WL 500492 (4th Cir. 2002); *Devito v. Chicago Park District*, 270 F.3 532, 534 (7th Cir. 2001); *Milton v. Scrivner, Inc.*, 53 F.3d 1118, 1124 (10th Cir. 1995)). Other authorities support a part-time schedule as an accommodation (EEOC, *Technical Assistance Manual; Waggoner*, 169 F.3d at 485).

An employer may be required to modify its internal policies to grant an adjusted work schedule as an accommodation (EEOC, *Reasonable Accommodation and Undue Hardship; see also* 42 U.S.C. § 12111(9)(B); *Barnett*, 535 U.S. at 405–06). However, policies contained in collective bargaining agreements may alter the employer's obligation to accommodate (*Barnett*, 535 U.S. at 394; *Kralik v. Durbin*, 130 F.3d 76, 81 (3d Cir. 1997)).

Modifying Workplace Policies

It is an accommodation to modify a workplace policy when required by an individual's disability-related limitations, so long as it does not cause the employer undue hardship (ADA, 42 U.S.C. § 12111(9)(B); *Barnett*, 535 U.S. at 397–98; 29 CFR § 1630.2 (o) (1); EEOC, *Technical Assistance Manual*; EEOC, *Psychiatric*

Disabilities; EEOC, *Reasonable Accommodation and Undue Hardship*). In fact, it is often necessary to modify workplace rules to implement accommodations. For example, a retail employer that bars cashiers from drinking beverages at checkout stations may need to modify this policy to accommodate an employee who needs to drink beverages to regulate insulin levels or to combat dry mouth, a side effect of his psychiatric medication (EEOC, *Psychiatric Disabilities*; EEOC, *Reasonable Accommodation and Undue Hardship*). Scheduling and leave accommodations often require policy modifications (EEOC, *Psychiatric Disabilities*; EEOC, *Reasonable Accommodation and Undue Hardship*; but see *Waggoner*, 169 F.3d at 484–85; *Jackson*, 22 F.3d at 279–80).

Courts are divided on whether employers are required to reasonably accommodate employee violations of a workplace conduct policy if the misconduct is a result of an underlying disability.[16] However, it is agreed upon that an employer may "hold a disabled employee to ... the same standards of conduct as a nondisabled employee" if "such standards are job-related and consistent with business necessity" (*Den Hartog*, 129 F.3d at 1086; *Walsted*, 113 F. Supp. 2d at 1342 n.7).

Job Restructuring

Job restructuring may also be a reasonable accommodation (29 C.F.R. pt. 1630, App. 1630.2(o), 1630.9; EEOC, *Technical Assistance Manual*; EEOC, *Reasonable Accommodation and Undue Hardship*). Job restructuring includes reallocating or redistributing marginal job functions that an employee is unable to perform because of a disability, and altering when or how a function is performed (29 C.F.R. pt. 1630, App. 1630.2(o), 1630.9; *Benson v. Northwest Airlines, Inc.*, 62 F.3d 1108 (8th Cir. 1995); *Kiphart v. Saturn Corp.*, 251 F.3d 573 (6th Cir. 2001)). When a function is removed from the disabled employee's job, the employer can require the employee to perform a different function in its place (EEOC, *Reasonable Accommodation and Undue Hardship*). State vocational rehabilitation agencies and other organizations with expertise in job analysis may provide technical assistance for job restructuring.

An employer is not required to reallocate the essential functions of a job as a reasonable accommodation (29 C.F.R. pt. 1630, App. 1630.2(o); *Webb v. Clyde L.*

---

[16]Compare *Jones v. American Postal Workers Union*, 192 F.3d 417, 429 (4th Cir. 1999); *Martinson v. Kinney Shoe Corp.*, 104 F.3d 683, 686 n. 3 (4th Cir. 1997); *Hamilton v. Southwestern Bell Tel. Co.*, 136 F.3d 1047, 1052 (5th Cir. 1997); *Pernice v. City of Chicago*, 237 F.3d 783, 785 (7th Cir. 2001); *Palmer v. Circuit Court*, 117 F.3d 351, 353 (7th Cir. 1997); *Harris v. Polk County*, 103 F.3d 696, 697 (8th Cir. 1996); *Newland v. Dalton*, 81 F.3d 904, 906 (9th Cir. 1996) *with Nielsen v. Moroni Feed Co.*, 162 F.3d 604, 608 (10th Cir. 1998); *Ward v. Massachusetts Health Research Inst., Inc.*, 209 F.3d 29, 38 (1st Cir. 2000); *Teahan v. Metro-North C. R. Co.*, 951 F.2d 511, 515 (2d Cir. 1991); *Salley v. Circuit City Stores*, 160 F.3d 977, 981 (3d Cir. 1998); *Walsted v. Woodbury County*, 113 F. Supp. 2d 1318, 1342 (D. Iowa 2000); *Humphrey*, 239 F.3d at 1139-40; *Den Hartog v. Wasatch Acad.*, 129 F.3d 1076, 1086 (10th Cir. 1997). And see *EEOC Enforcement Guidance, Psychiatric Disabilities*, pages 29-32, questions 30 & 31.

Choate Mental Health and Development Center, 230 F.3d 991 (7th Cir. 2000); *Bratten v. SSI Services, Inc.*, 185 F.3d 625 (6th Cir. 1999); *Mole v. Buckhorn Rubber Products, Inc.*, 165 F.3d 1212 (8th Cir.), cert. denied, 528 U.S. 821 (1999); *Gonzagowski v. Windall*, 115 F.3d 744 (10th Cir. 1997); *Kuehl v. Wal-Mart Stores, Inc.*, 909 F.Supp. 794 (D. Colo. 1995); *McDonald v. State of Kansas Dept. of Corrections*, 880 F.Supp. 1416 (D. Kan. 1995)), or to create a new job encompassing only those tasks the disabled employee is able to perform(*Hoskins v. Oakland County Sheriff's Dept.*, 227 F.3d 719 (6th Cir. 2000); *Sutton v. Lader*, 185 F.3d 1203 (11th Cir. 1999)). These principles create tension in the context of employer requirements that an employee rotate through several core job functions, one of which the employee cannot perform.[17] While the employer may argue that each job function is essential, the employee may argue that having one person do all of the functions is not essential.

Adjusting Supervisory Methods

Supervisors may be required to adjust their methods of supervision as a reasonable accommodation (EEOC, *Psychiatric Disabilities*; EEOC, *Reasonable Accommodation and Undue Hardship*). For example, an individual with a disability may request that the supervisor modify the mode of communicating assignments, instructions, or training – e.g., to use face-to-face communications, or emails – to ensure effective communication (EEOC, *Psychiatric Disabilities*). Or, in the case of an employee who has a supervisor with the habit of calling last-minute staff meetings that interfere with previously scheduled therapy appointments, it might be a reasonable accommodation for the employee to request that the supervisor provide 48 hours notice of meetings whenever possible (EEOC, *Reasonable Accommodation and Undue Hardship*).

Adjusting the level of supervision or structure may also enable the employee to perform the essential job functions (EEOC, *Psychiatric Disabilities*). For example, an otherwise qualified individual with a disability who experiences problems in concentration may request more detailed day-to-day guidance, feedback, or structure in order to perform his job (EEOC, *Psychiatric Disabilities; see also Kennedy v. Dresser Rand Co.*, 193 F.3d 120 (2d. Cir. 1999)).

It is common for employees with disabilities, including mental health disabilities, to experience problems at work when a positive and supportive supervisor is replaced by an unsupportive or negative supervisor. However, according to the EEOC, an employer is not required to provide an employee with a new supervisor as a reasonable accommodation (EEOC, *Reasonable Accommodation and Undue*

---

[17]Compare *Barnett*, 228 F.3d at 1117; *Kiphart*, 251 F.3d at 585-586; *Benson*, 62 F.3d at 1112; with *Anderson v. Coors Brewing Co.*, 181 F.3d 1171, 1176 (10th Cir. 1999); Laurin v. Providence Hosp., 150 F.3d 52, 59 (1st Cir. 1998); Malabarba v. Chicago Tribune Co., 149 F.3d 690, 700-01 (7th Cir.1998).

Hardship; see also Weiler v. Household Finance Corp., 101 F. 3d 519 at 526 (7th Cir. 1996); *Kennedy*, 193 F.3d at 122–23).

At the same time, an employee with a disability is protected from disability-based discrimination by supervisors, including disability-based harassment (EEOC, *Reasonable Accommodation and Undue Hardship; Fox*, 247 F.3d at 176–77; *Flowers*, 247 F.3d at 223–35). If the supervisor is creating a hostile working environment on the basis of disability, then the employer may have an obligation to remedy the situation, which may include providing a new supervisor.

Barring a finding by the employer of supervisor harassment, when an employee with disability is no longer able to work with a particular supervisor, often the only alternative from a disability law perspective is for the worker to seek a transfer to a vacant position.

Modified or More Training

Additional or modified training or instruction for an employee with disabilities may be a reasonable accommodation (*Walsted*, 113 F. Supp. 2d at 1334; *Kells v. Sinclair Buick-GMC Truck, Inc.*, 210 F.3d 827, 833 (8th Cir. 2000); EEOC, *Reasonable Accommodation and Undue Hardship*). Similarly, employers must provide reasonable accommodation if needed for the employee to access training programs (EEOC, *Reasonable Accommodation and Undue Hardship*). Needed accommodations may include ensuring accessible training sites, providing training materials in alternate formats, and modifying the manner in which training is provided (EEOC, *Technical Assistance Manual*). For example, it may be an accommodation to provide a tape recorder for an employee with limited concentration to review training sessions (EEOC, *Psychiatric Disabilities*). The obligation extends to in-house training, optional training, and training provided by outside entities (EEOC, *Reasonable Accommodation and Undue Hardship*).

An employer may be required to provide a temporary job coach to assist in the training of a qualified individual with a disability as a reasonable accommodation, barring undue hardship (EEOC, *Psychiatric Disabilities*; 29 C.F.R. pt. 1630 app. § 1630.9; *E.E.O.C. v. Hertz Corp.*, No. 96–72421, slip op. at 5 (E.D. Mich. Jan. 6, 1998)). An employer also may be required to allow a job coach paid by a public or private social service agency to accompany the employee at the job site as a reasonable accommodation (*E.E.O.C. v. Hertz Corp.*, No. 96–72421, slip op. at 5; *see also E.E.O.C. v. Dollar General Corp.*, 252 F. Supp. 2d 277, 291 (M.D.N.C. 2003)).

Working at Home

Working at home is a reasonable accommodation when the essential functions of the position can be performed at home and a work-at-home arrangement would not cause an undue hardship for the employer (EEOC, *Reasonable Accommodation*).

Courts have been mixed in their assessments of this accommodation.[18] Cases that reject working at home as a reasonable accommodation focus on evidence that personal contact, interaction, supervision, and coordination are needed for a specific position (*Whillock v. Delta Air Lines*, 926 F. Supp. 1555, 1564, (N.D. Ga. 1995), aff'd, 86 F.3d 1171 (11th Cir. 1996); *Misek-Falkoff v. International Business Machines Corp.*, 854 F. Supp. 215, 227–28 (S.D. N.Y. 1994)). However, advances in technology may make telecommuting an increasingly viable option (Gabel and Mansfield 2003). Indeed, one study found that about one-fifth of the adult workforce do some type of telework (International Telework Association and Council 2001). In addition, more than 80% of full-time American employees work off-site themselves or work with others who telecommute (Richman et al. 2001).

The ADA may require an employer to permit an employee to work at home, even if the employer does not allow other employees to do so (EEOC, *Factsheet on Work At Home/Telework as a Reasonable Accommodation* (2005)). The employee must be able to perform the "essential functions" of the job at home (*Mason v. Avaya Communications, Inc.*, 357 F.3d 1114, 1122–1123 (10th Cir. 2004)). Factors include the employer's ability to adequately supervise the employee and the employee's need to work with certain equipment or tools that cannot be replicated at home (29 C.F.R. § 1630.2(o)(1)(ii), (2)(ii) (1997); EEOC, *Reasonable Accommodation and Undue Hardship; Vande Zande*, 44 F.3d at 545). For some jobs, such as those requiring face-to-face contact with customers, essential functions can only be performed at the work site, and working at home would not be effective or required. Employees in other kinds of jobs, such as telemarketing or proofreading, may be able to perform the essential functions of their positions at home (*Humphrey*, 239 F.3d at 1136). An employer may be required to reassign marginal job functions that cannot be done at home (EEOC, *Telework*).

An employer should not deny a request to work at home as a reasonable accommodation solely because a job involves some contact and coordination with other employees. Frequently, meetings can be conducted effectively by telephone and information can be exchanged quickly through email. If the employer determines that some job duties must be performed in the workplace, then the employer and employee need to decide whether working part-time at home and part-time in the workplace would be effective (EEOC, *Telework*).

Where an employer offers telework generally, it must allow employees with disabilities an equal opportunity to participate in such a program. Further, it may be required to waive certain requirements or otherwise modify its telework program for someone with a disability (EEOC, *Telework*).

As with any accommodation, the employer may deny a request to work at home if it can show that another accommodation would be effective, or if working at home will cause undue hardship (EEOC, *Reasonable Accommodation and Undue Hardship*).

---

[18]Compare *Humphrey*, 239 F.3d 1128, *Langon v. Department of Health and Human Servs.*, 959 F.2d 1053 (D.C. Cir. 1992); *Anzalone v. Allstate Insurance Co.*, 5 AD Cas. (BNA) 455 (E.D. La. 1995); *Carr v. Reno*, 23 F.3d 525 (D.D.C. 1994), with *Vande Zande v. Wisconsin Dep't of Admin.*, 44 F.3d 538 (7th Cir. 1995).

Environmental Changes

Individuals with mental health conditions may be acutely sensitive to external stimulation. Physical changes to the workplace may be effective accommodations for some individuals with mental health disabilities. For example, installing room dividers, partitions, or other soundproofing or visual barriers between workspaces may accommodate individuals who have disability-related limitations in concentration. Moving an individual away from noisy machinery or reducing other workplace noise that can be adjusted (e.g., lowering the volume or pitch of telephones) are similar reasonable accommodations. Permitting an individual to wear headphones to block out noisy distractions also may be effective (EEOC, *Psychiatric Disabilities*). Individuals with disabilities may also request changes to the ventilation system or relocation of a workspace away from harmful external stimuli (EEOC, *Psychiatric Disabilities*).[19]

Reassignment or Transfer to a Vacant Position

Reassignment to a vacant position should be considered as a reasonable accommodation when adjustments to an employee's present job are not possible or would impose an undue burden, or if both parties agree that a reassignment is preferable to accommodation in the present position (EEOC, *Technical Assistance Manual*; EEOC, *Reasonable Accommodation and Undue Hardship*). An employee seeking a transfer as an accommodation must be able to perform the essential functions of the open position, with or without reasonable accommodation (EEOC, *Reasonable Accommodation and Undue Hardship*; 29 C.F.R. app. § 1630.2(o); *Cravens v. Blue Cross and Blue Shield of Kansas City*, 214 F.3d 1011, 1016 (8th Cir. 2000)).

Reassignment should be made to an equivalent position (in terms of pay, status, or other relevant factors) that is vacant or will become vacant within a reasonable time (EEOC, *Technical Assistance Manual*; EEOC, *Reasonable Accommodation and Undue Hardship; Norville v. Staten Island Univ. Hosp.*, 196 F.3d 89, 99 (2d Cir. 1999)). If an equivalent position is not available, the employer must look for a vacant position at a lower level for which the employee is qualified (EEOC, *Technical Assistance Manual*; EEOC, *Reasonable Accommodation and Undue Hardship*). Reassignment does not include giving an employee a promotion (EEOC, *Reasonable Accommodation and Undue Hardship*). Reassignment may not be used to limit, segregate, or otherwise discriminate against an employee, nor may an employer reassign employees with disabilities only to certain undesirable positions, or only to certain offices or facilities (EEOC, *Technical Assistance Manual; Duda v. Board of Educ. of Franklin Park Pub. Sch. Dist.*, 133 F.3d 1054, 1060 (7th Cir. 1998)).

---

[19]But see *Buckles v. First Data Resources, Inc.*, 176 F.3d 1098, 1101 (8th Cir. 1999).

If there is no vacancy at either a comparable position or a lower position, then reassignment is not required (EEOC, *Psychiatric Disabilities*).

The employer has an affirmative obligation to help the employee identify appropriate job vacancies (EEOC, *Reasonable Accommodation and Undue Hardship; Barnett*, 228 F.3d at 1113; *Gile v. United Airlines*, 95 F.3d 492, 499 (7th Cir. 1996)). A vacant position means one that is available when the employee asks for reasonable accommodation, or that the employer knows will become available within a reasonable amount of time (EEOC, *Psychiatric Disabilities*; EEOC, *Reasonable Accommodation and Undue Hardship; Norville v. Staten Island Univ. Hosp.*, 196 F.3d at 99). What is a "reasonable amount of time" should be determined on a case-by-case basis considering relevant facts, such as whether the employer, based on experience, can anticipate that an appropriate position will become vacant within a short period of time (EEOC, *Technical Assistance Manual*; EEOC, *Reasonable Accommodation and Undue Hardship*). A position is considered vacant even if an employer has posted an announcement seeking applications (EEOC, *Reasonable Accommodation and Undue Hardship*).

The employer is not required to create a new position as a reasonable accommodation, or to bump an employee who currently holds the position (EEOC, *Technical Assistance Manual*; EEOC, *Reasonable Accommodation and Undue Hardship; Cravens v. Blue Cross & Blue Shield*, 214 F.3d at 1019; *Pond v. Michelin N. Am., Inc.*, 183 F.3d 592 (7th Cir. 1999); *Benson*, 62 F.3d at 1114; 29 C.F.R. pt. 1630 app. § 1630.2 (o); *White v. York Int'l Corp.*, 45 F.3d 357 (10th Cir. 1995)). However, it may be a reasonable accommodation for an employee with a disability to transfer to a position by "bumping" an employee with less seniority, if there is a collective bargaining agreement that generally permits employees to bump less senior employees (*Whitfield v. Pathmark Stores, Inc.*, 1999 WL 301649 at *6 (D. Del., 1999); *Norman v. Univ. of Pittsburgh*, 2002 WL 32194730, at *17 (W. D. Pa., 2002)).

According to the EEOC, reassignment means "something more" than simply allowing the employee with disability to apply and compete for vacancies alongside employees or job applicants without disabilities; reassignment generally means that the employee gets the vacant position if she is qualified for it (EEOC, *Reasonable Accommodation and Undue Hardship; Wood v County of Alameda*, 5 AD Cas. (BNA) 173, 184, 1995 WL 705139, * 14 (N.D. Cal. 1995); *Aka*, 156 F.3d at 1310–11; *States v. Denver*, 943 F.Supp. 1304, 1310–11 (D. Colo. 1996)). However, courts have disagreed over whether the ADA allows an employer to reject an applicant with disability for reassignment solely on the ground that another candidate is better qualified.[20]

Generally, it is considered unreasonable for an employer to violate a seniority system in order to provide a reassignment (EEOC, *Reasonable Accommodation and Undue Hardship; Barnett*, 535 U.S. at 394). However, an employee may show

---

[20]Compare *Smith*, 180 F.3d at 1165 with *EEOC v. Humiston-Keeling, Inc.*, 227 F.3d 1024, 1028 (7th Cir, 2000).

special circumstances that make the transfer nevertheless reasonable (EEOC, *Reasonable Accommodation and Undue Hardship; Barnett*, 535 U.S. at 405).

Protection from Unnecessary Medical Inquiries and Examinations

The ADA protects all employees and job applicants – with and without disabilities (*Roe v. Cheyenne Mountain Conference Resort*, 124 F.3d 1221, 1229 (10th Cir. 1997); *Conroy v. N.Y. State Dep't of Corr. Servs.*, 333 F.3d 88, 94–95 (2d Cir. 2003); *Cossette v. Minn. Power & Light*, 188 F.3d 964, 969 (8th Cir. 1999); *Griffin v. Steeltek, Inc.*, 160 F.3d 591, 593–94 (10th Cir., 1998); *Fredenburg v. Contra Costa County Department of Health Services*, 172 F.3d 1176, 1182 (9th Cir. 1999)) – from unnecessary medical inquiries (42 U.S.C. § 12112(d)(2); 28 C.F.R. § 42.513(a)). Further, any medical information obtained by an employer must be maintained as a separate and confidential file (not with the employee's regular personnel file) (42 U.S.C. § 12112(d)(4)(C).; 29 C.F.R. §§ 1630.14(c)(1); 1630.14(d)(1); 29 C.F.R. app. section 1630.14(c), (d)). Access to such information is strictly limited (29 C.F.R. § 1630.14(c)(1)). An employer may inform supervisors or managers about necessary restrictions on the work or duties of an employee and necessary accommodations.

Application Stage

The ADA prohibits employers from making disability-related inquiries on the application form, during the interview, or at any other pre-job offer stage of the application process, or from imposing medical exams prior to making a job offer (42 U.S.C. § 12112(d)(2)). For example, it is unlawful for employers to ask an applicant whether she has ever had a particular disability, has suffered from a mental health condition, or has received workers' compensation (*Doe v. Syracuse School District*, 508 F.Supp. 333, 335 (N.D.N.Y 1981); *Griffin v. Steeltek*, 160 F.3d at 592).

It is also unlawful for an employer to ask general questions that are likely to elicit disability-related information. For example, questions such as "What impairments do you have," or "Have you ever been hospitalized," are not permitted (EEOC, *Psychiatric Disabilities*). Questions about prescription drug use (e.g., "Do you take antidepressants?") and inquiries that are likely to lead to information about *past* illegal drug addiction or *past or current* alcoholism are also barred (EEOC, *Enforcement Guidance on Preemployment Disability-Related Inquiries and Medical Examinations* (1995)).

Employers cannot seek to obtain information from third parties that they cannot lawfully obtain from the applicant (EEOC, *Preemployment Inquiries*).

An employer is permitted to ask an applicant about his ability to perform tasks that are relevant to the job he is applying for (e.g., ability to lift 50 pounds) and, according to the EEOC, this inquiry is *not* limited to questions regarding "essential functions" of the job, provided all applicants are asked the same questions (EEOC, *Technical Assistance Manual*). Additionally, an employer can ask applicants to describe or demonstrate how they will perform a specific job function (EEOC, *Technical Assistance Manual*). However, an employer may not require an applicant to disclose whether she needs reasonable accommodation to perform the job, or to check a box indicating whether she can do an essential function "with __ or without __ reasonable accommodation" (EEOC, *Preemployment Inquiries*; EEOC, *Psychiatric Disabilities*).

An employer may ask about an applicant's nonmedical qualifications, such as educational background and required licenses (EEOC, *Preemployment Inquiries*), and may review an applicant's employment history and question the circumstances of any gaps or sudden departures of employment. This is permitted even if the applicant was unemployed or terminated due to disability-related reasons (EEOC, *Preemployment Inquiries*).

An applicant with work history gaps should be prepared to answer questions about time spent outside of formal employment. The applicant may explain truthfully that she took time off work to accomplish personal goals such as volunteering, traveling, or spending time with family. She may mention that while the time away from work was positive, she is excited to be returning to the workforce. It is also helpful to emphasize relevant qualifications and enthusiasm for the job for which she is applying. Such an applicant may wish to work with a vocational counselor to develop a résumé that focuses on experience rather than chronology.

Some disabilities may be obvious to the employer before or during the interview process. If the employer knows that an applicant has a disability, and has a reasonable belief that the individual will need reasonable accommodation, the employer may lawfully ask the applicant narrowly tailored questions about workplace accommodations in order to determine whether an applicant needs a reasonable accommodation to perform the functions of the job and what type of accommodations might be needed (EEOC, *Preemployment Inquiries*; EEOC, *Psychiatric Disabilities*).

The law does not bar asking about or testing for current illegal drug use (EEOC, *Preemployment Inquiries*). As part of such testing, an employer or its representative (usually a drug testing company) may ask an applicant or employee to disclose all prescribed drugs to determine whether a "positive" test result is caused by a lawful drug (EEOC, *Preemployment Inquiries*). These lawful drugs – e.g., antipsychotics, AZT – may reveal intimate medical information. Drug testing companies should reveal only the fact of a "positive" for illegal drugs, and no other information (*Pettus v. Cole*, 49 Cal.App.4th 402, 458 (1996); Cal. Civil Code § 56.10(a), (c)(8)(b); *see also* 45 C.F.R. Parts 160 and 164 (HIPAA privacy rules)). An employer performing its own drug testing should erect a firewall to ensure that information about lawful prescription drug usage is not disclosed (42 U.S.C. § 12112(d)(4)(C).; 29 C.F.R. §§ 1630.14(c)(1); 1630.14(d)(1); 29 C.F.R. app. section 1630.14(c), (d)).

Physical agility and physical fitness tests are generally permissible and not considered medical exams (EEOC, *Preemployment Inquiries*). According to the

EEOC, some psychological tests designed to measure personality traits, as opposed to mental disorders, are *not* considered medical examinations or inquiries, and thus are not regulated by the ADA.

A few jobs, mostly positions with disability service organizations and in the mental health system, are expressly labeled as "peer advocate" or "peer counselor" positions. The word "peer" in this kind of job posting means that the job is available only to individuals with a disability, and sometimes a particular kind of disability (e.g., a mental health condition). In this context, applying for the job is akin to disclosing a disability. Additionally, while the same rules arguably apply, an interview for a "peer" position will likely include questions that touch on the applicant's disability, such as: "How would you use your own experiences having a psychiatric condition to help others?" The applicant should be prepared with answers to these questions.

Note: How to Respond to Illegal Questions During an Interview or On an Application

As a practical matter, an applicant should become familiar with illegal questions that may be asked on application forms or during interviews. For example, a preinterview form asking the applicant to check boxes revealing her medical history is impermissible, but continues to be used by some employers. Unlawful questions may also be posed during the job interview (e.g., "Have you ever been hospitalized for mental illness?" or "How many days of work did you miss last year due to illness?").

If an employer asks the applicant an illegal question, either on a form or in the interview, the applicant has several imperfect options, and may: (1) answer the question honestly; (2) answer the question dishonestly or incompletely; (3) decline to answer, leaving a blank on the form, or stating that the question is illegal or inappropriate; or (4) sidestep the question by writing in "N/A" (Not Applicable) on the form, or making an analogous statement in the interview (e.g., stating "Oh, no, that wouldn't apply to me").

An applicant who answers an illegal question truthfully by disclosing a medical condition may risk being discriminated against in the hiring process, and proving such discrimination can be very difficult (*Doe v. Syracuse Sch. Dist.*, 508 F. Supp. at 337–38 & n. 4). When implementing this choice, the applicant should consider the simplest and least alarming manner to respond to the question. A detailed recitation of one's medical history is likely never the best course of action. A short, positive statement – e.g., "I had some health problems but am now completely recovered" – is best.

In the alternative, an applicant can choose to answer dishonestly by responding with "no" or "none." In this situation, the applicant may face a moral and ethical dilemma, but more importantly, she may face subsequent legal repercussions if the employer learns of this dishonesty. Court rulings vary with respect to whether lying in response to an illegal, preemployment medical question can justify an employer's decision to take action against the employee (*Downs v. Massachusetts Bay Transp. Auth.*, 13 F. Supp.2d 130, 140 (1998); *Kraft v. Police Commissioner of Boston*, 410 Mass. 155, 157, 571 N.E.2d 380 (1991); *Hartman v. City of Petaluma*, 841 F. Supp. 946, 949–50 (N.D. Cal. 1994); *Pinckard v. Metropolitan Government of Nashville*, 1 Fed. Appx. 359, 2001 WL 45285 (6th Cir. 2001) (unpublished)).

Where possible, a delicate but honest sidestep may be best. "N/A" is a truthful response to a medical question on an application form since the question does not apply to the applicant because it is illegal. Sidestepping an unlawful question may be more challenging during an in-person interview. However, it would also be truthful to make an analogous statement in response to an interview question about any prior health problems – "Oh, that wouldn't apply to me," and attempt to change the focus of the conversation to the ability to do the job. Another possibility is to restate the illegal question into a lawful question: "You asked how my health is. I expect you want to know if I can be here reliably. I haven't missed a day of work in years" (Kohlenberg 1998).

Post-offer Stage

Following an official job offer, the ADA allows employers to require medical examinations and/or medical questionnaires, including psychological examinations (42 U.S.C. § 12112(d)(3); EEOC, *Psychiatric Disabilities*). Employers are permitted to ask a variety of questions, even if they are unrelated to job performance, provided that the information is kept confidential and all entering employees in the same job category are subjected to the same inquiry (42 U.S.C. § 12112(d)(3); EEOC, *Preemployment Inquiries*). The ADA imposes no limits on the scope of such medical inquiries (EEOC, *Psychiatric Disabilities; see also Norman-Bloodsaw v. Lawrence Berkeley Laboratory*, 135 F.3d 1260, 1273 (9th Cir. 1998)). However, at least one state strictly limits such "post-offer" inquiries (Cal. Gov't Code § 12940(e)(3); Cal. Const. Art. 1, § 1).

To ensure that the ADA scheme functions as designed, post-offer examinations must be conducted as a "separate, second step of the selection process, after an individual has met all other job prerequisites (EEOC, *Technical Assistance Manual*)." The job offer must be "real" – the employer must evaluate "all relevant nonmedical information which it reasonably could have obtained and analyzed prior to giving the offer" (EEOC, *Preemployment Inquiries*).

If an employer uses the results of such examinations or inquiries to revoke the job offer, the employer may be required to prove that the individual is unqualified, that its reasons are "job-related and consistent with business necessity" (29 C.F.R. § 1630.14(b)(3); 29 C.F.R. app. § 1630.14(b); EEOC, *Preemployment Inquiries*; EEOC, *Psychiatric Disabilities*), or that the applicant poses a "direct threat" (EEOC, *Preemployment Inquiries*).

On the Job

The ADA prohibits an employer from requiring any medical examinations of an employee, or from making any disability-related inquiries of an employee, unless the examination or inquiry is "job-related and consistent with business necessity" (42 U.S.C. § 12112(d)(4)(A)).

The employer bears the burden of establishing the requisite showing, which is a question of fact (42 U.S.C. § 12112(d)(4)(A); *Fredenburg v. Contra Costa County Dep't of Heath Servs.*, 172 F.3d at 1182). Additionally, even if the inquiry is job-related and necessary, the employer may request only the particular information necessary to meet its business need (EEOC, *Psychiatric Disabilities*).

Examples of lawful, on-the-job medical inquiries include:

- An employer may seek "reasonable medical documentation" where an employee requests accommodation, and the disability or the need for accommodation is not obvious.
- If an employer has a "reasonable belief, based on objective evidence" that a medical condition is interfering with an employee's ability to perform the essential duties of the job, the employer may ask related medical questions or request a job-related "fitness for duty" examination.
- If an employer has a "reasonable belief, based on objective evidence" that an employee may pose a "direct threat" due to a medical condition, then the employer may ask limited medical questions or request an examination directed toward evaluating any threat.
- For certain safety-sensitive jobs, periodic medical inquiries or exams, if appropriately tailored, may be "job-related and consistent with business necessity," even without an individualized, objective basis for believing that there is some sort of medical problem interfering with the job.
- If an employee has a work-related injury, the employer may ask medical questions or request an examination to assess its liability under workers' compensation.

No Discrimination for Associating with Person with Disability, and No Retaliation or Interference for Engaging in Protected Activities

The ADA protects applicants and employees from discrimination on the basis of their association with a person with a disability (42 U.S.C. § 12112(b)(4)). Under these provisions, a person cannot be denied employment or terminated because they have a family member or a friend with a disability.

The employer has no obligation under the ADA to provide reasonable accommodations to an applicant or employee for the *associate*'s disability (*Den Hartog*, 129 F.3d at 1083–85; 29 C.F.R. § 1630 App.). Thus, an employee who needs time off from work to care for a disabled child will not find protection under the ADA's accommodation provisions (*Tyndall*, 31 F.3d at 214). Any protection would have to come from another law, such as the FMLA.

As with other civil rights laws, the ADA prohibits retaliation against an individual who has engaged in protected activities (42 U.S.C. 12203(a); *Maurey v. Univ. of S. Cal.*, 87 F. Supp. 2d. 1021, 1038 (C.D. Cal. 1999)). It is also "unlawful to coerce, intimidate, threaten, or interfere with any individual in the exercise or enjoyment of … any right granted or protected by this Act" (ADA, 42 U.S.C. § 12203(b); *Brown v. City of Tucson*, 336 F.3d 1181, 1188 (9th Cir. 2003); *Barnett*, 196 F.3d at 994).

Conclusion

Increased enforcement of nondiscrimination laws, and access to reasonable accommodations, are key to the inclusion of persons with psychiatric disabilities in the workplace. In many cases, reasonable accommodation may make the difference between job loss and a successful employment experience.

At the same time, nondiscrimination and reasonable accommodation are not sufficient by themselves to advance the work lives of adults living with mental disorders. The legal requirements of the ADA and state law nondiscrimination statutes must be implemented alongside additional initiatives, including access to health care and income supports, quality job training and placement, responsive supervision, and flexible and supportive workplace environments. Coordination between and among service providers specializing in different aspects of workplace integration must be enhanced.

By enshrining the principles of the disability rights movement, however – that a person may have a disability yet still be a qualified and independent participant in the workplace (with modest adjustments if requested) – the ADA provides the central model of respect and self-determination that must guide the struggle for workplace fairness, and a touchstone for a broad array of social programming.

References

Aka. v. Wash. Hosp. Ctr., 156 F.3d 1284 (D.C. Cir. 1998)
Albertson's, Inc. v. Kirkingburg, 527 U.S. 555 (1999)
Amy L. Allbright, 2003 Employment Decisions Under the ADA Title I - Survey Update, 25 MENTAL & PHYSICAL DISABILITY L. REP. 508, 509 (2003)
Association AB (1998) Study: employers win most ADA title I judicial and administrative complaints. Mental & Physical Disability L Rep 22, 403
Anderson v. Coors Brewing Co., 181 F.3d 1171 (10th Cir. 1999)
Anzalone v. Allstate Insurance Co., 5 AD Cas. (BNA) 455 (E.D. La. 1995)
Barnett v. U.S. Airways, 228 F.3d 1105 (9th Cir. 2000)

Bates v. UPS, 511 F.3d. 974 (9th Cir. 2007) (en banc)
Beck v. Univ. of Wis., 75 F.3d 1130 (7th Cir. 1996)
Belk v. Southwestern Bell Tel. Co., 194 F.3d 946 (8th Cir. 1999)
Benson v. Northwest Airlines, Inc., 62 F.3d 1108 (8th Cir. 1995)
Blanck PD (1996) Communicating the Americans with disabilities act, Transcending compliance:
    1996 follow-up report on sears. Roebuck and Co, Iowa City, IA
Blanck PD (1997) The economics of the employment provisions of the Americans with
    Disabilities Act: part I – workplace accommodations. DePaul L Rev 46(4):877–914
Blanck PD (1998) The emerging role of the staffing industry in the employment of persons with
    disabilities: a case report on Manpower Inc. Iowa City, Iowa
Bratten v. SSI Services, Inc., 185 F.3d 625 (6th Cir. 1999)
Brown v. City of Tucson, 336 F.3d 1181 (9th Cir. 2003)
Buckles v. First Data Resources, Inc., 176 F.3d 1098 (8th Cir. 1999)
Bultemeyer v. Fort Wayne Community School, 100 F.3d 1281 (7th Cir. 1996)
Carr v. Reno, 23 F.3d 525 (D.D.C. 1994)
Cehrs v. Northeast Ohio Alzheimer's Research Ctr., 155 F.3d 775 (6th Cir. 1998)
Chevron v. Echazabal, 536 U.S. 73 (2002)
Conroy v. N.Y. State Departmentt of Correctional Services, 333 F.3d 88 (2d Cir. 2003)
Cossette v. Minn. Power & Light, 188 F.3d 964 (8th Cir. 1999)
Cravens v. Blue Cross and Blue Shield of Kansas City, 214 F.3d 1011 (8th Cir. 2000)
Criado v. IBM Corp., 145 F.3d 437 (1st Cir. 1998)
Cripe v. City of San Jose, 261 F.3d 877 (9th Cir. 2001)
Den Hartog v. Wasatch Acadamy, 129 F.3d 1076 (10th Cir. 1997)
Devito v. Chicago Park District, 270 F.3 532 (7th Cir. 2001)
Doe v. Syracuse School District, 508 F.Supp. 333 (N.D.N.Y 1981)
Downs v. Massachusetts Bay Transportation Authority, 13 F.Supp.2d 130 (1998)
Duda v. Board of Education. of Franklin Park Public. School. District, 133 F.3d 1054 (7th Cir.
    1998)
Edwards DM, Allen M (1996) Stigma: perpetuating misperceptions. Winter Springs, FL: Quest
    Publishing Company, Inc. Retrieved from http://home.earthlink.net/~durangodave/html/writing/
    Stigma.htm
E.E.O.C. v. Dollar General Corp., 252 F. Supp. 2d 277 (M.D.N.C. 2003)
E.E.O.C. v. Hertz Corp., No. 96-72421 (E.D. Mich. Jan. 6, 1998)
EEOC v. Humiston-Keeling, Inc., 227 F.3d 1024 (7th Cir, 2000)
EEOC v. Newport News Shipbuilding & Drydock Co., 949 F. Supp. 403 (E.D. Va. 1996)
Feliberty v. Kember Corp., 98 F.3d 274 (7th Cir. 1996)
Fink PJ & Tasman A STIGMA AND MENTAL ILLNESS (1992), so I think the short form would
    be Fink & Tasman (1992)
Fjellestad v. Pizza Hut of Am., 188 F.3d 944 (8th Cir. 1999)
Fink PJ, Tasman A (1992) Stigma and mental illness. American Psychiatric Press, Washington,
    DC
Flowers v. South Regional Physician Services, 247 F.3d 229 (5th Cir. 2001)
Fox v. Gen. Motors Corp., 247 F.3d 169 (4th Cir. 2001)
Fredenburg v. Contra Costa County Department of Health Services, 172 F.3d 1176 (9th Cir. 1999)
Gabel J, Mansfield N (2003) The information revolution and its impact on the employment
    relationship: an analysis of the cyberspace workplace. American Business L J 40:301
Garcia-Ayala v. Lederle Parenterals, Inc., 212 F.3d 638 (1st Cir. 2000)
Globe B (1997) Survey Finds Stigma, Error Haunt Mental Illness
Gonzagowski v. Windall, 115 F.3d 744 (10th Cir. 1997)
Gile v. United Airlines, 95 F.3d 492 (7th Cir. 1996)
Granger B. Baron R.C. & Robinson, M.C.J. (1996). A national study on job accommodations for
    people with psychiatric disabilities: Final report. Matrix Research Institute, Philadelphia, PA
Grenier v. Cyanamid Plastics, Inc., 70 F.3d 667 (1st Cir. 1995)
Griffin v. Steeltek, Inc., 160 F.3d 591 (10th Cir., 1998)

Hamilton v. Southwestern Bell Tel. Co., 136 F.3d 1047 (5th Cir. 1997)

Hansen v. Henderson, 233 F.3d 521 (7th Cir. 2000)

Harris v. Polk County, 103 F.3d 696 (8th Cir. 1996)

Hartman v. City of Petaluma, 841 F. Supp. 946 (N.D. Cal. 1994)

Haschmann v. Time Warner Entertainment, 151 F.3d 591 (7th Cir. 1998)

Haysman v. Food Lion, 893 F. Supp. 1092 (S.D. Ga. 1995)

Hendricks-Robinson v. Excel Corp., 154 F.3d 685 (7th Cir. 1998)

Higgins v. New Balance Athletic Shoe, Inc., 194 F.3d 252 (1st Cir. 1999)

Hoskins v. Oakland County Sheriff's Dept., 227 F.3d 719 (6th Cir. 2000)

Hudson v. MCI Telecomm. Corp., 87 F.3d 1167 (10th Cir. 1996)

Huff, Charlotte (2000) Secret suffering: stigma can keep employees from disclosing mental condition. Fort Worth Star-Telegram, June 10

Humphrey v. Memorial Hosps. Ass'n, 239 F.3d 1128 (9th Cir. 2001)

International Telework Association & Council (2001) Telework America 2001 Survey

Jackson v. Veterans Administration, 22 F.3d 277 (11th Cir. 1994)

Jones v. American Postal Workers Union, 192 F.3d 417 (4th Cir. 1999)

Kells v. Sinclair Buick-GMC Truck, Inc., 210 F.3d 827 (8th Cir. 2000)

Kennedy v. Dresser Rand Co., 193 F.3d 120 (2d. Cir. 1999)

Kimbro v. Atlantic Richfield Co., 889 F.2d 869 (9th Cir. 1989)

Kiphart v. Saturn Corp., 251 F.3d 573 (6th Cir. 2001)

Kong D (October 1, 1997) Widespread misperception on mental illness; Survey finds stigma and errors linger. Boston *Globe*, p. B1

Kraft v. Police Commissioner of Boston, 410 Mass. 155 (1991)

Kralik v. Durbin, 130 F.3d 76 (3d Cir. 1997)

Krocka v. Riegler, 958 F. Supp. 1333 (N.D. Ill. 1997)

Kuehl v. Wal-Mart Stores, Inc., 909 F.Supp. 794 (D. Colo. 1995)

Kohlenberg B (1998) Talking about your disability in job interviews. San Francisco, CA: Kohlenberg & Associates. Retrieved from: http://www.bkohlenberg.com/disclosure.htm

Lamb v. Qualex, Inc., 33 Fed. Appx. 49, 2002 WL 500492 (4th Cir. 2002)

Langon v. Department of Health and Human Servs., 959 F.2d 1053 (D.C. Cir. 1992)

Laurin v. Providence Hosp., 150 F.3d 52 (1st Cir.1998)

Loulseged v. Akzo Nobel Inc., 178 F.3d 731. (5th Cir. 1999)

Lyons v. Legal Aid Society, 68 F.3d 1512 (2d Cir. 1995)

Malabarba v. Chicago Tribune Co., 149 F.3d 690 (7th Cir.1998)

Martinson v. Kinney Shoe Corp., 104 F.3d 683 (4th Cir. 1997)

Mason v. Avaya Communications, Inc., 357 F.3d 1114 (10th Cir. 2004)

Maurey v. Univ. of S. Cal., 87 F. Supp. 2d. 1021 (C.D. Cal. 1999)

McDonald v. State of Kansas Dept. of Corrections, 880 F.Supp. 1416 (D. Kan. 1995)

Mengine v. Runyon, 114 F.3d 415 (3d Cir. 1997)

Mental Health Liaison Group (2000) Responding to the mental health needs of americans: a briefing document for candidates and policymakers. Retrieved from: http://www.mhlg.org/10-13-00.pdf

Micari v. Trans World Airlines, Inc., 43 F. Supp. 2d 275 (E.D.N.Y. 1999)

Miller v. Ill. Dep't of Corr., 107 F.3d 483 (7th Cir. 1997)

Milton v. Scrivner, Inc., 53 F.3d 1118 (10th Cir. 1995)

Misek-Falkoff v. International Business Machines Corp., 854 F. Supp. 215 (S.D. N.Y. 1994)

Mole v. Buckhorn Rubber Products, Inc., 165 F.3d 1212 (8th Cir.), cert. denied, 528 U.S. 821 (1999)

Morton v. United Parcel Service, 272 F.3d 1249 (9th Cir. 2001)

Moses v. Am. Nonwovens, Inc., 97 F.3d 446 (11th Cir. 1996)

Murphy v. United Parcel Service, 527 U.S. 516 (1991)

Myers v. Hose, 50 F.3d 278 (4th Cir. 1995)

National Institute of Mental Health (2003) The numbers count: mental disorders in America [Electronic Version]. http://www.nimh.nih.gov/health/publications/the-numbers-count-mental-disorders-in-america/index.shtml. Accessed 15 June 2009

National Organization of Disability (2001) Employment rates of people with disabilities. Retrieved from: http://www.nod.org/index.cfm?fuseaction=feature.showFeature&FeatureID=110&C:\CFusion8\verity\Data\dummy.txt

Nesser v. Trans World Airlines, Inc., 160 F.3d 442 (8th Cir. 1998)

New Freedom Commission on Mental Health (2003) Achieving the promise: transforming mental health care in America. Final report. Rockville, MD. US Department of Health and Human Services, Publication SMA-03-3832. http://www.mentalhealthcommission.gov/reports/FinalReport/toc.html. Accessed 15 June 2009

Newland v. Dalton, 81 F.3d 904 (9th Cir. 1996)

Nielsen v. Moroni Feed Co., 162 F.3d 604 (10th Cir. 1998)

Norman v. Univ. of Pittsburgh, 2002 WL 32194730 (W. D. Pa., 2002)

Norman-Bloodsaw v. Lawrence Berkeley Laboratory, 135 F.3d 1260 (9th Cir. 1998)

Norville v. Staten Island University Hospital, 196 F.3d 89 (2d Cir. 1999)

Nowak v. St. Rita High School, 142 F.3d 999, 1003 (7th Cir. 1998)

Nunes v. Wal-Mart Stores, 164 F.3d 1243, 1247 (9th Cir. 1999)

Palmer v. Circuit Court, 117 F.3d 351 (7th Cir. 1997)

Parry JW (2000) 1999 Employment decisions under the ADA title I – survey update. Mental & Physical Disability L Rep 24, 348

Pernice v. City of Chicago, 237 F.3d 783 (7th Cir. 2001)

Pettus v. Cole, 49 Cal. App. 4th 402 (1996)

Pinckard v. Metropolitan Government of Nashville, 1 Fed. Appx. 359, 2001 WL 45285 (6th Cir. 2001) (unpublished)

Pond v. Michelin North America, Inc., 183 F.3d 592 (7th Cir. 1999)

President's New Freedom Commission on Mental Health (2003) Interim report to the President. http://www.mentalhealthcommission.gov/reports/interim_toc.htm. Accessed 15 June 2009

Prewitt v. United States Postal Service, 662 F.2d 292 (5th Cir. 1981)

Prilliman v. United Air Lines, Inc., 53 Cal. App. 4th 935 (1997)

Ralph v. Lucent Techs., Inc., 135 F.3d 166 (1st Cir. 1998)

Rascon v. U.S. West Communications, 143 F.3d 1324 (10th Cir. 1998)

Richman A, Noble K, Johnson A (2001) When the workplace is many places: the extent and nature of off-site work today. WFD Consulting, Watertown, MA

Rodal v. Anesthesia Group of Onondaga, 369 F.3d 113 (2d Cir. 2004)

Roe v. Cheyenne Mountain Conference Resort, 124 F.3d 1221 (10th Cir. 1997)

Rogers v. International Marine Terminals, Inc., 87 F.3d 755 (5th Cir. 1996)

Salley v. Circuit City Stores, 160 F.3d 977 (3d Cir. 1998)

Schmidt v. Safeway Inc., 864 F. Supp. 991 (D. Or. 1994)

Smith v. Midland Brake, Inc., 180 F.3d 1154 (10th Cir. 1999)

Snead v. Metro. Prop. & Cas. Ins. Co., 237 F.3d 1080 (9th Cir. 2001)

States v. Denver, 943 F.Supp. 1304 (D. Colo. 1996)

Stewart v. Brown County, 86 F.3d 107 (7th Cir. 1996)

Stewart v. Happy Herman's Cheshire Bridge, Inc., 117 F.3d 1278 (11th Cir. 1997)

Stoddard S, Jans L, Ripple J, Kraus L (1998) Chartbook on work and disability in the United States. U.S. National Institute on Disability and Rehabilitation Research, Washington, DC

Stubbs v. Marc Center, 950 F. Supp. 889 (C.D. Ill. 1997)

Sutton v. Lader, 185 F.3d 1203 (11th Cir. 1999)

Sutton v. United Airlines, 527 U.S. 471 (1999)

Tate v. Farmland Industries, Inc., 268 F.3d 989 (10th Cir. 2001)

Taylor v. Phoenixville Sch. Dist., 184 F.3d 296 (3d Cir. 1999)

Taylor v. Principal Fin. Group, Inc., 93 F.3d 155 (5th Cir. 1996)

Teahan v. Metro-North C. R. Co., 951 F.2d 511 (2d Cir. 1991)

Toyota Motor Mfg., Ky. v. Williams, 534 U.S. 184 (2002)

Treanor v. MCI Telecomm. Corp., 200 F.3d 570 (8th Cir. 2000)

Tyndall v. National Education Centers, 31 F.3d 209 (4th Cir. 1994)

U.S. Airways v. Barnett, 535 U.S. 391 (2002)

US Department of Health and Human Services (1999) Mental health: a report of the surgeon general. Rockville, MD: Department of Health and Human Services, Substance Abuse and Mental Health Services Administration, Center for Mental Health Services, National Institutes of Health, National Institute of Mental Health, 1999

US Equal Employment Opportunity Commission. (1995). Compliance manual. Washington, DC: Equal Employment Opportunity Commission, 8 FEP Manual (BNA) 405:7276–78, 7281

US Equal Employment Opportunity Commission. (2000). Enforcement guidance: disability-related inquiries and medical examinations of employees under the americans with disabilities act. Washington, DC: Equal Employment Opportunity Commission, publication 915.002. http://www.eeoc.gov/policy/docs/guidance-inquiries.html

US Equal Employment Opportunity Commission. (1997). Enforcement guidance on the americans with disabilities act and psychiatric disabilities. Washington, DC: Equal Employment Opportunity Commission, publication 915.002. http://www.eeoc.gov/policy/docs/psych.html

US Equal Employment Opportunity Commission (1995) Enforcement guidance: preemployment disability-related questions and medical examinations. Washington, DC: Equal Employment Opportunity Commission, publication 915.002. http://www.eeoc.gov/policy/docs/preemp.html

US Equal Employment Opportunity Commission (1999) Enforcement guidance: Reasonable accommodation and undue hardship under the americans with disabilities act. Washington, DC: Equal Employment Opportunity Commission, publication 915.002. http://www.eeoc.gov/policy/docs/accommodation.html

US Equal Employment Opportunity Commission (1992) Technical assistance manual on the employment provisions (Title I) of the ADA. Equal Employment Opportunity Commission, Washington, DC

US Equal Employment Opportunity Commission (2005) Work at home/telework as a reasonable accommodation. Washington, DC: Equal Employment Opportunity Commission. http://www.eeoc.gov/facts/telework.html

Vande Zande v. Wisconsin Department of Administration, 44 F.3d 538 (7th Cir. 1995)

Waggoner v. Olin Corp., 169 F.3d 481 (7th Cir. 1999)

Wahl O (1999) Mental health consumers' experience of stigma. The Schizophrenia Bulletin 25:467–478

Wahl OF (1999) Study examines consumer experiences with stigma. The Bell, December

Wahl OF (1998) Violence stereotype hurts people with mental health needs. The Bell, September

Wahl OF (Spring 2001) Consumer experiences of stigma. Community Connections (newsletter of the Mental Health Association of New York State)

Walsted v. Woodbury County, 113 F. Supp. 2d 1318 (D. Iowa 2000)

Walton v. Mental Health Association of Southeastern Pa., 168 F.3d 661 (3d Cir. 1999)

Ward v. Massachusetts Health Research Institute., Inc., 209 F.3d 29 (1st Cir. 2000)

Watkins v. J & S Oil Co., 164 F.3d 55 (1st Cir. 1998)

Webb v. Clyde L. Choate Mental Health and Development Center, 230 F.3d 991 (7th Cir. 2000)

Weiler v. Household Finance Corp., 101 F. 3d 519 (7th Cir. 1996)

Whillock v. Delta Air Lines, 926 F. Supp. 1555, 1564 (N.D. Ga. 1995), aff'd, 86 F.3d 1171 (11th Cir. 1996)

White v. York International Corp., 45 F.3d 357 (10th Cir. 1995)

Whitfield v. Pathmark Stores, Inc., 1999 WL 301649 (D. Del., 1999)

Wood v County of Alameda, 5 AD Cas. (BNA) 173, 184, 1995 WL 705139 (N.D. Cal. 1995)

World Health Organization (2001) The World Health Report 2001 – mental health: new understanding, new hope. http://www.who.int/whr/2001/en/

Chapter 2
Investing in the Mental Health of the Labor Force: Epidemiological and Economic Impact of Mental Health Disabilities in the Workplace

Carolyn S. Dewa and David McDaid

By the year 2020, depression will emerge as one of the leading causes of disability globally, second only to ischemic heart disease (World Health Organization 2001). Thus, governments are taking notice of the mental health of workers. For example, the European Ministers of Health have endorsed a detailed action plan calling for employers to "create healthy workplaces by introducing measures such as exercise, changes to work patterns, sensible hours, and healthy management styles," and also to "include mental health in programs dealing with occupational health and safety" (World Health Organization 2005). More recently the European Union's Pact for Mental Health and Wellbeing identifies the workplace as one of its four areas for action (Commission of the European Communities 2008).

Part of this interest is associated with the growing awareness that mental and emotional health problems are associated with staggering social and economic costs that create a heavy burden for the workplace. In its recent report, the Canadian Standing Senate Committee on Social Affairs, Science, and Technology (The Standing Senate Committee on Social Affairs 2006) raised prevention, promotion, and treatment of mental illness as critical issues to be addressed. The Committee went on to identify the workplace as one of the prime areas in which to begin. They asserted, "It is in the workplace that the human and economic dimensions of mental health and mental illness come together most evidently" (p. 171). In another example of the increased prominence of the workplace, in the UK the government commissioned an independent review of the health of the working population and requested recommendations to bring about positive change (Black 2008). Not only did their review highlight the impact of poor mental health to the economy and business, it also showed that "the scale of numbers on [long-term disability benefits] represents a historic failure of healthcare and employment support" (p. 13). In a detailed response, the English government subsequently committed itself to the development of a "National Mental Health and Employment Strategy to ensure that Government

C.S. Dewa (✉)
Work & Well-being Research & Evaluation Program,
Centre for Addiction and Mental Health, Toronto, ON, Canada M5S 2S1
e-mail: Carolyn_Dewa@camh.net

I.Z. Schultz and E.S. Rogers (eds.), *Work Accommodation and Retention in Mental Health*, 33
DOI 10.1007/978-1-4419-0428-7_2, © Springer Science+Business Media, LLC 2011

is doing all it can to support their particular needs" (p. 6) (Department of Work and Pensions and Department of Health 2008).

The business sector is also taking notice and seeking ways to decrease the impact of mental disorders among its employees. For example, business leaders have formed consortiums such as the Canadian-based Global Business and Economic Roundtable on Addiction and Mental Health to develop strategies to curb disability costs. One product of the alliances was the 2006 Business and Economic Plan for Mental Health and Productivity (Global Business and Economic Roundtable on Addiction and Mental Health 2006) that focuses on workers' mental health.

In this chapter we further consider the saliency of this concern. We begin by defining labor market participation, employment, unemployment, and discouraged workers, and how these factors reflect the health of the economy and how, in turn, the health of the economy could be affected by a mentally healthy work force.

Investing in a Healthy Labor Force

In 2008, approximately 66% of people in the US 16 years of age or older participated in the labor force. Of these 154 million people, about 73% of men and 60% of women who are 16 years of age or older were in the labor force. Of those in the labor force, about 6% of working age people were unemployed.

Labor force participation is one of the key indicators used to measure the economy's health. The US Bureau of Labor Statistics (2009) defines the labor force as being comprised of people who are 16 years of age or over, and who either are employed or are actively seeking work. This means that the rate does not count people who are either retired, do not wish to work, cannot work, or have given up looking for work. The last group is referred to as discouraged workers. These are people who would like to work, but are no longer actively looking for employment; they are discouraged and no longer believe that they can find a job. Discouraged workers are people who move from the category of "unemployed" to "discouraged." Thus, labor force participation can decrease as the number of workers who become unable to work or who no longer seek work grows. The latter group increases as it becomes more difficult to find jobs and people give up hope of finding one. In some countries, many of these people will not be counted as unemployed and instead be in receipt of some form of disability benefit, as for instance in Great Britain in 2007, where there were more people registered as disabled due to mental health problems than there are people unemployed (Department of Work and Pensions 2007). Reduced labor market participation is a negative sign of the health of the population. Decreasing disability among workers and increasing employment opportunities for people who would like to work benefits the economy and renews hope.

Employment and unemployment rates are other important economic indicators. The former is the proportion of the labor force that find work; the latter indicates the proportion of people who are still looking. Both are related to the country's gross domestic product (GDP), or the total value of goods and services produced in a country. Thus, GDP is affected by productivity, which is defined as the rate at which goods and services are produced. Labor productivity is a critical component of productivity. Labor productivity is defined as the output per unit of labor input. Total labor productivity is affected by two factors: (1) the number of hours worked, and (2) the productivity of those employed. Thus, the GDP is related to the productivity of workers. The GDP can rise as the productivity of the labor force rises. Productivity can be affected if workers who are employed are not able to work at their full capacity because of a disability or underemployment. The GDP can also decrease if the number of workers employed decreases, and with it, the total number of work hours decreases, another possibility associated with mental disorders.

Economists have noted a relationship between the growth in GDP and the unemployment rate. This relationship has been referred to as Okun's Law. Recalling that the GDP is affected by the two factors: (1) the number of hours worked, and (2) the productivity of those working, we see that in the short run there is a relationship between factors (1), (2), and the unemployment rate.

If the annual increase in GDP is not sustained, there is a risk of an increase in the rate of unemployment. Assuming the size of the labor force does not significantly change, investments in the health of the working population that lead to increased productivity and decreased disability can support sustained economic growth and result in a stable unemployment rate.

On the other hand, in the short run, if we assume that the number of hours worked remains constant but all the workers become more productive, the growth in GDP can also lead to an increase in unemployment (as fewer workers are needed to produce the same amount of goods and services). This suggests that strategies to create a more healthy and thereby more productive workforce could also result in fewer employment opportunities for people who want to work. This underscores the need for strategies that consider the short-run implications of current investments that may make workers more healthy and productive, but also ensure that there are opportunities for employment.

Consequently, in addition to the issues of human rights and justice, there also may be an economic argument for a global focus on the mental health of workers. This holds true if: (1) mental disorders decrease the proportion of the population who are not in the labor force, (2) mental disorders decrease the employment opportunities for people who would like to work, or (3) mental disorders affect the ability of workers to do their jobs. The remainder of this chapter will examine the relationship of mental disorders and these factors.

In the next sections, the prevalence and effects of mental disorders on the working population and society will be discussed. For the purposes of this chapter, "mental disorders" include mood disorders, anxiety disorders, psychotic disorders, substance use disorders, and traumatic brain injuries (TBI). To provide a context, the prevalence of mental disorders in the general population will be described. Then the chapter will go on to explore what is known about the prevalence of mental disorders in the working population and the differences in the prevalence of these disorders with regard to occupation, age, sex, physical conditions, work environments, and work-related stress. This is followed by a discussion of how mental disorders affect productivity and who bears the immediate losses. To conclude, there will be a discussion of who should pay for interventions for mental disorders.

Prevalence of Mental Disorders in the General Population

In North America, the prevalence of mental disorders in the adult population during a 1-year period ranges from 12% (Statistics Canada 2003) to 26% (Kessler et al. 2005). Based on the US National Comorbidity Survey Replication, Kessler and colleagues (2005) reported that 26% of the adult population had at least one mental disorder during a 1-year period. Of people who had a mental disorder, 40% experienced a mild episode, whereas 37% had a moderate episode, and 22% had a serious episode. Kessler et al. (2005) identified anxiety disorders as the most prevalent class of disorders, experienced by 18% of the adult population. Next were mood disorders (10%) and substance disorders (4%). A review of epidemiological surveys in the EU also reported that annually, 27% of the population had a mental disorder with anxiety and depression-related disorders as the principal categories of disorders (Wittchen and Jacobi 2005).

In addition, in a review of the literature, Thurman and colleagues (1999) noted the annual incidence of TBI in the US population ranges from 132 to 367 per 100,000 people (Thurman et al. 1999). The mortality rate for TBI is approximately 20 per 100,000 (Thurman et al. 1999). The majority of the TBI are for minor or moderate injuries (Finkelstein et al. 2006). In Europe, a systematic review covering 14 countries reported an overall incidence per year of 235 cases per 100,000 people and an average fatality rate per year of 15 per 100,000 (Tagliaferri et al. 2006).

In contrast, the Canadian Community Health Survey (CCHS) 1.2 indicated that at least 12% of the Canadian population between the ages of 15 and 64 years suffer from a mental disorder or substance dependence during a 1-year period. As in the US study, the major categories of disorders from which people suffer are anxiety (i.e., generalized anxiety disorder, panic disorder), mood (i.e., major depressive episode), and substance abuse disorder (Offord et al. 1996).

Most of the research on mental disorders in the general population focuses on depression. In North America, 4–7% of people experience a major depressive episode during the year (Kessler et al. 2005; Murphy et al. 2000; Newman and Bland 1998; Offord et al. 1996; Statistics Canada 2003). Rates ranging between 3% and 10% have been observed in different EU countries (Kessler et al. 2005; Wittchen and Jacobi 2005).

It should also be noted that mental disorders often occur in combination with one another. Kessler and colleagues (2005) observed that during a 1-year period, 11% of people have two or more mental disorders. They reported significant relationships among mood, anxiety, and substance disorders. In addition, it has been noted that depression is associated with mild traumatic brain injury (Langlois et al. 2006; Vanderploeg et al. 2007). Deb and colleagues (1999) found that 1 year following a TBI, 13% of people have a diagnosis of depression and 7% have a diagnosis of panic disorder.

Prevalence of Mental Disorders in the Working Population

As in the general population, Sanderson and Andrews (2006) report the most prevalent mental disorders in the working population are depression (a mood disorder) and simple phobia (an anxiety disorder). Estimates from the US, the Netherlands, Australia, and Canada indicate that about 2–7% of the workforce experience depression (Kessler and Frank 1997; Laitinen-Krispijn and Bijl 2000; Lim et al. 2000; Wang et al. 2006). In addition, 5–6% of workers have simple phobia (Kessler and Frank 1997; Laitinen-Krispijn and Bijl 2000; Wang et al. 2006). Overall, about 10% of the working population has at least one mental disorder (Dewa et al. 2007).

Variations in Prevalence Rates

Differences by Sex and Age. There are differences in the prevalence of mental disorders by sex and age (Marcotte et al. 1999; Stewart et al. 2003). Among the employed population, the prevalence of a recent major depressive disorder among women is almost double that among men (10% vs. 6%) (Cohidon et al. 2009; Marcotte et al. 1999). However, men are twice as likely as women to experience a TBI (Langlois et al. 2006).

In addition, there is a higher prevalence of mental disorders among middle-aged workers (i.e., 40–45 years) than among either younger (20–24 years) or older workers (50+ years) (Godin et al. 2009; Marcotte et al. 1999).

Differences by Health Status. There is also evidence of an association between mental illness and physical conditions. In a population-based sample of Canadian workers, it was observed that approximately 31% of respondents experienced chronic work stress either alone or in combination with chronic physical condition and/or a psychiatric disorder (Dewa et al. 2007). About 46% indicated they had at least one chronic physical condition either alone or accompanied by chronic work stress or a psychiatric disorder. Finally, 11% had a mental disorder. A European six-country population-based study reported lower physical health status associated with workers with major depression (Alonso et al. 2004a).

Differences by Occupation. There is also evidence that there are differences in the prevalence of the different types of mental disorders among occupation groups. Results from the Ontario Health Survey Mental Health Supplement suggest there are relatively higher rates of comorbid mental disorders among professionals, middle management, and unskilled clerical workers (Dewa and Lin 2000). There have also been reports in the literature indicating higher prevalence of mental disorders among particular groups, including night security guards and secretaries (Alfredsson et al. 1991; Garrison and Eaton 1992). A French study reported higher prevalence rates of a mental disorder among people who were office or manual workers, as compared with those who were farmers, tradespeople, or managers/ professionals (Cohidon et al. 2009). TBI has a higher prevalence in military personnel serving in Iraq and Afghanistan as well as in rescue workers (Langlois et al. 2006).

Differences by Work Characteristics. Data are accumulating regarding the negative impact of work stress on both mental and physical health (Bourbonnais et al. 1998; Clumeck et al. 2009; Ibrahim et al. 2001). For instance, high-strain work has been observed to be related to a higher risk of self-reported mental health problems (Amick et al. 1998; Parent-Thirion et al. 2007). A two- to four-fold increase in risk of depression has been associated with inadequate employment (Grzywacz and Dooley 2003) and poor psychosocial working conditions (Clumeck et al. 2009); while there is a twofold increase in risk of any psychiatric condition among those experiencing work-related stress (Bourbonnais and Mondor 2001). Also, job strain and work stress are associated with emotional exhaustion, worse health status, and development of chronic physical conditions (Bourbonnais et al. 1998; Ibrahim et al. 2001; Shields 2004). In addition, job characteristics seem to have different impacts on men and women. For example, there is a high proportion of men with depression who have high job strain, whereas there is a high proportion of women with depression who have low social support in the workplace (Godin et al. 2009).

There is also a significant body of work examining the relationship between work satisfaction and mental health status. A recent meta-analysis based on 500 reported studies found a strong correlation between job satisfaction and mental health status (Faragher et al. 2005). There appears to be a significant relationship between job insecurity and lower mental health status (Korten and Henderson 2000); job insecurity is associated with four times the odds for depression and three times the odds for anxiety (D'Souza et al. 2003). The Fourth European Working Conditions Survey in 2005 surveyed 30,000 workers in all EU member states plus Croatia, Norway, Turkey, and Switzerland (Parent-Thirion et al. 2007). It found that the combination of high level of job strain and high job insecurity may increase the risk of depression by 14 times compared to those who have control over active, secure jobs. Long working hours were associated with depression in women, while sickness absence is positively associated with monotonous work, not learning new skills, and low control over work, as well as nonparticipation at work.

It appears that compared to the general population, the prevalence of mental disorders in the working population is lower. However, there is a significant proportion of workers who are affected by mental disorders. Mental disorders also may be creating barriers to employment for people. In addition, mental disorders are associated with a variety of factors, including demographic characteristics, occupation, and job characteristics. Taken together, this suggests that the picture of mental disorders among workers is complex. Thus, we might expect their effects on productivity may also be complex, as will be the measures needed to address them.

How Mental Disorders Affect Productivity

In North America, the estimated annual societal costs of mental disorders is US$83.1 billion (Greenberg et al. 2003). Between 30% and 60% of the societal cost of depression is related to losses associated with decreased work productivity (Greenberg et al. 2003; Stephens and Jourbert 2001). Recent Canadian estimates indicate productivity losses related to mental disorders total about C$17.7 billion annually (Lim et al. 2008). It has been reported that workplace absenteeism related to mental health problems accounts for about 7% of the total payroll and is one of the principal causes of absences (Watson Wyatt Worldwide 2000). Conservatively, the annual costs of mental disorders have been estimated to be EUR$320 billion (2004 prices) in the EU 25 plus Norway, Iceland, and Switzerland (Andlin-Sobocki et al. 2005). This estimate included nearly EU$3 billion for brain trauma injuries. Overall, 60% of costs were due to productivity losses. Estimates from the US indicate the lifetime costs of traumatic brain injury totaled $60 billion in 2000 (Finkelstein et al. 2006). About 85% of these losses were related to decreased productivity. Together, these estimates provide a strong incentive to promote worker mental health and to prevent injury.

The annual burden of mental disorders is primarily associated with costs arising from unemployment, decreased productivity, and disability. These will be the focus of the subsequent discussion. There are at least four main groups who bear the immediate costs. They are the public sector (i.e., government), employers, workers, and their families. Insurance companies represent a fifth stakeholder in countries where insurance companies play a major role in paying for disability benefits (e.g., the US and Canada).

Barrier to Labor Force Participation and Unemployment

There is evidence that certain disorders serve as barriers to labor force participation as well as factors contributing to unemployment in the working age population. It is important to note that because of the way that the data are reported in the literature, it is difficult to distinguish between the proportion of people who are not in the labor force and those who are unemployed. As a result, in this section, there will be a departure from the traditional economic definition of unemployment.

Individuals with mental disorders are less likely to be working (Bowden 2005; Ettner et al. 1997; Marwaha and Johnson 2004; Mechanic et al. 2002; Patel et al. 2002; Waghorn and Lloyd 2005), due to the inability either to obtain or to retain employment (Lerner et al. 2004). In their review of the literature, Marwaha and

Johnson (2004) reported that UK studies show 70–80% unemployment rates among people with schizophrenia, while other European studies indicate the range is from 80% to 90%. Employment rates of people with severe mental disorders, including schizophrenia, are lower in most countries than those of people with severe physical disabilities (Kilian and Becker 2007).

In addition, the literature suggests that mood disorders as well as anxiety and alcohol disorders are associated with a greater likelihood of not working (Alonso et al. 2004b; Ettner and Grywacz 2001; Marcotte et al. 1999, 2000). There are estimates that 6% of adults not working have an anxiety disorder, 7% have an affective disorder, and 10% have both (Comino et al. 2003). Moreover, there is evidence that depression is associated with significantly lower labor market participation for men, but not necessarily for women (Marcotte et al. 1999).

Among individuals with TBI, there are also low employment rates. A review of return-to-work rates following a traumatic brain injury showed that preinjury employment rates of 61–75% drop to 10–70% postinjury (Yasuda et al. 2001). For individuals with severe TBI, postinjury unemployment rates were 78% compared to 4% preinjury (Yasuda et al. 2001). Similar trends were reported by Mailhan and colleagues (2005). Results from a multinational population-based study also indicate that mild traumatic brain injury is associated with a higher probability of underemployment and not working (Vanderploeg et al. 2007).

Although a large proportion of the burden of not working falls on individuals with mental disorders and their families, costs to governments in most industrialized countries are also substantial, due both to losses in income tax revenues and also to increased use of the public "safety net," leading to higher expenditures on unemployment benefits as well as on disability insurance and welfare programs.

A report by the UK's Office of the Deputy Prime Minister reported psychiatrists were reluctant to encourage their patients to seek work, fearing they would not be successful and would have difficulties in re-obtaining benefits (Office of the Deputy Prime Minister 2004). This highlights the role that publicly funded benefits play in inadvertently creating disincentives to working. These incentives can generate more benefit-related expenditures. In addition to increased use of unemployment benefits, unemployment may be associated with increased disability and social welfare benefit enrollment. For example, it has been observed that as disability benefit levels decrease, employment rates increase and unemployment benefits decrease (Westerhout 2001).

In Europe, the state (with some contributions from employers and employees) bears the primary responsibility for long-term disability benefits through its social welfare budget. Poor mental health substantially impacts this budget. Rather than registering for lower and sometimes time-limited unemployment benefits, most Europeans with a mental illness-related long-term labor market absence register for the higher disability benefits or disability-related retirement pensions. For example, in Ireland 22% of people with mental health problems in 2002 were employed; only 3% were described as unemployed. The remainder were on disability benefits and deemed economically inactive (Central Statistics Office 2002).

Because disability benefits far outweigh expenditures on unemployment benefits, reducing the overall expenditure levels for these benefits has become a major priority for European governments. The share of disability benefits paid to those with mental health problems has been increasing: between 1990 and 2003 Finland's short-term sickness absence levels for formally diagnosed mental health problems increased by 93%. Twenty percent of all sickness benefits and 42% of all disability pensions were paid out for people with mental health problems; overall, around 50% of all people recorded as suffering from depression are on long-term disability pensions (Jarvisalo et al. 2005). In the US, up to 25% of Social Security Disability Insurance recipients qualify on the basis of mental illness (Beneficiaries in Current Payment Status 2004).

Although disability benefits can provide an important safety net for those unable to work, they also can act as a disincentive to returning to the labor market. In a recent study that included ten western European countries, disability benefit expenditure levels were negatively associated with the labor force participation of people with schizophrenia. In Italy, where expenditure was lowest, the highest rates of labor participation were observed (Kilian and Becker 2007). In many European countries individuals must be registered as disabled to obtain higher rates of disability and associated benefits; but once labeled as "economically inactive" rather than unemployed, individuals may jeopardize their future job prospects.

The difference between disability benefit levels and potential employment earnings is just one factor that is considered in the decision to seek employment; there are also bureaucratic regulations to overcome. Individuals also fear that if they do return to the labor market (and hence lose part, if not all, of their disability benefits), then it may take a considerable period of time to reclaim disability benefits if they subsequently lose their jobs (Organisation for Economic Cooperation and Development 2003).

Early Retirement

Another way that mental health disorders affect labor force participation is through early retirement (e.g., Brown et al. 2005; Harkonmaki et al. 2006; Karpansalo et al. 2005). It has been observed that workers with poor mental health functioning are more likely to plan early retirement (Harkonmaki et al. 2006). Dewa et al.'s (2003) results reflected a similar pattern, with older workers more likely to retire or to terminate their employment rather than return to work after a depression-related short-term disability. Karpansalo and colleagues (2005) reported that employed men with depression retired almost 2 years earlier than those without. In Austria, by 2006 almost 27% of all early retirements were due to mental health problems, an increase from 21% in 2001 (Biffl and Leoni 2009). Another estimate from the Organisation of Austrian Social Insurance Funds suggested that nearly 29% of all early retirements were due to mental health problems in 2007 (Hauptverband der Österreichischen Sozialversicherungsträger 2008).

The rise in early retirement can be problematic for both public and private pension plans. The Organization of Economic Cooperation and Development (OECD) (2004) warned member countries about the dangers of rising early retirement rates. As the workforce ages and there are fewer young workers to replace those retiring, more is drawn from pension plans than contributed to them. In the absence of new additions to the labor pool, the remaining workforce will have to pay higher premiums and work for a longer time period to sustain the pension system.

Absenteeism, Presenteeism, and Short-term Disability

Even among those who are employed, workplace productivity costs associated with mental illness are substantial; depression has been shown to be associated with more work loss and work cutback days than most chronic medical conditions (Dewa and Lin 2000; Grzywacz and Ettner 2000). A European six-country study indicated that workers with neurological problems experience more work loss than workers with major depression (Alonso et al. 2004a). In the Netherlands, it has been observed that excess impairment days are significantly associated with anxiety, mood, and substance disorders, as well as with accidental injury, including TBI (Buist-Bouwman et al. 2005).

Work-related productivity losses associated with mental disorders have been characterized as presenteeism and absenteeism. Presenteeism has been defined as showing up to work but working with impaired functioning. Absenteeism has been defined as an absence from work (e.g., sick days). In Sweden, more than two-thirds of the costs of all mental health problems are due to lost productivity (Institute of Health Economics 1997); while in the Netherlands, the total costs of employee absence and long-term disability related to mental problems have been estimated to be 0.5% of GDP or €1.44 billion per annum (Jarvisalo et al. 2005). One English study estimated the costs of adult depression in 2002 to be about €15.46 billion; the vast majority of costs were due to absenteeism and premature mortality (Thomas and Morris 2003). Similarly, a review of studies looking at the costs of schizophrenia reported that costs associated with lost productivity typically accounted for 60–80% of total cost (Knapp et al. 2004).

Presenteeism days represent a significant proportion of the burden of mental disorders at work (Dewa and Lin 2000; Kessler et al. 2003; Lim et al. 2000; Sanderson and Andrews 2006). It is estimated that for a 2-week period, US workers lose an average of 4 h per week due to depression-related presenteeism; this translates into US$36 billion (Stewart et al. 2003). In contrast, the average depression-related absenteeism productivity loss is about 1 h per week, equivalent to US$8.3 billion (Stewart et al. 2003). Thus, if mental health affects workers via presenteeism, GDP can be reduced even if the unemployment rate does not change.

In addition, because individuals are reluctant to be labelled with a mental disorder, much absenteeism and illness may be labelled as "stress" rather than given a psychiatric diagnosis. In the UK, "stress" (generally taken to refer to mental health problems such as anxiety or depression) is the most significant cause of absence. The latest data from the Labour Force Survey in Great Britain indicate that almost 49% of all self-reported work-related illness (excluding absence due to injuries) now are due to stress, depression, or anxiety problems (Health and Safety Executive 2008). The European Working Conditions Observatory (Houtman 2004) reported that in Germany, the long-term sickness absences due to mental health problems (including stress) increased by 74% between 1995 and 2002. More generally, across the EU stress was experienced by an average of 22% of working Europeans (Parent-Thirion et al. 2007).

In North America, many mid- to large-size companies provide disability benefits to workers. In their survey, Watson Wyatt (1997) found 53% of the firms they surveyed self-administered their short-term disability benefits, while 45% depended on third-party administration (e.g., insurance carriers), with the remainder covered by government programs. In the countries where companies provide these benefits, employers bear a significant part of the disability burden for individuals who remain employed. In contrast to publicly funded disability benefits, the employer-sponsored short- and long-term disability benefits not only provide income support during the absence, they also guarantee a position is waiting for the worker when s/he returns to work. In these terms, the worker is still a member of the company. As part of the benefit, the worker has a case manager, who is either part of the company's occupational health department or works for the disability management company, following his/her progress.

Mental and nervous disorder disability claims have increased. The Health Insurance Association of America (HIAA) (1995) reported that between 1989 and 1994, disability claims doubled. HIAA (1995) also found respondent companies spend between $360 and $540 million on disability claims related to this group of disorders. Over half of short-term mental or nervous disorder disability claims are attributed to depression (Conti and Burton 1994; Dewa et al. 2002; Health Insurance Association of America 1995). In Canada, mental illness-related short- and long-term disability accounts for up to a third of claims and for about 70% of the total costs, translating into $15 to $33 billion annually (Dewa et al. 2002). It should be noted that because there are no universal definitions for short-term and long-term disability and because criteria differ by disability insurance plans, the above esti-mates under-report the true costs associated with disability claims. The burden of disability is further compounded by administrative costs incurred by the employer in managing the claim through the occupational health and human resources departments.

Workers are also likely to bear a substantial portion of the decreased productivity cost, through reduced income (Wang et al. 2004) and failure to obtain promotions or raises in their salaries (Nicholas 1998; Scheid 1999). Slower economic growth due to lower worker productivity may also lead to declines in government tax revenues, including both individual and corporate income.

Decreased Productivity Related to Spillover Effects on Coworkers and Supervisors

Workers do not work in isolation; they have the potential to affect their work environments. As they struggle with the symptoms of their disorders, coworkers and supervisors will also be affected. For example, Hoge and colleagues (2005) found members of the armed forces who had a mental illness were more likely to have misconduct and legal problems than those who did not. In addition, Smith et al. (2002) observed that workers with depression experienced workplace conflict. Furthermore, studies have shown that workers suffering from depression are more likely to have difficulty focusing on tasks and meeting their quotas (Lerner et al. 2003; Wang et al. 2004). The impairment in functioning can also affect coworkers who may need to undertake the additional workload to meet a deadline or quota.

Much of the burden of "spillover" effects is again likely to fall on the employer. Some economists have postulated there is a "friction" period or a limited amount of time that a company takes to adjust to work disruptions related to a worker's disability (Koopmanschap and van Ineveld 1992). Assuming the workplace will be able to find substitutes for a worker, the friction costs are those incurred during this period as the company readjusts and the workplace returns to "normal." This raises the question of how long it takes for a workplace to return to "normal." For example, employers incur costs when the most experienced workers are lost, leaving the most inexperienced, and decreasing the potential overall workplace productivity (Dewa et al. 2002). Where individuals work as part of a team, the costs to a firm may be greater if the team is no longer capable of functioning as effectively. Although the more experienced workers can be replaced (incurring recruitment costs), it may take many years for a new generation of workers to acquire expertise and in-house knowledge – something not easily replaced. There are also the costs of hiring temporary workers until a permanent human or technological replacement can be made; the wages for temporary labor may be higher than that for permanent employees. Thus a friction period for an employer may not end when the absent worker's position is filled. It may extend to when the replacement can acquire all the skills and knowledge necessary to be fully productive in this position. However, replacement costs will be influenced by existing economic conditions – during recessions when available workers outnumber available jobs, such costs may be much reduced (Marcotte 2004).

The symptoms associated with mental disorders also may expose coworkers and managers to undesirable work situations. This, in turn, may make colleagues more vulnerable to increased stress and increase their risk of experiencing a mental disorder (D'Souza et al. 2005). In addition, if spillover from a colleague's mental disorder makes another worker less productive, or the overall firm less profitable, the other worker's job opportunities or wages may suffer as a result. There has been little research on mental disorders' impacts on coworker productivity, and careful analysis of this potential spillover effect is required.

Decreased Productivity Related to Spillover Effects on Families

Family members may also experience spillover effects from their relative's mental illness. Individuals who stop working because of mental health problems may require family member support. In turn, these family members may have to give up the opportunity of paid employment to provide such support. They may also incur additional out-of-pocket expenses to financially support a relative, and may themselves suffer from both physical and mental health problems as a result of caregiver burden, which again can entail significant costs to the health system. In one Italian study of the costs associated with schizophrenia, it was estimated that approximately 29% were due to the lost employment, forgone employment opportunities, and leisure time costs of family care providers (Tarricone et al. 2000).

An Australian study estimated that approximately 9% of people with bipolar disorders have caregivers who gave up work to care full-time. The lost wages of these 9,000-plus caregivers was estimated in 2003 to be almost A$200 million (Access Economics & SANE Australia 2003). One US-based study suggested that the costs of lost employment to family caregivers of people with schizophrenia alone was approximately 18% of the illness's total costs (Rice and Miller 1996).

Conclusion

At the outset of this chapter, we posed the question of whether there is sufficient evidence to make an economic argument for investing in the mental health of workers. We proposed that such would be the case if: (1) mental disorders decrease the proportion of the population who are not in the labor force, (2) mental disorders decrease the employment opportunities for people who would like to work, or (3) mental disorders affect the ability of workers to do their jobs. The evidence presented in this chapter indicates that mental disorders do affect productivity and that they have a number of effects, including decreased labor force participation, increased unemployment, and decreased ability to work. Furthermore, economies around the world are all facing similar challenges related to mental disorders in their work force, and these disorders affect not only the workers with the disorders but coworkers and families as well. Thus, there is support for the contention that the economic impact of mental disorders calls for a global focus on prevention and reduction of associated occupational disability.

In addition, the immediate costs of the effects of mental disorders are distributed across multiple stakeholders. These costs are interrelated; an attempt to decrease the burden for one group of stakeholders will inevitably affect other stakeholders. Thus, effectively addressing the issue requires all stakeholders to identify how they can contribute to the solution. In most industrialized countries, this could mean redefining the responsibilities of the public healthcare system, employers, and employees (Klepper and Salber 2005; McClanathan 2004; McDaid et al. 2005).

A clear challenge for all systems is to identify ways to provide incentives to encourage all stakeholders to invest in workplace mental health promotion. There have been comparatively few incentives for employers and workers individually to take the first step; although, for example, much work has now been undertaken in England to help make a business case for investing in mental well-being in the workplace (Black 2008; Dewe and Kompier 2008; NICE 2009). Perhaps each is waiting for government to make the first move. Given what is at stake, it is imperative that someone begins to move. But government cannot do it alone; there is a role for all stakeholders, just as there potentially are benefits for all. Employers, employees, the healthcare system, and the insurers must work with government to promote good workplace health. For its part, government can enrich the incentives for collaboration through financial incentives, such as the careful use of subsidies and regulations.

In addition to the desire to decrease the costs of mental health problems to the workplace, there may be other factors influencing workplace mental health policy development. In Europe, great emphasis is placed on the EU's 10-year objectives, set out in its Lisbon agenda (Lisbon European Council 2000), of increased productivity and economic development. EU policy makers are now taking steps to help facilitate a greater focus on workplace mental health, working across sectors and in partnership with national governments, employers, trade unions, and workplace health promotion professionals to address these issues (McDaid 2008). The goal of promoting social inclusion in the workforce of vulnerable groups, including people with mental health problems, is also an integral part of this approach (Commission of the European Communities 2008). As the costly role played by mental health disorders among workers and the working population becomes increasingly clear, North America may be confronted with the same choices. Rather than dwell on the costs, it will, hopefully, embrace the opportunities to build a stronger economy, and in the process, a healthy, more inclusive, and better supported workforce.

Acknowledgments Dr. Dewa gratefully acknowledges the support provided by her CIHR/PHAC Applied Public Health Chair. The Centre for Addiction and Mental Health receives funding from the Ontario Ministry of Health and Long Term Care to support research infrastructure. This article does not necessarily reflect their views. Any remaining errors are the sole responsibility of the authors.

References

Access Economics & SANE Australia (2003) Bipolar disorder: costs. An analysis of the burden of bipolar disorder and related suicide in Australia. SANE Australia, Melbourne
Alfredsson L, Akerstedt T, Mattsson M, Wilborg B (1991) Self-Reported health and well being amongst night security guards: a comparison with the working population. Ergonomics 34(5):525–530
Alonso J, Angermeyer MC, Bernert S, Bruffaerts R, Brugha TS, Bryson H, de Girolamo G, Graaf R, Demyttenaere K, Gasquet I, Haro JM, Katz SJ, Kessler RC, Kovess V, Lepine JP, Ormel J, Polidori G, Russo LJ, Vilagut G, Almansa J, Arbabzadeh-Bouchez S, Autonell J, Bernal M,

Buist-Bouwman MA, Codony M, Domingo-Salvany A, Ferrer M, Joo SS, Martinez-Alonso M, Matschinger H, Mazzi F, Morgan Z, Morosini P, Palacin C, Romera B, Taub N, Vollebergh WA (2004a) Disability and quality of life impact of mental disorders in Europe: results from the European Study of the Epidemiology of Mental Disorders (ESEMeD) project. Acta Psychiatr Scand Suppl (420):38–46

Alonso J, Angermeyer MC, Bernert S, Bruffaerts R, Brugha TS, Bryson H, de Girolamo G, Graaf R, Demyttenaere K, Gasquet I, Haro JM, Katz SJ, Kessler RC, Kovess, V, Lepine JP, Ormel J, Polidori G, Russo LJ, Vilagut G, Almansa J, Arbabzadeh-Bouchez S, Autonell J, Bernal M, Buist-Bouwman MA, Codony M, Domingo-Salvany A, Ferrer M, Joo SS, Martinez-Alonso M, Matschinger H, Mazzi F, Morgan Z, Morosini P, Palacin C, Romera B, Taub N, Vollebergh WA (2004b) Prevalence of mental disorders in Europe: results from the European Study of the Epidemiology of Mental Disorders (ESEMeD) project. Acta Psychiatr Scand Suppl (420):21–27

Amick BC 3rd, Kawachi I, Coakley EH, Lerner D, Levine S, Colditz GA (1998) Relationship of job strain and iso-strain to health status in a cohort of women in the United States. Scand J Work Environ Health 24(1):54–61

Andlin-Sobocki P, Jonsson B, Wittchen HU, Olesen J (2005) Cost of disorders of the brain in Europe. Eur J Neurol (12 Suppl 1):1–27

Beneficiaries in Current Payment Status (2004) http://www.ssa.gov/policy/docs/statcomps/di_asr/2004/sect01b.pdf. Accessed 1 Sept 2006

Biffl G, Leoni T (2009) Arbeitsplatzbelastungen, arbeitsbedingte Krankheiten und Invalidität. Österreichisches Institut für Wirtschaftsforschung. http://www.wifo.ac.at/wwa/servlet/wwa.upload.DownloadServlet/bdoc/S_2009_INVALIDITAET_35901$.PDF

Black C (2008) Working for a healthier tomorrow. TSO, London

Bourbonnais R, Comeau M, Vezina M, Dion G (1998) Job strain, psychological distress, and burnout in nurses. Am J Ind Med 34(1):20–28

Bourbonnais R, Mondor M (2001) Job strain and sickness absence among nurses in the province of Quebec. Am J Ind Med 39(2):194–202

Bowden CL (2005) Bipolar disorder and work loss. Am J Manage Care 11(3 Suppl):S91–S94

Brown J, Reetoo KN, Murray KJ, Thom W, Macdonald EB (2005) The involvement of occupational health services prior to ill-health retirement in NHS staff in Scotland and predictors of re-employment. Occup Med 55(5):357–363

Buist-Bouwman MA, de Graaf R, Vollebergh WA, Ormel J (2005) Comorbidity of physical and mental disorders and the effect on work-loss days. Acta Psychiatr Scand 111(6):436–443

Bureau of Labor Statistics (2009) Labor force statistics from the current population survey. http://www.bls.gov/cps/lfcharacteristics.htm. Accessed 28 Sept 2009

Central Statistics Office (2002) Quarterly national household survey statistical release, disability in the labour force second quarter. Central Statistics Office, Dublin

Clumeck N, Kempenaers C, Godin I, Dramaix M, Kornitzer M, Linkowski P, Kittel F (2009) Working conditions predict incidence of long-term spells of sick leave due to depression: results from the Belstress I prospective study. J Epidemiol Community Health 63(4):286–292

Cohidon C, Imbernon E, Gorldberg M (2009) Prevalence of common mental disorders and their work consequences in France, according to occupational category. Am J Ind Med 52(2):141–152

Comino EJ, Harris E, Chey T, Manicavasagar V, Penrose Wall J, Powell Davies G, Harris MF (2003) Relationship between mental health disorders and unemployment status in Australian adults. Aust NZ J Psychiatry 37(2):230–235

Commission of the European Communities (2008) European pact for mental health and wellbeing. Commission of the European Communities. http://ec.europa.eu/health/ph_determinants/life_style/mental/docs/pact_en.pdf. Accessed 1 Oct 2009

Conti DJ, Burton WN (1994) The economic impact of depression in a workplace. J Occup Med 36(9):983–988

D'Souza RM, Strazdins L, Clements MS, Broom DH, Parslow R, Rodgers B (2005) The health effects of jobs: status, working conditions, or both? Aust NZ J Psychiatry 29(3):222–228

D'Souza RM, Strazdins L, Lim LL, Broom DH, Rodgers B (2003) Work and health in a contemporary society: demands, control, and insecurity. J Epidemiol Community Health 57(11):849–854

Deb S, Lyons I, Koutzoukis C, Ali I, McCarthy G (1999) Rate of psychiatric illness 1 year after traumatic brain injury. Am J Psychiatry 156(3):374–378

Department of Work and Pensions (2007) Incapacity benefit payments in Great Britain. Department of Work and Pensions, London

Department of Work and Pensions and Department of Health (2008) Improving health and work: changing lives. Department of Work and Pensions, London

Dewa CS, Goering P, Lin E, Paterson M (2002) Depression-related short-term disability in an employed population. J Occup Environ Med 44(7):628–633

Dewa CS, Hoch JS, Lin E, Paterson M, Goering P (2003) Pattern of antidepressant use and duration of depression-related absence from work. Br J Psychiatry 183:507–513

Dewa CS, Lin E (2000) Chronic physical illness, psychiatric disorder and disability in the workplace. Soc Sci Med 51(1):41–50

Dewa CS, Lin E, Kooehoorn M, Goldner E (2007) Association of chronic work stress, psychiatric disorders, and chronic physical conditions with disability among workers. Psychiatr Serv 58(5):652–658

Dewe P, Kompier M (2008) Wellbeing and work: future challenges. Foresight, Government Office for Science, London

Ettner SL, Frank RG, Kessler R (1997) The impact of psychiatric disorders on labor market outcomes. Ind Labor Relat Rev 51(1):64–81

Ettner SL, Grywacz J (2001) Worker perceptions of how jobs affect health. J Occup Health Psychol 6(2):101–113

Faragher EB, Cass M, Cooper CL (2005) The relationship between job satisfaction and health: a meta-analysis. Occup Environ Med 62(2):105–112

Finkelstein EA, Corso PS, Miller TR (2006) The incidence and economic burden of injuries in the United States. Oxford University Press, Oxford

Garrison R, Eaton WW (1992) Secretaries, depression and absenteeism. Women Health 18(4):53–76

Global Business and Economic Roundtable on Addiction and Mental Health (2006) On the road to mental health in the workplace. Global Business and Economic Roundtable on Addiction and Mental Health, Toronto

Godin I, Kornitzer M, Clumeck N, Linkowski P, Valente F, Kittel F (2009) Gender specificity in the prediction of clinically diagnosed depression. Results of a large cohort of Belgian workers. Soc Psychiatry Psychiatr Epidemiol 44(7):592–600

Greenberg PE, Kessler R, Birnbaum HG, Leong SA, Lowe SW, Bergland PA, Corey-Lisle PK (2003) The economic burden of depression in the United States: how did it change between 1990 and 2000? J Clin Psychiatry 64(12):1465–1475

Grzywacz J, Ettner SL (2000) Lost time on the job: the effect of depression versus physical health conditions. Econ Neurosci 2(6):41–46

Grzywacz JG, Dooley D (2003) "Good jobs" to "bad jobs": replicated evidence of an employment continuum from two large surveys. Soc Sci Med 56(8):1749–1760

Harkonmaki K, Lahelma E, Martikainen P, Rahkonen O, Silventoinen K (2006) Mental health functioning (SF-36) and intentions to retire early among ageing municipal employees: the Helsinki Health Study. Scand J Public Health 34(2):190–198

Hauptverband der der österreichischen Sozialversicherungsträger (2008) Statistisches Handbuch der österreichischen Sozialversicherung 2008. Hauptverband der der österreichischen Sozialversicherungsträger, Vienna

Health and Safety Executive (2008) Estimated days (full-day equivalent) off work and associated average days lost per (full-time equivalent) worker and per case due to a self-reported work-related illness or workplace injury. Health and Safety Executive. http://www.hse.gov.uk/statistics/lfs/0708/swit1.htm. Accessed 1 Oct 2009

Health Insurance Association of America (1995) Disability claims for mental and nervous disorders. Health Insurance Association of America, Washington, DC

Hoge CW, Toboni HE, Messer SC, Bell N, Amoroso P, Orman DT (2005) The occupational burden of mental disorders in the US military: psychiatric hospitalizations, involuntary separations, and disability. Am J Psychiatry 162(3):585–591

Houtman ILD (2004) Work-related stress. European Foundation for the Improvement of Living and Working Conditions, Dublin

Ibrahim SA, Scott FE, Cole DC, Shannon HS, Eyles J (2001) Job strain and self-reported health among working women and men: an analysis of the 1994/5 Canadian National Population Health Survey. Women Health 33(1–2):105–124

Institute of Health Economics (1997) Läkemedlen och sjukvårdskostnaderna: några utvecklingslinjer. Institute of Health Economics.Lund, Sweden

Jarvisalo J, Andersson B, Boedeker W, Houtman I (eds) (2005) Mental disorders as a major challenge in the prevention of work disability: experiences in Finland, Germany, the Netherlands and Sweden. The Social Insurance Institution, Helsinki, Finland

Karpansalo M, Kauhanen J, Lakka TA, Manninen P, Kaplan GA, Salonen JT (2005) Depression and early retirement: prospective population based study in middle aged men. J Epidemiol Community Health 59(1):70–74

Kessler R, Frank RG (1997) The impact of psychiatric disorders on work loss days. Psychol Med 27:861–873

Kessler RC, Chiu WT, Demler O, Merikangas KR, Walters EE (2005) Prevalence, severity, and comorbidity of 12-month DSM-IV disorders in the National Comorbidity Survey Replication. Arch Gen Psychiatry 62(6):617–627

Kessler RC, Ormel J, Demler O, Stang PE (2003) Comorbid mental disorders account for the role impairment of commonly occurring chronic physical disorders: results from the National Comorbidity Survey. J Occup Environ Med 45(12):1257–1266

Kilian R, Becker T (2007) Macro-economic indicators and labour force participation of people with schizophrenia. J Ment Health 16(2):211–222

Klepper B, Salber P (2005) The business case for reform. Employers must demand needed change to create a sustainable system of care. Mod Healthc 35(41):22

Knapp M, Mangalore R, Simon J (2004) The global costs of schizophrenia. Schizophr Bull 2:297–293

Koopmanschap MA, van Ineveld BM (1992) Towards a new approach for estimating indirect costs of disease. Soc Sci Med 34(9):1005–1010

Korten A, Henderson S (2000) The Australian national survey of mental health and well-being. Common psychological symptoms and disablement. Br J Psychiatry 177:325–330

Laitinen-Krispijn S, Bijl RV (2000) Mental disorders and employee sickness absence: the NEMESIS study. Netherlands Mental Health Survey and Incidence Study. Soc Psychiatry Psychiatr Epidemiol 35(2):71–77

Langlois JA, Rutland-Brown W, Wald MM (2006) The epidemiology and impact of traumatic brain injury: a brief overview. J Head Trauma Rehabil 21(5):375–378

Lerner D, Adler DA, Chang H, Lapitsky L, Hood MY, Perissinotto C, Reed J, McLaughlin TJ, Berndt ER, Rogers WH (2004) Unemployment, job retention, and productivity loss among employees with depression. Psychiatr Serv 55(12):1371–1378

Lerner D, Amick BC 3rd, Lee JC, Rooney T, Rogers WH, Chang H, Berndt ER (2003) Relationship of employee-reported work limitations to work productivity. Med Care 41(5):649–659

Lisbon European Council. Lisbon European Council presidency conclusions [Internet]. 2000 [cited 2007 Apr 2]. Available from: http://ve.evint/vedocs/cns.Data/docs/pressData/en/ec/odocr1.en.htm.

Lim D, Sanderson K, Andrews G (2000) Lost productivity among full-time workers with mental disorders. J Ment Health Policy Econ 3(3):139–146

Lim KL, Jacobs P, Ohinmaa A, Schopflocher D, Dewa CS (2008) A new population-based measure of the economic burden of mental illness in Canada. Chron Dis Can 28(3):92–98

Mailhan L, Azouvi P, Dazord A (2005) Life satisfaction and disability after severe traumatic brain injury. Brain Inj 19(4):227–238

Marcotte DE (2004) Essential to understand the relationship between mental illness and work. Healthc Pap 5(2):26–31

Marcotte DE, Wilcox-Gok V, Redmon DP (2000) The labor market effects of mental illness. The case of affective disorders. In: Research in human capital and development. Stamford: JAI Press

Marcotte DE, Wilcox-Gök V, Redmon P (1999) Prevalence and patterns of major depressive disorder in the United States labor force. J Ment Health Policy Econ 2:121–131

Marwaha S, Johnson S (2004) Schizophrenia and employment: a review. Soc Psychiatry Psychiatr Epidemiol 39:337–349

McClanathan A (2004) Behavioral health benefits: key to a healthy workforce. Empl Benefits J 29(1):11–16

McDaid D (2008) Mental health reform: Europe at the cross-roads. Health Econ Policy Law 3(Pt 3):219–228

McDaid D, Curran C, Knapp M (2005) Promoting mental well-being in the workplace: a European policy perspective. Int Rev Psychiatry 17(5):365–373

Mechanic D, Bilder S, McAlpine DD (2002) Employing persons with serious mental illness. Health Aff 21(5):242–253

Murphy JM, Laird NM, Monson RR, Sobol AM, Leighton AH (2000) Incidence of depression in the Stirling County Study: historical and comparative perspectives. Psychol Med 30(3):505–514

Newman SC, Bland RC (1998) Incidence of mental disorders in Edmonton: estimates of rates and methodological issues. J Psychiatr Res 32:273–282

NICE (2009) Promoting mental wellbeing at work. National Institute for Health and Clinical Excellence, London

Nicholas G (1998) Workplace effects on the stigmatization of depression. J Occup Environ Med 40(9):793–800

Office of the Deputy Prime Minister (2004) Mental health and social exclusion. Social Exclusion Unit report. Office of the Deputy Prime Minister, London

Offord DR, Boyle MH, Campbell D, Goering P, Lin E, Wong M, Racine YA (1996) One-year prevalence of psychiatric disorder in Ontarians 15 to 64 years of age. Can J Psychiatry 41(9):559–563

Organisation for Economic Co-operation and Development (2003) Transforming disability into ability: policies to promote work and income security for disabled people. OECD, Paris

Parent-Thirion A, Fernández Macías E et al (2007) Fourth European Working Conditions survey. European Foundation for the Improvement of Living and Working Conditions, Dublin

Patel A, Knapp M, Henderson J, Baldwin D (2002) The economic consequences of social phobia. J Affect Disord 68(2–3):221–233

Rice DP, Miller LS (1996) The economic burden of schizophrenia: conceptual and methodological issues and cost estimates. In: Moscarelli M, Rupp A, Sartorius N (eds) Handbook of mental health economics and health policy, volume 1, schizophrenia. Wiley, London

Sanderson K, Andrews G (2006) Common mental disorders in the workforce: recent findings from descriptive and social epidemiology. Can J Psychiatry 51(2):63–75

Scheid TL (1999) Employment of individuals with mental disabilities: business response to the ADA's challenge. Behav Sci Law 17:73–91

Shields M (2004) Stress, health and the benefit of social support. Health Rep 15(1):9–38

Smith JL, Rost KM, Nutting PA, Libby AM, Elliott CE, Pyne JM (2002) Impact of primary care depression intervention on employment and workplace conflict outcomes: is value added? J Ment Health Policy Econ 5(1):43–49

Statistics Canada (2003) Canadian community health survey: mental health and well-being. http://www.statcan.ca/Daily/English/030903/d030903a.htm. Accessed 8 Sept 2003

Stephens T, Jourbert N (2001) The economic burden of mental health problems in Canada. Chron Dis Can 22(1):18–23

Stewart WF, Ricci JA, Chee E, Hahn SR, Morganstein D (2003) Cost of lost productive work time among US workers with depression. J Am Med Assoc 289(23):3135–3144

Tagliaferri F, Compagnone C et al (2006) A systematic review of brain injury epidemiology in Europe. Acta Neurochir 148(3):255–268

Tarricone R, Gerzeli S, Montanelli R, Frattura L, Percudani M, Racagni G (2000) Direct and indirect costs of schizophrenia in community psychiatric services in Italy. The GISIES study. Interdisciplinary Study Group on the Economic Impact of Schizophrenia. Health Policy 51(1):1–18

The Standing Senate Committee on Social Affairs, Science and Technology (2006) Out of the shadows at last transforming mental health, mental illness and addiction services in Canada. The Senate, Ottawa

Thomas C, Morris S (2003) Cost of depression among adults in England in 2000. Br J Psychiatry 183:514–519

Thurman DJ, Alverson C, Browne D, Dunn KA, Guerrero J, Johnson R, Johnson V, Langlois JA, Pilkey D, Sneizek JE, Toal S (1999) Traumatic brain injury: a report to congress. Centers for Disease Control and Prevention, Atlanta

Vanderploeg RD, Curtiss G, Luis CA, Salazar AM (2007) Long-term morbidities following self-reported mild traumatic brain injury. J Clin Exp Neuropsychol 29(6):585–598

Waghorn G, Lloyd C (2005) The employment of people with mental illness. AeJAMH 4(2 Suppl):1–43

Wang J, Adair CE, Patten SB (2006) Mental health and related disability among workers: a population-based study. Am J Ind Med 49:514–522

Wang PS, Beck AL, Berglund P, McKenas DK, Pronk NP, Simon GE, Kessler RC (2004) Effects of major depression on moment-in-time work performance. Am J Psychiatry 161(10):1885–1891

Watson Wyatt Worldwide (2000) Staying at Work 2000/2001 – the dollars and sense of effective disability management. Watson Wyatt Worldwide, Vancouver

Westerhout E (2001) Disability risk, disability benefits, and equilibrium unemployment. Int Tax and Public Finance 8(3):219–243

Wittchen HU, Jacobi F (2005) Size and burden of mental disorders in Europe – a critical review and appraisal of 27 studies. Eur Neuropsychopharmacol 15(4):357–376

World Health Organization (2005) Mental health action plan for Europe. Facing the challenges, building solutions. World Health Organization, Copenhagen

World Health Organization (2001) The World Health Report 2001 mental health: new understanding, new hope. WHO, Geneva

Wyatt W (1997) Staying @ work. Watson Wyatt Canada, Toronto

Yasuda S, Wehman P, Targett P, Cifu D, West M (2001) Return to work for persons with traumatic brain injury. Am J Phys Med Rehabil 80(11):852–864

Chapter 3
Stigma, Discrimination, and Employment Outcomes among Persons with Mental Health Disabilities

Marjorie L. Baldwin and Steven C. Marcus

Introduction

In major cultures around the world mental disorders are common and associated with substantial levels of disability. According to projections of the World Health Organization, depression-related disorders will be the single leading cause of global disease burden by the year 2020 (Murray and Lopez 1996). The 12-month prevalence of serious mental disorders in the U.S. is 6.5%, with higher rates among women than men (Kessler et al. 2008). Persons with mental disorders comprise the largest single diagnostic category of persons receiving SSDI or SSI (McAlpine and Warner 2002), and tend to be among the most severely disabled recipients (Estroff et al. 1997).

Some of the most important factors contributing to the disease burden associated with mental disorders are the poor outcomes this group experiences in the labor market. Persons with mental disorders have significantly lower wages and employment rates than nondisabled persons or persons with physical disorders (Baldwin 1999). Recent estimates indicate serious mental disorders were associated with a loss of $193 billion in personal earnings in the U.S. in 2002, three-fourths attributed to lower wages among persons with mental disorders who are employed, one-fourth attributed to lost wages among those who are not (Kessler et al. 2008). In part due to their poor prospects in the labor market, persons with mental health disorders have lower socioeconomic status, on average, and greater risk of living in poverty, than persons with physical disorders (Dohrenwend et al. 1992).

The low wages and employment rates of persons with mental disorders are surely related to the social, emotional, and cognitive limitations characteristic of mental health disorders, but stigma and discrimination may also play a part. Studies that rank health conditions by the degree of stigma they elicit consistently find that mental disorders generate some of the strongest negative attitudes, with little change over

M.L. Baldwin (✉)
Department of Economics, W.P. Carey School of Business,
Arizona State University, PO Box 873806, Tempe, AZ 85287-3806, USA
e-mail: Marjorie.Baldwin@asu.edu

I.Z. Schultz and E.S. Rogers (eds.), *Work Accommodation and Retention in Mental Health,*   53
DOI 10.1007/978-1-4419-0428-7_3, © Springer Science+Business Media, LLC 2011

the last three decades (Tringo 1970; Albrecht et al. 1982; Royal and Roberts 1987; Yuker 1987; Westbrook et al. 1993). Persons with mental health disorders experience stigma comparable to persons with AIDS or ex-convicts. There is indirect evidence that stigma translates to discrimination in the workplace: mental disorders are the health condition most frequently cited in employment discrimination charges filed under the Americans with Disabilities Act (Moss et al. 1999).

To the extent that labor market discrimination limits employment opportunities and reduces the wages of persons with mental disorders, it imposes substantial costs on an already burdened population. Some of the costs are obvious: decreased earning potential and denied access to employment-related health insurance coverage. Other costs are subtler. Employment is a source of role identification, social acceptance, and a measure of success in most societies. Denial of access to the labor market, or segregation into poorly paid, second tier jobs, can add to the sense of failure and stigma already associated with mental disorders (Priebe et al. 1998; Strauss and Davidson 1997).

It is important, therefore, to understand the magnitude of the problem. To what extent are the low wages and employment rates of persons with mental disorders explained by functional limitations? What part can reasonably be attributed to employer discrimination? How can these two components be measured? How does the relative importance of functional limitations versus discrimination vary across different types of mental disorders? Here we begin to answer the questions with a review of the literature on mental health disorders and work disability, a description of methods used by economists to estimate the impact of discrimination, and a summary of our research that applies these methods to persons with mental health disorders.

Background

Mental Disorders and Work Disability

Results of numerous studies show mental disorders can have profound effects on an individual's ability to function in the labor market (Druss et al. 2000; Dewa and Lin 2000; Yelin and Cisternas 1997; Ormel et al. 1994). Using data from the 1994 to 1995 National Health Interview Disability Survey, Druss and colleagues (2000) examined the association between functional disability and the presence of a mental health disorder, a general medical condition, or combined mental/medical conditions. Functional limitations were examined across three domains: social, cognitive, and physical. Respondents reporting mental health disorders were more likely than those with general medical disorders to report difficulties in social and cognitive functioning, and to report difficulties in multiple functional domains. The authors concluded that general medical conditions primarily affect physical functioning, whereas mental health conditions lead to deficits in higher order social and cognitive skills that may be particularly important for success in the workplace.

Using data for a sample of employed persons from the 1990 to 1991 Ontario Health Survey (Mental Health Supplement), Dewa and Lin (2000) examined the association between chronic physical disorders, psychiatric disorders, and work disability. Results show that both chronic physical disorders and psychiatric disorders are associated with increased levels of work disability (as measured by days of work absences and days of reduced productivity), but the effects of physical and mental disorders are different. Chronic physical disorders increase both measures of work disability; whereas a mental disorder alone has no significant impact on days absent from work, but significantly increases days of reduced productivity.

Ormel and colleagues (1994) examined the relationship between psychiatric status and disability using data collected by the World Health Organization Collaborative Study on Psychological Problems in General Health Care. Results suggested that psychiatric disorders are associated with substantial levels of disability in major cultures around the world, including cognitive, motivational, and emotional impairments that affect the highest level functional capacities of human beings. The authors concluded that moderate levels of psychiatric disorders are more disabling than moderate levels of physical disorders.

All types of mental disorders have a substantial impact on outcomes in the labor market, but recent studies indicate that different categories of mental disorders are associated with different measures of disability. Bassett et al. (1998), for example, analyzed functional disability and its relationship to diagnosed mental disorders, health care utilization, and receipt of disability benefits, using data from the 1980 to 1981 Eastern Baltimore Health Survey. They found that mood disorders were associated with elevated utilization of general medical or general mental health visits; schizophrenia, mood disorders, substance use, and anxiety disorders were associated with significant increases in utilization of specialized mental health visits; and schizophrenia and cognitive disorders were associated with greater likelihood of receiving disability payments. Dewa and Lin (2000) reported that affective disorders have a stronger association with days of reduced productivity than do other mental disorders (see also Chap. 2, Dewa and McDaid 2010).

Taken together, studies show that specific diagnostic categories of mental disorders are significant predictors of health care utilization, reduced productivity, and receipt of disability benefits. The association between specific diagnoses of mental disorders and high health care and disability costs may reinforce employer prejudices regarding persons with mental disorders and deter some employers from hiring persons with mental conditions.

Yet research suggests that success in the workplace, and the self-esteem it accords, are particularly important for persons with mental disorders (Priebe et al. 1998; Strauss and Davidson 1997; Ormel et al. 1994). Based on results from the WHO data, Ormel and colleagues (1994) argued that psychiatric disorders cause impairments in functioning that can lead to reduced social interaction and loss of self-esteem, leading to further psychiatric distress. Priebe and colleagues (1998) studied employed and unemployed persons with schizophrenia or schizoaffective disorder and found that employed persons exhibited significant advantages, not only with respect to finances, but also on measures of global and psychological well-being. They concluded that work is an important source of role identity, social interaction, and structured time that is associated with improved quality of life and recovery for persons with serious mental disorders.

Mental Disorders and Stigma

Persons with serious mental disorders are frequently unable to obtain good jobs because of prejudice among key members in their communities: employers, coworkers, or customers. Numerous studies have documented the public's widespread endorsement of stigmatizing attitudes toward persons with mental illness (e.g., Westbrook et al. 1993; Corrigan et al. 2000; Sartorius 2004; Schulze 2007; Jenkins and Carpenter-Song 2009). These attitudes have an adverse impact on the probability of obtaining and keeping good jobs (Brockington et al. 1993; Hamre et al. 1994; Wahl 1999; Baldwin and Marcus 2006; Jenkins and Carpenter-Song 2009). Thus stigma (that is, negative attitudes toward a group) leads to discrimination or unfair actions against the disfavored group.

Social cognitive theories provide an especially promising approach to understanding stigma against those with mental illness. Prominent among these is attribution theory. As developed by Weiner, attribution theory is fundamentally a model of human motivation and emotion based on the assumption that individuals search for causal understanding of everyday events (Weiner 1980, 1983, 1985, 1993, 1995). When encountering successful or unsuccessful outcomes people ask themselves why one outcome occurred vs. another (for example, "Why did I get a pay raise?" "Why can't a person with mental illness care for himself?"), and then attempt to explain those outcomes in terms of attributes perceived in themselves or others. Weiner explores controllability and stability as attributes often used as the basis for causal explanations (Weiner 1985, 1993, 1995); Corrigan and Kleinlein (2004) extend the model to consider other attributes commonly associated with mental disorders.

Controllability attributions are particularly fruitful for outlining the relationship between stigmatizing attitudes toward persons with serious mental disorders and the discriminatory actions that may follow. Controllability refers to the amount of perceived influence an individual may be able to exert over a situation or a disability (Weiner 1985, 1993, 1995). Weiner's theory suggests that controllability attributions are associated with positive or negative emotional responses and corresponding behavior. Persons who are perceived as unable to control a negative outcome (e.g., cancer patients) are reacted to with pity and helping behaviors (Weiner et al. 1988; Dooley 1995; Menec and Perry 1998). Conversely, persons who are perceived to be able to control a negative outcome (e.g., persons with alcohol or drug dependence) are more likely to be held responsible and reacted to with anger and punishing behaviors (Weiner et al. 1988; Dooley 1995; Graham et al. 1992, 1997; Rush 1998). Punitive responses to controllability attributions may involve withholding opportunities for competitive jobs or good housing (Pescosolido et al. 1999; Corrigan 1998).

Research suggests that mental behavioral disorders are viewed as more controllable than physical disorders (Weiner et al. 1988; Lin 1993). If the symptoms of mental disorders are perceived as controllable, punishing behaviors like discrimination may be elicited. If, however, mental disorders are perceived as biological and beyond individual control, reactions may tend more toward pity and helping behavior (Weiner et al. 1988; Menec and Perry 1998; Lin 1993). The complex links between attributions of mental disorders, negative emotional responses, and punishing behaviors have not yet been examined in depth, but this model potentially describes the cognitive/emotional antecedents to discriminatory behavior.

In addition to controllability, other common attributions of mental disorders, including dangerousness, incompetence, and stability, can be integrated into the model. The perception that persons with mental disorders are dangerous is widespread and often fueled by media reports (Angermeyer and Matschinger 1996; Link et al. 1999; Pescosolido et al. 1999; Brockington et al. 1993). If the perceived danger is believed to be under the individual's control, reactions of anger and associated punishing behavior are likely. Even if the individual is perceived as unable to control threatening behaviors, perceptions of dangerousness may lead directly to fear. Most people respond to threats of any kind with apprehension; fear, in turn, is associated with avoidance behaviors (Johnson-Dalzine et al. 1996). Studies have shown that fearful reactions to media representations of mental disorders or substance abuse lead to the desire to "stay away" from individuals with mental disorders (Madianos et al. 1987; Levey et al. 1995; Angermeyer and Matschinger 2003). In the labor market, avoidance behavior can manifest itself as lost opportunities, for example employers refusing to hire persons with serious mental disorders.

A significant proportion of the public also views persons with mental disorders as incompetent, that is, incapable of making sound decisions or achieving major life goals in such important areas as work, relationships, and independent living (Rahav et al. 1984). When incompetence is a perceived attribute of mental disorders the response may be pity and helping behavior, but not helping that leads to independence. In the labor market perceptions of incompetence may lead to anger and punishment, if a team member, for example, is perceived to be compromising the productivity of the work group.

Note the apparent irony between the perceptions of controllability and incompetence: people with serious mental disorders are often viewed as able to make the life choices that lead to the disorders but unable to make other life choices. This attitude is manifest in the impact of public education programs that focus on mental disorders as "brain disorders." Although such programs seek to counter the stereotype that people with mental disorders are responsible for their disorders (e.g., "they do not choose schizophrenia, it is genetically inherited"), these messages may unintentionally convey the idea that people with mental disorders are not responsible for any of their behavior (i.e., they are incompetent).

Stability refers to the perceived persistence of an outcome over time (Weiner 1985, 1995). Some negative outcomes remain potent over time (e.g., recovery from a mental disorder may never be observed), while others wax and wane, and still others are resolved completely. Research suggests that stability attributions modify the intensity of emotional and behavioral responses to a stigmatizing attribute (Weiner et al. 1982; Roesch and Weiner 2001). Public perceptions of a disorder as stable (i.e., the prognosis for a psychotic disorder is unlikely to improve even with treatment) can modify or amplify the effects of controllability, dangerousness, and incompetence attributions on pity and anger. Stable attributions such as "he will never get better," for example, may lead to decreased helping behavior from others.

The attribution theory develops a model of stigma in which emotional responses and behavioral reactions stem from public perceptions and stereotypes of mental disorders. In particular, attributions of controllability, stability, dangerousness, contagion, and incompetence generate emotions of pity, anger, or fear, leading to helping or punishing behaviors by members of the community. In the labor market, punishing behaviors become discriminatory when members of a stigmatized group receive lower wages, or have lower probabilities of employment, than what is expected based on their relative productivity.

Research on disability-related discrimination supports the hypothesis that stigma is an important factor contributing to poor employment outcomes for persons with serious mental disorders. Relative to general medical disorders, mental disorders elicit stronger stigma rankings, lower employability rankings, and larger wage gaps (Baldwin and Johnson 1994). Evidence on employer hiring practices is consistent with studies showing stronger stigma against mental disorders than against general medical disorders. Among the businesses interviewed in a recent survey, 68% make a special effort to hire minorities, 41% make a special effort to hire persons with general medical disorders, but only 33% make a special effort to hire persons with mental disorders (Scheid 2000).

Workers' self-reports of job discrimination are consistent with the evidence that stigma plays an important role in determining employment outcomes. Surveys of mental health consumers indicate they experience stigma from many sources, including relatives and mental health providers, but discrimination and stigma are most commonly experienced in the process of seeking employment, or in the workplace itself (Gaebel and Baumann 2003). Respondents in a focus group of consumers with schizophrenia indicate that stigma presents a major obstacle in gaining access to important social roles, including employment (Schulze and Angermeyer 2003).

Still, it is irrefutable that mental disorders impose significant functional limitations on many persons, even among those who are responsive to medication. The challenge for economists studying the effects of disability-related discrimination in the workplace is to disentangle the effects of avoidance and punishing behaviors associated with stigma from the effects of functional limitations on worker productivity. The following section describes the econometric methods that have been developed to identify these separate effects on employment outcomes.

Estimating the Effects of Discrimination

Definition

According to economic theory, relative employment and wage rates reflect differences in worker productivity and differences in the desirability of different jobs. All else equal, more productive workers are more likely to be employed and earn higher average wages than less productive workers, providing workers with incentives to acquire "human capital" (education, training, or work experience that enhances productivity). Dangerous or unpleasant jobs must pay higher average wages, all else equal, than more pleasant jobs, thus providing incentives for workers to accept employment offers in less attractive jobs. Wage and employment differences associated with education or danger are not considered discriminatory; labor market discrimination occurs only when there are differences in relative employment rates or wages (or other employment outcomes) that cannot be explained by differences in worker productivity or job characteristics. Formally, *labor market discrimination* occurs when *two groups of workers* with *equal average productivity* have different average wages or different opportunities for employment *in similar jobs*.

This economists' definition of discrimination differs in important respects from the colloquial understanding of the term. First, the economic definition of discrimination relates to groups, not individuals. We expect random variations in wages and employment prospects across individuals, so economic models cannot determine whether the poor employment experiences or outcomes of any one individual are the result of discrimination or random events. Thus, discrimination is identified across groups where we are able to compare average wages and employment rates, and the random events or differences among individuals cancel out within each group.

A second important feature of the economists' definition of discrimination is that it compares groups of workers with equal average productivity doing similar jobs. To separate out the effects of discrimination, therefore, the econometric models control for differences in the average productivity and job characteristics of workers in the two comparison groups. In studies of discrimination against African Americans or women, this is managed by including variables for workers' education, experience, occupation, and so forth, in the estimating equations. The problem of controlling for productivity differences is even more complex in studies of discrimination against workers with disabilities because the functional limitations associated with a disability also reduce worker productivity in many jobs. Thus, measures of functional limitations must be included as additional control variables in the employment or wage models. To study discrimination against persons with mental disorders it is particularly important to have data that include good measures of cognitive and emotional limitations.

We have analyzed the effects of stigma and discrimination on employment outcomes of persons with mental disorders using methods economists have applied to women, African Americans, Hispanics, and persons with physical disabilities. Empirically, the estimate of discrimination is the wage or employment differential that remains after controlling for between-group differences in functional limitations, worker characteristics (such as education and work experience), and job characteristics (such as occupation and union membership). The various methods include using a binary variable (i.e., a two-level measure such as "yes/ no") to measure the impact of minority group membership on outcomes, or a decomposition approach which was introduced to the economics literature by Oaxaca (1973). Either method requires data with good measures of productivity-related characteristics to be consistent with the definition of labor market discrimination noted above.

Ideally, studies of disability-related discrimination would utilize large, nationally representative data sets that include clinically valid information about persons with mental disorders and key labor market outcomes (wages, employment status, etc.). To control for differences in average productivity of persons with and without mental disorders, the data should also include measures of: (a) human capital (education, work experience, functional limitations, other health conditions); (b) job characteristics (union membership, industry, occupation, public or private sector); (c) demographics (gender, race, geographic region); and (d) functional limitations associated with mental disorders (emotional, social, cognitive limitations).

Unfortunately, the required elements noted above are not typically available together in the same data set. Surveys of earnings and income (e.g., *Survey of Income and Program Participation*; *National Longitudinal Survey*) have good measures of employment outcomes and job characteristics, but lack clinical information about mental disorders and detailed health data. Health-related surveys (e.g., *National Health Interview Survey*; *Medical Expenditure Panel Survey*) present exactly the opposite problem. Researchers conducting empirical studies of disability-related discrimination use whichever data are best suited to their particular problem and include appropriate caveats regarding missing data elements.

Models Predicting Discrimination

The binary variable approach is one straightforward method of estimating the effects of discrimination on employment outcomes. Essentially, a binary variable identifying persons with mental disorders is included in an employment model with controls for human capital, job characteristics, and functional limitations. The model is estimated on data that include persons with and without a mental disorders. The model compares the predicted probability of employment (or other employment outcomes) for the disadvantaged group (i.e. persons with a mental disorder) with their predicted probability of employment without the stigmatizing effect of a mental disorder. The difference in predicted probabilities is an econometric estimate of discrimination (see Appendix for additional detail and formulae for the estimation).

One limitation to the binary approach is estimating the model from a pooled sample, which implicitly assumes the underlying employment functions are identical for the disadvantaged and comparison groups. That is, the model assumes persons with and without mental disorders can expect the same relationships between probabilities of employment and an additional year of education, work experience, or any other characteristic in the model. In fact, persons with mental disorders may benefit more (or less) from additional education or experience than persons without mental disorders, so the entire structure of the employment function may differ between the two groups.

Another approach, called the decomposition approach, has two important advantages over the binary variable approach. First, it captures underlying differences in employment functions for the disadvantaged group and the controls, because

Table 3.1 Hypothetical example predicting employment rates

	Number of persons (n)	Observed employment rate	Education		Predicted employment rate (conditional on education)
			High school graduate	College graduate	
Persons without mental disorders	1,000	95%	25% ($n=250$)	75% ($n=750$)	95% ($n=950$)
Persons with mental disorders	100	78%	50% ($n=50$)	50% ($n=50$)	90% ($n=90$)

employment functions are estimated separately for the two groups. Second, it provides details on the specific differences in characteristics that, in addition to the possibility of discrimination, may contribute to differences in employment outcomes (see Appendix for further descriptions and formulae).

The rationale underlying the decomposition formula is quite straightforward. Consider the hypothetical example of decomposition results comparing persons with and without mental disorders in Table 3.1. The observed employment rate for persons without mental disorders is 17 percentage points higher than the rate for persons with mental disorders (95% vs. 78%). Part of the employment gap clearly is explained by differences in educational attainment of the two groups (75% vs. 50% college graduates). Suppose employment rates are 100% for college graduates but only 80% for high school graduates. We would expect 95% of persons without mental disorders to be employed (all 750 college graduates and 200 high school graduates, 950 persons total). Similarly, we would expect 90% of persons with mental disorders to be employed (all 50 college graduates and 40 high school graduates, or 90 persons total). Thus five percentage points of the gap in employment rates are "explained" by differences in education. The remaining 12 percentage points are "unexplained" and may be associated with discrimination against persons with mental disorders.

More likely, some part of the unexplained gap is associated with omitted variables. In this example, the only control variable is education; differences in job experience, functional limitations, etc., have not been taken into account. We would expect differences between persons with and without mental disorders in these variables as well, so the actual discrimination effect would be smaller than 12 points. Note that in this simplified example employment structures for the two groups are identical (100% of college graduates and 80% of high school graduates employed). Thus the binary model would yield similar overall results.

Limitations of the Empirical Models

The example makes clear the main limitation of both empirical models of discrimination (binary variable and decomposition). To the extent that important measures of productivity-related characteristics are omitted from the models, the estimates will overstate the impact of discrimination on employment outcomes. Estimates of discrimination are, therefore, an "upper bound" and subject to the criticism that additional variables could explain more of the gap in outcomes.

Another limitation of the empirical methods is that one cannot infer the source of potential discrimination from the output. Typically, discrimination is attributed to stigma and prejudice, operationalized as the desire to avoid or punish members of an undesirable group. Although some empirical evidence is supportive of this hypothesis, it has not been subjected to rigorous study. Another possibility is that discrimination is motivated by misconceptions about the productive capacities of a group. Economists call this "statistical discrimination," which occurs when individuals with the same measured productive characteristics are treated differently on a systematic basis because of their group affiliation (Ehrenberg and Smith 1994).

Economists make no claim to having solved all the problems in analyzing the complex issue of labor market discrimination. However, while not perfect, these methods are a good first step in estimating the potential effects of discrimination on employment outcomes of persons with mental disorders, and in helping identify to what extent 3 accommodations, without additional sanctions against discrimination, can be expected to alleviate these disparate outcomes.

Empirical Estimates

We have applied the methods for estimating employment discrimination to several national data sets. Here we report results from applying the decomposition model described above to the 1999 Medical Expenditure Panel Survey (MEPS) (Baldwin and Marcus 2006). The MEPS is a nationally representative data set designed to gather detailed information on health conditions, utilization of health care, and its costs. The MEPS allows us to identify persons with mental disorders by DSM-IV diagnostic categories, and provides satisfactory measures of the emotional and cognitive limitations associated with mental disorders. Although not designed to be an employment survey, the MEPS includes some demographic and employment information. Our employment models control for functional limitations, demographic and family characteristics, human capital, and non-wage incomes.

As is typical for health surveys, the MEPS data do not contain the comprehensive measures of worker productivity we would like to have in an "ideal" model of discrimination. The most glaring omission is that the MEPS does not have a measure of work experience, a key aspect of human capital. This is a particular problem for us, because persons with mental disorders often have sporadic work histories reflecting the cyclic nature of the disorders. Whenever key variables that reflect productivity differences between two groups are missing from the employment models, estimates of discrimination are likely biased upward, and results must be interpreted with this caveat in mind.

Table 3.2 summarizes results for persons with psychotic disorders (e.g., schizophrenia, manic-depressive psychosis; ICD-9 = 295–298); anxiety disorders

Table 3.2 Decompositions of employment differentials from the MEPS

	Mental disorders		
	Mood disorders	Anxiety disorders	Psychotic disorders
Expected employment rate	0.733	0.765	0.513
Difference relative to persons with no mental disorder	0.163	0.130	0.382
Explained difference	79%	88%	72%
Unexplained difference	21%	12%	28%

Source: Medical Expenditure Panel Survey, Household Component, 1999

(e.g., obsessive-compulsive disorder, panic disorder; ICD-9 = 300); and mood disorders (e.g., major depression, dysthymia; ICD-9 = 311) compared to persons with no mental disorder.

The expected employment rate for persons with mood disorders is 73%, compared to 90% for persons with no mental disorder. Most of the differential is explained by differences in functional limitations, non-wage incomes, and gender (more than two-thirds of persons with mood disorders are female, compared to only half of the group with no mental disorder, and females have lower employment rates than males), leaving only 20% unexplained and possibly associated with stigma and disconnection. Results for persons with anxiety disorders are similar, with an even smaller (12%) unexplained difference. Persons with psychotic disorders, typically the most severe of mental disorders, have extremely low expected employment rates (51%), largely explained by differences in functional limitations, non-wage incomes, and education. Gender is not a relevant explanatory factor for this group, because the prevalence of psychotic disorders is roughly equal for men and women. Still, almost one-third of the employment differential is unexplained and possibly attributed to discrimination, a result consistent with the high levels of stigma associated with psychotic disorders.

Summary

Mental health disorders are a leading cause of work disability. Persons with serious mental disorders have poorer employment outcomes than persons without disabilities or persons with general medical disorders. The poor employment outcomes relate, in part, to the fairly severe emotional and cognitive limitations that can affect the productivity of workers with mental disorders. Mental disorders also elicit punishing and avoidance behaviors associated with stigma. Studies that rank health conditions by the degree of stigma they elicit consistently find mental disorders generate some of the strongest stigma, with little change in attitudes over the last three decades (Tringo 1970; Royal and Roberts 1987; Westbrook et al. 1993). Poor employment outcomes among persons with mental disorders may also relate, therefore, to the discriminatory actions that accompany stigma.

Economists have developed empirical methods to distinguish the relative importance of productivity-related differences vs. discrimination in determining the poor employment outcomes for a disadvantaged group. Applying the results to persons with mental disorders compared to persons without, we find the unexplained difference in employment rates vary from 12 percent (anxiety disorders) to 28 percent (psychotic disorders) of the overall gap in employment rates. The larger part of the employment differential is explained and attributed primarily to differences in functional limitations, non-wage incomes, and education (the latter applies only to psychotic disorders).

Results of analyses such as this can be used to inform employment policies and practices for persons with mental disorders. For example, workplace accommodations are designed to mitigate the effects of functional limitations so that persons with disabilities are able to perform the essential functions of a job for which they are qualified. Other chapters in this handbook describe the types of accommodations that may be needed to compensate for the emotional and cognitive limitations associated with mental disorders. Our results suggest, however, that accommodations alone are unlikely to improve employment rates for persons with mental disorders; rather, job accommodations must be accompanied by policies that address other deterrents to employment. For example, policies that ensure persons with mental disorders are able to retain their Social Security disability benefits and Medicare coverage when they attempt to enter the workforce can address the disincentive effects of non-wage incomes. Policies that enforce appropriate accommodations in colleges and universities can help address the education gap between persons with psychotic disorders and their non-disabled peers.

Conclusions

The report of the President's New Freedom Commission on Mental Health (2003) states that protecting and enhancing the rights of persons with mental disorders should be one of the primary goals of improved mental health policies. The Commission finds that many persons with mental disorders are underemployed, and identifies workplace discrimination as an important contributing cause. The report states that the ADA "has not fulfilled its potential to prevent discrimination in the workplace. Workplace discrimination, either overt or covert, continues to occur" (p. 34). Without a better understanding of the role of stigma and discrimination, and the complexity of other factors that contribute to the poor employment outcomes of persons with mental disabilities in the competitive labor market, public policies to address this issue are likely to continue to fail.

Appendix

Binary Model

Define the binary variable M_i such that:

$M_i = 0$ if the ith individual is in the comparison group (C)

$M_i = 1$ if the ith individual is a person with a mental disorder (MD)

Also define a vector of variables, Z_i, that includes controls for human capital, job characteristics, demographics, and functional limitations. Parameters of the employment function for the pooled sample can be estimated using a logistic regression model (where the dependent variable equals one if employed, zero otherwise). Predicted probabilities of employment, $\hat{P}$, for each group are:

$$\hat{P}_C = \left[\frac{1}{n_C} \sum_{i=1}^{n_C} F(\hat{\beta} Z_i) \right] (\text{comparsion group})$$

$$\hat{P}_{MD} = \left[\frac{1}{n_{MD}} \sum_{i=1}^{n_{MD}} F(\hat{\beta} Z_i + \hat{\alpha} M_i) \right] (\text{group with mental disorders}) \tag{3.1}$$

The expression $\hat{\beta}$ represents estimated coefficients of the control variables in the employment function, that is, determinants of the relationships between an additional year of education, experience, etc., and the predicted employment rate. The expressions $F(\hat{\beta} Z_i)$ and $F(\hat{\beta} Z_i + \hat{\alpha} M_i)$ are predicted probabilities of employment for the ith individual in the comparison and disadvantaged groups, respectively. So the predicted probabilities for each group $(\hat{P}_C, \hat{P}_{MD})$, are simply the averages across the (n_C, n_{MD}) numbers of individual probabilities in the group.

If discrimination reduces the probability of employment for persons with mental disorders beyond what would be expected given their human capital, job characteristics, and functional limitations, then $\hat{\alpha} < 0$. In this case, the estimated effect of discrimination is the difference $\tilde{P}_{MD} - \hat{P}_{MD}$, where

$$\tilde{P}_{MD} = \left[\frac{1}{n_{MD}} \sum_{i=1}^{n_{MD}} F(\hat{\beta} Z_i) \right],$$

the probability of employment for the disadvantaged group without the stigmatizing penalty associated with a mental disorder.

Decomposition Model

Employment functions are estimated separately for persons with and without mental disorders with the same control variables as described above, except that the binary variable identifying the disadvantaged group is omitted. The predicted employment rate for each group $(\hat{P}_j)$ can be expressed as:

$$\hat{P}_j = \left[\frac{1}{n_j} \sum_{i=1}^{n_j} F(\hat{\beta} Z_{ij}) \right],\tag{3.2}$$

where n_j is the number of persons in the group ($j = C$ or MD). After some simple algebraic manipulation, the difference in employment rates can be decomposed into an "explained" part attributed to differences in human capital, job characteristics, or functional limitations; and an "unexplained" part attributed, in some part, to discrimination against persons with mental disorders:

$$\hat{P}_C - \hat{P}_{MD} = \left[\frac{1}{n_C} \sum_{i=1}^{n_C} F(\hat{\beta}_C Z_{iC}) - \frac{1}{n_{MD}} \sum_{i=1}^{n_{MD}} F(\hat{\beta}_C Z_{iMD}) \right]$$
$$+ \left[\frac{1}{n_{MD}} \sum_{i=1}^{n_{MD}} F(\hat{\beta}_C Z_{iMD}) - \frac{1}{n_{MD}} \sum_{i=1}^{n_{MD}} F(\hat{\beta}_{MD} Z_{iMD}) \right]\tag{3.3}$$

To see this, note that the first term on the right hand side of (3.3) reflects differences in employment rates attributed to differences in productivity-related characteristics (differences in the vectors Z_j) between the comparison group and persons with mental disorders (the "explained" part of the difference in employment rates). The second term reflects differences in employment rates attributed to differences in the structure of employment functions (the coefficients β) for the two groups (the "unexplained" part).

References

Albrecht GL, Walker VG, Levy JA (1982) Social distance from the stigmatized: a test of two theories. Soc Sci Med 16:1319–1328

Angermeyer MC, Matschinger H (1996) The effect of violent attacks by schizophrenic persons on the attitude of the public towards the mentally ill. Soc Sci Med 43(12):1721–1728

Angermeyer MC, Matschinger H (2003) The stigma of mental illness: effects of labeling on public attitudes towards people with mental disorders. Acta Psychiatr Scand 108:304–309

Baldwin ML (1999) The effects of impairments on employment and wages: estimates from the 1984 and 1990 SIPP. Behav Sci Law 17(1):7–27

Baldwin ML, Johnson WG (1994) Labor market discrimination against men with disabilities. J Hum Resour 29(1):1–19

Baldwin ML, Marcus SC (2006) Perceived and measured stigma among persons with serious mental disorders. Psychiatr Serv 57(3):388–392

Bassett SS, Chase GA, Folstein MF, Regier DA (1998) Disability and psychiatric disorders in an urban community: measurement, prevalence, outcomes. Psychol Med 28:509–517

Brockington IF, Hall P, Levings J, Murphy C (1993) The community's tolerance of the mentally ill. Br J Psychiatry 162:93–99

Corrigan PW (1998) The impact of stigma on severe mental illness. Cogn Behav Pract 5:201–222

Corrigan PW, Kleinlein P (2004) The impact of mental illness stigma. In: Corrigan PW (ed) On the stigma of mental illness: implications for research and social change. American Psychological Association, Washington, DC, pp 11–44

Corrigan PW, River LP, Lundin RK, Wasowski KU, Campion J, Mathisen J et al (2000) Stigmatizing attributions about mental illness. J Commun Psychol 28:91–103

Dewa CS, Lin E (2000) Chronic physical illness, psychiatric disorder and disability in the workplace. Soc Sci Med 51:41–50

Dewa C, McDaid D (2010) Investing in the mental health of the labor force: epidemiological and economic impact of mental health disabilities in the workplace. In: Schultz IZ, Sally Rogers E (eds) Handbook of work accommodation and retention in mental health. Springer, New York, pp 33–52

Dohrenwend BP, Levav I, Shrout PE, Schwartz S, Naveh G, Link BG et al (1992) Socioeconomic status and psychiatric disorders: the causation-selection issue. Science 255(5047):946–952

Dooley PA (1995) Perceptions of the onset controllability of AIDS and helping judgments – an attributional analysis. J Appl Soc Psychol 25(10):858–869

Druss BG, Marcus SC, Rosenheck RA, Olfson M, Tanielian T, Pincus HA (2000) Understanding disability in mental and general medical conditions. Am J Psychiatry 157(9):1485–1491

Ehrenberg RG, Smith RS (1994) Modern labor economics: theory and public policy, 5th edn. Harper Collins, New York, NY

Estroff S, Patrick D, Zimmer C, Lachicotte J (1997) Pathways to disability income among persons with severe, persistent psychiatric disorders. Milbank Q 75(4):495–532

Gaebel W, Baumann AE (2003) Interventions to reduce the stigma associated with severe mental illness: experiences from the open the doors program in Germany. Can J Psychiatry 48(10):657–662

Graham S, Hudley C, Williams E (1992) Attributional and emotional determinants of aggression among African-American and Latino young adolescents. Dev Psychol 28(4):731–740

Graham S, Weiner B, Zucker GS (1997) An attributional analysis of punishment goals and public reactions to O. J. Simpson. Pers Soc Psychol Bull 23(4):331–346

Hamre P, Dahl A, Malt U (1994) Public attitudes to the quality of psychiatric treatment, psychiatric patients, and prevalence of mental disorders. Norweigan J Psychiatry 74:1464–1480

Jenkins JH, Carpenter-Song EA (2009) Awareness of stigma among persons with schizophrenia. J Nerv Ment Dis 197(7):520–529

Johnson-Dalzine P, Dalzine L, Martin-Stanley C (1996) Fear of criminal violence and the African-American elderly: assessment of crime prevention strategy. J Negro Edu 65:462–469

Kessler RC, Heeringa S, Lakoma MD, Petukhova M, Rupp AE, Schoenbaum M et al (2008) Individual and societal effects of mental disorders on earnings in the United States: results from the National Comorbidity Survey Replication. Am J Psychiatry 165(6):703–711

Levey S, Howells K, Cowden E (1995) Dangerousness, unpredictability and the fear of people with schizophrenia. J Forensic Psychiatry 6(1):19–39

Lin Z (1993) An exploratory study of the social judgments of Chinese college students from the perspectives of attributional theory. Acta Psychologica Sinica 25:155–163

Link BG, Phelan JC, Bresnahan M, Stueve A, Pescosolido BA (1999) Public conceptions of mental illness: labels, causes, dangerousness, and social distance. Am J Public Health 89(9):1328–1333

Madianos MG, Madianou D, Vlachonikolis J, Stefanis CN (1987) Attitudes towards mental illness in the Athens area: implications for community mental health intervention. Acta Psychiatr Scand 75(2):158–165

McAlpine D, Warner LA (2002) Barriers to employment among persons with mental illness: a review of the literature. University of Illinois at Urbana-Champaign: Disability Research Institute. http://www.als.uiuc.edu/dri

Menec VH, Perry RP (1998) Reactions to stigmas among Canadian students: testing an attribution-affect-help judgment model. J Soc Psychol 138(4):443–453

Moss K, Ullman M, Starrett BE, Burris S, Johnsen MC (1999) Outcomes of employment discrimination charges filed under the Americans with Disabilities Act. Psychiatr Serv 50:1028–1035

Murray CJL, Lopez A (1996) Global health statistics: a compendium of incidence, prevalence and mortality estimates for over 2,000 conditions. Harvard School of Public Health, Cambridge

New Freedom Commission on Mental Health (2003) Achieving the promise: transforming mental health care in America. Final report (DHHS Pub. No. SMA-03–3832). Mental Health Commission, Rockville, MD

Oaxaca RL (1973) Male–female wage differentials in urban labor markets. Int Econ Rev 14:693–709

Ormel J, VonKorff M, Ustun TB, Pini S, Korten A, Oldehinkel T (1994) Common mental disorders and disability across cultures: results from the WHO Collaborative Study on Psychological Problems in General Health Care. J Am Med Assoc 272(22):1741–1748

Pescosolido B, Monahan J, Link B, Stueve A, Kikuzawa S (1999) The public's view of the competence, dangerousness, and need for legal coercion of persons with mental health problems. Am J Public Health 89(9):1339–1345

Priebe S, Warner R, Hubschmic T, Eckle I (1998) Employment, attitudes toward work, and quality of life among people with schizophrenia in three countries. Schizophr Bull 24(3):469–477

Rahav M, Struening EL, Andrews H (1984) Opinions on mental illness in Israel. Soc Sci Med 19(11):1151–1158

Roesch SC, Weiner B (2001) A meta-analytic review of coping with illness: do causal attributions matter? J Psychosom Res 50(4):205–219

Royal GP, Roberts MC (1987) Students' perceptions of and attitudes toward disabilities: a comparison of twenty conditions. J Clin Child Psychol 16:122–132

Rush LL (1998) Affective reactions to multiple social stigmas. J Soc Psychol 138(4):421–430

Sartorius N (2004) Diminishing the stigma of schizophrenia. Adv Schizophrenia Clin Psychiatry 1:50–54

Scheid TL (2000) Compliance with the ADA and employment of those with mental disabilities. In: Blanck PD (ed) Employment, disability, and the Americans with disabilities act: issues in law, public policy, and research. Northwestern University Press, Evanston, IL, pp 146–173

Schulze B (2007) Stigma and mental health professionals: a review of the evidence on an intricate relationship. Int Rev Psychiatry 19(2):137–155

Schulze B, Angermeyer MC (2003) Subjective experiences of stigma. A focus group study of schizophrenic patients, their relatives and mental health professionals. Soc Sci Med 56(2):299–312

Strauss JS, Davidson L (1997) Mental disorders, work, and choice. In: Bonnie RJ, Monahan J (eds) Mental disorder, work disability, and the law. University of Chicago Press, Chicago, pp 105–130

Tringo JL (1970) The hierarchy of preference toward disability groups. J Spec Educ 4:295–306

Wahl OF (1999) Mental health consumers' experience of stigma. Schizophr Bull 25(3):467–478

Weiner B (1980) A cognitive (attribution) – emotion – action model of motivated behavior: an analysis of judgments of help-giving. J Pers Soc Psychol 39(2):186–200

Weiner B (1983) Some methodological pitfalls in attributional research. J Educ Psychol 75(4):530–543

Weiner B (1985) An attributional theory of achievement motivation and emotion. Psychol Rev 92(4):548–573

Weiner B (1993) On sin versus sickness. A theory of perceived responsibility and social motivation. Am Psychol 48(9):957–965

Weiner B (1995) Judgments of responsibility: a foundation for a theory of social conduct. Guilford Press, New York

Weiner B, Graham S, Chandler C (1982) Pity, anger, and guilt: an attributional analysis. Pers Soc Psychol Bull 8(2):226–232

Weiner B, Perry RP, Magnusson J (1988) An attributional analysis of reactions to stigmas. J Pers Soc Psychol 55(5):738–748

Westbrook MT, Legge V, Pennay M (1993) Attitudes towards disabilities in a multicultural society. Soc Sci Med 36:615–624

Yelin EH, Cisternas MG (1997) Employment patterns among persons with and without mental conditions. In: Bonnie RJ, Monahan J (eds) Mental disorder, work disability and the law. University of Chicago Press, Chicago, pp 25–51

Yuker HE (1987) The disability hierarchies: comparative reactions to various types of physical and mental disabilities. Hofstra University, Mimeo

Part II
Mental Health Disabilities and Work Functioning

Chapter 4
Vocational Capacity among Individuals with Mental Health Disabilities

E. Sally Rogers and Kim L. MacDonald-Wilson

Introduction

Decades have passed since individuals with mental health disabilities, particularly those with serious mental illness, were widely considered unable to sustain work. Many individuals were housed in institutional settings, but those in the community were often labeled unemployable. With the advent of deinstitutionalization in the 1960s and 1970s, individuals with more severe mental health problems were served using a "train and place" model of vocational rehabilitation services, delivered primarily in sheltered workshops. This approach dictated that individuals be placed in work-like settings in order to be "trained," sometimes for years at a time, prior to being "placed" in employment. With the recent evolution of supported employment, the prevailing approach has become "place and train," introducing the concept of rapid entry into employment with wraparound supports. The cornerstone of supported employment is the philosophy that the majority of individuals with mental health disabilities who want to work, can work. But what have these last few decades of community-based vocational rehabilitation services revealed in terms of the capacity of individuals with mental health disabilities to enter into the workplace and maintain employment? We explore in this chapter the empirical findings and knowledge regarding the assessment and prediction of vocational capacity among individuals with significant mental health problems.

E.S. Rogers(✉)
Center for Psychiatric Rehabilitation, Boston University, 940 Commonwealth Avenue West, Boston, MA 02215, USA
e-mail: erogers@bu.edu

I.Z. Schultz and E.S. Rogers (eds.), *Work Accommodation and Retention in Mental Health*, DOI 10.1007/978-1-4419-0428-7_4, © Springer Science+Business Media, LLC 2011

Employment Rates of Individuals with Mental Health Disabilities

Estimates of employment rates vary, but research suggests that without vocational interventions only about one-third of individuals with all disabilities are employed (Harris Interactive 2000). For those with more severe disabilities, the employment rate may hover around 20%. Large, population-based studies suggest unemployment rates among individuals with mental health disabilities ranging from about 39 to 68% in the U.S. (McAlpine and Warner 2001), 61 to 73% in England (Crowther et al. 2001), and approximately 84% in Australia (Waghron, Chant and White ford 2002). In the U.S., mental disorders account for a large and growing segment of the Social Security Disability rolls (Jans et al. 2004), a large portion of private long-term disability claims and costs (Salkever et al. 2000), and a substantial proportion of those served by state-federal vocational rehabilitation programs (Jans et al. 2004). Of people on disability payments through Social Security, approximately 25–33% are receiving benefits due to a mental disorder (Jans et al. 2004). A number of studies indicate that people with mental health disabilities have fewer successful closures in the U.S. state-federal vocational rehabilitation system when compared to people with other disabilities (National Institute on Disability and Rehabilitation Research 1997; Andrews et al. 1992; Marshak et al. 1990). Although vocational rehabilitation services for people with mental health disabilities can triple employment outcomes (Cook et al. 2005; Bond et al. 2008; Rosenthal et al. 2007), the resulting employment rates still leave the majority of people with mental health disabilities unemployed.

What factors account for the ability of individuals with moderate to severe mental health disabilities to successfully obtain and retain employment? A review of existing empirical information suggests that clinical, demographic, cognitive, and service-related variables affect these outcomes. We describe relevant studies in each of these categories.

Clinical Predictors of Work Functioning and Capacity

The bulk of the literature examining prediction of work capacity has focused on whether and how well psychiatric diagnoses and symptoms are able to predict the ability of individuals with mental health problems to obtain and retain employment. Numerous small- and large-scale studies have been conducted to examine these predictors. Taken together, the results are both equivocal and suggestive, and have been refined over time (MacDonald-Wilson et al. 2001). Part of the difficulty in drawing conclusions from studies of this kind (both earlier studies and those conducted more recently) is the variation in the populations studied, the variety of predictors examined, and the precision with which important clinical variables such as psychiatric diagnosis, symptomatology, and social and cognitive skills (to name a few) are assessed. This variation makes it difficult to draw conclusions across studies and to conduct systematic reviews.

Early Studies Examining Clinical Predictors

Early studies examining clinical predictors of vocational capacity and outcomes for individuals with mental health disabilities reported little relationship between future work performance and psychiatric diagnosis or assessments of symptoms (Ciardiello et al. 1988; Moller et al. 1982; Schwartz et al. 1975; Strauss and Carpenter 1972, 1974). Despite prevailing rehabilitation practices at the time, these studies indicated that there was no set or pattern of symptoms that were consistently related to work performance. For example, an early review (Anthony and Jansen 1984) of vocational capacity and outcomes for persons with psychiatric or mental health disabilities concluded that psychiatric symptoms, diagnostic category, and standardized psychometric assessments (intelligence, aptitude, and personality tests) were poor predictors of future work performance. The authors drew a fairly controversial conclusion at that time: that there was little or no correlation between a person's symptoms and their functional or vocational skills. This conclusion ran counter to the typical vocational rehabilitation service at the time, which relied heavily on both clinical and paper-and-pencil assessments to determine the vocational capacity of its applicants. Anthony and Jansen (1984) concluded that the best predictors of future work performance were: (1) ratings of a person's work adjustment skills made in a simulated work environment and (2) the ability to get along in the workplace or function socially. They concluded that the best demographic predictor of future work performance was a person's prior employment history.

Recent Studies Examining Clinical Predictors of Vocational Outcomes

Studies conducted since the 1980s have uncovered a modest relationship between psychiatric symptoms, work performance, and vocational outcomes, especially for those individuals receiving vocational rehabilitation services (Brekke et al. 1997; Taylor and Liberzon 1999; Hodel et al. 1998; Bryson et al. 1998; Lysaker et al. 1995a; Gold et al. 1999). Several researchers have indicated that psychiatric diagnosis and symptoms predicted poorer outcomes, poorer role functioning, and less likelihood of being employed (Massel et al. 1990; Coryell et al. 1990; Tsuang and Coryell 1993), particularly for individuals diagnosed with schizophrenia. Rogers and colleagues concluded that there was a small but significant relationship between measures of symptoms and vocational outcomes among persons with mental health disabilities in vocational programs (Rogers et al. 1997; Anthony et al. 1995), with negative symptoms (e.g., withdrawal) being a better predictor of vocational functioning than positive symptoms (e.g., hallucinations).

Several recent systematic reviews and large-scale studies have examined clinical predictors and have drawn conflicting conclusions. This includes three meta-analyses or systematic reviews (Michon et al. 2005; Tsang et al. 2000; Wewiorski and Fabian 2004) which have been conducted on vocational outcomes. The first, by Wewiorski and Fabian (2004) examined 17 studies of vocational outcomes for individuals with severe psychiatric disabilities. The authors were significantly constrained by the limited number of common predictor and outcome variables that were contained in these various studies, and were able to examine only a handful of demographic variables (age, gender, race) and one clinical variable (diagnosis). They examined the ability of these factors to predict employment status and the ability to attain and retain employment at 3 and 6 months after the initial assessment. They concluded that across available studies, individuals with schizophrenia were less likely to attain or retain employment. Individuals with an affective disorder, when compared to individuals classified as having an "other" disorder (not a diagnosis of schizophrenia or affective disorder) were more likely to be employed.

In a comprehensive review of predictors of work outcome in 35 well-designed research studies published between 1985 and 1997, Tsang et al. (2000), found that, of available clinical predictors, mixed results were apparent for diagnosis, substance abuse, cognitive functioning, and previous functioning when predicting work outcome; and that social skills, work history, and premorbid functioning were the most consistent predictors of work outcome for people with mental health disabilities. Measures of negative symptoms tended to be more strongly related to vocational outcome. Social skills were the most consistent and strongest predictors and the factor most frequently identified among all others, which included medical aspects such as symptoms (ten studies) and diagnosis (three studies), demographic characteristics (nine studies), psychological aspects such as cognitive functioning (four studies) and ego functioning (one study), other premorbid functioning variables (six studies), and environmental aspects (three studies).

Michon et al. (2005) conducted a systematic review of vocational outcomes using 16 articles representing eight vocational studies which focused on individuals receiving psychiatric vocational rehabilitation (PVR) services. The sample sizes in the studies ranged from a low of 60 to a high of 907, with the average age of the participating subjects being 28–37 (somewhat younger than in other research in this area). They examined the influence of past functioning, work history, diagnosis, severity of symptoms, and psychiatric history, and determined that those variables were largely outweighed by measures of work performance. They concluded that positive employment outcomes were "most clearly and strongly" related to better work performance as measured at the beginning of a vocational program. In addition, participants' work-related self-efficacy and social functioning in the PVR program were associated with better outcomes. Psychiatric diagnosis and history were not associated with outcomes. This review involved somewhat restricted samples, in that all participants were involved in a rehabilitation program.

Waghorn, Chant and Whiteford (2002), using population-based data of individuals diagnosed with a psychotic disorder from a national survey in Australia, examined demographic and clinical factors that predict employment. They assessed "self-reported course of illness" as a predictor and measured it in a fairly rigorous way, but through self-reporting. They examined both current employment status and ability to retain employment ("durable" employment), and suggested that having a chronic deteriorating course of a psychiatric illness and having a family history of a psychiatric disorder other than schizophrenia predicted poorer employment status. A good premorbid work adjustment and not having a lifetime cannabis dependence were predictors of positive work outcomes. Those with schizophrenia were more likely to be unemployed.

Cook et al. (2008) examined a host of demographic (age, race, gender, education) and clinical variables (diagnosis, substance abuse, Social Security status, work history, hospitalization history, motivation to work, symptoms, co-occurring developmental disability, and physical health) in a large dataset of individuals participating in best practice vocational programs ($N = 1,273$). This is the largest study conducted to date in the U.S. of best and evidence-based practices of vocational rehabilitation services for individuals with mental health disabilities, and was funded by the U.S. Substance Abuse and Mental Health Services Administration (SAMHSA). The Employment Intervention Demonstration Project (EIDP), as it is called, has resulted in significant knowledge about the vocational outcomes of individuals with mental health disabilities (Cook et al. 2005). The authors examined the patterns of employment over 24 months (Cook et al. 2008a) and determined that study participants receiving a best practice vocational intervention without a diagnosis of schizophrenia had the best outcomes (almost 35% employed at the peak of employment rates), followed by experimental subjects with schizophrenia (22%). Participants not receiving the best practice vocational intervention fared the worst (approximately 14–18% employment rates). The authors concluded that those with better work histories and greater work motivation and those with a diagnosis other than schizophrenia (e.g., depressive or bipolar disorders) were more likely to work. There was a negative association of work outcomes with depressive symptoms. Participants with higher excitement scores (a component of psychiatric symptoms) were more likely to be working. Most important to note is that those with a diagnosis of schizophrenia in the experimental condition outperformed their counterparts with other diagnoses not receiving the experimental intervention.

Razzano et al. (2005) examined clinical predictors of employment outcomes using the same data from the multisite EIDP and found that, even when they controlled for demographic variables and study condition (that is, receipt of best practice vocational intervention or not), several variables were associated with the participants' ability to work in competitive settings and to work at a meaningful level of intensity (i.e., more than 40 hours per month). Those included poor self-rated functioning (similar to the findings of Waghorn, Chant and Whiteford 2002), negative psychiatric symptoms, and recent hospitalizations, which were "consistently associated" with poorer vocational outcomes (Razzano et al. 2005). In addition, comorbid conditions (i.e., co-existing physical health problems, substance use disorders, mental retardation, or head injury) were associated with poorer vocational outcomes.

An important study of vocational outcomes for individuals receiving state vocational rehabilitation (VR) services was recently reported using measures beyond those typically utilized in VR closure studies. Daniels (2007), in a study of state VR recipients with all disabilities ($N = 5,305$), reported that higher levels of self-esteem, internal locus of control, and fewer functional limitations were related to better vocational outcomes.

Demographic Predictors of Work Functioning and Capacity

As mentioned above, three systematic reviews (Tsang et al. 2000; Wewiorski and Fabian 2004; Michon et al. 2005) and several large-scale studies (e.g., Bromet 2005; Chant 2002; Cook et al. 2005; Daniels 2007) have examined demographic predictors of vocational outcomes and have drawn equivocal findings. Wewiorski and Fabian (2004), in examining 17 studies of vocational outcomes for individuals with severe mental health disabilities, uncovered little evidence that gender was related to obtaining or retaining work, but Caucasian race was. This finding was confirmed by Daniels (2007); Cook and her colleagues found evidence that race, gender and age were related to employment outcome (2008a, b). (Note, however, that the Daniels study (2007) was a large-scale secondary analysis of state VR data, but did include all disabilities.) On the contrary, analyses conducted by Bromet (2005), in a study of $N = 79,967$ recipients of state VR services, suggested that race was not a significant predictor of outcome for individuals with mental health disabilities. In their systematic review, Tsang et al. (2000) discovered no consistent demographic predictors of vocational outcomes, except in terms of work history. Age, gender, marital status, and race/ethnicity were mixed in their predictive value for work outcomes among individuals with mental health disabilities. Michon et al. (2005), in a systematic review of eight vocational rehabilitation intervention studies, concluded that education level was related to positive employment outcomes. Education has been linked to better vocational outcomes in several studies, and several authors have also suggested that being younger is associated with better work outcomes (Cook et al. 2008b; Chant 2002; Daniels 2007). Cook and colleagues, in examining demographic predictors of employment outcomes in the multisite EIDP study, also found that greater work motivation was associated with better work outcomes.

Cognitive Predictors of Work Outcomes

Recently, literature has appeared on the relationship between cognitive functioning and vocational outcomes, particularly among individuals diagnosed with schizophrenia or other psychotic disorders. Measures of cognitive functioning include such factors as vigilance, memory, executive functioning, verbal fluency, and visuo-motor functioning (Green et al. 2000; Brekke et al. 1997). In a review of recent literature and meta-analyses, Green and colleagues (2000) concluded that measures of memory, vigilance, and executive functioning were significantly related to community outcome, which included occupational functioning. Visuospatial processing in particular was related to increased work functioning (Brekke et al. 1997). Verbal memory accounted for significant variance in work performance in a vocational rehabilitation program, according to findings of another study (Bryson et al. 1998). Other studies and reviews have documented the importance of executive functioning and other cognitive processes to vocational performance (Lysaker et al. 1995b; McGurk and Wykes 2008; McGurk and Mueser 2006). A global screening measure of cognitive functioning called the RBANS (Repeatable Battery for the Assessment of Neuropsychological Status) successfully discriminated between employed and unemployed participants on the total score and four of five index scores, including immediate memory, attention and delayed memory indices, and language (Gold et al. 1999). (Review the chapter by Guilmette and Giuliano (2010), in this book, for findings on individuals with moderate to severe brain injuries, and the chapter by Fraser et al. (2010), also in this book, on individuals with mild cognitive disorders).

McGurk and Wykes (2008) summarized the empirical evidence documenting cognitive impairments among individuals with severe mental health disabilities, including those of attention, memory, verbal learning, and executive functioning. They describe a consistent pattern which has emerged from the literature suggesting that cognitive impairments limit work performance and are predictive of poorer vocational outcomes. There is also evidence that impaired cognitive functioning may compromise the ability of individuals to benefit from vocational services (Bond et al. 2008). McGurk and Wykes' (2008) review of the research evidence also suggests that cognitive remediation interventions (such as McGurk's "Thinking for Work" program) improve both cognitive functioning and work outcomes over and above what is possible through vocational rehabilitation services alone. Much research remains to be done to better understand the relationship between cognitive impairments and vocational functioning among individuals with mental health disabilities and what can be done to improve both using cognitively based interventions.

Work Outcomes Following Rehabilitation Interventions

Several studies have been conducted suggesting that vocational interventions have definitive beneficial effects on the vocational outcomes of individuals with mental health disabilities (e.g., Cook et al. 2005; Bond et al. 2008; Fabian 1992; Jacobs et al. 1992; Lysaker et al. 1995; Bryson et al. 1999; Michon et al. 2005; Rogers et al. 1991a). Cook and colleagues, in the EIDP multisite study, concluded that study participants receiving a best practice vocational intervention significantly benefited from these services in terms of work outcomes (Cook et al. 2005). Evidence has also accrued from numerous randomized trials that receipt of supported employment services using the Individual Placement and Support model in particular (Drake and Bond 2008) is highly predictive of positive outcomes (Bond et al. 2008). In analyses of recipients of state VR services, Bromet's data suggested that receipt of vocational rehabilitation services (e.g., job search assistance) predicted positive work outcomes (2005). This finding was confirmed by

Rosenthal et al. (2007) who, using state VR data, concluded that receiving job placement services was the most critical variable differentiating those who become employed from those who do not. Taken together, these results suggest that receipt of vocational rehabilitation services of various kinds, but particularly targeted job development, job finding assistance, or evidence-based supported employment services, is a strong predictor of positive work outcomes for individuals with mental health disabilities.

Nature of Mental Health Conditions

There are several characteristics of mental health disabilities that contribute to the inability to find robust and consistent predictors of work outcomes. Many mental health disorders are episodic and recurrent; that is, exacerbations in symptoms and deterioration in functioning may recur over time. However, some individuals experience only one episode of impairment and may return to premorbid levels of functioning. Unfortunately, researchers and clinicians are unable to predict the course of the illness or disability for one individual at any specific point in time. In addition, for those whose condition is episodic, it is difficult to predict when periods of exacerbation may occur. Thus, assessing the capacity to work in any "slice-in-time" may not provide representative information of that person's functioning longitudinally. Indeed, there is evidence that work evaluations may be better predictors of current work performance than future performance, particularly if there are changes in clinical status (Liberman 1990; Massel et al. 1990).

In addition to variability in functioning over time, there is also considerable variability within diagnostic groups in terms of manifestation of the symptoms of the illness and limitations in functioning. Harding (1988) refers to this as the "hidden heterogeneity" of mental health disabilities. This may be one of the reasons that specific diagnoses are not highly correlated with vocational outcomes. In addition, the reliability of psychiatric diagnosis has historically been poor, although advances in the American Psychiatric Association's Diagnostic and Statistical Manual of Mental Disorders (DSM-IV-TR 2000) and related structured clinical interviews (e.g., First et al. 1997) have improved the reliability of diagnoses. Therefore, interpretation of the results of studies correlating diagnosis with vocational outcome should be done cautiously, considering the source of the diagnosis. There is also significant variability across diagnoses that are grouped under the Social Security Administration category of "mental impairments." Studies on diagnosis or symptoms and work performance may involve only one specific diagnostic group, groups of diagnoses that may be categorized as psychotic or nonpsychotic, or unspecified psychiatric diagnoses grouped generically. These issues create problems in generalizing findings between and across studies examining different diagnoses or categories of "mental impairments." Finally, difference in measures, methods, and the samples used in these various studies make cross-study comparisons difficult.

Assessments of Vocational Capacity

The areas of functioning affected by mental health disabilities have often been described as lying in the social/interpersonal, emotional, and cognitive domains (Ikebuchi et al. 1999; MacDonald-Wilson et al. 2002; Wallace 1986). There are numerous measures of mental health symptoms and mood states, but few that focus on emotional, social, or interpersonal functioning vis-à-vis work. Hodel and colleagues (1998) suggested that individuals diagnosed with schizophrenia and other psychotic disorders may lack skills in emotion recognition, processing, regulation, and expression (Taylor and Liberzon 1999; Hodel et al. 1998). Emotional and interpersonal problems on the job, more than the quality of work, are often cited as the reasons that people with mental health disabilities leave jobs (Becker et al. 1998; Tsang and Pearson 1996; Wallace et al. 1999). Instruments measuring social and interpersonal functioning are varied and may focus on different aspects of this domain, such as activities, role functioning, social adjustment, or interpersonal skills; however, none of these foci alone are entirely useful when examining social and interpersonal skills in relation to work. Assessments may also be global or specific in nature (MacDonald-Wilson and Nemec 2008; Dickerson 1997; Scott and Lehman 1998; Wykes 1998). Furthermore, the literature on social/interpersonal skills has suggested that such interactions have a cognitive component, as in the receiving-processing-sending model of social skills (Wallace et al. 1999), which complicates interventions designed to improve social and interpersonal skills in the workplace. While measures of specific interpersonal skills in a work setting may relate to work outcome (Anthony 1994), a major limitation of many social and interpersonal assessments is that they have not been well correlated with work outcomes. In many respects, the components of emotional, social, and interpersonal functioning in the workplace are less well defined and understood when compared to physical and cognitive functioning. But, taken together, there appear to be few well-validated measures of social or interpersonal skills in relation to work or that can predict work outcomes (MacDonald-Wilson and Nemec 2008).

More recent literature has appeared on cognitive functioning (McGurk and Mueser 2006) as noted above, and particularly in relation to people with diagnoses of a psychotic disorder or schizophrenia. Measures of cognitive functioning are more well-developed and psychometrically sound than measures in other domains, such as social functioning. In addition, recent studies have correlated some cognitive functions with vocational outcomes (Green et al. 2000; McGurk and Wykes 2008).

Despite the evidence that social/interpersonal, psychiatric, and cognitive functioning are related to vocational capacity and that they may predict work outcomes, we have few standardized and sound yet manageable assessment approaches or batteries to assess work functioning and capacity. Additional study is needed with large and representative samples, various measures of the same domains and functions, greater rigor, and especially studies of the predictive power of such instruments for work outcomes.

Promising Assessment Methods of Vocational Capacity

In the past two decades new approaches to the assessment of work capacity have been developed and tested on individuals with mental health disabilities. These include a method of vocational evaluation known as situational assessment. Situational assessment techniques involve direct observations of a person in a real or simulated work environment and use of behavioral rating scales to assess skill performance over a period of time. These behaviors are known as general work or work adjustment skills. For people with mental health disabilities, these skills are often in the social/interpersonal, "work adjustment," or cognitive domains. A number of older studies pointed to the relevance and usefulness of situational assessment techniques for people with mental health disabilities (Rehabilitation Services Administration 1995; Hursh et al. 1988; Tashjian et al. 1989) and have suggested that situational assessment is the best method to determine vocational potential (Hursh et al. 1988; Bond and Friedmeyer 1987; Tashjian et al. 1989). A more recent study (McGuire et al. 2007) found a situational assessment approach useful for predicting outcomes among individuals with schizophrenia, but not necessarily job acquisition. Given the inability of most paper-and-pencil measures, psychological assessments, and measures of social skills to predict work outcomes, the use of situational assessments are a logical step. To date, however, situational assessment measures have proven difficult to implement, as described below.

The Standardized Assessment of Work Behaviour (Griffiths 1973, 1975, 1977; Watts 1978) is one of the first vocational situational assessment measures developed to assess work capacity among individuals with mental health disabilities. Using 25 behavioral items designed to assess a broad range of work behaviors, items are rated on a five-point continuum, from "strength" (e.g., looks for more work) to "deficit" (e.g., waits to be given work), and are based upon observed work behaviors over a 10-day period. Using the initial work of Griffiths, Rogers and her colleagues developed the Work Adjustment and Interpersonal Skills Scales and situational assessment procedures (Rogers et al. 1991b). The instrument contains two separate scales consisting of 21 work adjustment skills and 14 interpersonal skills measured in a work context by a trained vocational observer. The scale and assessment approach has demonstrated modest success in predicting work outcome. Similarly, the Work Behavior Inventory (WBI), a 36-item work performance assessment instrument specifically designed for people with severe mental illness (Bryson et al. 1999), is intended for use in a real work situation. It consists of five subscales: Work Habits, Work Quality, Personal Presentation, Cooperativeness, and Social Skills, all of which are rated on a five-point scale from "consistently inferior" to "consistently superior."

The Vocational Cognitive Rating Scale is a 16-item instrument designed to measure cognitive skills in the workplace for individuals with mental health disabilities (Greig et al. 2004). Items are measured on a five-point scale similar to the WBI. Ratings are made by an observer and focus on attention, initiative, and other cognitive-related work skills, specifically for individuals with disorders affecting cognitive functioning. The scale has demonstrated promise in its ability to predict vocational outcomes.

Unfortunately, none of these situational assessment measures have been widely adopted, to our knowledge. Perhaps they are seen as too costly to use (at least in the short term), too labor intensive, and too unwieldy to arrange because they require vocationally trained staff and work environments in which to observe those being assessed. At least one study (Zarate et al. 1998) suggests that situational assessment procedures may be valid with very short observation periods, which could render them more cost-effective. While these approaches to measuring vocational capacity are promising, additional research is needed to determine the optimal length of assessment time, the ability to predict work outcomes, and any costs or savings resulting from using these methods.

Conclusions

While considerable progress has been made in the past few decades in the assessment of factors related to work outcomes for persons with mental health disabilities, prediction of vocational capacity and outcomes continues to be fraught with methodological problems. Psychiatric symptoms and diagnoses, while necessary to establish the presence of a medical impairment (necessary for the Social Security disability determination process), have not been demonstrated to be the only important factors for assessing capacity to work. Reviews of the literature suggest that there are currently no global, self-reporting, clinical or traditional measures of vocational capacity that can predict work outcomes or account for a major portion of the variance in vocational functioning for individuals with mental health disabilities. Situational assessment approaches hold promise, but are often more time-consuming, resource-intensive, and logistically difficult to arrange because of the need for a real or simulated work setting in which to conduct such assessments. Furthermore, none of the approaches or measures described in this article, including the situation-based measures, sufficiently address the issue of the day-to-day fluctuations in functioning or the hidden heterogeneity that is characteristic of individuals with mental health disabilities.

There are many design challenges inherent in studying vocational capacity among individuals with mental health disabilities. For example, longitudinal studies are often needed to examine the ability of a measure or set of measures to validly predict later work function, but such studies are time-consuming and costly, and loss of study participants at follow-up is a real concern. The other major design consideration is what population or populations to sample for such studies. Studying only those individuals who receive state VR services limits the population and the knowledge that can be gained about work outcomes. Including individuals who have been adjudicated as disabled is problematic, because only a small minority of those individuals return to work. Finding individuals who have experienced a work disability due to mental health impairments and who have not accessed public mental health services or federal income support programs might shed light on these issues, but is a considerable challenge. Despite this, vocational rehabilitation personnel continue to need reliable and valid measures of vocational capacity and an understanding of those factors that predict work functioning. Studies are needed to test existing and new measures of work capacity, to test promising situational assessment methods, and to continue to examine predictors of work outcomes.

Implications for Vocational Rehabilitation Personnel

Taken together, we can conclude the following from our current state of knowledge regarding prediction of work capacity among individuals with mental health disabilities:

- The ability to predict vocational capacity from demographic factors such as gender, race, or age is equivocal; whatever factors may be operating in the local economy regarding these characteristics may also operate for individuals with mental health disabilities (for example, the workforce may favor younger over older employees, more educated employees over less educated employees). However, our state of knowledge in terms of predicting vocational capacity for individuals with mental health disabilities using demographic factors is not strong and should be avoided.
- Individuals with a psychotic or schizophrenia-spectrum disorder may experience more difficulty obtaining, performing in, and retaining employment.
- Individuals who are more symptomatic in terms of their mental health condition or who are experiencing a more difficult course of their illness (a deteriorating course or a course with exacerbations and relapses) will have more difficulty achieving good work outcomes.
- Individuals with comorbid conditions (physical, substance abuse, or developmental disabilities) may fare worse in terms of work outcomes.
- Individuals with better social skills, a higher sense of self-efficacy, higher work motivation, better "premorbid" functioning, and a prior work history may fare better in terms of work outcomes.
- Individuals with significant difficulties in cognitive functioning, including difficulties in memory, attention, and executive function, may fare worse in terms of work outcomes.
- Individuals receiving direct vocational assistance through job development, job placement, and general vocational rehabilitation services and best practice supported employment services will experience better vocational outcomes.
- Individuals with cognitive impairments who receive cognitive remediation in conjunction with vocational rehabilitation fare better in terms of vocational rehabilitation outcomes.

These last two points are critical and highly instructive for vocational rehabilitation personnel. Our state of the knowledge is such that we cannot accurately predict which individuals with a mental health disability have the capacity for work. We do know that individuals with multiple predictive factors (for example, individuals with a fluctuating or deteriorating course, who are symptomatic, and who have diagnoses such as schizophrenia and cognitive deficits) will have more difficulty achieving positive work outcomes. However, we also know from existing research that vocational rehabilitation interventions, ranging from state VR, to best practice vocational services, to the evidence-based Individual Placement and Support model, will all be highly successful in improving outcomes among individuals with mental health difficulties (Bond et al. 2008; Bromet 2005; Cook et al. 2005; Krupa 2010; Rosenthal et al. 2007). Further, cognitive remediation provided in conjunction with vocational rehabilitation services will improve outcomes over and above vocational rehabilitation alone (McGurk and Wykes 2008). Given the evidence that a deteriorating course of illness and more severe symptoms are associated with poorer outcomes, vocational rehabilitation providers should strive to integrate clinical and vocational services, as recommended by best practice guidelines (Drake and Bond 2008; Rogers et al. 2005), to ensure that individuals have the best psychopharmacology and psychiatric treatment available, as well as access to the best practice in vocational and cognitive interventions, to maximize their vocational outcomes.

References

Andrews H, Barker J, Pittman J, Mars L, Struening EL, LaRocca N (1992) National trends in vocational rehabilitation: a comparison of individuals with physical disabilities and individuals with psychiatric disabilities. J Rehabil 58(1):7–18

Anthony WA (1994) Characteristics of people with psychiatric disabilities that are predictive of entry into the rehabilitation process and successful employment outcomes. Psychosoc Rehabil J 17(3):3–13

Anthony WA, Jansen MA (1984) Predicting the vocational capacity of the chronically mentally ill: research and policy implications. Am Psychol 39(5):537–544

Anthony WA, Rogers ES, Cohen M, Davies RR (1995) Relationships between psychiatric symptomatology, work skills, and future vocational performance. Psychiatr Serv 46(4):353–358

Becker DR, Drake RE, Bond GR, Xie H, Dain BJ, Harrison K (1998) Job terminations among persons with severe mental illness participating in supported employment. Community Ment Health J 34(1):71–82

Bond GR, Drake RE, Becker DR (2008) An update on randomized controlled trials of evidence based supported employment. Psychiatr Rehabil J 31(4):280–290

Bond GR & Friedmeyer MH (1987) Predictive validity of situational assessment at a psychiatric rehabilitation center. Rehabil Psychol 32(2):99–112

Brekke JS, Raine A, Ansel M, Lencz T, Bird L (1997) Neuropsychological and psychophysiological correlates of psychosocial functioning in schizophrenia. Schizophr Bull 23(1):19–28

Bromet ES (2005) The relationship between vocational rehabilitation services, demographic variables and outcomes among individuals with psychiatric disabilities. Dissertation, The Ohio State University

Bryson G, Bell MD, Kaplan E, Greig T (1998) The functional consequences of memory impairments on initial work performance in people with schizophrenia. J Nerv Ment Dis 186(10):610–615

Bryson GJ, Bell MD, Lysaker PH, Zito W (1999) The Work Behavior Inventory: a scale for the assessment of work behavior for people with severe mental illness. Psychiatr Rehabil J 20:47–55

Ciardiello JA, Klein ME & Sobkowski S (1988) Ego functioning and vocational rehabilitation. In: Ciardiello JA, Bell MD (eds) Vocational rehabilitation of persons with prolonged psychiatric disorders. The John Hopkins University Press, Baltimore, MD, pp 196–207

Cook JA, Leff HS, Blyler CR, Gold PB, Goldberg RW, Mueser KT et al (2005) Results of a multisited randomized trial of supported employment interventions with severe mental illness. Arch Gen Psychiatry 62(5):506–512

Cook JA, Blyler CR, Burke-Miller JK, McFarlane WR, Leff HS, Mueser KT et al (2008a) Effectiveness of supported employment for individuals with schizophrenia: results of a multisite, randomized trial. Clin Schizophr Relat Psychoses 2:37–46

Cook JA, Blyler CR, Leff HS, McFarlane WR, Goldberg RW, Gold PB et al (2008b) The employment intervention demonstration program: major findings and policy implications. Psychiatr Rehabil J 31(4):291–295

Coryell W, Keller M, Lavori P, Endicott J (1990) Affective syndromes, psychotic features, and prognosis. Arch Gen Psychiatry 47:651–657

Crowther RE, Marshall M, Bond GR, Huxley P (2001) Helping people with severe mental illness to obtain work: systematic review. Br Med J 322(7280):204–208

Daniels TM (2007) Determinants of employment outcomes among persons with disabilities in public vocational rehabilitation. Dissertation, Columbian College of Arts and Sciences of the George Washington University

Dickerson FB (1997) Assessing clinical outcomes: the community functioning of persons with serious mental illness. Psychiatr Serv 48(7):897–902

Drake RE, Bond GR (2008) Special issue: supported employment: 1998 to 2008. Psychiatr Rehabil J 21:274–276

Fabian ES (1992) Longitudinal outcomes in supported employment: a survival analysis. Rehabil Psychol 37(1):23–35

First MB, Spitzer RL, Gibbon M, Williams JB (1997) Structured clinical interview for DSM-IV axis I disorders, research version, patient edition (SCID I/P). Biometrics Research Press, New York State Psychiatric Institute, New York

Fraser RT, Strand D, Johnson C (2010) Employment interventions for persons with mild cognitive disorders. In: Schultz IZ, Sally Rogers E (eds) Handbook of work accommodation and retention in mental health. Springer, New York

Gold JM, Queern C, Iannone VN, Buchanan RW (1999) Repeatable battery for the assessment of neuropsychological status as a screening test in schizophrenia, I: sensitivity, reliability, and validity. Am J Psychiatry 156(12):1944–1950

Green MF, Kern RS, Braff DL, Mintz J (2000) Neurocognitive deficits and functional outcome in schizophrenia: are we measuring the "right stuff"? Schizophr Bull 26(1):119–136

Greig TC, Nicholls SS, Bryson GJ, Bell MD (2004) The vocational cognitive rating scale: a scale for the assessment of cognitive functioning as work for clients with severe mental illness. J Vocat Rehabil 21(2):71–81

Griffiths RD (1973) A standardized assessment of the work behaviour of psychiatric patients. Br J Psychiatry 123(575):403–408

Griffiths RD (1975) The accuracy and correlates of psychiatric patients' self-assessment of their work behavior. Br J Soc Clin Psychol 14:181–189

Griffiths RD (1977) The prediction of psychiatric patients' work adjustment in the community. Br J Soc Clin Psychol 16:165–173

Guilmette TJ, Giuliano AJ (2010) Brain injury and work performance. In: Schultz IZ, Sally Rogers E (eds) Handbook of work accommodation and retention in mental health. Springer, New York

Harding CM (1988) Course types in schizophrenia: an analysis of European and American studies. Schizophr Bull 14(4):633–643

Harris Interactive (2000) N.O.D./Harris survey of Americans with disabilities (Study No. 12384). Harris Interactive, New York

Hodel B, Brenner HD, Merlo MCG, Teuber JF (1998) Emotional management therapy in early psychosis. Br J Psychiatry 172(Suppl 33):128–133

Hursh NC, Rogers ES, Anthony WA (1988) Vocational evaluation with people who are psychiatrically disabled: results of a national survey. Vocat Eval Work Adjust Bull 21(4):149–155

Ikebuchi E, Iwasaki S, Sugimoto T, Miyauchi M, Liberman R (1999) The factor structure of disability in schizophrenia. Psychiatr Rehabil Skills 3(2):220–230

Jacobs H, Wissusik D, Collier R, Stackman D, Burkeman D (1992) Correlations between psychiatric disabilities and vocational outcome. Hosp Community Psychiatry 43:365–369

Jans L, Stoddard S, Kraus L (2004) Chartbook on mental health and disability in the United States. An InfoUse report. U.S. Department of Education, National Institute on Disability and Rehabilitation Research, Washington, DC

Krupa T (2010) Approaches to improving employment outcomes for people with serious mental illness. In: Schultz IZ, Sally Rogers E (Eds) Handbook of work accommodation and retention in mental health. Springer, New York

Liberman R (1990) Psychiatric symptoms and the functional capacity for work (final report). Los Angeles: UCLA Clinical Research Center for Schizophrenia and Psychiatric Rehabilitation, UCLA School of Medicine and VA Medical Center. Social Security Administration Grant No. 10-P-98193-9004

Lysaker PH, Bell MD, Goulet JL (1995a) The Wisconsin Card Sorting Test and work performance in schizophrenia. Schizophr Bull 11:45–51

Lysaker PH, Bell MD, Zito WS, Bioty SM (1995b) Social skills at work: deficits and predictors of improvement in schizophrenia. J Nerv Ment Dis 183(11):688–692

MacDonald-Wilson K, Nemec PB (2008) Assessment in psychiatric rehabilitation. In: Bolton B, Parker R (Eds) Handbook of measurement and evaluation in rehabilitation, 4th edn. Pro Ed, Austin, TX, pp 527–568

MacDonald-Wilson K, Rogers ES, Anthony WA (2001) Unique issues in assessing work function among individuals with psychiatric disabilities. J Occup Rehabil 11(3):217–232

MacDonald-Wilson KL, Rogers ES, Massaro JM, Lyass A, Crean T (2002) An investigation of reasonable workplace accommodations for people with psychiatric disabilities: quantitative findings from a multi-site study. Community Ment Health J 38(1):35–50

Marshak L, Bostick D, Turton L (1990) Closure outcomes for clients with psychiatric disabilities served by the vocational rehabilitation system. Rehabil Couns Bull 33:247–250

Massel HK, Liberman R, Mintz J, Jacobs HE, Rush TV, Giannini CA et al (1990) Evaluating the capacity to work of the mentally ill. Psychiatry 53:31–43

McAlpine DD, Warner L (2001) Barriers to employment among persons with mental illness: a review of the literature (publication from Institute for Health, Health Care Policy, and Aging Research: http://www.dri.uiuc.edu/research/p0-04c/final_technical_reportp01-04c). Accessed 15 February 2010

McGuire AB, Bond GR, Evans JD, Lysaker P, Kim HW (2007) Situational assessment in psychiatric rehabilitation: a reappraisal. J Vocat Rehabil 27(1):49–55

McGurk SR, Mueser KT (2006) Cognitive and clinical predictors of work outcomes in clients with schizophrenia receiving supported employment services: 4-year follow-up. Adm Policy Ment Health Ment Health Serv Res 33(5):598–606

McGurk SR, Wykes T (2008) Cognitive remediation and vocational rehabilitation. Psychiatr Rehabil J 31(4):350–359

Michon HW, van Weeghel J, Kroon H, Schene AH (2005) Person-related predictors of employment outcomes after participation in psychiatric vocational rehabilitation programmes: a systematic review. Soc Psychiatry Psychiatr Epidemiol 40:408–416

Moller H, von Zerssen D, Werner-Eilert K, Wuschenr-Stockheim M (1982) Outcome in schizophrenic and similar paranoid psychoses. Schizophr Bull 8:99–108

National Institute on Disability and Rehabilitation Research (1997) Trends in labor force participation among persons with disabilities, 1983–1994 (report 10). Disability statistics report. U.S. Department of Education, Office of Special Education, Office of Special Education and Rehabilitative Services

Razzano LA, Cook JA, Burke-Miller JK, Mueser KT, Pickett-Schenk SA, Grey DD et al (2005) Clinical factors associated with employment among people with severe mental illness: findings from the employment intervention demonstration program. J Nerv Ment Dis 193(11):705–713

Rehabilitation Services Administration (1995) The provision of vocational rehabilitation services to individuals who have severe mental illnesses: Program administrative review (Final report). Office of Special Education and Rehabilitative Services, U.S. Department of Education, Washington, DC

Rogers ES, Anthony WA, Toole J, Brown MA (1991a) Vocational outcomes following psychosocial rehabilitation: a longitudinal study of three programs. J Vocat Rehabil 1(3):21–29

Rogers ES, Sciarappa K, Anthony WA (1991b) Development and evaluation of situational assessment instruments and procedures for persons with psychiatric disability. Vocat Eval Work Adjust Bull 24(2):61–67

Rogers ES, Anthony WA, Cohen M, Davies RR (1997) Prediction of vocational outcome based on clinical and demographic indicators among vocationally ready clients. Community Ment Health J 33(2):99–112

Rogers E, Razzano L, Rutkowski D, Courtenay C (2005) Vocational rehabilitation practices and psychiatric disability. In: Dew D, Allan GM (eds) Report from the study group; 30th Institute on Rehabilitation Issues, innovative methods for providing services to individuals with psychiatric disabilities. Rehabilitation Services Administration, US Department of Education, Washington, DC, pp 49–79

Rosenthal DA, Dalton JA, Gervey R (2007) Analyzing vocational outcomes of individuals with psychiatric disabilities who received state vocational rehabilitation services: a data mining approach. Int J Soc Psychiatry 53(4):357–368

Salkever D, Goldman HH, Purushothaman M, Shinogle J (2000) Disability management, employee health and fringe benefits, and long-term disability claims for mental disorders: an empirical exploration. Milbank Q 78(1):79–113

Schwartz C, Myers J, Astrachan B (1975) Concordance of multiple assessments of the outcome of schizophrenia. Arch Gen Psychiatry 32:1221–1227

Scott JE, Lehman AF (1998) Social functioning in the community. In: Mueser KJ, Tarrier N (eds) Handbook of social functioning in schizophrenia. Allyn & Bacon, Needham Heights, MA, pp 1–19

Strauss JS, Carpenter WT (1972) The prediction of outcome of schizophrenia: I. Characteristics of outcome. Arch Gen Psychiatry 27:739–746

Strauss JS, Carpenter WT (1974) The prediction of outcome of schizophrenia: II. Relationships between predictor and outcome variables. Arch Gen Psychiatry 31:37–42

Tashjian MD, Hayward BJ, Stoddard S, Kraus L (1989) Best practice study of vocational rehabilitation services to severely mentally ill persons. (Vol. 1: Study findings). Policy Study Associates, Inc., Washington, DC

Taylor SF, Liberzon I (1999) Paying attention to emotions in schizophrenia. Br J Psychiatry 174:6–8

Tsang H, Lam P, Ng B, Leung O (2000) Predictors of employment outcome for people with psychiatric disabilities: a review of the literature since the mid'80s. J Rehabil 66(2):19–31

Tsang H, Pearson V (1996) A conceptual framework for work-related social skills in psychiatric rehabilitation. J Rehabil 62(3):61–67

Tsuang M, Coryell W (1993) An 8-year follow-up of patients with DSM-III-R psychotic depression, schizoaffective disorder, and schizophrenia. Am J Psychiatry 150:1182–1188

Waghorn D, Chant D and Whiteford H (2002) Clinical and non-clinical predictors of vocational recovery for Australians with psychotic disorders. Journal of Rehabilitation, 68(4):40–51

Wallace CJ (1986) Functional assessment in rehabilitation. Schizophr Bull 12(4):604–624

Wallace CJ, Tauber R, Wilde J (1999) Teaching fundamental workplace skills to persons with serious mental illness. Psychiatr Serv 50(9):1147–1149

Watts FN (1978) A study of work behaviour in a psychiatric rehabilitation unit. Br J Soc Clin Psychol 17(1):85–92

Wewiorski NJ, Fabian ES (2004) Association between demographic and diagnostic factors and employment outcomes for people with psychiatric disabilities: a synthesis of recent research. Ment Health Serv Res 6(1):9–20

Wykes T (1998) Social functioning in residential and institutional settings. In: Mueser J, Tarrier N (eds) Handbook of social functioning schizophrenia. Allyn & Bacon, Needham Heights, MA, pp 1–19

Zarate R, Liberman R, Mintz J, Massel HK (1998) Validation of a work capacity evaluation for individuals with psychiatric disorders. J Rehabil 64:28–34

Chapter 5
Employment and Serious Mental Health Disabilities

Terry Krupa

Serious mental illness (SMI) includes a range of mental disorders characterized by symptoms and impairments that are severe enough, and of a long enough duration, to significantly interfere with the capacity to function and adapt in important domains of daily life (Kelly 2002). SMI is not diagnostic specific, although it often includes mental disorders in which psychosis is a prevalent feature (for example, schizophrenia). A broader range of mental disorders would be included in the categorization of serious mental illness if they had a significant and long-standing impact on daily activities and social participation.

Difficulties with employment are a feature of serious mental illness. Unemployment among people with SMI is exceptionally high, with reported rates in the 70–90% range (Marwaha and Johnson 2004). The problems of work and SMI are common around the world, although employment rates are sensitive to regional differences with variations in labor and welfare structures (Marwaha and Johnson 2004; Warner 1994).

This chapter begins with a discussion of the employment status and patterns of people with SMI. It then moves on to describe the benefits of employment for people with SMI, with a particular focus on the relationship between work and recovery. Following this is a brief review of factors that influence the employment of people with SMI and considers how these factors are being addressed in the mental health field with a view to improving employment outcomes. The chapter concludes with a brief discussion of the philosophies underpinning contemporary efforts to improve employment outcomes.

T. Krupa (✉)
School of Rehabilitation Therapy, Queen's University, Louise D. Acton Building,
Room 200, 31 George Street, Kingston, ON, Canada, K7L 3N6
e-mail: terry.krupa@queensu.ca

I.Z. Schultz and E.S. Rogers (eds.), *Work Accommodation and Retention in Mental Health*,
DOI 10.1007/978-1-4419-0428-7_5, © Springer Science+Business Media, LLC 2011

Employment Marginalization and Serious Mental Illness

The concept of "unemployment" is an imprecise representation of the work status and patterns of people with serious mental illness. Unemployment is a concept often used to refer to the work status of individuals who are willing and able to work but are not in paid employment. In some jurisdictions unemployment is also defined by evidence of active efforts to secure work. The concept of unemployment fails to capture the extent to which the patterns of work involvement of people with SMI are characterized by marginalization. This idea of marginalization suggests that people with SMI find themselves at the fringes of society's labor and economic markets, and that this outsider position is perpetuated through systematic processes.

This fringe status of people with SMI in the area of employment is expressed in many ways. For example:

- People with SMI generally lack ties to the workforce. These ties include connections to important networks of people and social structures that might enable them to secure and sustain employment.
- While there are growing efforts to build a case for the economic and social benefits of improving the employment outcomes for people with mental illness (Wilson et al. 2000), these efforts have largely focused on people who are employed or returning to work after a defined period of absence. People with SMI typically have histories of long periods of time away from employment.
- To deal with their daily experience of lack of structure and meaningful activity, people with SMI may be engaged in an array of work activity programs developed within the mental health system. These programs largely operate parallel to, not integrated within the community labor force. It appears that these programs contribute to a weakened sense of responsibility for ensuring access to community employment on the part of employers (Krupa et al. 2010).
- The jobs that people with SMI do get are often low paying, with limited potential for career advancements and improvements in their social status (Baron and Salzer 2002). This problem can lead to a situation where working is not personally meaningful and financially unviable.

Ultimately, this outsider status compromises their access to fair and equitable employment, and serves to undermine their own belief in their rights to and potential for community employment.

Since the employment patterns of people with SMI are so characterized by marginalization, focusing exclusively on employment rates provides a very limited view of their work status. While improving employment rates is an important goal, this outcome alone will not tell us about the extent to which people with SMI are integrated within their work settings, the extent to which they are able to manage and enjoy other aspects of daily life while engaged in employment, or the extent to which their economic status is actually improving. Goering et al. (2004) have proposed a model for conceptualizing productivity and severe mental illness that includes a range of outcomes to provide a more comprehensive understanding of an individual's work status than employment rates alone. These outcomes are summarized in Table 5.1,

Table 5.1 Employment outcomes and serious mental illness (Goering et al. 2004)

A range of work-related activities	Aspects of the work experience	Work status in the broader social context

Involvement in:
• Work for pay
• Unpaid work
• Education/training
• Parenting
• Vocational programs
	Hours of participation
	Continuity/tenure
	Integration with
	  community
	Time use/life balance
	Personal satisfaction
	Personal growth/change
		Social-economic status
		Promotion/advancement
		Pay level
		Cost/benefit
		Reliance on government
		  income benefits

and include attention to the range of work activities engaged in, aspects of the experience of working, and outcomes that assist with interpreting work status within the broader social context.

The Benefits of Working

Contrary to this outsider status is the important evidence that people with SMI want to work in community employment, and that given the appropriate opportunities and supports, and access to the rights they are due as citizens, we can expect that they will experience success in employment (Cook and Razzano 2000; Crowther et al. 2001). We can also expect that they will experience the same benefits that people in the general population receive from employment, including: the potential to develop and use their abilities; the development of a socially valued identity; a chance to contribute meaningfully to their societies, communities, and families; increased income to meet their basic needs and to plan for their futures; and access to opportunities and events that enrich their quality of life.

In addition to these benefits, there is growing evidence that employment is important to recovery from serious mental illness. Recovery does not necessarily mean a cure, but rather the idea that people can come to live meaningful and satisfying lives that are not primarily defined by their mental illnesses (Anthony 2004). There are several ways that employment might facilitate recovery, and only a few are offered here. Observing others with SMI who are successful at employment may spark the hope that a meaningful life beyond the experience of mental illness is possible, and encourage an individual to invest their energies towards this goal. The desire to work may engage the individual in efforts to learn to actively cope with the mental illness, and to seek out and integrate knowledge about health practices that could sustain employment. Having people with SMI demonstrate success in working provides evidence that employment is indeed achievable, and may secure the commitment of service providers, families, and friends to provide their ongoing support and assistance.

Factors Contributing to Employment Marginalization

A range of factors influences the employment experiences of people with serious mental illness, and there has been a concerted effort to address these factors with a view to facilitating better outcomes.

Developmental Factors

Adolescence and young adulthood is a critical time period for developing a work identity; gaining experience with work expectations, relationships, and behaviors; and completing education and training associated with adult work. It is also the time period when young people are developing autonomy in daily living and community participation – an important foundation for employment success in adulthood. Unfortunately, the experience of SMI typically begins during this very stage in life, disrupting career planning, work experiences, and academic achievement, and often leading to an ongoing reliance on family (Woodside et al. 2007). Experiencing mental illness during young adulthood has also been associated with limited capacity for understanding and reasoning about employment-related problems (Van Ormer 2001).

Current best practices in employment focus on providing people with SMI opportunities to gain direct work experience while providing ongoing support, with a view to gaining on-the-job experience (Kirsh et al. 2005). Counseling and training that focuses on preparing people for looking for jobs, preparing résumés, and interviewing, address an individual's lack of experience with these activities. There has also been an emphasis on supporting people to achieve levels of education and training that will enhance their career potential (Murphy et al. 2005; Baron and Salzer 2002). In addition, early intervention services have now been widely replicated, with a view to ensuring that the education and career planning of young people experiencing their first episodes of illness is not derailed, and that the work disability and social marginalization that has been associated with SMI can be prevented (Bertolote and McGorry 2005).

Illness-Related Factors

While there is no doubt that illness factors influence work participation, the nature of this relationship is complex and not well understood. For example, the extent to which diagnosis predicts employment has been the source of much debate. While schizophrenia has widely been considered to be the most disabling serious mental illness, a diagnosis of schizophrenia has not routinely been found to predict participation in employment, although it may impact aspects of the employment experience, such as the number of hours that an individual works (Razzano et al. 2005).

At this time there is a general agreement that diagnosis should not be used to judge who is likely to (or unlikely to) work.

There is a growing body of research examining how specific symptoms and impairments influence employment. Some of these findings have been counterintuitive – suggesting that an increase in symptoms does not always relate to poorer outcomes. For example, the evidence regarding the impact of positive symptoms of psychosis, such as hallucinations and delusions, has not consistently demonstrated worse employment outcomes (Razzano et al. 2005). Those symptoms grouped together as "negative symptoms," including general feelings of apathy, avolition, blunted affect, social withdrawal, and limited feelings of pleasure in daily life activities, have been associated with worse employment outcomes (Razzano et al. 2005). Similarly, ongoing symptoms of depression have been associated with prolonged and severe work impairment (Judd et al. 2007). Profound negative and depressive symptoms can deprive an individual of the drive and energy for work, and interfere with task and social performance on the job. They can also weaken the emotional connection to work-related goals and activities, even when the work itself is highly valued.

Cognitive symptoms have also consistently been associated with poorer employment outcomes. SMI can lead to disturbances in several cognitive functions, such as personal insight, working memory, processing speed, attention, and executive functioning, that are fundamental to the task and interpersonal demands of contemporary work settings (McGurk and Mueser 2004; McGurk and Meltzer 2000). Similarly, lower levels of general functioning, whether clinical or self-rated, have also been associated with poorer employment outcomes in SMI (Razzano et al. 2005; Cook and Razzano 2000). General functioning typically refers to the ability to live with autonomy in the community, the degree of self-distress experienced by an individual, the extent to which behaviors are consistent with social norms and expected social role participation, and the quality of interpersonal relationships (Phelan et al. 1994).

Within a recovery-oriented mental health system there has been growing interest in addressing illness-related issues in a manner that is likely to support improved employment outcomes. The following are only a few examples:

- The integration of clinical and employment services is now considered best practice in community mental health, to ensure that experiences of mental illness will be addressed in a manner that is likely to promote employment (Bond et al. 2008).
- Structured interventions to improve personal illness management in daily life have been widely disseminated (Hasson-Ohayon et al. 2007), although their impact on employment has yet to be established. Qualitative research is revealing how individuals with SMI who are successful in employment actively cope with the symptoms of their illness on the job (Krupa 2004; Cunningham et al. 2000). Education and training in coping skills is a mainstay of employment support.
- Best practices in employment support highlight the importance of eliciting and honoring the preferences of individuals with serious mental illness, with a view

to limiting the impact of motivational challenges. Best practice also includes careful job matching between individual strengths and job demands (Bond et al. 2008; Kirsh 2000).

- Interventions aimed at remediation and compensation for impairments in cognitive functioning in SMI have been developed, and some positive associations with improved employment outcomes have been established (Velligan et al. 2006). To address difficulties in interpersonal relationships, social skills training and social problem solving initiatives are being implemented (Tsang and Pearson 2001). A complementary strategy has involved the development of positive social networks for individuals with SMI, who often experience deterioration of their social networks over time (Rollins et al. 2002).

Treatment-Related Factors

Unfortunately, the treatment meant to improve the health and well-being of people with SMI has also been associated with creating barriers to their employment. Medication side effects such as drowsiness, fatigue, tremors, and disturbed movement patterns can have a direct impact on work performance; while others, like weight gain, can impact employment by compromising an individual's self-esteem. Evidence-based practice in psychopharmacology highlights the importance of a collaborative approach with individuals in treatment to ensure that side effects and their impacts are identified and decisions about treatments are jointly developed (Mueser et al. 2003).

Mental health systems have been slow to champion the employment of people with SMI as a priority. Employment supports are not routinely accessible within mental health service systems (Hanrahan et al. 2006). Certainly this state of affairs can be attributed somewhat to the funding limitations that constrain the development of mental health services and supports. In addition, the development of employment-focused support has been hindered by traditionally held assumptions that SMI has a downward and deteriorating course that is inconsistent with employment (Harding et al. 1992). Well-established conceptual frameworks that link the experience of stress to the exacerbation of acute illness further hinder an investment in employment initiatives when work is understood as a primary source of stress (Krupa 2004).

Contemporary efforts to design recovery-oriented mental health systems have focused on shifting the paradigm of service delivery away from an institutional, segregation, and care orientation, to a community- and empowerment-oriented perspective that supports growth potential and full citizenship. Within this new paradigm, employment is viewed as a valued social resource – a resource to which people with SMI should have full access (Nelson et al. 2001). While the relationship between the experience of stress and poor mental health is acknowledged, the broad and uncritical application of stress reduction is being challenged. Marrone and Golowka (2000), for example, poignantly challenge the field to consider that the poverty, social isolation, and lack of meaning that is part of unemployment might be well beyond the stress experienced at work. Indeed, research examining the impact of employment on mental illness has not, to date, demonstrated that working leads to the worsening of mental illness. Buoyed by research evidence indicating that people with SMI can experience improvements in the course of their illnesses, level of functioning, and community well-being (Harding et al. 1992), there has been growing attention to ensuring that employment is addressed within the treatment and service plans of all individuals with serious mental illness, and that best practices in employment support are widely disseminated and accessible to all people with SMI.

Societal and Environmental Factors

The ability of people with SMI to secure and maintain employment is closely linked to the societal and environment context. The stigma of mental illness has been recognized as perhaps the most profound societal barrier to full participation in employment. Stigma refers to the negative attitudes and labeling that lead to a disposition to discriminatory behaviors and practices (Corrigan and Lam 2007). In the case of employment and mental illness, this means the tendency of societal attitudes to devalue people with SMI as workers and to create a social situation that unfairly constrains their ability to secure and maintain work. More damaging perhaps, is the fact that the devaluation that is at the core of stigma and discrimination becomes internalized by people with mental illness, and compromises their sense of entitlement to valued social resources such as employment (Corrigan and O'Shaughnessy 2007).

A recently developed model of stigma in the workplace has identified a range of underlying assumptions and attitudes that can lead to employment discrimination (Krupa et al. 2010). These include the following assumptions: people with SMI lack the competencies necessary to fulfill the task and social demands of employment; people with SMI are prone to violence and dangerous behavior in the workplace; mental illnesses are not legitimate illnesses and therefore are not entitled to accommodations; people with mental illness will be made more ill by employment; and employing people with mental illness will weaken workplace productivity.

Efforts are being directed to address the assumptions and attitudes that perpetuate the stigma of mental illness and limit full employment. One approach is to challenge these assumptions through the education of people who can influence the employment process, such as employers, human resources staff, and work supervisors. A second approach, with the strongest impact on stigma, is to assertively create opportunities for meaningful and positive contacts between people with mental illness and key employment people (Corrigan et al. 2007). In addition, attention is now being paid to the issue of disclosure of mental illness at work, with the emerging development of specific disclosure practices that reduce potential for stigma in the workplace (MacDonald-Wilson 2005).

Another environmental factor to be considered is the characteristics of the workplace itself. Many features of employment have been linked to poor mental health in the general public, but aspects of the workplace climate and the organization and structure of jobs may pose a particular disadvantage for those with serious mental illness. There have been efforts to link important aspects of SMI to job demands, with a view to understanding how jobs can be constructed to promote success (Mechanic 1998). For example: early work hours may conflict with sleep and fatigue issues secondary to medications; congested work environments may be difficult to negotiate when cognitive functions such as concentration are impacted; anxiety on the job may be increased by supervision that is overly critical; and brief on-the-job training may not be conducive to deep learning when an individual has limited work experience or a lengthy history of unemployment. In such circumstances, the implementation of reasonable accommodations in the workplace can be used to respect both an individual's need for job refinements and the employer's concern for a productive workplace. Flexible working hours, the use of privacy screens, the delivery of balanced performance reviews, and longer training periods and mentors might be reasonable accommodations to deal with the case scenarios identified above.

Another significant barrier to the full employment participation of people with SMI is their reliance on government disability income supports. On the positive side, in countries where disability income support is available, individuals with SMI are provided with some modicum of financial security, albeit typically at poverty levels. On the negative side, the evidence suggests that these pensions are linked to poorer employment outcomes (Honkonen et al. 2007; Rosenheck et al. 2006). These income structures are believed to act as a disincentive to employment in a variety of ways: for example, producing fear that financial security will be lost if employment is secured, encouraging a self-view that is primarily characterized as "disabled" and thus unable to work, and offering complex administrative and policy structures that are difficult to understand and navigate (Leff and Warner 2006). Choosing disability benefits over payment from employment can also be seen as a rational economic choice for people with SMI who will incur high costs associated with the treatment of mental illness, but may not be able to access employment that offers the salary or the benefits to offset these costs.

Approaches to address the constraints posed by government disability income supports are directed to both the individual with SMI and to the structures of income support. At the individual level, providing counseling that increases the awareness of opportunities and policies related to income benefits, as well as assistance in navigating complex income systems, can help to ensure informed decision-making for employment (Tremblay et al. 2006). Advocacy efforts focus on identifying problematic aspects of disability structures and lobbying recommendations for change. For example, advocates in some regions might lobby for the separation of guaranteed coverage for medications and other treatments from receipt of government disability income as a means to improve access to treatment, reduce poverty levels, and improve employment outcomes among people with serious mental illness.

Conclusion: Advancing a Capability- and Opportunity-Based Approach to Employment and Serious Mental Illness

This chapter has presented the employment status of people with SMI as characterized by employment marginalization – a situation where people find themselves located outside of the community-based work force. This fringe status is continued by multiple interacting factors that systematically disadvantage people with SMI in securing and maintaining employment. The chapter provides a brief review of a variety of employment interventions and initiatives directed to addressing factors that have been linked to a range of poor employment outcomes experienced by people with SMI.

Considered independently, each employment intervention or support is unlikely to have much impact on employment marginalization. For example, cognitive remediation intervention may have a positive effect on problems of concentration, but for an individual with SMI who is marginalized from the workforce, this improvement may not translate to employment success without consideration of other factors. Provided within an integrated and comprehensive approach, the potential for full employment participation improves dramatically.

This comprehensive approach to employment support is consistent with a capabilities approach, applied to people with mental illness by Ware et al. (2007). From a capabilities approach, participation in employment in the community workforce is considered a right of people with SMI as citizens. This right is meaningless, however, when employment remains a hypothetical ideal. The potential for participation in community employment is increased when individuals with SMI are provided the range of supports and resources they need to maximize their capabilities, and the opportunities within the world of work to exercise and grow these capacities, and ultimately to both flourish and make valued social contributions.

References

Anthony WA (2004) The recovery effect. Psychiatr Rehabil J 27(4):303–304
Baron R, Salzer M (2002) Accounting for unemployment among people with mental illness. Behav Sci Law 20:585–599
Bertolote J, McGorry P (2005) Early intervention and recovery for young people with early psychosis: consensus statement. Br J Psychiatry 187(Suppl 48):116–119
Bond GR, Drake RE, Becker DR (2008) An update on randomized controlled trials of evidence-based supported employment. Psychiatr Rehabil J 31(4):280–290
Cook JA, Razzano LA (2000) Vocational rehabilitation for people with psychiatric disability. Am Rehabil 20:2–12
Corrigan PW, Lam C (2007) Challenging the structural discrimination of psychiatric disabilities: lessons learned from the American disability community. Rehabil Educ 21(1):53–58
Corrigan PW, O'Shaughnessy JR (2007) Changing mental illness stigma as it exists in the real world. Aust Psychol 42(2):90–97
Corrigan PW, Larson JE, Kuwabara SA (2007) Mental illness stigma and the fundamental components of supported employment. Rehabil Psychol 52(4):451–457

Crowther RE, Marshall M, Bond GR, Huxley P (2001) Helping people with serious mental illness to obtain work. Psychiatr Serv 49:196–201

Cunningham K, Wolbert R, Brockmeier MB (2000) Moving beyond the illness: factors contributing to gaining and maintaining employment. Am J Community Psychol 28:481–494

Goering P, Durbin J, Krupa T, Kirsh B, Dewa C (2004) Developing a model of meaningful activity for persons with severe mental illness. Annual conference of the American Public Health Association, Washington, DC, 6–10 Nov 2004

Hanrahan P, Heiser W, Cooper AE, Oulvey G, Luchins DJ (2006) Limitations of system integration in providing employment services for persons with mental illness. Adm Policy Ment Health 33:244–252

Harding CM, Zubin J, Strauss JS (1992) Chronicity in schizophrenia: revisited. Br J Psychiatry 161(Suppl 18):27–37

Hasson-Ohayon I, Roe D, Kravetz S (2007) A randomized controlled of the effectiveness of the Illness Management and Recovery Program. Psychiatr Serv 58:1461–1466

Honkonen T, Stengård E, Virtanen M, Salokangas R (2007) Employment predictors for discharged schizophrenia patients. Soc Psychiatry Psychiatr Epidemiol 42:372–380

Judd LL, Schettler PJ, Solomon DA, Maser JD, Coryell W, Endicott J, Akiskal HS (2007) Psychosocial disability and work role function compared across the long-term course of bipolar I, bipolar II and unipolar major depressive disorders. J Affect Disord 18:49–58

Kelly T (2002) A policy maker's guide to mental illness. htttp://www.heritage.org/Research/HealthCare/BG12522ES.cfm. Accessed 29 July 2008

Kirsh B (2000) Factors associated with employment for mental health consumers. Psychiatr Rehabil J 24:13–22

Kirsh B, Cockburn L, Gewurtz R (2005) Best practice in occupational therapy: program characteristics that influence vocational outcomes for people with serious mental illness. Can J Occup Ther 72(5):265–280

Krupa T (2004) Employment, recovery and schizophrenia: integrating health and disorder. Psychiatr Rehabil J 28(1):8–15

Krupa T, Kirsh B, Cockburn L, Gewurtz R (2010) Understanding the stigma of mental illness in employment. Work 33:413–425

Leff J, Warner R (2006) Social inclusion of people with mental illness. Cambridge University Press, Cambridge

MacDonald-Wilson KL (2005) Managing disclosure of psychiatric disabilities to employers. J Appl Rehabil Couns 36(4):11–21

Marrone J, Golowka E (2000) If work makes people with mental illness sick, what do unemployment, poverty and social isolation cause? Psychiatr Rehabil J 23(2):187–193

Marwaha S, Johnson S (2004) Schizophrenia and employment: a review. Soc Psychiatry Psychiatr Epidemiol 39:337–349

McGurk SR, Meltzer HY (2000) Neurocognition as a determinant of employment status in schizophrenia. Schizophr Res 45(3):175–184

McGurk SR, Mueser KT (2004) Cognitive functioning, symptoms and work in supported employment: a review and heuristic model. Schizophr Res 63:1–27

Mechanic D (1998) Cultural and organizational aspects of application of the Americans with Disabilities Act to persons with psychiatric disabilities. Milbank Q 76:5–23

Mueser KT, Torrey WC, Lynde D, Singer P, Drake RE (2003) Implementing evidence-based practices for people with severe mental illness. Behav Modif 27:387–411

Murphy AA, Mullen MG, Spagnolo AB (2005) Enhancing individual placement and support: promoting job tenure by integrating natural supports and supported education. Am J Psychiatr Rehabil 8:37–61

Nelson G, Lord J, Ochocka J (2001) Shifting the paradigm in community mental health: towards empowerment and community. University of Toronto Press, Toronto

Phelan M, Wykes T, Goldman H (1994) Global function scales. Soc Psychiatry Psychiatr Epidemiol 29:205–211

Razzano LA, Cook JA, Burke-Miller JK, Mueser KT, Pickett-Schenk SA, Grey DG, Goldberg R et al (2005) Clinical factors associated with employment among people with serious mental illness: findings from the employment intervention demonstration program. J Nerv Ment Dis 193:705–713

Rollins AL, Mueser KT, Bond GR, Becker DR (2002) Social relationships at work: does the employment model make a difference? Psychosoc Rehabil J 26:51–61

Rosenheck R, Leslie D, Keefe R, McEvoy J, Swartz M, Perkins D, Stroup S, Hsiao JK, Lieberman J (2006) Barriers to employment for people with schizophrenia. Am J Psychiatry 163:411–417

Tremblay T, Smith J, Xie H, Drake RE (2006) Effect of benefits counseling services on employment outcomes for people with psychiatric disabilities. Psychiatr Serv 57:816–821

Tsang HW, Pearson V (2001) Work-related social skills training for people with schizophrenia in Hong Kong. Schizophr Bull 27:139–148

Van Ormer EA (2001) Developmental status as a predictor of outcomes in a vocational rehabilitation program. Lynch School of Education, Boston College, Dissertations and Theses

Velligan DI, Kern RS, Gold JM (2006) Cognitive rehabilitation for schizophrenia and the putative role of motivation and expectancies. Schizophr Bull 32:474–485

Ware NC, Hopper K, Tugenberg T, Dickey B, Fisher D (2007) Connectedness and citizenship: redefining social integration. Psychiatr Serv 58:469–474

Warner R (1994) Recovery from schizophrenia: psychiatry and political economy. Routledge, London

Wilson M, Joffe RT, Wilkerson B (2000) The unheralded business crisis in Canada: depression at work. Global business and economic roundtable on addiction and mental health, Toronto, 20 July 2000

Woodside H, Krupa T, Pocock K (2007) A conceptual model to guide rehabilitation and recovery in early psychosis. Psychiatr Rehabil J 32(2):367–373

Chapter 6
Depression and Work Performance: The Work and Health Initiative Study

Debra Lerner, David Adler, Richard C. Hermann, William H. Rogers, Hong Chang, Pamella Thomas, Annabel Greenhill, and Katherine Perch

Introduction

Depression, a chronic, episodic condition affecting at least 4.9% of the working age population (Blazer et al. 1994), causes substantial functional limitation and social role disability (Wells 1985, 1997; Wells et al. 1991). Since Wells et al. (1989) first reported on the disabling impact of depression in the 1980s, evidence of its human and economic burdens has continued to accumulate (Druss et al. 2000; Hirschfeld et al. 2002; Lin et al. 2000; Simon et al. 1998). As a result, there is a greater awareness of the adverse impact of depression on employment and work productivity. Thus, it is not an overstatement to say that depression is a major threat to the public's health, quality of life, and economic well-being.

In the United States (US) alone, depression is estimated to cost between $36.6 and $51.5 billion annually in lost productivity (Greenberg et al. 2003; Kessler et al. 2006; Stewart et al. 2003). As research by our team and others (Lerner and Henke 2008) has shown, employment problems such as job loss, premature retirement, on-the-job functional limitations, at-work productivity loss (known as "presenteeism") and time lost from work (absenteeism) are hiding these enormous costs. Between 2000 and 2004, we conducted a longitudinal observational study of workers with depression. This study was designed to provide in-depth information about the range and magnitude of workers' employment issues and the mechanisms underlying employment outcomes. To accomplish this, we enrolled adults with depression and a group of depression-free controls. Participants had to be employed at baseline and have no plans to leave the labor market. Workers with serious physical and mental chronic health problems (other than depression) were excluded. To locate a broad sample of working people, we screened more than 14,000 patients in primary care office waiting rooms in Massachusetts. The final sample consisted of 286 adults who met the Diagnostic and Statistical Manual of Mental Disorders (DSM-IV) criteria

D. Lerner (✉)
Institute for Clinical Research and Health Policy Studies, Tufts Medical Center,
800 Washington Street, 02111, Boston, MA, USA
e-mail: Dlerner@tuftsmedicalcenter.org

I.Z. Schultz and E.S. Rogers (eds.), *Work Accommodation and Retention in Mental Health*,   103
DOI 10.1007/978-1-4419-0428-7_6, © Springer Science+Business Media, LLC 2011

for major depression and/or dysthymia and 193 depression-free controls. Participants were surveyed every 6 months for 18 months. The outcomes measured included job loss, presenteeism, and absenteeism (the latter two were evaluated using the Work Limitations Questionnaire (WLQ) Lerner 2001, 2002, 2003).

By any metric, depression had an adverse impact on employment and work productivity. Based on data from the WLQ, measuring four dimensions of presenteeism, depressed subjects experienced at-work difficulty performing job tasks between 19.5 and 36.2% of the time in the 2 weeks prior to the baseline assessment. The amount of presenteeism was four times the level observed within the control group. In addition, at baseline the depression group was absent a mean of 1.8 days in the prior 2 weeks, three times the mean number of the control group's baseline absence days ($p \leq 0.001$ on all comparisons). After baseline, at each follow-up, the depression group still had significantly worse presenteeism and absenteeism ($p \leq 0.001$ on all comparisons). With regard to job loss, the depression group had a 15% job loss rate at 6 months vs. 3.5% in the control group. By the final 18 month follow-up, few subjects in the depression group who had lost their jobs had found new ones (Lerner et al. 2004b). Finally, while this was not a treatment study, we did not observe any significant differences among workers with depression who were in mental health treatment compared to those not in mental health treatment.

The Mechanisms Underlying Work Loss

It is important to understand the mechanisms by which depression leads to employment problems and loss of productivity. Currently, the research literature has little to say on the topic of mechanisms per se, but has addressed the vulnerabilities of certain subgroups of the population with depression (Lerner and Henke 2008).

Generally, workers with more severe depression symptoms do worse than those whose symptoms are moderate or mild. Dooley et al. (2000) found that depression severity at baseline predicted employment at 2-year follow-up. Greco et al. (2004) found that symptom severity had a significant important impact on work functioning, concluding that depression symptom severity had a greater influence on employment than physical symptom severity. Three studies examining absenteeism rates found that individuals with more severe depression symptoms were more likely than others to report days of lost work due to depression (Kessler and Frank 1997; Lerner et al. 2004a; Pflanz and Ogle 2006). Further, Simon et al. (2005) found that symptom improvement predicted reduced absenteeism. Similarly, Berndt et al. (1998) found that at-work performance of chronically depressed patients improved with reduction of depression severity.

Other research has shown that depressed workers with physical and/or mental comorbidities do worse at work than those who have depression and no comorbidities. For example, we found that depression severity and physical health status were both highly related to presenteeism (Lerner et al. 2004a).

In order to better understand the mechanisms underlying employment outcomes, we have focused on the role of specific depression symptoms. Using depression symptom data from our observational study, we found that symptoms such as difficulty concentrating, distractibility, fatigue, and difficulty sleeping had the strongest relationships to presenteeism.

Compelling evidence regarding the role of symptoms and symptom severity has also been obtained from intervention research. Generally, studies show that subgroups of adults with depression who receive high quality depression care (diagnosis and treatment) have better employment outcomes compared to those whose care is suboptimal. Depression treatment studies suggest that improving the quality of depression care reduces depression symptom severity, which contributes to improved work outcomes (Rost et al. 2004; Schoenbaum et al. 2002). The most definitive study to date testing the effect of high quality depression treatment vs. usual care on employment outcomes reported that treatment reduced absenteeism but not presenteeism (Wang et al. 2007).

While symptom severity, specific symptoms, and the quality of medical care appear to influence outcomes, new research suggests the work itself also affects outcomes. In our longitudinal observational study, we found that recovery from depression did not necessarily result in an adequate improvement in the ability to work (Adler et al. 2006). Specifically, 6 months after baseline, 17% of the subjects in the depression group either had a clinically significant reduction in symptoms or were in remission, 61% had no clinically significant change, and 22% had worsened. At each follow-up, subjects with the best clinical outcomes (the improved group) still had worse WLQ scores than controls. Thus, functional losses persisted even after clinical improvement (Fig. 6.1).

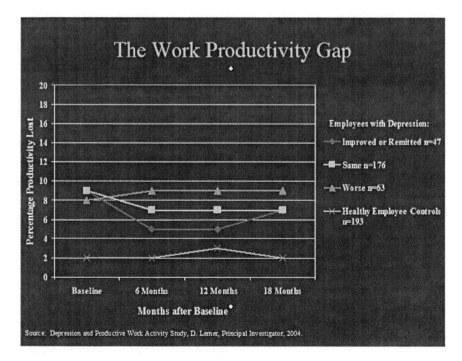

Fig. 6.1 The work productivity gap

Next, we examined the importance of occupational demands to outcome at 6 months after baseline among depressed workers. We found that multiple dimensions of presenteeism, measured with the WLQ, were explained by occupational demands (measured objectively with the US O*NET occupational classification system (O*Net Resource Center 2008), which serves as the nation's primary source of occupational information). Presenteeism was higher among workers whose occupations required more judgment and decision-making, and/or interpersonal interactions.

Using data from the entire observation period of 18 months, which consisted of a minimum of 1,080 observations with up to four data points per subject, we modeled the impact of psychosocial work characteristics on both presenteeism and absenteeism. Both presenteeism and absenteeism were explained mainly by depression symptom severity ($p \leq 0.0001$ on all comparisons). Poorer physical health (PCS-12 scores; Ware 1996) also influenced the outcomes ($p = 0.001-0.02$). Importantly, several dimensions of presenteeism were explained by low control at work (p-values $0.01-0.047$) and psychologically demanding work ($p = 0.005-0.046$). Thus, results suggest that in addition to improving depression diagnosis and treatment, changing conditions within the work environment may help workers with depression to function better and reduce productivity loss.

Service Gaps

This new research inevitably leads to questions about the adequacy of services to address depression among workers. Currently, the US does not have an integrated system of medical and vocational services aimed at helping workers who develop illnesses such as depression, or those who experience relapses, to continue working.

The medical care system, and typically primary care, remains the predominant source of help for working age US adults who have mental and/or chronic physical health problems. The primary care system (where most workers receive mental health care) has advanced in depression recognition and treatment, but problems remain. In one epidemiological study of more than 25,000 primary care patients, only 54% reporting depressive symptoms on a screener had been recognized as having psychological problems by their primary care provider, and only 15% had been given a diagnosis of depression. Additionally, many adults with depression, even when insured, do not seek care because they lack sufficient information (Ustiin et al. 1995). Low health literacy and the fear of job loss or reassignment from jobs are barriers to participating in screening and/or treatment. Stigma and health behaviors inhibit care-seeking and treatment (Katon and Schulberg 1992; Rost et al. 1994; U.S. Department of Health and Human Services and AHCPR 1993; Wells and Sturm 1995). Many primary care physicians and nurses have neither the time for adequate patient education or follow-up nor the infrastructure to systematically monitor treatment adherence or outcomes (Rost et al. 1994; U.S. Department of Health and Human Services and AHCPR 1993).

In medical care, routine assessment of work difficulties rarely, if ever, occurs. Many physicians lack the training, skills, tools, and time necessary to detect and treat patients' health-related employment and productivity problems. And workers, unaccustomed to discussing work issues with their doctors, often do not access care for this reason. Partly because of this situation, physicians often become involved in a patient's employment problems only when there is a disability or Workers Compensation claim involved (Lerner et al. 2005). Finally, access to medical rehabilitation is limited and most workers with depression will not qualify for such services.

The US occupational health and safety system, comprised of state and federal agencies and private sector occupational health personnel and worksite health clinics, mainly addresses physical health hazards in the work environment. Services for workers with depression are generally not part of the occupational health and safety infrastructure. In fact, most states do not recognize mental health claims in their workers compensation system unless it is clearly precipitated by specific compensable physical trauma. Unable to work, employees eventually lose employer-based medical coverage, which would have provided the necessary treatment needed to improve sufficiently to return to the workplace.

On the employment services side, various job training services are available through state and federal government programs. However, these generally assist clients with job preparation and job placement. They are not oriented to helping workers with depression and/or other chronic health problems, nor are they oriented to helping adults who are already in the labor market to keep working.

In the US, state and federal vocational rehabilitation programs provide employment support to individuals with disabilities. However, services are available mainly to adults with the most severe health problems, most of whom have little or no work experience.

Insurance programs, such as the US Social Security disability programs and private disability insurance, provide income replacement. The few return to work services, such as the Ticket to Work program under Social Security, if provided, are limited in scope.

The Americans with Disabilities Act (Thomas and Gostin 2009) has generally helped more with hiring than retention. The latest amendments broaden the definition of disability and conceivably could help more workers with depression, but are as yet largely untested.

The New Intervention Model

The extant data on depression and employment services led us to conclude that new care models integrating depression symptom treatment and work-focused interventions were important to helping workers with depression remain engaged and productive in their work. We developed a new multimodal care model aimed at preventing depression-related losses in job performance and productivity. Named the Work and Health Initiative (WHI), this model was influenced by emerging perspectives on work productivity such as the "high road to productivity" and health and productivity management concepts (Appelbaum et al. 2000; Ichniowski et al. 1996; Lowe et al. 2003; Perry et al. 1996).

Problem	Modality
Poor detection of depression and impaired ability to work.	Web-based depression screening. Uses validated brief survey tools to identify employees ruling in for major depression and/or dysthymia and/or work limitations, report results to employees and instruct employees about next steps.
Barriers to obtaining depression treatment and recovery.	Care Coordination. 1) Educate about depression and its work impact, and role of treatment; 2) if PHQ ≥ 10 with no medication or if on ineffective dosage, encourage use of treatment & adherence; 3) monitor symptoms, treatment, adherence, and side effects; & 4) report to treating physician on depression, treatment, and work issues.
Features of the work and work behavior interfering with functioning at work.	Work Modification/Coaching.1) Review performance deficits; 2) identify modifiable elements of work, work stressors, resources and constraints; 3) coach in strategies to change work behavior and/or work conditions; & 4) monitor impact.
Thoughts, feelings and behaviors interfering with functioning at work.	CBT Strategies. 1) Identify work and other functional problems related to dysfunctional cognitive-behavioral patterns; 2) teach CBT strategies; & 3) monitor progress and provide reinforcement.

Fig. 6.2 Summary of WHI model

According to these models, productivity within work organizations is achieved by investing in employee health, well-being, and a good quality of working life.

To overcome some of the care delivery issues mentioned earlier and pave the way for future dissemination, we chose to provide WHI services in the context of existing Employee Assistance Programs (EAPs). EAPs focus on both work and mental health issues, often playing an intermediary role between workers and other external services. The WHI may be the first program integrating mental health and vocational care for workers using this existing community-based resource.

The WHI, which is provided through EAPs to employees within companies that contract for its services, consists of several components (Fig. 6.2). The first component is a web-based screening tool to detect depression and work limitations. Once an employee views a study advertisement, either online or in hard copy, he or she visits the URL listed in the advertisement to take the web-based screener. If he/she is determined to be eligible for services, he/she is provided care by an EAP counselor who has been trained to provide the program's three care components. The WHI program consists of: (1) a medical care coordination component (based on the Collaborative Care model for depression (Gilbody et al. 2006)); (2) a work modification and coaching component (a new method addressing personal and environmental barriers to effective functioning and well-being at work); and (3) a cognitive- behavioral therapy (CBT) strategies component (based on the work of Ludman et al. 2007).

Workplace-Based Depression Screening and Medical Care Coordination

As described above, the WHI integrates web-based health screening and medical care coordination activities to improve the quality of care for workers with depression (Katon 2003). The screening component of the WHI responds to the US Preventive

Services Task Force (2002) report finding that screening improves the detection of depression and, when linked to effective treatment and follow-up, leads to more positive outcomes. The WHI's medical care coordination component is based on principles of the Collaborative Care model, which has been described as "a structured approach to care based on chronic disease management principles and a greater role for nonmedical specialists, such as nurse care managers, working in conjunction with the primary care physician" (Gilbody et al. 2006, p 2314). An extensive meta-analysis found that collaborative care improves initiation and adherence to antidepressant treatment and sustains improvements in symptoms and functioning for up to 5 years (Gilbody et al. 2003). To support recovery from depression, WHI counselors use a combination of educational and counseling techniques, monitoring, and follow-up. To strengthen coordination between the employee, the medical care system, and the EAP, the EAP counselor will periodically send updates to the employee's treating physician. These updates are sent when the employee has given the counselor permission to contact his or her physician. The updates are hard copy reports displaying the results of standardized assessments such as the PHQ-9 (Personal Health Questionnaire-9) for depression, the WLQ for work limitations, and treatment adherence. Assessment results are displayed graphically and summarized.

Work Modification and Coaching Component

This WHI component is aimed at reducing personal and/or environmental barriers to effective functioning at work. Conceptually, it draws upon principles articulated in research on return to work programs for work-injured employees (Goetzel et al. 2001), employment support for severely and chronically mentally ill populations (Handler et al. 2003; Kopelowicz and Liberman 2003; Lehman et al. 2002; Loisel and Durand 2001), and disability (Altman et al. 2001; Ellwood et al. 2000; Lehman et al. 2002) (depicting social role performance as the result of a complex personal and environmental interaction). Its approach to intervention reflects discoveries regarding the complex relationships between exposure to work stressors and health outcomes (Marmot et al. 1996; NIOSH Working Group 1999; Sparks et al. 2001; Stansfeld et al. 1995). In this WHI component, we emphasize the importance of obtaining an accurate assessment of the employee's work limitations using the WLQ and interviewing techniques. Next, the counselor elicits information from the employee regarding the personal and environmental barriers that are interfering with effective functioning. Then, drawing upon resources available inside and outside of the workplace, the employee is guided towards initiating changes aimed at improving functioning. The recommended changes are designed to be normative and safe, as opposed to formal job accommodations. The recommended approaches are both behavioral and environmental, depending on the nature of the problem and the employee's preferences. For example, employees may be having difficulty working because they feel disorganized and/or distracted. After careful assessment of his or her limitations, the work situation, and the employee's preferences, the counselor may suggest strategies such as using organizing tools, checking in with coworkers for feedback and advice, or turning off email and telephones at designated times. If disorganization and/or distraction are related to cognitive/emotional issues such as ruminating and wandering thoughts, the counselor may suggest a CBT strategy.

Work-Focused Cognitive Behavioral Therapy (CBT) Strategies Component

Depression is associated with multiple problems relative to cognition, motivation, behavior, and interpersonal functioning that may interfere with work functioning. Thus, individuals with the disorder will, conceivably, benefit from developing a new repertoire of compensatory and coping strategies, and identifying supports. The WHI incorporates CBT strategies originally developed by Beck (Beck 1979, 1991, 1995); Lewinsohn et al. (1984); and adapted by others (Fava et al. 1998; Jacobson et al. 1996; Jarrett et al. 2001; Katon et al. 1999; Ludman et al. 2000, 2001; Paykel et al. 1999; Scott et al. 2000). CBT involves strategies to change cognitions (e.g. distorted and negative thoughts) and maladaptive behaviors contributing to impaired functioning. Studies report that CBT is an effective, time-limited alternative to more traditional therapeutic approaches to depression treatment (Brewin 1996), and that it improves patients' symptoms and functional outcomes (Hirschfeld et al. 2002; Keller et al. 2000). Additionally, randomized controlled trials (RCTs) of enhanced primary care have shown that CBT interventions reduce disability (Katon et al. 2001, 2002; Ludman et al. 2000; Rost et al. 2001; Wells et al. 2000). CBT has also been studied within employee populations showing moderate positive effect on mental health, quality of working life, and absenteeism (Marhold et al. 2001, 2002; van der Klink et al. 2001, 2003; Wang et al. 2007). Recent studies using time-limited telephone-based CBT strategies have demonstrated efficacy in depression treatment (Ludman et al. 2007; Lynch et al. 1997; Mohr et al. 2005). These strategies are educational, goal oriented, and time-limited collaborative endeavors that attempt to help individuals learn skills that enable more adaptive problem solving.

Through EAPs, trained counselors provide short-term, work-focused CBT. Based on an eight-session telephone psychotherapy program, employees and their counselors use a workbook as a tool for guiding employees through the process of change, leading towards improved symptoms and the ability to function more effectively at work. The approach used in the WHI is adapted from the manual *Creating a Balance* (Simon et al. 2006), which has been shown to improve depressive symptomatology in telephone-based treatment interventions. EAP counselors help employees learn to use active behavioral and cognitive coping strategies both at home and on the job.

The WHI Infrastructure

The WHI care model is embedded in an infrastructure consisting of an EAP counselor training program and a clinical information system.

EAP Counselor Training

EAPs offer a convenient, low-cost, and highly confidential service to managers and employees, including a range of personal, mental health, and work life services. The goals of EAPs typically are aligned with those of the WHI, but EAP counselors may not have, or routinely use, all of the knowledge and skills necessary for the WHI.

The WHI requires basic knowledge of depression symptoms and treatment options, coordination of care skills, the study's work modification and coaching strategies, the CBT component, methods for applying standardized assessment tools (e.g. the PHQ-9 and WLQ), and training in the use of the web-based WHI information system. Also, many counselors are not accustomed to proactive outreach as a means for eliciting employee engagement. In the majority of EAP contacts, the employee makes the first contact.

Moreover, the WHI requires a high level of counselor adherence to specific protocols, including routine assessment of counselor fidelity and impact. In the provision of the WHI, there are concerns with achieving both structural and content fidelity. Structural fidelity refers to the degree to which counselors adhere to the program design features (e.g. number of visits, assessments completed, etc.). Content fidelity refers to the degree to which counselors perform the standardized functions of the WHI (e.g. use of the three care components when appropriate). Thus, WHI program counselors require both training and tools to support excellence.

A four-part WHI training program was developed that consists of: (1) a ten-chapter counselor manual; (2) in-depth, in-person, and remote training; (3) weekly clinical supervision and ongoing mentoring; and (4) performance monitoring that uses a counselor performance dashboard, consisting of a graphic representation of key performance metrics.

The WHI counselor manual is a detailed document for implementing the protocol. The minimum 50-h counselor training program involves didactic sessions, roleplay techniques, readings, and problem-focused discussions. Training covers intervention techniques, including CBT and motivational interviewing, telephone outreach strategies, use of structured assessment methods, and electronic record keeping. Pre- and post-training examinations are administered to determine if learning objectives were met.

Training is reinforced by counselor supervision sessions with the Clinical Management team, occurring weekly for 60–90 min. The counselor performance dashboard allows the Clinical Managers to monitor key performance indicators and to identify training needs. These needs are addressed through individualized mentoring and supplemental trainings. The WHI information system, described in the next section, supplies much of the required quantitative and qualitative performance data.

Web-Based WHI Information System

A secure web-based information system was developed to perform four important functions: (1) employee health screening for potential WHI participants; (2) WHI program enrollment (including obtaining informed consent, if required); (3) process of care documentation (including housing the tools and information required to provide care); and (4) outcomes evaluation. The system reflects the growing importance of efficient, inexpensive, and practical information systems among EAPs and employers, insurers, and federal and state agencies, all of which have a stake in monitoring care access, quality, outcome, and/or cost. A critical concern in developing this system was protection of privacy. It was important for employees to feel sure that the confidential personal information they were being asked to provide would be shared only with their assigned counselor. This level of privacy requires technological approaches (e.g. use of a password-protected clinical management website, secure databases, and encryption of personal identifiers); counselor training (e.g. protocols regarding the use of information); employer preparation (e.g. restricted access to any information collected other than de-identified summary statistics); review of protocols by an appropriate entity such as a university-based institutional review board; and a communications strategy that reinforces all of the protections in place and the commitment to privacy.

As part of the WHI information system, the study's self-administered health screening tool includes state-of-the-art questionnaires such as the PHQ-9 (depression), the 8-item WLQ short form (health-related job performance and work productivity deficits), and the SF-12 (short form health status measure). The anonymous screening tool takes 5–10 min to complete, depending on the number of questions the employee is required to answer. Employees completing the screener receive immediate e-feedback about their results. This summary report contains graphical displays of depression and work impairment scores, the company's average scores (based on the aggregate respondent pool), and national norms from our archival database. A brief narrative reports results and recommendations. Employees reporting suicidal ideation are given instructions for contacting a health professional.

After screening, eligible employees can voluntarily enroll in the WHI on the website. Once an employee is enrolled and entered into the WHI, the WHI counselor portion of the website is automatically populated with data from the screener and baseline questionnaires. At each contact and/or visit, the counselor enters data into the system, enabling the counselor to record and manage all aspects of the care process, such as scheduling, generating physician reports, documenting treatments and responses, and monitoring employee progress using standard assessment forms. At the end of the 4-month treatment period, the information system prompts the EAP counselor and WHI program staff to instruct the participating employee to complete a final post-intervention questionnaire, also administered over the web. The system combines the data from each source and time period into an analytic database for outcomes assessment.

WHI Test Results

The WHI's development has been facilitated by two research grants. With sponsorship from the US Centers for Disease Control and Prevention, the first pilot study (Grant Number: R01DP000101) of the WHI was conducted within a national manufacturing company. The National Institute of Mental Health supported the second pilot (Grant Number: 5R34MH72735-3), a small randomized study, which was conducted with the participation of a public sector (state government) employer. Both employee populations consisted of a range of occupations. The government employer was mainly located in a rural state in the northeast, while the manufacturing employer operated in many locations (urban and rural) throughout the US. In the remainder of this chapter, the data from each of these tests are presented.

Unmet Need for Care

Both pilots confirmed that an unmet need for work-focused care, observed in prior research, persists (Table 6.1). At baseline, approximately 20–30% of each employee sample had major depression at the start of the study, 30–44% had dysthymia and 23–38% had both major depression and dysthymia (referred to as double depression). Approximately one-fourth (28%) of the manufacturing company sample and 50% of the government employee sample were on antidepressant medication.

Table 6.1 Baseline characteristics of WHI Scale scores indicating the mean percentage of time limitations that occurred in the prior 2 weeks in the treatment vs. comparison group

	Manufacturing company sample treatment group ($n=79$)	Government employee sample treatment group ($n=52$)	Government employee sample usual care group ($n=27$)
Mean age	44.6	45.6	45.9
% ≥50 years old	35.0	34.0	44.0
% Male	50.0	23.1	18.5
% White	93.4	100.0	96.3
% Professional, technical, managerial	76.3	57.7	66.7
% Sales, support, service	26.3	38.5	33.3
% Repairs, const., prod., trans., other	9.2	1.9	0
% Major depression	27.6	29.8	19.2
% Dysthymia	32.9	40.4	44.4
% Both	38.2	23.1	33.3
Mean PHQ-9 severity (0–27)	13.5	13.1	12.2
% On antidepressant	28.9	55.8	51.9
Mean WLQ scores – Presenteeism			
Time management	45.4	45.6	43.7
Physical tasks	15.4	23.3	18.3
Mental-interpersonal tasks	35.3	37.3	38.5
Output tasks	39.1	40.7	39.1
Percentage productivity lost	9.7	10.3	10.1
Mean WLQ scores – Absenteeism			
Days missed	1.3	1.7	1.1
Percentage productivity lost	13.2	16.6	11.9

With regard to presenteeism, the mean percentage of time in the prior 2 weeks in which employees had limitations ranged from: 43.7 to 45.6% for Time Management, 15.4 to 23.3% for Physical Tasks, 35.3 to 38.5% for Mental-Interpersonal Tasks and 39.1 to 40.7% for Output Tasks. These rates translated into an at-work productivity loss of 10% on average. The mean number of days absent in the 2 weeks prior to baseline was 1.1–1.7, resulting in an average productivity loss of 11.9–16.6%.

Insight into Underlying Mechanisms Linking Depression to Work Loss

Table 6.2 identifies the specific problems experienced by study subjects at work and the strategies implemented to improve work outcomes under the Work Modification and Coaching portion of the WHI. This information is new and important both because it indicates how depression expresses itself in the context of the work, and because it identifies potentially useful strategies. Many of the workers in each study had multiple problems during their participation.

Table 6.2 Top ranking work issues (of 14) and related counselor interventions (of 9) for two samples

Work issues	Manufacturing company sample	Rank	Government employee sample	Rank
Difficulty concentrating, mind wandering, ruminating thoughts	30 (39.5%)	1	8 (20.5%)	4
Irritated with others, difficulty in work relationships	22 (30.3%)	2	7 (18.0%)	5
Loss of feelings of competency, loss of self-confidence	17 (22.4%)	3	10 (25.6%)	3
Trouble completing work and/or meeting deadlines	16 (21.1%)	4	14 (35.9%)	1
Loss of enjoyment in work	9 (11.8%)	5	11 (28.2%)	2
Work strategies				
Behavioral (e.g. relaxation, pleasurable activities, balancing work-home, getting advice)	23 (30.3%)	1	6 (22.2%)	3
Cognitive (e.g. manage negative, self-defeating, distracting thoughts and feelings)	16 (21.1%)	2	8 (29.6%)	2
Interpersonal (e.g. change frequency and types of social interactions at work)	13 (17.1%)	3	4 (14.8%)	4
Organizational (e.g. planning, scheduling, prioritizing work tasks)	13 (17.1%)	3	13 (48.1%)	1

Adherence to the Process of Care

In the government employer pilot study, counselor adherence to the WHI protocol was measured extensively. Adherence is important at the testing stage and will be important to disseminating the WHI. We measured adherence as the number of protocol steps completed divided by the steps required. Generally, counselor adherence to the protocol was high. For example, adherence to the steps in the medical care coordination procedure ranged from 73 to 100%. Adherence to the steps in the work coaching component and CBT component ranged from 79 to 95% and 76 to 100%, respectively.

In addition to these metrics, we monitor visit frequency and other key parameters. From the first to the second test, the mean number of counselor visits per client increased, though to a nonsignificant degree, from 4.5 to 5.1 ($p=0.31$), suggesting improved counselor performance. The number of CBT manual chapters completed increased from a mean of 2.4–3.9 ($p=0.001$). Also, the average time spent by counselors per visit (preparing, treating, documenting, and following up) declined from 87.5 min per visit to 70.1 min ($p<0.001$); an indicator of improved efficiency.

Outcomes

Data from the manufacturing company pilot have been analyzed (Table 6.3). Because it was a small pilot study, it did not include a control group (as does the government employer study). Comparing the pre- and post-intervention scores on the major variables indicated that statistically significant improvements were achieved on seven of the nine primary study endpoints. Work limitations related to time management decreased by 17.3 points, limitations in performing mental and interpersonal job tasks

Table 6.3 Means and mean change scores from baseline to follow-up in the WHI

WLQ scores[a]	WHI treatment		
	Baseline N=76	Follow-up n=76	Change
Time management	45.37	28.11	−17.27 ($p=0.000$)
Physical tasks	15.42	11.86	−3.57 ($p=0.221$)
Mental-interpersonal tasks	35.29	22.96	−12.33 ($p=0.000$)
Output tasks	39.09	22.69	−16.40 ($p=0.000$)
At-work productivity loss	0.10	0.06	−0.04 ($p=0.000$)
Days missed	1.322	0.966	−0.356 ($p=0.246$)
Productivity loss due to absences	0.132	0.087	−0.050 ($p=0.125$)
PHQ-9 depression severity[b]	13.49	6.99	−6.50 ($p=0.000$)

[a]Work Limitations Questionnaire scores range from 0 to 100 (least to most limited)
[b]PHQ-9 scores range from 0 to 27 (least to highest severity)
Note: Estimates generated from multiple regression. Models were adjusted for age, gender, marital status, WLQ at-work productivity loss score at baseline, and occupation

decreased by 12.3 points, and limitations in performing output tasks declined by 16.4 points. Overall at-work productivity loss declined by 0.04 points.

These results are encouraging. In our previous observational study (Lerner et al. 2010), which had similar eligibility criteria, we found that WLQ Scale score changes ranged from 5.0 to 11.7, while PHQ-9 scores improved an average of 2.2 points. However, further assessment of the WHI is essential. The randomized clinical trial, which has just been completed, will answer additional questions about the program's performance and potential value.

Conclusions

Depression is draining both workers of their opportunity to participate fully and productively in the labor market, and employers of much-needed productivity. This large and important problem is not being addressed by the existing services system, which rarely provides integrated health and vocational supports. The WHI, which achieves this integration, is an innovation in services as well as a practical solution to a large and difficult problem. Work-focused interventions conceivably could occur at the level of the worker and/or the workplace. Organization level interventions, such as increasing worker autonomy through job redesign, may also be beneficial, but are not yet a part of the approach. At present, the organizational innovation is changing the role of the EAP and its relationship to employees and the medical care system.

In addition to quality medical care for their persistent depressive symptoms, employees with depression need other services. New intervention models such as the WHI that help employees remain engaged in their work, participating fully and productively, are urgently needed.

References

Adler DA, McLaughlin TJ, Rogers WH, Chang H, Lapitsky L, Lerner D (2006) Job performance deficits due to depression. Am J Psychiatry 163(9):1569–1576
Altman BM (2001) Disability definitions, models, classification schemes, and applications. In: Albrecht GL, Seelman KD, Bury M (eds) Handbook of disability studies. Sage, Thousand Oaks, CA, pp 97–122
Appelbaum E, Bailey T, Berg P, Kalleberg AL (2000) High-performance work systems and worker outcomes manufacturing advantage: why high-performance work systems pay off. Cornell University Press, Ithaca, NY
Beck A (1979) Cognitive therapy and the emotional disorders. International University Press, New York
Beck A (1991) Cognitive therapy. A 30-year retrospective. Am Psychol 46(4):368–375
Beck J (1995) Cognitive therapy: basics and beyond. Guildford, New York
Berndt ER, Finkelstein SN, Greenberg PE, Howland RH, Keith A, Rush AJ et al (1998) Workplace performance effects from chronic depression and its treatment. J Health Econ 17(5):511–535

Blazer DG, Kessler RC, McGonagle KA, Swartz MS (1994) The prevalence and distribution of major depression in a national community sample: the national comorbidity survey. Am J Psychiatry 151(7):979–986

Brewin CR (1996) Theoretical foundations of cognitive-behavior therapy for anxiety and depression. Annu Rev Psychol 47:33–57

Dooley D, Prause J, Ham-Rowbottom KA (2000) Underemployment and depression: longitudinal relationships. J Health Soc Behav 41(4):421–436

Druss BG, Rosenheck RA, Sledge WH (2000) Health and disability costs of depressive illness in a major US corporation. Am J Psychiatry 157(8):1274–1278

Ellwood DT, Blank RM, Blasi J, Kruse D, Niskanen WA, Lynn-Dyson K (2000) A working nation: workers, work, and government in the new economy. Russell Sage Foundation, New York

Fava GA, Rafanelli C, Grandi S, Canestrari R, Morphy MA (1998) Six-year outcome for cognitive behavioral treatment of residual symptoms in major depression. Am J Psychiatry 155(10):1443–1445

Gilbody S, Bower P, Fletcher J, Richards D, Sutton AJ (2006) Collaborative care for depression: a cumulative meta-analysis and review of longer-term outcomes. Arch Intern Med 166(21):2314–2321

Gilbody S, Whitty P, Grimshaw J, Thomas R (2003) Educational and organizational interventions to improve the management of depression in primary care: a systematic review. J Am Med Assoc 289(23):3145–3151

Goetzel RZ, Guindon AM, Turshen IJ, Ozminkowski RJ (2001) Health and productivity management: establishing key performance measures, benchmarks, and best practices. J Occup Environ Med 43(1):10–17

Greco T, Eckert G, Kroenke K (2004) The outcome of physical symptoms with treatment of depression. J Gen Intern Med 19(8):813–818

Greenberg PE, Leong SA, Birnbaum HG, Robinson RL (2003) The economic burden of depression with painful symptoms. J Clin Psychiatry 64(Suppl 7):17–23

Handler J, Doel K, Henry A, Lucca A (2003) Rehab rounds: implementing supported employment services in a real-world setting. Psychiatr Serv 54(7):960–962

Hirschfeld RM, Dunner DL, Keitner G, Klein DN, Koran LM, Kornstein SG et al (2002) Does psychosocial functioning improve independent of depressive symptoms? A comparison of nefazodone, psychotherapy, and their combination. Biol Psychiatry 51(2):123–133

Ichniowski C, Kochan TA, Levine D, Olson C, Strauss G (1996) What works at work: overview and assessment. Ind Relat 35:299–333

Jacobson NS, Dobson KS, Truax PA, Addis ME, Koerner K, Gollan JK et al (1996) A component analysis of cognitive-behavioral treatment for depression. J Consult Clin Psychol 64(2):295–304

Jarrett RB, Kraft D, Doyle J, Foster BM, Eaves GG, Silver PC (2001) Preventing recurrent depression using cognitive therapy with and without a continuation phase: a randomized clinical trial. Arch Gen Psychiatry 58(4):381–388

Katon W (2003) The Institute of Medicine "Chasm" report: implications for depression collaborative care models. Gen Hosp Psychiatry 25(4):222–229

Katon W, Ludman E, Simon G, Lin E, Von KM, Walker E et al (2002) The depression helpbook. Bull, Boulder, CO

Katon W, Rutter C, Ludman EJ, Von Korff M, Lin E, Simon G et al (2001) A randomized trial of relapse prevention of depression in primary care. Arch Gen Psychiatry 58(3):241–247

Katon W, Schulberg H (1992) Epidemiology of depression in primary care. Gen Hosp Psychiatry 14(4):237–247

Katon W, Von Korff M, Lin E, Simon G, Walker E, Unutzer J et al (1999) Stepped collaborative care for primary care patients with persistent symptoms of depression: a randomized trial. Arch Gen Psychiatry 56(12):1109–1115

Keller MB, McCullough JP, Klein DN, Arnow B, Dunner DL, Gelenberg AJ et al (2000) A comparison of nefazodone, the cognitive behavioral-analysis system of psychotherapy, and their combination for the treatment of chronic depression. N Engl J Med 342(20):1462–1470

Kessler R, Akiskal HS, Ames M, Birnbaum H, Greenberg P, Rm A et al (2006) Prevalence and effects of mood disorders on work performance in a nationally representative sample of US workers. Am J Psychiatry 163(9):1561–1568

Kessler R, Frank RG (1997) The impact of psychiatric disorders on work loss days. Psychol Med 27(4):861–873

Kopelowicz A, Liberman RP (2003) Integrating treatment with rehabilitation for persons with major mental illnesses. Psychiatr Serv 54(11):1491–1498

Lehman AF, Goldberg R, Dixon LB, McNary S, Postrado L, Hackman A et al (2002) Improving employment outcomes for persons with severe mental illnesses. Arch Gen Psychiatry 59(2):165–172

Lerner D, Amick BC, III, Rogers WH, Malspeis S, Bungay K, Cynn D (2001) The Work Limitations Questionnaire. Med. Care. 39: 72–85

Lerner D, Reed JI, Massarotti E, Wester LM, Burke TA (2002) The Work Limitations Questionnaire's validity and reliability among patients with osteoarthritis. J. Clin. Epidemiol 55: 197–208

Lerner D, Amick BC, III, Lee JC, Rooney T, Rogers WH, Chang H, Berndt ER (2003) Relationship of employee-reported work limitations to work productivity. Med Care. 41: 649–659

Lerner D, Adler DA, Chang H, Berndt ER, Irish JT, Lapitsky L et al (2004a) The clinical and occupational correlates of work productivity loss among employed patients with depression. J Occup Environ Med 46(Suppl 6):S46–S55

Lerner D, Adler DA, Rogers WH, Chang H, Lapitsky L, McLaughlin T et al (2010) Work performance of employees with depression: the impact of work stressors. Am J Health Promot 24(3):205–213

Lerner D, Adler DA, Chang H, Lapitsky L, Hood MY, Perissinotto C et al (2004b) Unemployment, job retention, and productivity loss among employees with depression. Psychiatr Serv 55(12):1371–1378

Lerner D, Allaire SH, Reisine ST (2005) Work disability resulting from chronic health conditions. J Occup Environ Med 47(3):253–264

Lerner D, Henke RM (2008) What does research tell us about depression, job performance and work productivity? J Occup Environ Med 50(4):401–410

Lewinsohn PM, Antonuccio DA, Steinmetz J, Teri L (1984) The coping with depression course: a psychoeducational intervention for unipolar depression. Castalia, Eugene, OR

Lin EH, VonKorff M, Russo J, Katon W, Simon GE, Unutzer J et al (2000) Can depression treatment in primary care reduce disability? A stepped care approach. Arch Fam Med 9(10):1052–1058

Loisel P, Durand MJ (2001) Disability prevention: the new paradigm of management of occupational back pain. Dis Manag Health Outcomes 9(7):351–360

Lowe GS, Schellenberg G, Shannon HS (2003) Correlates of employees' perceptions of a healthy work environment. Am J Health Promot 17(6):390–399

Ludman E, Bush T, Lin E, Von Korff M, Simon G, Katon W et al (2001) Behavioral factors associated with medication adherence and symptom outcomes in a primary care-based depression prevention intervention trial. Ann Behav Med 23(Suppl):S034

Ludman E, Simon GE, Tutty S, Von Korff M (2007) A randomized trial of telephone psychotherapy and pharmacotherapy for depression: continuation and durability of effects. J Consult Clin Psychol 75(2):257–266

Ludman E, Von Korff M, Katon W, Lin E, Simon G, Walker E et al (2000) The design, implementation, and acceptance of a primary care-based intervention to prevent depression relapse. Int J Psychiatry Med 30(3):229–245

Lynch DJ, Tamburrino MB, Nagel R (1997) Telephone counseling for patients with minor depression: preliminary findings in a family practice setting. J Fam Pract 44(3):293–298

Marhold C, Linton SJ, Melin L (2001) A cognitive-behavioral return-to-work program: effects on pain patients with a history of long-term versus short-term sick leave. Pain 91(1–2):155–163

Marhold C, Linton SJ, Melin L (2002) Identification of obstacles for chronic pain patients to return to work: evaluation of a questionnaire. J Occup Rehabil 12(2):65–75

Marmot M, Feeney A (1996) Work and health: implications for individuals and society. In: Blane D, Brunner E, Wilkinson R (eds) Health and social organization towards a health policy for the 21st century. Routledge, London, pp 235–254

Mohr DC, Hart SL, Julian L, Catledge C, Honos-Webb L, Vella L et al (2005) Telephone-administered psychotherapy for depression. Arch Gen Psychiatry 62(9):1007–1014

NIOSH Working Group (1999) Stress at work. US Department of Health and Human Services, Cincinnati, OH

O*Net Resource Center (2008) O*NET Research and Technical Reports. Retrieved on 30 Jan 2008, from http://www.onetcenter.org/research.html

Paykel ES, Scott J, Teasdale JD, Johnson AL, Garland A, Moore R et al (1999) Prevention of relapse in residual depression by cognitive therapy: a controlled trial. Arch Gen Psychiatry 56(9):829–835

Perry M, Hill R, Davidson C (1996) Workplace reform and the quality of working life: a tale of three workplaces. Labour Market Bulletin 51–70

Pflanz SE, Ogle AD (2006) Job stress, depression, work performance, and perceptions of supervisors in military personnel. Mil Med 171(9):861–865

Rost K, Humphrey J, Kelleher K (1994) Physician management preferences and barriers to care for rural patients with depression. Arch Fam Med 3(5):409–414

Rost K, Nutting P, Smith J, Werner J, Duan N (2001) Improving depression outcomes in community primary care practice: a randomized trial of the quEST intervention. Quality enhancement by strategic teaming. J Gen Intern Med 16(3):143–149

Rost K, Smith JL, Dickinson M (2004) The effect of improving primary care depression management on employee absenteeism and productivity. A randomized trial. Med Care 42(12):1202–1210

Schoenbaum M, Unutzer J, McCaffrey D, Duan N, Sherbourne C, Wells KB (2002) The effects of primary care depression treatment on patients' clinical status and employment. Health Serv Res 37(5):1145–1158

Scott J, Teasdale JD, Paykel ES, Johnson AL, Abbott R, Hayhurst H et al (2000) Effects of cognitive therapy on psychological symptoms and social functioning in residual depression. Br J Psychiatry 177:440–446

Simon G, Katon W, Rutter C, Von Korff M, Lin E, Robinson P et al (1998) Impact of improved depression treatment in primary care on daily functioning and disability. Psychol Med 28(3):693–701

Simon G, Ludman E, Tutty S (2006) Creating a balance: a step by step approach to managing stress and lifting your mood. Trafford, Victoria, BC (copyright, Group Health Coorperative)

Simon G, Von Korff M, Lin E (2005) Clinical and functional outcomes of depression treatment in patients with and without chronic medical illness. Psychol Med 35(2):271–279

Sparks K, Faragher B, Cooper CL (2001) Well-being and occupational health in the 21st century workplace. J Occup Organ Psychol 74:489–509

Stansfeld SA, North FM, White I, Marmot MG (1995) Work characteristics and psychiatric disorder in civil servants in London. J Epidemiol Community Health 49(1):48–53

Stewart WF, Ricci JA, Chee E, Hahn SR, Morganstein D (2003) Cost of lost productive work time among US workers with depression. J Am Med Assoc 289(23):3135–3144

Thomas VL, Gostin LO (2009) The Americans with disabilities act: shattered aspirations and new hope. J Am Med Assoc 301(1):95–97

US Department of Health and Human Services and AHCPR (1993) Depression in primary care: vol 1: detection and diagnosis. Diane, Darby, PA

US Preventive Services Task Force (2002) Screening for depression: recommendations and rationale. Ann Intern Med 136(10):760–764

Ustiin TB, Sartorius N. (1995) Mental illness in general health care: an international study. Wiley, New York

van der Klink J, Blonk RW, Schene AH, van Dijk FJ (2001) The benefits of interventions for work-related stress. Am J Public Health 91(2):270–276 van der Klink J, Blonk RW, Schene AH, van Dijk FJ (2003) Reducing long term sickness absence by an activating intervention in adjustment disorders: a cluster randomised controlled design. Occup Environ Med 60(6):429–437

Wang PS, Simon GE, Avorn J, Azocar F, Ludman EJ, McCulloch J et al (2007) Telephone screening, outreach, and care management for depressed workers and impact on clinical and work productivity outcomes: a randomized controlled trial. J Am Med Assoc 298(12):1401–1411

Ware JE, Jr., Kosinski M, Keller SD (1996) A 12-Item Short-Form Health Survey: construction of scales and preliminary tests of reliability and validity. Med Care 34(3):220–33

Wells K (1985) Depression as a tracer condition for the national study oSf medical care outcomes: background review. Rand, Santa Monica, CA

Wells K (1997) Caring for depression in primary care: defining and illustrating the policy context. J Clin Psychiatry 58(Suppl 1):24–27

Wells K, Rogers W, Burnam A, Greenfield S, Ware JE Jr (1991) How the medical comorbidity of depressed patients differs across health care settings: results from the Medical Outcomes Study. Am J Psychiatry 148(12):1688–1696

Wells K, Sherbourne C, Schoenbaum M, Duan N, Meredith L, Unutzer J et al (2000) Impact of disseminating quality improvement programs for depression in managed primary care: a randomized controlled trial. J Am Med Assoc 283(2):212–220

Wells K, Stewart A, Hays RD, Burnam A, Rogers W, Daniels M et al (1989) The functioning and well-being of depressed patients: results from the Medical Outcomes Study. J Am Med Assoc 262(7):914–919

Wells K, Sturm R (1995) Care for depression in a changing environment. Health Aff 14(3):78–89

Chapter 7
Anxiety Disorders and Work Performance

Jaye Wald

Scope of the Problem

Anxiety disorders are among the most prevalent group of psychiatric illnesses in the workforce (American Psychiatric Association (APA) 2000; Sanderson and Andrews 2006).The lifetime prevalence of anxiety disorders (as a group) in the general population is approximately 14% (APA 2000; ESEMeD/MHEDEA Investigators 2004a; Kessler et al. 2005). These disorders typically follow a chronic course and are accompanied by substantial functional impairment and an elevated suicide risk (Khan et al. 2002; Spitzer et al. 1995; van Balkom et al. 2008; Yonkers et al. 2003). More than half of the persons diagnosed with an anxiety disorder will have a comorbid psychiatric condition, the most common being another anxiety disorder, major depressive disorder, and substance use disorders (van Balkom et al. 2008). Chronic medical conditions (e.g., cardiovascular disease, chronic pain, asthma) are also common among people with anxiety disorders (Demyttenare et al. 2008; Roy-Bryne et al. 2008; Zvolensy and Smits 2008). Despite significant advances in evidence-based interventions, these disorders tend to go unrecognized and untreated (Lim et al. 2000; Schonfeld et al. 1997).

Several studies have shown that reduced work performance, as reflected by work absenteeism (missed work days or cut-back days), presenteeism (decreased work productivity at work), and unemployment, are particularly common consequences of these disorders (El-Guebaly et al. 2007; ESEMeD/MHEDEA Investigators 2004b; Laitinen-Krispijn and Bijl 2000; Lim et al. 2000; Waghorn et al. 2005; Waghorn and Chant 2005). A population survey in Canada found that 30% of 831 individuals with an anxiety disorder with no comorbidity did not work, or were disabled from working, in the previous week (el-Guebaly et al. 2007). Another population study reported that a sample of full-time employed individuals with an anxiety disorder had an average of 1.7 work loss days and 4.5 cut-back days in the

J. Wald (✉)
Department of Psychiatry, University of British Columbia, 2255 Wesbrook Mall, Vancouver, BC, Canada V6T 2A1
e-mail: jwald@interchange.ubc.ca

I.Z. Schultz and E.S. Rogers (eds.), *Work Accommodation and Retention in Mental Health*, 121
DOI 10.1007/978-1-4419-0428-7_7, © Springer Science+Business Media, LLC 2011

past month (Lim et al. 2000). Community-based and treatment studies also report high unemployment rates among persons with anxiety disorders, often ranging between 20 and 40% (e.g., Rapaport et al. 2005). These disorders may also disrupt vocational functioning in more subtle ways, such as underemployment (e.g., working part-time instead of full-time), reduced job seeking and advancement, and narrowed occupational choices (Waghorn et al. 2005; Waghorn and Chant 2005). In turn, the negative impact of these disorders on work function is reflected in a significant disease burden to society (e.g., workplace costs due to lost productivity, disability insurance costs, health care expenditures), and personal suffering to individuals (e.g., a reduced quality of life) (Greenberg et al. 1999; Mogotsi et al. 2000; Rapaport et al. 2005).

Several investigations suggest that anxiety disorders, as a combined group, have comparable rates of work impairment and disability to mood disorders, and higher rates than substance abuse disorders and many chronic medical conditions (Buist-Bowman et al. 2005; El-Guebaly et al. 2007; ESEMeD/MHEDEA Investigators 2004b; Lim et al. 2000). Studies have also found anxiety disorders comorbid with another psychiatric disorder or a serious medical disorder, are associated with greater levels of work impairment than found in anxiety disorders without comorbidity (Buist-Bowman et al. 2005; El-Guebaly et al. 2007; ESEMeD/MHEDEA Investigators 2004b; Lim et al. 2000; Waghorn et al. 2005). Even subthreshold forms of anxiety disorders, defined as having some symptoms but not meeting full diagnostic criteria, have been shown to still have a deleterious impact on work functioning, albeit to a lesser degree than those individuals with full-blown disorders (Amaya-Jackson et al. 1999; Goracci et al. 2007; Norman et al. 2007; Wittchen et al. 2000). Symptom severity has been shown to be a consistent predictor of the level of work impairment in the anxiety disorders (e.g., Diefenbach et al. 2007; Latas et al. 2004; Taylor et al. 2006). However, very little is known about other clinical, psychosocial, and work-related risk factors of work impairment in this population.

This chapter summarizes the current literature on the impact of anxiety disorders on work performance. The first section provides an overview of the DSM-IV anxiety disorders and summarizes the current knowledge on work impairment and disability for each disorder. The next section discusses the various ways in which anxiety disorders may interfere with work performance. Assessment methods for evaluating work performance are then reviewed. The last section addresses the issue of work rehabilitation, with an emphasis on the potential role of job accommodations.

A Review of the Anxiety Disorders, Work Impairment, and Disability

The Diagnostic and Statistical Manual of Mental Disorders, fourth edition, text revision (DSM-IV-TR; APA 2000), includes the following major anxiety disorders: Panic disorder (with or without agoraphobia) (PD), Posttraumatic stress disorder (PTSD), generalized anxiety disorder (GAD), obsessive compulsive disorder (OCD), social phobia or social anxiety disorder (SAD), and specific phobia (SP). Other anxiety disorders recognized by the DSM-IV include acute stress disorder (ASD), anxiety disorder not otherwise specified, anxiety disorder due to a medical condition, and substance-abuse anxiety disorder. ASD is briefly discussed along with PTSD, and the other latter disorders are considered residual disorders and have undergone little research, and thus are not discussed in this chapter.

Each of the major anxiety disorders is defined by specific symptom criteria; however, they share a number of common features, including a severe and persistent mood state of anxiety, often accompanied by various behavioral (e.g., avoidance of anxiety-provoking situations), emotional (e.g., episodes of intense fear or panic, irritability, low stress tolerance), cognitive (e.g., worry over a perceived future danger or threat, impaired concentration and memory), and physical symptoms (e.g., muscle tension, dizziness, sweating, fatigue), which cause significant distress and/or functional interference. However, there is also considerable variation within each disorder in terms of the severity and frequency of symptoms and functional difficulties that are experienced. Moreover, although these disorders may follow a relatively chronic course if untreated, symptom severity and associated functional impairment are not static; they tend to fluctuate over time and contexts (e.g., remitting and then relapsing), and are often exacerbated by situational stressors (e.g., Yonkers et al. 2003).

To date, most studies have looked at work performance in anxiety disorders as a group, and there is relatively little data on the individual disorders. The limited research suggests that PD, PTSD, SAD, and GAD may have the highest level of work impairment of the anxiety disorders (ESEMeD/MHEDEA Investigators 2004b; Stein et al. 2005); however, further research is needed to confirm these findings. Brief descriptions of the DSM-IV anxiety disorders and research findings on work impairment and disability respective to each of these conditions are given below.

Panic disorder with or without agoraphobia (PD). The hallmark feature of PD involves recurrent, unexpected panic attacks, or discrete periods of intense fear or discomfort in the absence of real dangers that develop abruptly, and are accompanied by a number of physical symptoms (e.g., racing heart, trembling or shaking, shortness of breath, feeling of choking, chest pain, dizziness) (APA 2000; see Taylor 2000, for a review). When the attacks are followed by a persistent concern about having more attacks, worry about the implications or consequences of the attacks (e.g., "I could go crazy"), or behavioral changes as a result of the attacks (e.g., avoidance of work activities), a PD diagnosis may be warranted. The lifetime prevalence of PD is approximately 1.5 – 3.5% (APA 2000). Over half of individuals with PD develop agoraphobia as a consequence of experiencing panic attacks, which involves anxiety and avoidance of places or situations from which escape might be difficult, embarrassing, or in circumstances in which help might not be available in the event of having a panic attack (e.g., being outside the home, being alone at home, crowds) (APA 2000, see Taylor 2000, for a review).

Only a few studies have specifically examined work functioning in PD. In a recent study of patients with PD with agoraphobia, 64% were employed, 18% were unemployed, and 81% of employed patients reported decreased work productivity (Latas et al. 2004). Similar findings were found by Ettigi et al. (1997), who reported that in a sample of patients with PD with or without agoraphobia, 57% were employed full-time, 18% were employed part-time, and 25% were unemployed. Employed patients in this sample reported working to only 56% of their work performance capacity. The presence of agoraphobic-related avoidance in PD, as compared to PD without agoraphobia, has been shown to be associated with more severe work impairment, a greater number of sickness absence days, and higher levels of unemployment (Latas et al. 2004; Sanderson et al. 2001; Yonkers et al. 2003). For example, in a study of treatment-seeking individuals, 41% of patients with PD with agoraphobia were employed full-time, whereas 53% of patients with PD without agoraphobia were employed full-time (Yonkers et al. 2003).

Posttraumatic stress disorder (PTSD). Survivors of severe traumatic events (e.g., road traffic collisions, criminal victimization, combat exposure, childhood physical and sexual abuse) are at risk of developing PTSD. This disorder develops when a person (a) experiences a traumatic event and reacts with intense fear, helplessness, or horror; and (b) develops particular symptoms that persist for at least a month (APA 2000). These symptoms include vivid re-experiencing of the trauma (e.g., intrusive memories, nightmares, flashbacks); avoidance of reminders of the trauma (e.g., memories, places, people); emotional numbing (e.g., feeling detached from others); and hyperarousal (e.g., irritabililty, sleep difficulties) (APA 2000; see Friedman et al. 2007, for a review). The lifetime prevalence of PTSD in the general population is 8% (APA 2000), thus making it a relatively common disorder. Within the initial month after the trauma, the presence of PTSD symptoms may be diagnosed as *ASD*. However, a diagnosis of ASD also requires the presence of various dissociative reactions (e.g., reduced awareness of one's surroundings, feeling detached from one's body), which develop immediately after the traumatic event (APA 2000; Harvey and Bryant 2002). The majority of people who develop ASD go on to develop PTSD (APA 2000; Harvey and Bryant 2002).

Numerous studies have demonstrated the negative impact of PTSD on work functioning (Breslau et al. 2004; ESEMeD/MHEDEA Investigators 2004b; Matthews 2005; Norman et al. 2007). A population study reported that 27% of a chronic PTSD sample were unable to work at all during the previous month in response to being upset by the traumatic event. This study also found that among the employed persons who were able to continue working, 84% reported at least one work-loss day and 90% had at least one cut-back day over the previous month (Breslau et al. 2004). Another investigation found that 42% of individuals with chronic PTSD had not returned to work 8 months posttrauma (Matthews 2005). A study of individuals that developed chronic PTSD from a work-related traumatic event showed that 34% were unable to return to work in any capacity following the trauma (McDonald et al. 2003). High rates of unemployment are also reported in PTSD treatment studies, with unemployment rates at the pretreatment assessment often ranging between 30 and 38% (e.g., Foa et al. 2005). Previous investigations have consistently shown that the severity of PTSD is correlated with, and is predictive of, work dysfunction (e.g., Matthews and Chinnery 2005; Taylor et al. 2006).

Other variables that have also been shown to contribute to work impairment in PTSD include depression, anger, impaired physical functioning (e.g., due to chronic pain from injuries sustained from the accident), a preoccupation with health, and various work-related factors (e.g., poor organization support, poor job satisfaction, and job insecurity) (Carlier et al. 1997; Evans et al. 2006).

Generalized anxiety disorder (GAD). GAD is characterized by chronic and excessive anxiety and worry about multiple life events (e.g., work and family responsibilities, health, finances, minor daily hassles) that lasts at least 6 months (APA 2000; see Dugas and Robichaud 2006, for a review). Individuals with GAD find it difficult to control their worrying, and experience several symptoms of anxious arousal, such as restlessness, fatigue, impaired concentration, muscle tension, irritability, and sleep disturbance. The lifetime prevalence of GAD in the general population is approximately 4 – 5% (APA 2000; Grant et al. 2005). The excessive and uncontrollable worrying associated with GAD tends to result in significant functional impairment (Grant et al. 2005; Hoffman et al. 2008).

The existing sparse research suggests that work impairment and disability are significant issues for persons with GAD (Grant et al. 2005; Hoffman et al. 2008). In a community sample of primary care patients, 67% of individuals diagnosed with GAD (without psychiatric comorbidity) and 81% of individuals diagnosed with GAD (with comorbid major depressive episode) were disabled from working in the past month due to their psychiatric problems (Wittchen et al. 2002). In a treatment-seeking sample of patients with GAD, only 46% were employed full-time (Yonkers et al. 2003). A population study found that a sample of full-time employed persons with GAD reported an average of 2.5 work-loss days and 5.9 cut-back days during the past month (Lim et al. 2000).

Obsessive compulsive disorder (OCD). OCD involves a spectrum of various obsessions (recurrent, intrusive, and frightening images, thoughts, or impulses) and/or compulsions (repetitive and ritualized behaviors or mental acts aimed at neutralizing or reducing the obsessions) (see Abramowitz et al. 2008, for a review; APA 2000). Examples of common obsessions include thoughts of harm (to self or others), thoughts of contamination or disease (e.g., fear of contracting a disease from a public place), or other perceived repugnant thoughts (e.g., involving sexual or aggressive content). Compulsive behaviors may include handwashing or checking, whereas compulsive mental thoughts can involve various acts such as praying, counting, or repeating words silently. Individuals with OCD recognize to varying extents that these symptoms are irrational and excessive, and are often highly distressed by them. Although less common than some of the other anxiety disorders, with a lifetime prevalence of approximately 2.5% in the general population (see Abramowitz et al. 2008, for a review; APA 2000), OCD, particularly the presence of compulsions and avoidance behaviors, can lead to high levels of functional impairment (see Abramowitz et al. 2008, for a review).

It is difficult to draw conclusions on the extent of work impairment and disability in OCD because of lack of data, but there is some evidence that OCD may be associated with less dysfunction than GAD, PTSD, or SAD. For example, a population study found that full-time employed individuals with OCD reported an average of 0.1 work-loss days and 3.0 cut back days in past month, which was lower than the rates reported for GAD and SAD (Lim et al. 2000). However, treatment studies of OCD and subthreshold OCD generally report much higher rates of work impairment. For example, in one study of treatment-seeking individuals with sub-threshold OCD, 46% were employed full-time, 10% employed part-time, 2% had temporary employment, and 6% were unemployed (Goracci et al. 2007). This study also reported that symptom severity also predicted reduced quality of life in work activities.

Social phobia or social anxiety disorder (SAD). This anxiety disorder involves a marked fear of one or more social or performance situations (e.g., public speaking, eating or drinking in front of other people, using public washrooms), in which a person fears being observed, humiliated, or negatively judged (APA 2000; see Stravynski 2007, for a review). Individuals with this disorder will avoid or escape these feared situations or will tolerate them with significant distress. The DSM-IV recognizes a form of this disorder that involves a pervasive fear of most social situations. This is known as generalized SAD, and it is associated with greater levels of impairment than with more circumscribed social fears (APA 2000; see Stravynski 2007, for a review). It is the most common anxiety disorder, with lifetime prevalence rates in the general population ranging from 7 to 13% (APA 2000; Wittchen and Fehm 2001).

This disorder can have a profound impact on the person's work functioning (Bruch et al. 2003; Fehm et al. 2000; Stansfield et al. 2008; Wittchen et al. 2000). A recent community sample examined work impairment between SAD with no comorbidity, SAD with comorbidity, and subthreshold SAD. They found that individuals with SAD and comorbidity had the highest rates of work impairment with 7% employed full-time, 16% employed part-time, 22% unemployed. The rates of part-time work or unemployment were slightly lower for those without a comorbid disorder and those with subthreshold SAD. In a population sample of full-time workers, individuals with SAD reported an average of 2.5 work-loss days and 3.5 cut-back days in the past month (Lim et al. 2000). Bruch et al. (2003) found that individuals with SAD experienced various difficulties in work adjustment, including underemployment, beliefs that involve negative views by supervisors, avoidance of interpersonally oriented jobs (particularly women with generalized SAD), and increased anxiety upon starting a new job.

Specific phobia (SP). A SP is a pronounced, excessive, and irrational fear of a specific object or situation. The DSM-IV classifies SPs into the following types: animal type (e.g., insects, dogs), natural environment type (e.g., water, storms), blood-injection-injury type, situational type (e.g., driving, flying, enclosed spaces), or other type (e.g., choking phobia) (APA 2000). People with an SP recognize that their fear is unreasonable, yet they will experience a marked anxiety response when confronted with the phobic situation and will try to avoid exposure to it whenever possible. SPs are relatively common, with a lifetime prevalence rate between 10 and 11% in the general population (APA 2000; see Pull 2008, for a review). Given the circumscribed nature of this disorder, it generally tends to be associated with less distress and impairment than the other anxiety disorders.

For example, Mataix-Cols et al. (2005) found that persons with an SP experienced less work impairment than those with SAD or those with agoraphobia. However, it is important to recognize that certain types of SPs (e.g., driving phobia, fear of flying) can have a significant impact on a person's vocational choices and functioning (e.g., a person with a driving phobia may turn down jobs that require driving to and from work, or to different work locations).

The Impact of Anxiety Disorders on Work Performance

Although there has been progress towards understanding the ways in which severe psychiatric disorders (e.g., schizophrenia) can impact work performance (e.g., McDonald et al. 2003), very little attention has been given to anxiety disorders (e.g., Fischler and Booth 1999, for a review). As a result, there is a lack of data on the actual types of work performance deficits that may be experienced by people with anxiety disorders. Furthermore, the relationships between specific symptoms and their functional impact on work performance has yet to be empirically investigated. However, symptom–function relationships are complex and difficult to tease apart; it is likely the case that multiple symptoms can lead to a particular work deficit. Conversely, a single symptom could possibly have numerous effects on work performance. For example, the occurrence of frightening thoughts, images, and physical sensations is likely to be highly distracting and contribute to a range of work performance deficits (e.g., reduced productivity, difficulties with time management, and difficulties performing tasks requiring concentration and other mental effort). Furthermore, the relationship between anxiety symptoms and work functioning is further complicated by the use of psychotropic medication and its side effects, as well as self-medication with other substances (e.g., herbal products, caffeine, alcohol), which is common among persons with anxiety disorders (Haslam et al. 2005a; Murray et al. 2006). With these caveats in mind for the reader, the following is a summary of different aspects of work performance that may be experienced by persons with anxiety disorders, which is based upon the limited available literature of anxiety disorders and the broader psychiatric disability literature. This information is also presented in Table 7.1.

Work attendance. As described in the previous section, individuals with anxiety disorders are likely to have difficulty with maintaining regular work attendance and working the required hours, likely because of various factors associated with the condition (e.g., feeling too anxious to go to work, leaving work early after becoming distressed, feeling ill due to medication side effects, taking time off to attend health care appointments). To illustrate, an employee who has PD with agoraphobia may arrive at work late, because of the fear of having a panic attack while driving or taking the bus. The person may also leave work following a panic attack or because of the marked apprehension and distress over the fear of having a panic attack. As another example, a worker who was physically assaulted and robbed on the job subsequently develops PTSD and experiences a lengthy work absence because of a fear of returning to work.

Table 7.1 Summary of possible work limitations and suggestions for job accommodations for persons with anxiety disorders

Dimensions of work performance	Possible functional limitations	Suggestions for job accommodations
Attendance	• Difficulty maintaining regular attendance (e.g., sick and cut-back days) • Difficulty working required hours	• Flexible work schedule (e.g., in start or end of workday, flexible and frequent breaks during workday) • Part-time work • Flex-time or job sharing • Working at home or telecommuting • Time off to attend health care appointments or breaks during workday to call healthcare professional
Productivity	• Reduced or inconsistent productivity and output • Difficulty handling workload • Difficulty completing work tasks • Increased mistakes and errors	• Modification of job duties (e.g., reduce workload, eliminate unnecessary work tasks) • Self-paced workload • Concrete, structured, and written tasks • Simple and time-limited projects and goals • Additional training and supervision (e.g., regular scheduled meetings with supervisor)
Time management	• Difficulty initiating, organizing, and completing tasks within a schedule • Difficulty performing at a consistent pace, or for an extended period • Difficulty working under pressure or deadlines	• Modification of job duties • Allow for self-paced schedule • Assign simple and time-limited projects • Schedule frequent breaks • Regular scheduled meetings with supervisor • Assistive devices (e.g., electronic diaries)
Cognitive	• Unable to sustain attention and concentration for extended periods (e.g., easily distracted by internal or external stimuli, such as noise) • Problems understanding work instructions and procedures • Problems learning and remembering work instructions and procedures • Problems dividing one's attention (e.g., multitasking when completing more complex tasks) • Problems making work-related decisions and solving problems	• Modification of job duties • Self-paced schedule • Single and time-limited projects • Written assignments and instructions • Schedule frequent breaks • Modification of work environment (reduce noise and distractions) • Assistive devices (e.g., electronic day-timers) • Job coach

(continued)

Table 7.1 (continued)

Dimensions of work performance	Possible functional limitations	Suggestions for job accommodations
Interpersonal	• Difficulty communicating, interacting, and socializing with supervisors, coworkers, and general public (e.g., telephone, face-to-face, during meetings) • Difficulty accepting and responding to feedback, criticism, or requests • Difficulty controlling one's irritability and anger around others • Difficulty working around other people • Difficulty working without supervision • Difficulty requesting assistance or disclosing limitations	• Natural supports (e.g., "buddy" system), or being paired up with a coworker) • Job coach (e.g., at supervision meetings) • Education for employers and coworkers • Structured system of feedback • Concrete expectations and written work agreements • Regular scheduled meetings with supervisor • Written goals and expectations • Modified supervisory style (e.g., provision of support and reinforcement) • Use other communication media (e.g., email vs. phone)
Emotional	• Difficulty coping with work-related stressors • Difficulty coping with emotional reactions at work • Problems coping with complex, unpredictable work situations, or change within the workplace	• Modification of work duties • Schedule frequent breaks • Natural supports • Advance notice of workplace changes and gradual transition/changes • Increase sense of control over job and work environment • Structured and predictable work environment • Written strategies and plans
Physical	• Difficulty coping with aspects of physical work environment and stimuli (e.g., noise, visual, smells) • Difficulty responding to safety issues and awareness of and taking precautions of workplace hazards • Difficulty with mobility and transportation (e.g., travelling to and from work, or within the workplace because of fears related to driving or passenger travel, using public transportation, using elevators, walking outside)	• Modification of work environment (e.g., quiet and private work area, open office space, wall partitions, headphones) • Assist in transportation and parking (e.g., near building entrance) • "Buddy system" (e.g., for unfamiliar or anxiety-provoking work areas) • Reduce unnecessary environmental stimuli and noise (e.g., coworkers talking around person, equipment, or noisy work area)

Work productivity. Research suggests that many individuals with anxiety disorder are likely to experience presenteeism: attending work while ill or symptomatic, which results in reduced work productivity (Sanderson et al. 2007). Decreased work productivity can be reflected in difficulties handling the workload and completing required work output and demands (e.g., Lerner et al. 2001). Anxiety symptoms may also interfere with the quality of work productivity and lead to increased mistakes and errors. As an example, a worker with GAD may be very worried about his or work performance and is often distracted by other concerns (e.g., other work-related issues, family, and home life), which results in decreased work productivity, which in turn perpetuates the worry cycle.

Time management. Another dimension of work performance that can be affected by anxiety symptoms involves time management skills. Specific limitations within this functional domain can include difficulties initiating, organizing, and completing tasks within a schedule (e.g., Lerner et al. 2001). Anxiety symptoms can also result in difficulties working at a consistent pace, or for an extended period without taking breaks. Individuals may also have difficulties working under pressure or deadlines. To illustrate, a worker with OCD may repeatedly check work tasks to ensure he or she has done them correctly, which in turn slows down the pace of work and results in the inability to complete work tasks within a deadline.

Cognitive. Symptoms of reduced concentration and memory are likely to contribute to a reduced ability to perform work activities requiring sustained attention (e.g., working on a computer, reading). These symptoms are also likely to interfere with understanding, learning, and remembering information, as well as problems making decisions and problem solving. These functional difficulties are likely to be further exacerbated as work demands increase in complexity (e.g., tasks that require divided attention, or multitasking) and unpredictability (e.g., new or rapidly changing work situations). To illustrate, an employee with PTSD describes being able to concentrate for only a few minutes at a time at work because of intrusive distressing memories, which in turn slows down the pace of work performance and work output, as well as possibly leading to increased work errors. Because of reduced ability to attend and concentrate on information, the worker may forget to complete work tasks or be unable to remember work instructions.

Interpersonal. Anxiety symptoms can also result in various interpersonal difficulties at work, such as problems communicating and interacting with supervisors, coworkers, and the general public. For example, irritability, a common anxiety symptom, can lead to difficulties controlling one's irritability and anger around other people at work, which in turn could result in inappropriate behavioral responses and outbursts of anger towards others. Another limitation may involve difficulties accepting and responding to feedback, criticism, or requests. Other interpersonal limitations may involve interrupting other coworkers (e.g., repeatedly asking them for assistance, seeking excessive reassurance), or conversely, not requesting assistance when needed. To illustrate, an employee with social phobia may be perceived as shy by other coworkers. He or she may tend to prefer to work alone, and avoid social and work-related interactions with coworkers and supervisors. The person is likely to be highly fearful and avoidant of being observed or negatively evaluated (e.g., the person may avoid interacting with others at breaks, eating lunch with other staff, or speaking at meetings).

Emotional. Linden and Muschalla (2007) found that persons with anxiety disorders also are likely to experience disorder-specific work anxieties (e.g., panic-related apprehension in PD, trauma-related emotional reactivity to trauma reminders in PTSD). Furthermore, individuals with anxiety disorders are also likely to have problems coping with their emotional responses at work. They may become easily overwhelmed even by relatively minor work hassles and demands, and are likely to have significant difficulties coping with more challenging work stressors (e.g., psychological and physical job demands, lack of job control, lack of supervisor and coworker support). Moreover, exposure to work-related stressors has been shown to further exacerbate anxiety disorder symptoms (Dewa et al. 2007; Melchior et al. 2007; Saint-Arnaud et al. 2006). For example, an employee with PD may experience unexpected panic attacks as well as cued panic attacks in response to work-related stressors, which in turn leads to constant apprehension and fear of the possibility of having future ones. When this disorder is accompanied by agoraphobia, the individual may dread being away from home, working alone (because of fear of having a panic attack and not having assistance), and attending staff meetings.

Environmental, mobility, and safety issues. Anxiety symptoms may also contribute to difficulties coping with various aspects of the physical work environment (e.g., noise, office space). Specifically, individuals may have difficulty working within a noisy workplace, as these stimuli can be distracting and anxiety-provoking. There may be specific functional issues with regards to mobility and transportation, such as difficulties travelling to and from work (e.g., driving, public transportation, parking), as well as getting around within the workplace itself (e.g., using elevators, escalators, stairwells). For example, a person with an SP of enclosed spaces (also known as claustrophobia) may be very anxious and reluctant to work in small offices without windows, or to use elevators or stairwells, but may otherwise have no other functional problems at work. Another concern is that anxiety disorders (as well as possibly medication effects) have been implicated in the occurrence of increased workplace accidents (Haslam et al. 2005b; Murray et al. 2006).

In summary, anxiety disorders are likely to impact job performance in a myriad of ways, which may vary considerably across individuals. There are likely both common and unique ways that anxiety disorders affect work functioning. The relationship between symptoms and functioning is complex, and work performance is likely influenced by a broad range of other clinical, psychosocial, and work-related factors. However, these general conclusions lack empirical validation, and both theory development and research in this area are still in their infancy. Thus again, with little empirical guidance, the following two sections address the practical implications of the impact of anxiety disorders on work performance – assessment and rehabilitation.

Assessment of Work Performance in the Anxiety Disorders

Previous studies on work performance in the anxiety disorders have primarily used generic assessment tools, such as the Medical Outcomes Study Short Form 12 (Ware et al. 1996) and Short Form 36 (Ware and Sherbourne 1992), the Brief Disability Questionnaire (Von Korff et al. 1996), the Liebowitz Self-Rated Disability Scale (Schneier et al. 1994), the Sheehan Disability Scale (Sheehan 1983), the Occupational Functioning Scale (Hannula et al. 2005), the Work and Social Adjustment Scale (Mundt et al. 2002), and the Work Productivity and Impairment Questionnaire (Reilly et al. 1993).

These types of instruments provide a global index of impairment, and are particularly useful for cross-comparisons across studies and populations. However, there is limited evidence on the psychometric properties of these measures for use in anxiety disorder samples (Mataix-Cols et al. 2005; Mundt et al. 2002; Sanderson and Andrews 2002). In a study that compared the use of the Short Form 12 (SF-12), the Brief Disability Questionnaire, and the number of disability days in the past month, the SF-12 was found to be most sensitive in detecting differences among functioning between the anxiety disorders (Sanderson et al. 2001). The Work and Social Adjustment Scale has been found to be a valid and reliable measure of general work impairment in OCD (Mundt et al. 2002) as well as agoraphobia, SAD, and SP (Mataix-Cols et al. 2005). The Sheehan Disability Scale has been used for assessing work disability in PD (Leon et al. 1992).

Although these measures provide a global indicator of the severity of work dysfunction, they do not provide information on the specific ways in which work functioning may be impacted. This type of information is required in developing individualized treatment and rehabilitation plans to remediate specific functional difficulties on the job. Although there have been recent advances in developing instruments to assess job performance deficits in other severe psychiatric disorders such as schizophrenia (e.g., the Work Behavior Inventory; Bryson et al. 1999), very little research has focused on the anxiety disorders, and there is a lack of validated work assessment tools for this population.

One promising self-report measure for assessing work performance in anxiety disorders is the Work Limitations Questionnaire (WLQ; Lerner et al. 2001). This self-report measure examines the percentage of time an employed person experienced difficulty with various job demands over the previous 2 weeks because of the severity of emotional and physical problems. There are four WLQ subscales, including time management (e.g., difficulty working the required number of hours); physical demands (e.g., difficulty with physical tasks, such as walking or lifting); mental-interpersonal demands (e.g., problems with concentration or controlling one's anger around others); and output demands (e.g., difficulties handling workload). This measure has been used for assessing work limitations across a variety of chronic physical health problems; recently, it has been expanded to depression (Adler et al. 2006; Burton et al. 2004; Lerner et al. 2004; Sanderson et al. 2007). Another new and useful instrument is the MINI Work Anxiety Interview, which consists of six questions to assess specific work-related anxieties for persons with anxiety disorders (Linden and Muschalla 2007).

As an assessment framework, it may be useful to describe a person's work limitations based upon the dimensions described in the above section. The evaluation of work performance also needs to be completed within a broader comprehensive assessment that includes a clinical interview and self-report inventories, in order to establish an accurate diagnosis. These methods should provide data on the frequency and severity of individual symptoms and the possible impact of these symptoms on work functioning. Relevant clinical, psychosocial, and other background information (e.g., work history, medication usage, other life stressors) needs to be obtained and integrated with other assessment findings. Other areas to assess include the person's perceptions of their illness and recovery expectations, as these factors have been shown to be robust predictors of return to work (Cole et al. 2002). Another key area of evaluation involves the psychosocial aspects of the person's work environment (e.g., Karasek et al. 1998). Factors such as job satisfaction, job control, workplace social support, psychological job demands, and job stressors have been shown to be associated with employee distress and work absence, and thus are important issues to address (Marchand et al. 2005; Schroer et al. 2005). Issues of perceived stigma and concerns regarding disclosure issues also need to be examined, as the presence of these factors can be significant return-to-work barriers (Baldwin and Marcus 2006). Medication and safety issues at work are also critical areas to assess (Haslam et al. 2005b). Ideally, a collateral report of work functioning can provide objective data on the person's work ability (e.g., interview or work performance ratings from a supervisor). The evaluation should also identify other ameloriating and complicating factors related to the person's work functioning, and identify ones that could possibly be modified. The end result of the assessment should provide a summary of the person's symptoms, work abilities, and limitations, which in turn are used for developing appropriate individualized recommendations for treatment and work rehabilitation strategies.

Work Rehabilitation of Anxiety Disorders: The Role of Job Accommodations

Current empirically supported psychosocial and pharmacological treatments for anxiety disorders are aimed at symptom reduction and do not specifically target the person's functioning. Thus, many individuals continue to experience work impairment after treatment (e.g., Diefenbach et al. 2007; Knekt et al. 2008). Another problematic issue is that there is no empirical evidence on how work impairment among persons with anxiety disorders can be most effectively rehabilitated. One promising strategy is the role of job accommodations. These workplace-based interventions can help individuals return to work as soon as possible after onset or relapse of the illness, thus minimizing extended sickness absence and its negative consequences (e.g., loss of work identity), including the secondary prevention of long-term work disability. Furthermore, many people with these conditions continue working despite the presence of symptoms and functional limitations. Thus, job accommodations can facilitate the likelihood of successful job retention and optimizing work performance. Moreover, the ability to remain engaged in the workforce can have a therapeutic effect on recovery from the disorder itself.

Currently there is a lack of empirically based practice guidelines for determining appropriate job accommodations for persons with anxiety disorders. Thus, for suggestions, it is necessary to draw upon the existing literature on established job accommodations for other types of severe psychiatric disorders (e.g., schizophrenia, bipolar disorder) (e.g., Coles 1996; Gates 2000; Gates et al. 1998; MacDonald-Wilson et al. 2003; Mancuso 1990; McCrum and Burnside 1997; Secker and Membry 2003). The types of strategies include modifications to the job (e.g., exchanging or removing nonessential job tasks), the work schedule (e.g., flexible scheduling), and the work environment (e.g., reducing noise and other distracting environmental stimuli, improving access to and with the workplace, modifying the office space). Other accommodations include job coach assistance, job sharing, and assistive technologies. Still other accommodation practices involve facilitating interpersonal processes and communication among the individual, supervisors, and coworkers: such as the building of natural supports (e.g., a "buddy system" of working with a coworker, or being accompanied by a coworker to anxiety-provoking work situations, such as meeting a new employee or taking the elevator); education and training for supervisors and coworkers; support and education for the individual on the disclosure of functional difficulties; and modification of supervision practices (e.g., adopting a more supportive supervisory style, regular supervision meetings) (see Fischler and Booth 1999, for a conceptual model of the impact of anxiety disorders on work performance and suggested job accommodations). Table 7.1 also provides examples of job accommodations that might alleviate specific types of functional limitations experienced by persons with anxiety disorders.

In addition to job accommodations, individuals with anxiety disorders require effective treatment and management of the disorder itself. Successful management of symptoms and effective coping skills are essential in facilitating work re-entry and retention. Furthermore, through treatment, individuals can learn to apply coping strategies on the job (Melchior et al. 2007). Access to other workplace-based interventions, such as cognitive behavioral stress management programs, has also been shown to be effective in reducing stress-related complaints and improving perceptions of quality of work (van der Klink et al. 2001). Organization-based education programs aimed at employers and employees can potentially be useful for providing education about anxiety disorders and their impact on work, for implementing job accommodations and other interventions, and for reducing social stigma and discrimination associated with mental illness (Gewurtz et al. 2006). Individuals with severe anxiety disorders, and those who have experienced a prolonged work absence, will likely need additional long-term and integrated support services that involve vocational, psychological, and psychiatric components. However, the extent of assistance needs for this population has not been established.

Although there is considerable evidence on the efficacy of individual placement and support services for severe mental illness (e.g., Becker et al. 2007; Bond et al. 2008), it is not known whether these types of programs would be beneficial for persons with anxiety disorders.

Conclusions and Future Directions

This chapter has summarized the existing limited knowledge on work performance in the anxiety disorders, and has suggested recommendations for assessment methods and work rehabilitation, with an emphasis on the potential role of job accommodations for this population.

The current literature suggests that, given the severity and nature of their symptoms, persons with anxiety disorders are likely to experience a myriad of work-related problems. Functional work areas that may be affected include attendance, work productivity, time management, and the interpersonal, cognitive, and emotional domains. Individuals with anxiety disorders may also experience functional difficulties because of specific aspects of the work environment (e.g., limitations with mobility and transportation, low tolerance to noise exposure, safety concerns because of possible increased accident risks). Empirical research is needed to confirm these conclusions, as well as to delineate more clearly the exact nature and types of work deficits across the anxiety disorders and for each individual disorder.

This chapter has also demonstrated that there is ample room for improving the research and practice of assessing work performance among persons with anxiety disorders. A review of the literature found that most studies rely on global measures of work impairment in anxiety disorders, and relatively few have been validated for this population. There is also a lack of specific tools to assess different dimensions of work performance, and there is a need for the development and validation of instruments that can assess interpersonal, cognitive, and emotional difficulties at work. Improved assessment methods are needed to gain a better understanding of the functional impact of the individual anxiety disorders, to accurately delineate the disease burden for each disorder, and to identify which disorders have the greatest need for work rehabilitation efforts.

The last section has described the potential benefits of job accommodations as an effective intervention to optimize the work functioning of individuals with anxiety disorders. The selection of job accommodations needs to be guided by assessment findings of the worker's symptoms and work limitations, as well as the availability of "reasonable" accommodations by the employer. Possible job accommodations may include modifications to the job, to the work schedule, and to the work environment. Job coach assistance, job sharing, and assistive technologies may also be helpful. Organizational initiatives to educate employees on anxiety disorders, to reduce stigma and discrimination, and to facilitate effective interpersonal processes and communication between the individual, supervisors, and coworkers are other critical components.

In summary, work impairment and disability are common and serious issues facing people who suffer from an anxiety disorder. Despite the evidence of work dysfunction, and its associated disease burden, it remains an understudied research area. This chapter has highlighted both the advances and challenges in this emerging field, and hopefully provides a foundation for future research and practice developments aimed at facilitating successful work participation among persons with anxiety disorders.

References

Abramowitz JS, McKay D, Taylor S (2008) Clinical handbook of obsessive-compulsive disorder and related problems. John Hopkins University Press, New York

Adler DA, McLaughlin TJ, Rogers WH, Chang H, Lapitsky L, Lerner D (2006) Job performance deficits due to depression. Am J Psychiatry 163:1569–1576

Amaya-Jackson L, Davidson JR, Hughes DS, Swartz M, Reynolds V, George KV, Blazer DG (1999) Functional impairment and utilization of services associated with post traumatic stress in the community. J Trauma Stress 12:709–724

American Psychiatric Association (2000). Diagnostic and statistical manual of mental disorders, 4th edn, text revision. Author, Washington, DC

Baldwin ML, Marcus SC (2006) Perceived and measured stigma among workers with serious mental illness. Psychiatr Serv 57:388–392

Becker D, Whitley R, Bailey EL, Drake RE (2007) Long-term employment trajectories among participants with severe mental illness in supported employment. Psychiatr Serv 58:922–928

Bond GR, Drake RE, Becker DR (2008) An update on randomized controlled trials of evidence-based supported employment. Psychiatr Rehabil J 31:280–290

Breslau N, Lucia VC, Davis GC (2004) Partial PTSD versus full PTSD: An empirical investigation of associated impairment. Psychol Med 34:1205–1204

Bruch MA, Fallon M, Heimberg RG (2003) Social phobia and difficulties in occupational adjustment. J Couns Psychol 50:109–117

Bryson G, Greig Bell T, Kaplan E (1999) The Work Behavior Inventory: Prediction of future work success of people with schizophrenia. J Psychiatr Rehabil 23:113–117

Buist-Bowman MA, de Graaf R, Vollebergh WAM (2005) Comorbidity of physical and mental disorders and the effect on work-loss days. Acta Psychiatr Scand 111:436–443

Burton WN, Pransky G, Conti DJ, Chen CY, Edington SW (2004) The associatioin of medical conditions and presenteeism. J Occup Environ Med 46:S38–S45

Carlier IV, Lamberts RD, Gersons (1997). Risk factors for posttraumatic stress symptomatology in police officers. J Nerv Ment Dis 185:498–506

Cole DC, Mondoch MV, Hogg-Johnson S (2002) Listening to injured workers: How recovery expectations predict outcomes – a prospective study. Can Med Assoc J 166:749–754

Coles A (1996) Reasonable accommodations in the workplace for persons with affective mood disorders: A case study. J Appl Rehabil Couns 27(4):40–44

Demyttenare K, Bonnewyn A, Bruffaerts R, De Graf R, Haro JM, Alonso J (2008) Comorbid painful physical symptoms and anxiety: Prevalence, work loss, and help seeking. J Affect Disord 109:264–272

Dewa CS, Lin E, Kooehoorn M, Goldner E (2007) Association of chronic work stress, psychiatric disorders, and chronic physical conditions with disability among workers. Psychiatr Serv 58:652–658

Diefenbach GJ, Abramowitz JS, Norberg MM, Tolin DF (2007) Changes in quality of life following cognitive-behavioral therapy for obsessive-compulsive disorder. Behav Res Ther 45:3060–3068

Dugas MJ, Robichaud M (2006) Generalized anxiety disorder: From science to practice. Routledge, New York

El-Guebaly N, Currie S, Williams J, Wang J, Beck CA, Maxwell C, Pattern SB (2007) Association of mood, anxiety, and substance use disorders with occupational status and disability in a community sample. Psychiatr Serv 58:659–667

ESEMeD/MHEDEA Investigators 2000 (2004a). Prevalence of mental disorders in Europe: Results from the European Study of the Epidemiology of Mental Disorders (ESEMed) project. Acta Psychiatr Scand 109 (Suppl 1):21–27

ESEMeD/MHEDEA Investigators 2000 (2004b). Disability and quality of life impact of mental disorders in Europe: Results from the European Study of the Epidemiology of Mental Disorders (ESEMed) project. Acta Psychiatr Scand 109 (Suppl 1):38–46.

Ettigi P, Allen M, Chirban JT, Jake JR, Reid WS (1997) The quality of life and employment in panic disorder. J Nerv Ment Dis 185:368–372

Evans S, Giosan C, Patt I, Spielman L, Difede J (2006) Anger and its association to distress and social/occupational functioning in symptomatic disaster relief workers responding to the September 11, 2001, World Trade Center Disaster. J Trauma Stress 19:147–152

Fehm L, Pelissolo A, Furmark T, Wittchen HU (2000) Size and burden of social phobia in Europe. Eur Neuropsychopharmacol 15:453–462

Fischler GL, Booth N (1999) Vocational impact of psychiatric disorders: A guide for rehabilitation professionals. Aspen Publishers, Inc., Gaithersburg, MD

Foa EB, Hembree EA, Cahill SP, Rauch AM, Riggs DS, Feeny NC, Yadin E (2005) Randomized trial of prolonged exposure for posttraumatic stress disorder with and without cognitive restructuring: Outcome at academic and community clinics. J Consult Clin Psychol 73:953–964

Friedman MJ, Keane TM, Resick PA (eds) (2007) Handbook of PTSD: Science and practice. Guilford Press, New York

Gates LB (2000) Workplace accommodation as a social process. J Occup Rehabil 10:85–98

Gates LB, Akabas SH, Oran-Sabia V (1998) Relationship accommodations involving the work group: Improving work prognosis for persons with mental health conditions. Psychiatr Rehabil J 21(3):264–272

Gewurtz R, Kirsh B, Jacobson N, Rappolt S (2006) The influence of mental illness on work potential and career development. Can J Community Ment Health 25:207–220

Goracci A, Martinucci M, Kaperoini A, Fagliolini A, Sbaragli C, Corsi E, Castrogiovanni P (2007) Quality of life and subthreshold obsessive-compulsive disorder. Acta Neuropsychiatr 19:357–361

Grant BF, Hasin DS, Stinson FS, Dawson DA, Ruan WJ, Goldstein RB, Smith SM, Saha TD, Huang B (2005) Prevalence, correlates, co-morbidity, and comparative disability of DSM-IV generalized anxiety disorder in the USA: Results from the National Epidemiologic Survey on alcohol and related conditions. Psychol Med 35:1747–1759

Greenberg PE, Stiglin T, Kessler RC, Finkelstein SN, Berndt ER, Davidson JR, Ballenger JC, Fyer AJ (1999) The economic burden of anxiety disorders in the 1990s. J Clin Psychiatry 60:427–435

Hannula JA, Lahtela K, Järvikoski, Salminen JK, Mäkelä P (2005) Occupational Functioning Scale (OFS) – An instrument for assessment of work ablility in psychiatric disorders. Nord J Psychiatry 60:372–378

Harvey AG, Bryant RA (2002) Acute stress disorder: A synthesis and critique. Psychol Bull 128:886–902

Haslam C, Atkinson S, Brown SS, Haslan RA (2005a) Anxiety and depression in the workplace: Effects on the individual and organization (a focus group investigation). J Affect Disord 88:209–215

Haslam C, Atkinson S, Brown SS, Haslan RA (2005b) Perceptions of the impact of depression and anxiety and the medication for these issues of safety in the workplace. Occup Environ Med 62:538–545

Hoffman DL, Dukes EM, Wittchen HU (2008) Human and economic burden of generalized anxiety disorder. Depress Anxiety 25:72–90

Karasek R, Brisson C, Kawakami N, Houtman I, Bongers P, Amick B (1998) The job content questionnaire (JCQ): an instrument for internationally comparative assessments of psychosocial job characteristics. J Occup Health Psychol 34:322–355

Kessler RC, Chiu, Demler O et al (2005). Prevalence, severity, and comorbidity of 12-month DSM-IV disorders in the National Comorbidity Survey Replication. Arch Gen Psychiatry 62:617–627

Khan A, Leventhal RM, Khan S, Brown WA (2002) Suicide risk in patients with anxiety disorders: A meta-analysis of the FDA database. J Affect Disord 68:183–190

Knekt P, Lindfors O, Laakesonen MA, Raitasalo R, Haaramo P, Jarvikoski A, The Helskinki Psychotherapy Study Group (2008) Efffectiveness of short-term and long-term psychotherapy on work ability and functional capacity – A randomized clinical trial on depressive and anxiety disorders. J Affect Disord 107:95–106

Laitinen-Krispijn S, Bijl RV (2000) Mental disorders and employee sickness absence: The NEMESIS study. Soc Psychiatry Psychiatr Epidemiol 35:71–77

Latas M, Starcevic V, Vucinic D (2004) Predictors of work disabilities in patients with panic disorder. Eur Psychiatry 19:280–284

Leon AC, Shear MK, Portera L, Klerman GL (1992) Assessing impairment in patients with panic disorder: The Sheehan Disability Scale. Soc Psychiatry Psychiatr Epidemiol 27:78–85

Lerner D, Amick B III, Rogers WH, Malspeis S, Bungay K, Cynn D (2001) The work limitations questionnaire. Med Care 39:72–85

Lerner D, Adler DA, Chang H, Berndt ER, Irish JT, Lapitsky L, Hood MY, Reed J, Rogers WH (2004) The clinical and occupational correlates of work productivity loss among employed patients with depression. J Occup Environ Med 46:S46–S55

Lim D, Sanderson K, Andrews G (2000) Lost productivity among full-time workers with mental disorders. J Ment Health Policy Econ 3:139–146

Linden M, Muschalla B (2007) Anxiety disorders and workplace-related anxieties. J Anxiety Disord 21:467–474

MacDonald-Wilson K, Rogers ES, Massaro J (2003) Identifying relationships between functional limitations, job accommodations, and demographic characteristics of persons with psychiatric disabilities. J Vocat Rehabil 18:15–24

Mancuso LL (1990) Reasonable accommodations for workers with psychiatric disabilities. Psychosoc Rehabil 14:3–19

Marchand A, Demers A, Durand P (2005) Does work really cause distress? The contribution of occupational tructure and work organization to the experience of psychological stress. Soc Sci Med 61:1–14

Mataix-Cols D, Cowley AJ, Hankins M, Schneider A, Bachofen M, Kenwright M, Gega L, Cameron R, Marks IM (2005) Reliability and validity of the Work and Social Adjustment Scale in phobic disorders. Compr Psychiatry 46:223–228

Matthews LR (2005) Work potential of road accident survivors with post-traumatic stress disorder. Behav Res Ther 43:475–483

Matthews L, Chinnery D (2005) Prediction of work functioning following accidental injury: The contribution of PTSD symptom severity and other established risk factors. Int J Psychol 40:339–348

McCrum BW, Burnside LK (1997) Organising for work: A job clinic for people with mental health needs. J Ment Health 6:503–514

McDonald H, Karlinsky H, Colotla V, Flamer S (2003) Posttraumatic stress disorder (PTSD) in the workplace: A descriptive study of workers experiencing PTSD resulting from work injury. J Occup Rehabil 13:63–77

Melchior M, Caspi A, Milne BJ, Danese A, Poulton R, Moffitt T (2007) Work stress precipitates depression and anxiety in young, working women and men. Psychol Med 37:1119–1129

Mogotsi M, Kaminer D, Stein DJ (2000) Quality of life in the anxiety disorders. Harv Rev Psychiatry 8:273–282

Mundt JC, Marks IM, Greist JH, Shear K (2002) The Work and Social Adjustment Scale: A simple accurate measure of impairment in functioning. Br J Psychiatry 180:461–464

Murray SA, Duxbury L, Higgins C (2006) The individual and organizational consequences of stress, anxiety, and depression in the workplace: A case study. Can J Community Ment Health 25:143–157

Norman SB, Stein MB, Davidson JRT (2007) Profiling posttraumatic functional impairment. J Nerv Ment Dis 195:48–53

Pull CB (2008) Recent trends in the study of specific phobias. Curr Opin Psychiatry 21:43–50

Rapaport MH, Clary C, Fayyad R, Endicott J (2005) Quality of life impairment in depressive and anxiety disorders. Am J Psychiatry 162:1171–1178

Reilly MC, Zbrozek AS, Dukes EM (1993) The validity and reproducibility of a work productivity and impairment instrument. Pharmacoeconomics 4:353–365

Roy-Byrne PP, Davidson KW, Kessler RC, Asmundson GJG, Goodwin RD, Kubzansky L, Lydiard B, Massie MJ, Katon W, Laden SK, Stein MB (2008) Anxiety disorders and comorbid medical illness. Gen Hosp Psychiatry 30:208–225

Saint-Arnaud L, Saint-Jean M, Damasse J (2006) Towards an enhanced understanding of factors involved in the return-to-work process of employees absent due to mental health problems. Can J Community Ment Health 25:303–315

Sanderson K, Andrews G (2002) Prevalance and severity of mental health – related disability and relationship to diagnosis. Psychiatr Serv 53:80–86

Sanderson K, Andrews G (2006) Common mental disorders in the workforce: Recent findings from descriptive and social epidemiology. Can J Psychiatry 51:63–75

Sanderson K, Andrews G, Jelsma W (2001) Disability measurement in the anxiety disorders: Comparison of three brief measures. J Anxiety Disord 15:333–344

Sanderson K, Tilse E, Nicholson J, Oldenburg B, Graves N (2007) Which presenteeism measures are mores sensitive to depression and anxiety. J Affect Disord 101:65–74

Schneier FR, Heckelman LR, Garkinkel R, Campeas R, Fallon BA, Gitow A, Street L, DelBene D, Liebowitz MR (1994) Functional impairment in social phobia. J Clin Psychiatry 55:322–331

Schonfeld WH, Verboncoeur CJ, Fifer SK, Lipschurtz RC, Lubeck DP, Bueshing DP (1997) The functioning and well-being of patients with unrecognized anxiety disorders and major depressive disorder. J Affect Disord 43:105–119

Schroer CAP, Janssen M, van Amelsvoort V, Bosma H, Swaen GMH, Nijhuis FJN, Eijk JV (2005) Organizational characteristics as predictors of work disability: A prospective study among sick employees of for-profit and not-for-profit organizations. J Occup Rehabil 15:435–445

Secker J, Membry H (2003) Promoting mental health through employment and developing healthy orkplaces: The potential of natural supports at work. Health Educ Res 2003(18):207–215

Sheehan DV (1983) The anxiety disease. Scribners, New York

Spitzer RL, Kroenke K, Linzer M, Hahn SR, Williams JB, deGruy FV III, Brody D, Davies M (1995) Health-related quality of life in primary care patients with mental disorders: Results from the PRIME-MD 1000 Study. J Am Med Assoc 274(19):1511–1517

Stansfield SA, Robertson-Blackmore E, Zagorski BM, Munce S, Stewart DE, Weller I (2008) Work characteristics and social phobia in a nationally representative employed sample. Can J Psychiatry 52:371–376

Stein MB, Roy-Byrne PP, Craske MG, Bystritky A, Sullivan G, Pyne JM, Katon W, Sherbourne CD (2005) Functional impact and health utility of anxiety disorders in primary care outpatients. Med Care 43:1164–1170

Stravynski A (2007) Fearing others: The nature and treatment of *social phobia*. Cambridge University Press, New York

Taylor S (2000) Understanding and treating panic disorder. John & Wiley Sons, New York

Taylor S, Wald J, Asmundson GJG (2006) Factors associated with occupational impairment in people seeking treatment for posttraumatic stress disorder. Can J Community Ment Health 25:289–301 van Balkom AJLM, van Boeijen CA, Boeke JP, van Oppen P, Kempe PT, van Dyck R (2008) Comorbid depression, but not comorbid anxiety disorders, predicts poor outcome in anxiety disorders. Depress Anxiety 25:408–415

Van der Klink J, Blonk RWB, Schene AH, van Dijk FJH (2001) The benefits of interventions for work-related stress. Am J Public Health 91:270–276

Von Korff M, Ustun TB, Ormel J et al. (1996). Self-report disability in an international primary care study of psychological illness. Journal of Clinical Epidemiology, 49:297–303

Waghorn G, Chant D, White P, Whiteford H (2005) Disability, employment, and work performance among people with ICD-10 anxiety disorders. Aust NZ J Psychiatry 39:55–66

Waghorn GR, Chant DC (2005) Employment restrictions among persons with ICD-10 anxiety disorders: Characteristics from a population survey. J Anxiety Disord 19:642–657

Ware JE, Kosinski M, Keller SD (1996) A 12-Item Short-Form Health Survey: Construction of scales and preliminary tests of reliability and validity. Med Care 34:220–233

Ware JE, Sherbourne CD (1992) The MOS 36-item short form health survey (SF-36). I. Conceptual framework and item selection. Med Care 30:473–483

Wittchen HU, Fehm L (2001) Epidemiology, patterns of comorbidity, and associated disabilities of social phobia. Psychiatr Clin North Am 24:617–641

Wittchen HU, Fuetsch M, Sonntag H, Muller N, Liebowitz M (2000) Disability and quality of life in pure and comorbid social phobia. Findings from a controlled study. Eur Psychiatry 15:46–58

Wittchen HU, Kessler RC, Beesdo K, Krause P, Hofler M, Hoyer J (2002) Generalized anxiety and depression in primary care: Prevalence, recognition, and management. J Clin Psychiatry 63(Suppl 8):24–32

Yonkers KA, Bruce SE, Dyck IR, Keller MB (2003) Chronicity, relapse, and illness – course of panic disorder, social phobia, and generalized anxiety disorder. Depress Anxiety 17:173–179

Zvolensy MJ, Smits JA (eds) (2008) Anxiety in health behaviors and physical illness. Springer Science and Business Media, New York

Chapter 8
Brain Injury and Work Performance

Thomas J. Guilmette and Anthony J. Giuliano

Among the many causes of brain injury, traumatic brain injury (TBI) is the most common cause of disability and death in young people (Coronado et al. 2006). Moreover, one of the most socially relevant and personally meaningful domains of post-injury disability is significantly reduced work capacity and performance, often during the period of psychosocial development in which young people are entering the workforce. The purpose of this chapter is to provide an overview of the current state of our knowledge regarding vocational outcomes following TBI. The focus will be selectively organized around the vocational implications of moderate and severe TBI, and the consequences and outcomes of mild TBI (mTBI) will not be addressed in detail (see Fraser et al. 2010, this volume for additional information about mTBI). This chapter will review the definition and epidemiology of TBI; the terminology and conventions related to classification of injury severity; the physical, cognitive, and emotional-behavioral consequences of TBI; and the vocational outcomes associated with moderate and severe TBI, including the variables and processes which may predict, mediate, and moderate various kinds of work outcomes. The chapter will summarize the conclusions of recent review articles, the selective findings of some methodologically stronger individual studies, the relationships among sex and ethnicity/race and employment following TBI, and the trends and limitations of extant literature in this area.

Definition and Mechanisms

TBI is a subtype of acquired brain injury and typically occurs as a result of external mechanical forces that impact the head. In most TBIs, linear or angular acceleration-deceleration forces and/or blunt trauma produce the more common closed head

T.J. Guilmette (✉)
Department of Psychology, Providence College, 118 Albertus Magnus Hall,
One Cunningham Square, Providence, RI 02918, USA
e-mail: tguilmet@providence.edu

I.Z. Schultz and E.S. Rogers (eds.), *Work Accommodation and Retention in Mental Health*, 141
DOI 10.1007/978-1-4419-0428-7_8, © Springer Science+Business Media, LLC 2011

injuries, while projectile forces that penetrate the dura, the outermost covering of the brain inside the skull, produce open head injuries. TBIs may also be the consequence of multiple mechanisms, such as blast injuries which include several forces (i.e., a "blast wave" or over-pressurization, flying debris/projectile forces, and body displacement/blunt trauma as a result of being thrown into other objects). Orthopedic and other internal organ injuries are also a common co-occurring consequence of these same forces, and represent simultaneous targets of emergency medical attention and sources of morbidity or mortality. Injury severity occurs across a broad continuum, ranging from very mild transient injuries to catastrophic injuries that result in death or severe disability.

Epidemiology

Numerous methodological differences among studies complicate any synthesis of the extant epidemiological literature on TBI. The most recent working estimates indicate that approximately 120 per 100,000 people in the United States experience a fatal or nonfatal hospitalized TBI each year (Kraus and Chu 2005). Between 1995 and 2001, approximately 1.4 million people were treated in an emergency department (ED) for a TBI in the United States. Of those receiving treatment in an ED, 50,000 (or 3.6% of) people died as a result of their injuries, 235,000 (or 16.8%) were hospitalized, and 1.1 million (79.6%) were treated and discharged (Langlois et al. 2004). Over the past 25 years, those patients admitted to a hospital after sustaining a TBI demonstrated injury severity patterns that were 80% mild, 10% moderate, and 10% severe (Kraus and Chu 2005). However, rates of mild, moderate, and severe TBI vary as a function of treatment setting: Level 1 trauma centers are more likely to receive and treat people with more severe injuries than other kinds of institutions (Thurman and Guerrero 1999). Moreover, the incidence of mTBI is not fully known, largely because people who sustain a mTBI often do not seek treatment after their injury (Kraus and Chu 2005).

While TBIs occur across all ages of the lifespan, there are ages of peak risk and incidence which contribute to a developmental understanding of the psychosocial, particularly vocational, consequences of TBI. The highest rates of injury are found among 15–24 year olds, followed by adults over the age of 75 (Thurman et al. 2007), thus representing a bimodal age-related incidence distribution. Many transitional age youth (ages 16–25) experience TBI during the formative and dynamic developmental period when vocational skills and identity are just beginning to be developed (see Arnett 2000).

The worldwide literature provides reliable evidence for differential incidence rates among males and females. TBI is much more frequent in men than women in the United States, and is marked by a rate ratio of 1.6:2.8 (Kraus and Chu 2005). As with other race- and ethnicity-related injury data, racial or ethnic differences in rates of TBI remain undetermined, largely due to variable recording practices. One working estimate is derived from the National Traumatic Brain Injury Model

Systems database, which reports that 33% of their participants are members of minority groups (Department of Education, National Institute on Disability and Rehabilitation Research 2008). Other data also indicate that the rate of TBI is higher among families at lower socioeconomic levels (Kraus et al. 1986; Sosin et al. 1989). This should be kept in mind when attempting to understand vocational outcomes associated with TBI, as socioeconomic status has been identified as an important predictor of adult employment in adolescents in general (Leventhal et al. 2001) and following TBI in particular (Hoofien et al. 2002). Lastly, it is well-established that blood alcohol concentration (BAC) is reliably associated with both risk and severity of TBI (Kraus et al. 1989; Smith and Kraus 1988). Nearly 56% of adults who sustain a TBI showed a positive BAC test result, and almost 50% of those with a positive BAC test result were at or above the legal level of 0.10% (Kraus et al. 1989).

Causes of TBI vary as a function of age and developmental phase. The most frequent cause of fatal and nonfatal TBI is transportation accidents involving motor vehicles, motorcycles, bicycles, watercraft, and others. Falls are the second most frequent cause of TBI, but tend to be associated with both younger and older age. Other common causes include assault-related injuries, some of which involve firearms and result in penetrating head injuries, and sports-related injuries (Kraus and Chu 2005).

Injury Severity: Terminology and Classification Schemes

While there are no universally accepted classification criteria for injury severity, the most common schemes use three key variables to broadly classify TBIs as *mild, moderate, or severe*: the Glasgow coma scale (GCS), duration of unconsciousness, and duration of posttraumatic amnesia (PTA) or confusion (see Table 8.1). Over the past 30 years, altered consciousness has been indexed by depth of coma and length of unconsciousness. The clinical convention has been to grade depth of coma with the GCS (Teasdale and Jennett 1974). The GCS is widely used by emergency medical technicians at the scene of an accident, emergency department personnel, and researchers. The GCS is used to evaluate three components of arousal and responsiveness: (1) the level of stimulation required to elicit eye-opening; (2) the best motor response; and (3) the best verbal response. GCS scores range from 3 to 15.

Table 8.1 Classification of traumatic brain injury severity

Classification	GCS	Length of unconsciousness	PTA
Mild	13–15	<30 min, if at all	<24 h
Moderate	9–12	30 min–24 h	1–7 days
Severe	3–8	>24 h	>7 days

GCS, Glasgow coma scale; PTA, posttraumatic amnesia

The conventional severity classification ranges for GCS scores are: mild = 13–15; moderate = 9–12, and severe = 3–8.

Duration of unconsciousness refers to the length of time that an individual is observed to be in a state marked by a "lack of awareness of self or the environment" (Jennett 1996). Loss of consciousness may be brief or not at all, such as might occur in a sports-related mTBI (e.g., a concussion), or it may be protracted. Mild injuries are generally regarded as those involving 30 min or less of unconsciousness, while moderate-to-severe injury is defined by a loss of consciousness greater than 30 min (Mild Traumatic Brain Injury Committee 1993). Duration of unconsciousness can be influenced by a variety of factors, including the severity of the primary brain injury, level of intoxication, disruption of cardiopulmonary functions, and/or suppression of neurophysiology due to sedating medications (often used emergently to manage neurotrauma).

The third variable most commonly used to grade injury severity is PTA. Some individuals are unable to recall events prior to the TBI (called "retrograde" amnesia), while others may not be able to remember new events that occur after the TBI, such as being cared for by emergency clinicians or visited by family members (called "anterograde" amnesia). PTA is a period of anterograde amnesia marked by an inability to encode and recall new experiences or information that follows recovery of consciousness. Patients who sustain moderate or severe TBIs often experience both kinds of amnesia, although PTA, or anterograde amnesia, is more common than retrograde amnesia and is usually of longer duration. PTA may last from seconds to months, and estimates of duration are often confounded by intoxication, medication effects, and/or recall biases.

The Special Circumstances of mTBI

mTBI, sometimes referred to as concussion, is characterized by a limited alteration in mental status at the time of injury (as noted in text and Table 8.1 above), and is by far the most common type of acquired brain injury. In spite of the prevalence of mTBI, the long-term effects of and recovery from this level of injury remain sources of controversy in the literature. However, a number of meta-analytic studies over the last decade have consistently revealed that, unlike moderate and severe brain injuries, in the vast majority of uncomplicated mTBIs (i.e., injuries with no evidence of brain trauma on neuroimaging) there is full recovery of cognitive and general functioning within weeks to, at most, one to three months after the trauma (Belanger et al. 2005; Binder 1997; Dikmen et al. 1995; Iverson 2005; Larrabee 1997; Schretlen and Shapiro 2003).

In summarizing the scientific literature, McCrea (2008) made the distinction that outcome in moderate to severe TBI was most associated with injury severity, whereas in mTBI it was "most often predicted by noninjury-related factors, e.g., premorbid psychosocial issues, psychological comorbidities, postinjury stressors, substance abuse, litigation" (p. 15). Thus, it is clear that mTBI is "a different animal"

(McCrea 2008, p. 129) when compared to moderate to severe injuries, and ideally should be examined as a separate diagnostic category with regard to outcome. Consequently, we have chosen to report the results of individual studies that include only participants with moderate to severe TBIs. We will, however, summarize the results of review articles that include mTBI participants, because most review articles do not routinely exclude this subset of individuals.

Consequences of Moderate to Severe TBI

The consequences of moderate and severe TBIs may be temporary, prolonged, or permanent, and include a wide range of physical, cognitive, and emotional–behavioral impairments. These problems may contribute differentially to specific psychosocial and/or vocational disabilities depending upon the nature of the work role (e.g., specific task demands), residual symptom severity, and environmental supports or compensations available to an individual.

The most obvious physical consequences of moderate or severe TBIs include a range of motor impairments and movement disorders involving one or both lower and/or upper extremities that interfere with mobility and motor control and coordination of everyday tasks. Motor impairments may themselves range from mild to severe, and are more likely to occur as a result of a severe TBI. These include paresis (weakness) or plegia (paralysis), spasticity (increased muscle tone and exaggerated reflexes), and ataxia (loss of muscle coordination). Post-injury movement disorders may include slowness or reduced movement (hypokinesia) or excessive involuntary movements (hyperkinesia). Tremors and dystonias are among the most common movement disorders and sometimes occur together (Krauss and Jankovic 2002). Tremors manifest as rhythmic, oscillatory movements; dystonias are characterized by sustained muscle contractions that cause twisting or repetitive movements, and/or abnormal postures or positions of any body part.

Other common physical problems include impairments in static or dynamic balance; dizziness (which is typically caused by a peripheral injury to the vestibular system, as opposed to brain injury); and sensory and cranial nerve impairments (such as visual deficits, double vision or diplopia, and ocular abnormalities; deficits in smell, taste, hearing, sensation; facial movement, swallowing, tongue movements, and neck strength). Posttraumatic headaches, or new headache problems that manifest within the first week after injury, are common consequences of injuries to the head or neck, or both. While posttraumatic headaches often resolve within 3 months, for some patients they persist beyond 3–6 months, and are regarded as chronic conditions. Depending upon their severity, chronic headaches themselves can be quite disabling, and are frequently associated with considerable psychological distress, most often depression (Martelli et al. 1999; Ohayon and Schatzberg 2003). Similarly, other pain problems, particularly chronic pain (i.e., pain problems that persist beyond 6 months) regardless of its source(s), are associated with avoidance behaviors and reduced activity, increased psychological

distress, and considerable functional disability (Zasler et al. 2005). Sleep and
fatigue problems also represent common consequences of TBI, and tend to
exacerbate other problems (such as headaches, cognitive problems, and mood
disorders). Patients most often report feelings of fatigue, low endurance/tolerance
for exertion, and a sense of exhaustion. Research over the past 10 years has documented that approximately 16–32% of patients who sustain TBIs report significant
problems with fatigue at 1 year post-injury (Bushnik et al. 2008), 21–68% at
2 years post-injury (Bushnik et al. 2008), and 37–73% at 5 years post-injury
(Hillier et al. 1997; Olver et al. 1996). Lastly, approximately 5% of patients who
sustain closed TBIs and 30–50% of those with open TBIs (i.e., when the dura has
been penetrated) develop posttraumatic epilepsy or seizures that, particularly if
they remain uncontrolled with medication, often interfere significantly with daily
functioning (e.g., retaining a driver's license, competitive employment) (Frey
2003; Tucker 2005).

Moderate and severe TBIs often produce a range of neurocognitive impairments, though their pattern and level of severity is fairly individualized and
difficult to predict. As a group, those who sustain a severe TBI are very likely to
demonstrate some degree of persistent neurocognitive impairment (Dikmen et al.
2003; Schretlen and Shapiro 2003). Cognitive deficits are often regarded as the
most salient and functionally relevant consequences following TBI, and have
been shown to be more predictive of persisting disability than most physical
symptoms except for quadriplegia. The most common cognitive impairments
include a slowed rate of information processing (which often makes a patient
vulnerable to confusion and secondary agitation), reduced sustained attention/
concentration, working memory, new learning and memory, executive function
(e.g., initiation, planning, organization, inhibition, problem-solving, performance
regulation), and language/communication, particularly discourse comprehension
and language pragmatics (e.g., more fragmentation of speech and comprehension, difficulty initiating/maintaining conversation topic). All of these limitations
have implications for functioning in a work setting. Moreover, lack of awareness
of physical, cognitive, or behavioral deficits and their functional consequences
often complicates identification, remediation, and rehabilitation of these problems. Nearly 45% of people who sustain a moderate to severe TBI have been
observed to demonstrate reduced awareness (Flashman and McAllister 2002;
Flashman et al. 2005).

Lastly, moderate and severe TBIs often result in permanent impairments in neurobehavioral, neuropsychiatric, emotional, and/or social functioning. Neurobehavioral
and emotional changes following TBI can include personality changes, problems
with emotion regulation, apathy, disinhibition, and onset of new psychiatric symptoms or disorders such as depression, anxiety, psychosis, and/or secondary atten-
tion-deficit/hyperactivity disorder (ADHD). These symptoms can cause a wide
range of difficulties in work settings and seriously interfere with employability. The
onset of new psychiatric disorders tends to be complicated by overlapping ages of
peak risk; that is, it is well-established that primary psychiatric disorders, including
mood and psychotic disorders, have their peak age of onset in adolescence and

young adulthood (Kessler et al. 2005; New Freedom Commission on Mental Health 2003), precisely the same period of peak risk for the occurrence of a TBI.

The prevalence of depression following TBI has varied estimates, and ranges from 11 to 77% (Silver et al. 2001). While rates of depression generally decrease over time post-injury, chronic (14%) and late onset (10%) depression have been reported at 3 years post-injury (Hibbard et al. 2004), and some studies have documented that the prevalence of depressive symptoms remains notably elevated over eight (61%; Hibbard et al. 1998) and 30 (27%; Koopnen et al. 2002) years post-injury. While less common than depression, the post-injury onset of anxiety disorders ranges from 0 to 42%, depending on anxiety disorder subtype (e.g., generalized anxiety disorder: 8–24%; panic disorder: 1–9%; posttraumatic stress disorder: up to 42% in some studies). While cause and comorbidity remain unclear, some research has documented a relationship between TBI and psychosis in some patients (Corcoran et al. 2005), particularly paranoid symptoms (Achte et al. 1991). Secondary ADHD has also been shown to emerge following moderate and particularly severe TBI, especially in youth (Max et al. 2004). More generally, TBIs can cause changes in personality that are marked by impulsivity, emotional lability, apathy, lack of interest, episodic but intense irritability, and aggressive outbursts (O'Shanick and O'Shanick 2005).

Numerous studies have documented the negative impact of moderate and severe TBI on psychosocial outcomes. Among the most disruptive problems that add excess morbidity and interferes with rehabilitation and community reintegration is substance abuse. Silver and his colleagues (2001) reported that the prevalence of alcohol and drug abuse disorders was 25 and 11%, respectively, in people with a history of TBI, compared to 10 and 5% in people without a known history of TBI. Not surprisingly, for the majority of people, post-injury substance abuse problems represent a gradual return to use patterns that were established before the injury, marked by increasing rates of use over the first 3 years post-injury (Bombardier et al. 2003; Corrigan et al. 1995). Return to pre-injury alcohol use patterns tends to be faster than with drug use patterns, though there is a subgroup of individuals in whom new onset drinking problems emerge (about 15%; Corrigan et al. 2007).

Social strain and reduced social support are also common consequences of TBI, and loneliness and social isolation are frequently reported by people who sustain a moderate or severe TBI. Across all family relationships, reports of increased burden over time, frustration and reactivity to personality and behavioral changes (that exceed neurological–physical or cognitive changes), and increased coping and adjustment demands are quite common (Cavallo and Kay 2005). Limited community mobility and transportation costs/problems, particularly driving restrictions secondary to sensory and motor impairments; cognitive deficits (e.g., distractibility, poor judgment); and/or self-regulation and performance regulation problems (e.g., impulsivity) represent ongoing challenges and significantly disrupt community integration and vocational options. In fact, driving tends to be a highly desired personal goal that may be unattainable, which itself becomes a source of ongoing distress (Kneipp and Rubin 2007).

Overview of Research Literature: Trends, Limitations, and Challenges

The scientific literature examining the relationship between TBI and work capabilities is far from definitive, as is evident from estimates for returning to work after TBI, which are highly variable and range from 10 to 70% (Nightingale et al. 2007). The National Traumatic Brain Injury Model Systems database reported a 28% return-to-work (RTW) rate after 1 year post-injury compared with 63% of their participants who were employed pre-injury (Department of Education, National Institute on Disability and Rehabilitation Research 2008).

There are multiple difficulties in conducting research in this area that limit the generalizations that one can make from the available studies. One of the most problematic issues is that employment is a very complex psychosocial phenomenon with multiple determinants, even within a non-brain-injured population. For example, having a job is dependent on the local economy and the availability within the community of a position that fits the potential employee's capabilities and job skills. The brain injury survivor also has to be motivated to work. Motivation for work can be influenced by numerous factors such as age, financial needs, pending litigation, family dynamics, disability compensation, and other variables directly related to the brain injury such as damage to select cortical, especially frontal, areas resulting in impaired drive and initiation (e.g., abulia). In addition, challenges associated with how best to define work "outcome" contribute to substantial heterogeneity in the literature: whether it should include both full-time or part-time, supported and competitive, or sustained and intermittent; the extent to which it should be equivalent to the survivor's premorbid level of employment; and whether school matriculation at any level should be considered comparable to work with regard to productivity. RTW rates reported in published studies vary markedly because the operational definitions of employment differ, i.e., narrow and restrictive versus broad and inclusive, with the latter potentially inflating estimates of work outcomes (Kendall et al. 2006).

Other issues that make comparisons difficult across, and firm conclusions from, naturalistic studies include limited sample characterization on one or more of the following key variables: the level of injury severity of study participants (e.g., mixing mTBI survivors with more severely injured sample participants); the length of time since injury; the survivor's participation in rehabilitation services (particularly specialized vocational services); access to transportation; and other demographic, pre-injury (e.g., employment status, substance use), and environmental factors. Historically, the research on prediction of RTW has focused narrowly on demographic and TBI severity variables; results of these studies generally reveal that older age and greater injury severity are associated with worse psychosocial outcomes, including RTW (Dawson et al. 2007). However, what has become increasingly clear to researchers in and reviewers of this area of study is that vocational outcome is multiply determined and its study requires a multifactorial approach. If researchers do not include certain variables, such as premorbid and contemporary environmental factors, despite their relative ease of acquisition, then it cannot be determined whether those variables have any value in predicting work performance.

Nightingale et al. (2007) concluded that prediction studies of employment following TBI should examine a minimum of 17 variables. They suggest that studies include information about level of education, pre-injury employment/job stability, psychosocial background, premorbid intelligence, history of learning disability, prior psychological and/or psychiatric problems, history and type of substance abuse, prior brain injury, developmental history, injury severity, early post-injury cognitive status related to attention and processing speed, memory, executive functioning, and other post-injury variables such as behavioral functioning, neurophysical functioning, functional/disability status, and community participation. In their review of nearly 2,000 articles on RTW after TBI, they did not find one article that addressed all these variables.

Summary of Recent Reviews

Four recent reviews have ably summarized the literature on RTW following TBI, although as noted above there are substantial methodological issues that limit firm conclusions. Starting in 2004, Ownsworth and McKenna reviewed 85 studies published between 1980 and December 2003 that examined the relationship between TBI and vocational outcome. After independent ratings of methodological quality, they selected 50 studies to include in their review. The most consistent predictors of employment were premorbid occupational status, functional status at discharge, perceptual ability, executive functioning (e.g., concept formation, divided and selective attention, mental flexibility, mental programming, and planning), involvement in vocational rehabilitation services, and emotional status. The authors did not find a consistent relationship between injury severity and employment outcome. They reported that "a number of studies found a relationship between longer periods of coma and poor employment outcome ... however, other studies found a considerable number of people who never lost consciousness but who were unable to return to work" (p. 772). It is obvious from their review that they included studies with mTBI and, as noted above, studies that include participants with mTBI tend to complicate the picture, as vocational outcome following mTBI may be determined by different sets of variables than in the case of moderate and severe TBI.

In a selective review of the literature on TBI and employment, Wehman et al. (2005) concluded the following: rates of employment for TBI survivors change over time; self-awareness and acceptance of disability have an impact on employment outcome; duration of coma and GCS scores are related to RTW; an age of 40 or older at the time of injury adversely affects postinjury employment; some persons with severe TBI can be competitively employed, in part because the severity of post-injury disability and functional status is not perfectly associated with injury severity; and community-based vocational rehabilitation services can be successful in assisting TBI survivors to return to work in some capacity (see also Malec and Degiorgio 2002; Targett and Wehman 2010, this volume).

Nightingale and colleagues (2007) identified only 27 out of 1948 studies of TBI and employment outcome that met methodologically rigorous standards, and which examined three domains that they considered critical to operationalize in prediction models: pre-injury functioning, injury severity parameters, and post-injury factors. Their review found limited evidence for the prognostic value of age and education (e.g., generally low correlations with vocational outcome); in contrast, they concluded that other pre-injury variables have not been adequately studied with the exception of pre-injury employment, which was a significant factor in every study in which it was examined. Injury severity was of limited use in predicting employment if it was studied within a homogeneous severity sample, likely because of restricted ranges on the injury parameters. Early post-injury variables have not been consistently studied, although there is emerging evidence that level of executive functioning may have some positive prognostic significance. Overall, these reviewers documented that the current state of the research literature on RTW prediction variables is mixed and largely inconclusive. They summarized that "there is insufficient evidence to reliably and validly identify variables with prognostic significance for RTW, and a comprehensive, methodologically rigorous study needs to be conducted" (p. 138).

Summarizing their MEDLINE review of TBI and RTW factors, Shames et al. (2007) identified injury severity and lack of self-awareness as the most significant predictors of return to work. They further concluded that "there appears to be a complex interaction between premorbid characteristics, injury factors, post-injury impairments, personal factors, and environmental factors in TBI patients. This influences RTW in ways that make predicting outcomes, at best, only moderately accurate" (p. 1,392).

Kendall et al. (2006) selected 26 studies to review pertaining to TBI and employment outcomes with specific analysis of the role of vocational rehabilitation and RTW. They found, consistent with other research, that there was a strong positive relationship between pre-injury and post-injury employment. They also noted that the majority of moderately to severely injured TBI participants in the studies they reviewed returned to productive employment in some capacity, usually within 1–2 years, especially those who participated in vocational rehabilitation programs. Although not a review article, Hammond and colleagues (2004) found similar results. They reported that the majority (79%) of TBI survivors' employability did not change from year 1 to year 5 post-injury. In addition, the National Traumatic Brain Injury Model Systems database reports only a 4% increase in employment from 1 to 2 years post-injury (Department of Education, National Institute on Disability and Rehabilitation Research 2008).

Summary of Individual RTW Studies

Table 8.2 lists individual studies that have examined RTW factors in TBI samples. Although some consistent findings emerge, it is important to note that many of the studies were conducted on samples with very different characteristics and at different

Table 8.2 Summary of individual return to work studies

Reference	Years post-injury	Number of participants	Percent employed	Conclusions
Devitt et al. 2006	7–24 ($M = 14$)	306	Not stated	Pre-injury behavior problems, male gender, post-injury cognitive and physical deficits, and lack of access to transportation were related to poorer outcome
Hanlon et al. 2005	1	24 with SAH 76 without SAH	42% 66%	TBI with SAH is associated with reduced employment rates
Johnstone et al. 2006	0.1–41 ($M = 8.9$)	TBI alone = 39 TBI & ortho = 26 TBI & seizure = 38 TBI & psych = 18 TBI & LD = 25	26% 23% 32% 11% 8%	Psychological factors and LD are associated with reduced employment rates
Johnstone et al. 2003 b	1	24 employed prior to injury	46%	No one who was unemployed at baseline was employed at one year post-injury
Keyser-Marcus et al. 2002	1–5	Year 1 = 451 Year 2 = 252 Year 3 = 187 Year 4 = 136 Year 5 = 120	Not stated	Pre-injury productivity and age were most reliably related to RTW, followed by educational level, discharge DRS, discharge FIM, and rehabilitation length of stay
Klonoff et al. 2006	1–7, post-rehabilitation	92 (54% TBI, 46% other)	74%	Younger age, higher education, non-right-hemisphere injury, and ability to drive were related to positive work outcome

(continued)

Table 8.2 (continued)

Reference	Years post-injury	Number of participants	Percent employed	Conclusions
Kreutzer et al. 2003	1, 2, and 3 or 4	186	Year 1 = 35% Year 2 = 37% Year 3 = 42%	Unemployment most related to being a minority group member, not completing high school, and being unmarried. Driving independence, age, length of unconsciousness, and DRS at one year were all related to job stability
Machamer et al. 2005	3–5	165	74%	RTW was most related to injury severity, neuro-psychological functioning at 1 month post-injury, and pre-injury work stability and earnings
Nakase-Richardson et al. 2007	1	171	30%	Individuals with greater confusion 1 month post-injury, older age (42 vs. 21), and fewer years of education (10 vs. 13) were less likely to be employed at 1 year
Walker et al. 2006	1	1,341	32–56%	Type of occupation influences RTW rates with the best outcomes for those in professional/managerial jobs and the worst with manual labor

SAH, subarachnoid hemorrhage; DRS, Disability Rating Scale; FIM, Functional Independence Measure times since injury. As noted above, none of the studies examined all or even most of the variables suggested by Nightingale et al. (2007), and methodological quality varies greatly. However, Table 8.2 largely includes only those studies that focused on complicated mild and moderate-to-severe TBIs, and deliberately excludes studies with very heterogeneous samples (e.g., mixing any mTBI with participants with moderate-to-severe TBIs).

A review of Table 8.2 reveals a RTW range following TBI from 8 to 74%, which is generally consistent with previous findings. However, the TBI samples at the extremes of the injury severity continuum include very different participants. For example, the 8% RTW rate was derived from a sample of persons with TBI and a history of learning disabilities who were receiving services from a state vocational rehabilitation program (Johnstone et al. 2006). This study compared TBI survivors with various concomitant disorders, such as orthopedic injuries, seizures, psychological issues, and learning disabilities. The generalizability of these results, however, is very limited, because the majority of the patients never completed the rehabilitation program and the sample in general was characterized by low average intelligence, largely male gender, and a wide range of time since injury (i.e., 40 years). Conversely, the 74% RTW rate reported by Klonoff and colleagues (2006) was derived from a sample of mixed TBI and other acquired brain injury etiologies, such as stroke, with relatively high levels of education (mean = 14.3 years), and following completion of a comprehensive neurorehabilitation program by all participants. In addition, their sample was notably selective, as nearly 99% were productive pre-injury. The other 74% RTW rate (Machamer et al. 2005) was reported with a sample that included complicated mild (57%) and moderate-to-severe TBIs (43%). A complicated mTBI is one in which the injury parameters of GCS scores and length of unconsciousness and PTA are the same as with uncomplicated mTBI (see text above and Table 8.1), except that there is evidence of brain injury-related abnormality on neuroimaging (e.g., on a head CT or MRI scan). Arguably, this sample overall would be less severely injured than one that was composed solely of greater than mild injuries.

Pre-injury factors were identified as being relevant in RTW rates in many of the individual studies reviewed. Specifically, the three most salient pre-injury characteristics related to successful employment following TBI were: (1) having been employed prior to the injury (Johnstone et al. 2006; Keyser-Marcus et al. 2002; Machamer et al. 2005); (2) higher levels of education (Keyser-Marcus et al. 2002; Kreutzer et al. 2003; Nakase-Richardson et al. 2007); and (3) younger age (Keyser-Marcus et al. 2002; Klonoff et al. 2006; Nakase-Richardson et al. 2007). Wehman et al. (2005) summarized that TBI survivors over the age of 40 have a worse RTW rate than those younger than 40; similarly, Keyser-Marcus et al. (2002) found that RTW rates at each of 5 post-injury years was lower for persons over 40. Lastly, in addition to the positive predictive value of pre-injury employment status, Walker and colleagues (2006) documented that occupational type may influence post-injury employment rates. More specifically, they reported that RTW rates differed at 1 year post-injury for persons with different occupational titles: professional/managerial (56%), skilled (40%), and manual labor (32%).

Although injury severity (e.g., GCS score, length of unconsciousness, length of PTA) was reported to be inconsistently associated with RTW in at least one review (Ownsworth and McKenna 2004), some individual studies have found increased severity to be related to reduced employment rates (Kreutzer et al. 2003; Machamer et al. 2005; Nakase-Richardson et al. 2007). The presence of a subarachnoid hemorrhage, an effect of the trauma to the brain resulting in arterial bleeding in the subarachnoid space, has also been associated with decreased employment levels (Hanlon et al. 2005). The presence of such an injury characteristic may signify a more severe injury than indexed by the more commonly used injury severity variables.

Greater injury severity is often associated with more cognitive and functional impairments in the months and years following injury. Consistent with this relationship, a number of studies have documented that postinjury functioning, whether measured by formal rating scales of disability (e.g., the Disability Rating Scale or the Functional Independence Measure), the need for longer rehabilitation hospital stays, or level of psychological and neuropsychological functioning, is related to RTW rates and employment stability postinjury (Devitt et al. 2006; Keyser-Marcus et al. 2002; Kreutzer et al. 2003; Machamer et al. 2005). Although cognitive disorders are a highly consistent finding in the postacute recovery phase of those persons who sustain moderate to severe TBIs, research has also revealed that global performance on neuropsychological tests is only marginally related to employability (Guilmette 2005; Sbordone and Guilmette 1999).

Turning again to Table 8.2, two studies revealed that driving independence was related to better vocational outcomes (Klonoff et al. 2006; Kreutzer et al. 2003). This finding likely reflects at least two processes: one is related to the pragmatic issue that not being able to drive simply makes it more difficult to get to a job. For example, Devitt et al. (2006) found that lack of access to transportation was related to poorer vocational outcome. Alternatively, or simultaneously, the inability to drive itself is a reflection of functional capacity, and is likely related to greater physical, cognitive, and/or behavioral impairments that directly limit job options as well.

Sex and Ethnicity Issues in Employment Following TBI

According to the National Traumatic Brain Injury Model Systems database, women sustain approximately one-fourth of TBIs in the US, and minorities account for approximately one-third of all cases (Department of Education, National Institute on Disability and Rehabilitation Research 2008). Because the majority of TBIs are sustained by Caucasian men and most study samples include participants with these characteristics, it remains unclear if predictors of psychosocial outcomes, such as employment status, are comparable for women and/or members of minority groups. Consequently, some researchers have specifically examined the effects of TBI on post-injury employment in women and non-Caucasian TBI survivors.

To date, the majority of studies report either the absence of sex differences in psychosocial outcomes or that men may have somewhat worse outcomes (Corrigan et al. 2007; Devitt et al. 2006). Given the limited number of studies to date, this issue remains unresolved. For example, in a sample of state vocational rehabilitation clients with TBI who were on average 9.6 years post-injury, 24% of men (13/55) were rated as being vocationally successful at closure compared to only 4% (1/23) of women (Bounds et al. 2003). However, the authors note that the men as a group showed a trend toward more severe TBI and more limited neuropsychological abilities, and had significantly higher rates of substance abuse than women, each of which could be conceived as a potentially confounding factor. In addition, their data also revealed that women were more likely than men to have their cases closed before rehabilitation services were initiated (73.9% vs. 56.4%, respectively), and that among closed cases, client refusal of services was higher for women (73%) than men (45.5%). Reasons for these gender disparities were unclear, but the researchers acknowledged that it was possible that more women were able to successfully find employment on their own, an outcome that was not included in the study.

Corrigan and colleagues (2007) merged datasets from the National Traumatic Brain Injury Model Systems and the South Carolina Traumatic Brain Injury Follow-up Registry to examine change in employment status from pre-injury to 1 year post-injury. They found, contrary to most prior studies, that women were more likely to decrease hours of employment or to stop working altogether compared to men, though this was not true for the oldest age cohort (55–64 years). The researchers also reported a significant sex/marital status interaction, with married women, compared to married men, being much more likely to work less or not at all. The processes underlying, and implications of, these findings were unclear and require further study.

Research has suggested that members of racial minorities may differ from whites with regard to vulnerability to TBI, causes and mechanisms of TBI, and outcome from TBI, including death (see Sherer et al. 2003). These differences may be the result of several factors, including the generally lower socioeconomic status of African-Americans compared to Caucasians, which can adversely affect increased risk exposure for certain types of injury; reduced prevention literacy; and reduced access to safer vehicles, communities, healthcare, and rehabilitation services. However, the literature on the effect of race and ethnicity on employment following TBI remains poorly understood. For example, Johnstone et al. (2003a) did not find a significant difference between the employment rates of African-Americans and Caucasians who had received state vocational rehabilitation services following TBI. However, like many state vocational rehabilitation samples, both client groups generally had low average IQs (e.g., 80–89) and achievement abilities, which may limit the generalizability of these findings.

Sherer and colleagues (2003) found that race was a significant predictor of productivity outcome at 1 year post-TBI with African-Americans and other racial minorities being about two times more likely to be defined as nonproductive than whites, but that the effect of race on employability was influenced substantially by other confounding factors, specifically pre-injury productivity, education level, and cause of injury (e.g., more frequent violent injuries). Using data from the National Traumatic Brain Injury Model Systems and after adjusting for pre-injury employment status, sex, Disability Rating Scale at discharge, marital status, cause of injury, and age, Aragano-Lasprilla et al. (2008) found that minorities were still 2.17 times more likely than whites to be unemployed at 1 year post-TBI. They suggested that lower levels of post-injury employment for African-Americans compared to whites is not the result of the TBI itself, but rather reflects the influence of race on job placement and a larger societal problem. Such societal and system-level variables are very infrequently addressed in studies of vocational outcome following TBI, and represent potential areas for more focused future study.

Conclusions

Most contemporary clinicians and researchers seem to agree that return to employment following TBI is likely the result of multiple, interacting factors, and that predicting RTW post-injury is at least highly difficult. There are substantial gaps in our knowledge about how different variables may directly or indirectly influence RTW and for whom, particularly at different developmental and post-injury stages. As a result, nearly all of our models demonstrate modest predictive accuracy at best. With the significant methodological limitations summarized above clearly in mind, we have listed below a set of provisional conclusions that may assist consumers, family members, and rehabilitation professionals in understanding some of the factors influencing the vocational outcomes of people who have sustained a moderate-to-severe TBI:

- Many persons with moderate-to-severe TBI can and do return to some level of community-based employment.
- Estimates of RTW rates vary markedly across studies, appear to hover around 30% on average over the years, and are likely related to a variety of rate-enhancing or rate-limiting sample characteristics.
- Though less frequently studied than demographic and injury severity factors, stable pre-injury employment, higher levels of education, and higher occupational categorization are generally associated with more favorable RTW rates.
- Younger TBI survivors (e.g., less than 40 years of age) tend to have higher employment rates post-injury than older survivors.
- While many studies document a significant negative relationship between injury severity and vocational outcome, level of disability or functioning at discharge is also an important predictor of post-injury employment status, despite being less studied.
- The relationship between injury severity and vocational outcome is likely mediated by level of cognitive impairment, particularly in executive functioning and self-awareness. Overall, regardless of injury severity, those with less impairment and less disability are more likely to enjoy favorable vocational outcomes.

- Most, but not necessarily all, gains with regard to employment occur within the first 1–2 years post-injury. If unemployed by that time post-injury, the likelihood of eventual RTW decreases. Overall, time since injury is probably less important than access to and participation in early and ongoing vocational rehabilitation services.
- Cognitive and neuropsychological impairments are more associated with unemployment post-injury than are physical limitations. Specifically, deficits in executive functioning (e.g., attention, planning and problem solving, conceptual functions, self-awareness, and self-regulation) increase the likelihood of unemployment.
- Post-injury psychological factors such as depression and substance abuse adversely affect RTW rates.
- Access to transportation and driving independence are associated with improved employment rates.
- Participation in rehabilitation programs, particularly those with specific vocational interventions and supports, increases RTW rates.
- Social, work, and physical environment and system level variables are significantly understudied (see last chapter on the need for integrated services by Schultz et al. 2010, this volume). Rehabilitation professionals should consider, in both research and practice, the new paradigm in rehabilitation research advanced by Gil et al. (2003) that focuses on the impact of these factors on vocational outcome.

References

Achte K, Jarho L, Kyykka T, Vesterinen E (1991) Paranoid disorders following ward brain damage: Preliminary report. Psychopathology 24:309–315

Aragano-Lasprilla JC, Ketchum JM, Williams K, Kreutzer JS, Marquez de la Plata CC, O'Neil-Pirozzi TM, Wehman P (2008) Racial differences in employment outcomes after traumatic brain injury. Arch Phys Med Rehabil 89:988–995

Arnett JJ (2000) Emerging adulthood: A theory of development from the late teens through the twenties. Am Psychol 55:469–480

Belanger HG, Curtiss G, Demery JA, Lebowitz BK, Vanderploeg RD (2005) Factors moderating neuropsychological outcomes following mild traumatic brain injury: A meta-analysis. J Int Neuropsychol Soc 11:215–227

Binder LM (1997) A review of mild head trauma. Part II: clinical implications. J Clin Exp Neuropsychol 19:432–457

Bombardier CH, Temkin NR, Machamer J, Dikmen SS (2003) The natural history of drinking and alcohol-related problems after traumatic brain injury. Arch Phys Med Rehabil 84:185–191

Bounds TA, Schoop L, Johnstone B, Unger C, Goldman H (2003) Gender differences in a sample of vocational rehabilitation clients with TBI. NeuroRehabilitation 18:189–196

Bushnik T, Englander J, Wright J (2008) The experience of fatigue in the first 2 years after moderate-to-severe traumatic brain injury: A preliminary report. J Head Trauma Rehabil 23:17–24

Cavallo MM, Kay T (2005) The family system. In: Silver JM, McAllister TW, Yudofsky SC (eds) Textbook of traumatic brain injury. American Psychiatric Publishing, Arlington, VA, pp 533–558

158 T.J. Guilmette and A.J. Giuliano

Corcoran C, McAllister TW, Malaspina D (2005) Psychotic disorders. In: Silver JM, McAllister TW, Yudofsky SC (eds) Textbook of traumatic brain injury. American Psychiatric Publishing, Arlington, VA, pp 213–229

Coronado VG, Johnson RL, Faul M, Kegler SR (2006) Incidence rates of hospitalization related to traumatic brain injury – 12 states. MMWR Morb Mortal Wkly Rep 55:201–204

Corrigan JD, Lamb-Hart GL, Rust E (1995) A programme of intervention for substance abuse following traumatic brain injury. Brain Inj 9:221–236

Corrigan JD, Lineberry LA, Komaroff E, Langlois JA, Selassie AW, Wood KD (2007) Employment after traumatic brain injury: Differences between men and women. Arch Phys Med Rehabil 88:1400–1409

Dawson DR, Schwartz ML, Winocur G, Stuss DT (2007) Return to productivity following traumatic brain injury: Cognitive, psychological, physical, spiritual, and environmental correlates. Disabil Rehabil 29:301–313

Department of Education National Institute on Disability and Rehabilitation Research (2008) 2008 Traumatic brain injury model systems of Care Presentation. Retrieved July 24, 2008, from http://www.tbindsc.org/2008%20TBIMS%20Slide%20Presentation.pdf

Devitt R, Colantonio A, Dawson D, Teare G, Ratcliff G, Chase S (2006) Prediction of long-term occupational performance outcomes for adults after moderate to severe traumatic injury. Disabil Rehabil 28:547–559

Dikmen SS, Machamer J, Winn HR, Temkin NR (1995) Neuropsychological outcome at one-year post head injury. Neuropsychology 9:80–90

Dikmen SS, Machamer JE, Powell JM, Temkin NR (2003) Outcome 3 to 5 years after moderate to severe traumatic brain injury. Arch Phys Med Rehabil 84:1449–1457

Flashman LA, McAllister TW (2002) Lack of awareness and its impact in traumatic brain injury. NeuroRehabilitation 17:285–296

Flashman LA, Amador X, McAllister TW (2005) Awareness of deficits. In: Silver JM, McAllister TW, Yudofsky SC (eds) Textbook of traumatic brain injury. American Psychiatric Publishing, Arlington, VA, pp 353–367

Fraser RT, Strand D, Johnson C (2010) Employment interventions for persons with mild cognitive disorders. In: Schultz IZ, Sally Rogers E (eds) Handbook of work accommodation and retention in mental health. Springer, New York

Frey LC (2003) Epidemiology of posttraumatic epilepsy: a critical review. Epilepsia 44:11–17

Gil CJ, Kewman DG, Grannon RW (2003) Transforming psychological practice and society: Policies that reflect this new paradigm. Am Psychol 58:305–312

Guilmette TJ (2005) Prediction of vocational functioning from neuropsychological data. In: Schultz IZ, Gatchel R (eds) Handbook of complex occupational disability claims: Early risk identification, intervention, and prevention. Springer, New York, pp 301–314

Hammond FM, Grattan KD, Sasser H, Corrigan JD, Rosenthal M, Bushnik T, Shull W (2004) Five years after traumatic brain injury: A study of individual outcomes and predictors of change in function. NeuroRehabilitation 19:25–35

Hanlon RE, Demery JA, Kuczen C, Kelly JP (2005) Effect of traumatic subarachnoid haemorrhage on neuropsychological profiles and vocational outcomes following moderate to severe traumatic brain injury. Brain Inj 19:257–262

Hibbard MR, Ashman TA, Spielman LA, Chun D, Charatz HJ, Melvin S (2004) Relationship between depression and psychosocial functioning after traumatic brain injury. Arch Phys Med Rehabil 85:S43–S53

Hibbard MR, Uysal S, Kepler K, Bogdany J, Silver J (1998) Axis I psychopathology in individuals with traumatic brain injury. J Head Trauma Rehabil 13:24–39

Hillier SL, Sharpe MH, Metzer J (1997) Outcomes 5 years post-traumatic brain injury (with further reference to neurophysical impairment and disability). Brain Inj 11:661–675

Hoofien D, Vakil E, Gilboa A, Danovick PJ, Barak O (2002) Comparison of the predictive power of SES variables, severity of injury, and age on long-term outcomes of brain injury: sample specific variables versus factors as predictors. Brain Inj 16:9–27

Iverson GL (2005) Outcome from mild traumatic brain injury. Curr Opin Psychiatry 18:301–317

Jennett B (1996) Clinical and pathological features of vegetative survival. In: Levin HS, Benton AL, Muizelaar JP, Eisenberg HM (eds) Catastrophic brain injury. Oxford University Press, New York

Johnstone B, Mount D, Gaines T, Goldfader P, Bounds T, Pitts O (2003a) Race differences in a sample of vocational clients with traumatic brain injury. Brain Inj 17:95–104

Johnstone B, Mount D, Schopp LH (2003b) Financial and vocational outcomes 1 year after traumatic brain injury. Arch Phys Med Rehabil 84:238–241

Johnstone B, Martin TA, Bounds TA, Brown E, Rupright J, Sherman A (2006) The impact of concomitant disabilities on employment outcomes for state vocational rehabilitation clients with traumatic brain injuries. J Vocat Rehabil 25:97–105

Kendall E, Muenchberger H, Gee T (2006) Vocational rehabilitation following traumatic brain injury: A qualitative synthesis of outcome studies. J Vocat Rehabil 25:149–160

Kessler RC, Berglund P, Demler O, Jin R, Merikangas KR, Walters EE (2005) Prevalence, severity, and comorbidity of 12-month DSM-IV disorders in the National Comorbidity Survey Replication. Arch Gen Psychiatry 62:617–627

Keyser-Marcus LA, Bricout JC, Wehman P, Campbell LR, Cifu DX, Englander J, High W, Zafonte RD (2002) Acute predictors of return to employment after traumatic brain injury: a longitudinal follow-up. Arch Phys Med Rehabil 83:635–641

Klonoff PS, Watt LM, Dawson LK, Henderson SW, Geherls JA, Wethe JV (2006) Psychosocial outcomes 1–7 years after comprehensive milieu-oriented neurorehabilitation: The role of pre-injury status. Brain Inj 20:601–612

Kneipp S, Rubin A (2007) Community re-entry issues and long-term care. In: Zasler ND, Katz DI, Zafonte RD (eds) Brain injury medicine. Demos Medical Publishing, New York, NY, pp 1085–1104

Koopnen S, Taiminen T, Portin R, Himanen L, Isoniemi H, Heinonen H et al (2002) Axis I and II psychiatric disorders after traumatic brain injury: a 30-year follow-up study. Am J Psychiatry 159:1315–1321

Kraus JF, Chu LD (2005) Epidemiology. In: Silver JM, McAllister TW, Yudofsky SC (eds) Textbook of traumatic brain injury. American Psychiatric Publishing, Arlington, VA, pp 3–26

Kraus JF, Fife D, Ramstein K, Conroy C, Cox P (1986) The relationship between family income to the incidence, external cause, and outcomes of serious brain injury, San Diego County, California. Am J Public Health 76:1345–1347

Kraus JF, Morgenstern H, Fife D et al (1989) Blood alcohol tests, prevalence of involvement, and outcomes following brain injury. Am J Public Health 79:294–299

Krauss JK, Jankovic J (2002) Head injury and posttraumatic movement disorders. Neurosurgery 50:927–939

Kreutzer JS, Marwitz JH, Walker W, Sander A, Sherer M, Bogner J, Fraser R, Bushnik T (2003) Moderating factors in return to work and job stability after traumatic brain injury. J Head Trauma Rehabil 18:128–138

Langlois JA, Rutland-Brown W, Thomas KE (2004) Traumatic brain injury in the United States: Emergency department visits, hospitalizations, and deaths. Centers for Disease Control and Prevention, National Center for Injury Prevention and Control, Atlanta, GA

Larrabee GJ (1997) Neuropsychological outcome, post concussion symptoms, and forensic considerations in mild closed head trauma. Semin Clin Neuropsychiatry 2:196–206

Leventhal T, Grabev JA, Brooks-Gunn J (2001) Adolescent transitions to young adulthood: Antecedents, correlates, and consequences of adult employment. J Res Adolesc 11:297–323

Machamer J, Temkin N, Fraser R, Doctor JN, Dikmen S (2005) Stability of employment after traumatic brain injury. J Int Neuropsychol Soc 11:807–816

Malec JF, Degiorgio L (2002) Characteristics of successful and unsuccessful completers of three postacute brain injury pathways. Arch Phys Med Rehabil 83:1759–1764

Martelli MF, Grayson RL, Zasler ND (1999) Posttraumatic headache: Neuropsychological and psychological effects and treatment implications. J Head Trauma Rehabil 14:49–69

Max JE, Lansing AE, Koele SL et al (2004) Attention deficit hyperactivity disorder in children and adolescents following traumatic brain injury. Dev Neuropsychol 25:159–177

McCrea M (2008) Mild traumatic brain injury and postconcussion syndrome: The new evidence base for diagnosis and treatment. Oxford University Press, New York

Mild Traumatic Brain Injury Committee, American Congress of Rehabilitation Medicine, Head Injury Interdisciplinary Special Interest Group (1993) Definition of mild traumatic brain injury. J Head Trauma Rehabil 8:86–87

Nakase-Richardson R, Yablon SA, Sherer M (2007) Prospective comparison of acute confusion severity with duration of post-traumatic amnesia in predicting employment outcome after traumatic brain injury. J Neurol Neurosurg Psychiatry 78:872–876

New Freedom Commission on Mental Health (2003) Achieving the Promise: Transforming Mental Health Care in America: Final report. DHHS Publication SMA-03-3832. Rockville, MD

Nightingale EJ, Soo CA, Tate RL (2007) A systematic review of early prognostic factors for return to work after traumatic brain injury. Brain Impair 8:101–142

Ohayon MM, Schatzberg AF (2003) Using chronic pain to predict depressive morbidity in the general population. Arch Gen Psychiatry 60:39–47

Olver JH, Ponsford JL, Curran CA (1996) Outcome following traumatic brain injury: a comparison between 2 and 5 years after injury. Brain Inj 10:841–848

O'Shanick GJ, O'Shanick AM (2005) Personality disorders. In: Silver JM, McAllister TW, Yudofsky SC (eds) Textbook of traumatic brain injury. American Psychiatric Publishing, Washington, DC, pp 245–258

Ownsworth T, McKenna K (2004) Investigation of factors related to employment outcome following traumatic brain injury: A critical review and conceptual model. Disabil Rehabil 26:765–784

Sbordone RJ, Guilmette TJ (1999) Ecological validity: Prediction of everyday and vocational functioning from neuropsychological test data. In: Sweet J (ed) Forensic neuropsychology: Fundamentals and practice. Swets & Zeitlinger Publishers, Lisse, The Netherlands, pp 227–254

Schretlen DJ, Shapiro AM (2003) A quantitative review of the effects of traumatic brain injury on cognitive functioning. Int Rev Psychiatry 15:341–349

Schultz IZ, Krupa T, Rogers ES (2010) Towards best practices in accommodating and retaining persons with mental health disabilities at work: Answered and unanswered questions. In: Schultz IZ, Sally Rogers E (eds) Handbook of work accommodation and retention in mental health. Springer, New York

Shames J, Treger I, Ring H, Giaquinto S (2007) Return to work following traumatic brain injury: Trends and challenges. Disabil Rehabil 29:1387–1395

Sherer M, Nick TG, Sander AM, Hart T, Hanks R, Rosenthal M et al (2003) Race and productivity outcome after traumatic brain injury: Influence of confounding factors. J Head Trauma Rehabil 18:408–424

Silver JM, Kramer R, Greenwald S, Weissman M (2001) The association between head injuries and psychiatric disorders: Findings from the New Haven NIMH Epidemiologic Catchment Area Study. Brain Inj 15:935–945

Smith GS, Kraus JF (1988) Alcohol and residential, recreational and occupational injuries: A review of the epidemiologic evidence. Am J Public Health 79:99–121

Sosin D, Sacks J, Smith S (1989) Head-injury associated deaths in the United States from 1979 to 1986. J Am Med Assoc 262:2251–2255

Targett PS, Wehman P (2010) Return to work after traumatic brain injury: a supported employment approach. In: Schultz IZ, Sally Rogers E (Eds) Handbook of work accommodation and retention in mental health. Springer, New York

Teasdale G, Jennett B (1974) Assessment of coma and impaired consciousness: A practical scale. Lancet 13:81–84

Thurman D, Guerrero J (1999) Trends in hospitalization associated with traumatic brain injury. J Am Med Assoc 282:954–957

Thurman DJ, Coronado V, Selassie A (2007) The epidemiology of TBI: Implications for public health. In: Zasler ND, Katz DI, Zafonte RD (eds) Brain injury medicine. Demos Medical Publishing, New York, NY, pp 45–55

Tucker GJ (2005) Seizures. In: Silver JM, McAllister TW, Yudofsky SC (eds) Textbook of traumatic brain injury. American Psychiatric Publishing, Washington, DC, pp 309–318

Walker WC, Marwitz JH, Kreutzer JS, Hart T, Novack TA (2006) Occupational categories and return to work after traumatic brain injury. Arch Phys Med Rehabil 87:1576–1582

Wehman P, Targett P, West M, Kregel J (2005) Productive work and employment for persons with traumatic brain injury: What have we learned after 20 years? J Head Trauma Rehabil 20:115–127

Zasler ND, Martelli MF, Nicholson K (2005) Chronic pain. In: Silver JM, McAllister TW, Yudofsky SC (eds) Textbook of traumatic brain injury. American Psychiatric Publishing, Washington, DC, pp 419–436

Chapter 9
Personality Disorders and Work

Susan L. Ettner

It is naive and foolish for you to believe that the world is populated entirely by people who are kind, well-intentioned, honest, and fair.

– Stuart Yudofsky (2005)

Background on Personality Disorders

The Diagnostic and Statistical Manual of Mental Disorders (DSM) divides psychiatric disorders into two categories, Axis I and Axis II disorders (American Psychiatric Association 2000). Axis I disorders are clinical disorders such as depression, anxiety, schizophrenia, and bipolar disorder. Axis II disorders are personality disorders (PDs), defined as "pervasive, inflexible, and enduring patterns of inner experiences and behavior that can lead to clinically significant distress or impairment in social, occupational, or other areas of functioning" and that reflect "inappropriate, ineffective, or painful ways of behaving and interacting" (American Psychiatric Association 2000). Although Axis II disorders were not officially recognized until about a quarter-century ago and continue to be less well known than their Axis I counterparts, they have become the focus of greater attention in recent years (Duggan et al. 2006; Gabbard 2005), due to their prevalence (Crawford et al. 2005; Grant et al. 2004; Lenzenweger 2006; Lenzenweger et al. 1997, 2007; Samuels et al. 2002) and associations with extensive use of psychiatric services (Bender et al. 2001; Soeteman et al. 2008) and a wide range of adverse outcomes (Bagge et al. 2004; Chen et al. 2004, 2006; Cramer et al. 2006; Daley et al. 2000; Goldstein et al. 2006; Johnson et al. 2000, 2006; King and Terrance 2006; Lavan and Johnson 2002; Leible and Snell 2004; Miller et al. 2007; Nestor 2002;

S.L. Ettner (✉)
Division of General Internal Medicine and Health Services Research, David Geffen School of Medicine at UCLA, 911 Broxton Plaza, Room 106, Box 951736, Los Angeles, CA, 90095-1736, USA
e-mail: settner@mednet.ucla.edu

I.Z. Schultz and E.S. Rogers (eds.), *Work Accommodation and Retention in Mental Health*, 163
DOI 10.1007/978-1-4419-0428-7_9, © Springer Science+Business Media, LLC 2011

Soeteman et al. 2005, 2008; Zaider et al. 2002). Soeteman and colleagues (2008) have argued that the burden of disease is even greater for those with PDs than for patients with HIV infection and type II diabetes.[1]

To be formally diagnosed with a PD, an individual must exhibit "an enduring pattern of inner experience and behavior that deviates markedly from the expectations of the individual's culture" (American Psychiatric Association 2000). This pattern must be demonstrated in at least two of the following: (1) cognition (i.e., ways of perceiving and interpreting oneself, other people, and events); (2) affectivity (i.e., the range, intensity, lability, and appropriateness of emotional response); (3) interpersonal functioning; and (4) impulse control. Furthermore, the pattern must be "inflexible and pervasive" across both personal and social situations; must lead to clinically significant distress or impairment in social, occupational, or other areas of functioning; must be stable and of long duration, with onset in adolescence or early adulthood; and cannot result from substance use, a medical condition, or another mental disorder (American Psychiatric Association 2000).

PDs are divided into three clusters (American Psychiatric Association 2000). Cluster A, which incorporates a cognitive dimension (Paris 2003), includes paranoid, schizoid, and schizotypal PDs. People with Cluster A disorders are often viewed as odd or eccentric, have abnormal cognitions or ideas, speak and act in strange ways, and have difficulty relating to others (American Psychiatric Association 2000). Cluster B, which corresponds to externalizing dimensions (Paris 2003), includes antisocial, borderline, histrionic, and narcissistic PDs. People with Cluster B disorders tend to act in dramatic, emotional, and erratic fashions, have difficulty with impulsive behavior, act out, and frequently violate social norms (American Psychiatric Association 2000). They are often hostile toward others and/ or self-abusive. Cluster C, which corresponds to internalizing dimensions (Paris 2003), includes avoidant, dependent, and obsessive-compulsive PDs. People with Cluster C disorders are often anxious, fearful, and excessively afraid of social interactions and of feeling out of control (American Psychiatric Association 2000). Table 9.1 provides a more detailed description of each PD.

Although competing theories persist regarding the relative roles of genetics and early childhood environment in determining PDs, most of the literature seems to suggest a confluence of "nature and nurture" (American Psychiatric Association 2000; Paris 1998; Yudofsky 2005). One point of agreement is that PDs tend to be difficult to treat and change, because they develop early in life and represent "lasting patterns of perceiving, relating to, and thinking about oneself and the environment." Thus PDs are generally assessed as lifetime diagnoses, despite evidence that change does occur over time for some individuals, especially in symptomatic behaviors (Lenzenweger 2006; McGlashan et al. 2005; Paris 2003; Sanislow and McGlashan 1998; Skodol et al. 2005).

The prevalence of PDs among community-dwelling adults depends on the population studied and the method of assessment, but has been estimated at about 9–16% in the United States (Crawford et al. 2005; Grant et al. 2004; Lenzenweger 2008; Reich et al. 1989; Samuels et al. 2002), 6.5% in Australia (Jackson and

---

[1] These findings were based on a treatment-seeking population, however, and may not generalize.

Table 9.1 Description of Axis II personality disorders (American Psychiatric Association 2000; The Merck Manuals Online Medical Library 2007)

Cluster A	*Individuals with this personality disorder...*
Paranoid	are extremely distrustful, suspicious of others, resentful of authority, vindictive, hypervigilant, blame-avoiding, excessively certain, and cognitively rigid. They often take legal action against others
Schizoid	are indifferent to social relationships, avoid interpersonal interactions, lack empathy, and have difficulty with emotional expression. Fantasizing is used as a defense mechanism
Schizotypal	demonstrate peculiarities of thinking, odd beliefs, and eccentricities of appearance, behavior, interpersonal style, and thought, such as belief in psychic phenomena or magical thinking
Cluster B	*Individuals with this personality disorder...*
Antisocial	have no superego or conscience and are unable to abide by societal rules. They lack regard for the moral or legal standards in the local culture, are willing to lie, have the potential for violence, and sometimes have a criminal record. They are reckless, irresponsible, and impulsive. They can be very charming but demonstrate a marked inability to get along with other people. They are aggressive and remorseless and enjoy humiliating and demeaning others
Narcissistic	demonstrate grandiosity, exhibitionism, and unrealistic self-evaluation. They have a need for constant approval/admiration and are preoccupied with success and hypersensitive to criticism. They exhibit a marked lack of empathy and are unable to see the viewpoints of others. Due to their sense of entitlement and envious nature, they can be very exploitative, although not with the same degree of deliberate intent that characterizes those with ASPD
Borderline	exhibit substantial emotional and interpersonal instability and experience rapid mood swings, including inappropriate, intense anger. They have a frantic fear of abandonment and react strongly to separations. Their relationships are often stormy, with verbal outbursts, and they rapidly switch between idealizing and devaluing others. They have an unstable self-image and lack identity, resulting in sudden changes in opinions and plans about career, sexual identity, values, and friends. They are inconsistent and impulsive, lack clear goals and direction, and perform poorly in unstructured work or school situations. They tend to undermine themselves at the moment goals are about to be realized, and are self-destructive, engaging in acts such as suicide attempts or self-mutilation

(continued)

Table 9.1 (continued)

Histrionic	demonstrate exaggerated displays of emotional reactions in everyday behavior. They are overly dramatic, attention-seeking, and vain. They are demanding and manipulative, throw frequent tantrums and need continual stimulation. They are sexually provocative and need to be the center of attention
Cluster C	*Individuals with this personality disorder…*
Obsessive-compulsive	become preoccupied with uncontrollable patterns of thought and action, to the extent that these patterns may interfere with their occupational and social functioning. They are perfectionistic, inflexible, unwilling to compromise and have a need for control. They tend to focus on minute detail and experience difficulty completing tasks. (Note that OCPD differs from obsessive-compulsive disorder, which is an Axis I diagnosis.)
Avoidant	experience marked social inhibition, feelings of inadequacy, apprehension, and mistrust, and extreme sensitivity to criticism. They are introverted, timid, and awkward. They wish to become involved with others but simultaneously fear such involvement. They are terrified by the thought of being embarrassed in front of others and avoid situations that inflict social discomfort on them, which in many cases leads to social withdrawal
Dependent	have a pervasive and excessive need to be taken care of by others. They fear separation and engage in clinging and submissive behavior, often denigrating themselves. They have a marked lack of decisiveness and self-confidence, to the point where they are unable to make any decisions or take an independent stand on their own

Burgess 2000), and 4.4% in Great Britain (Coid et al. 2006).[2,3] PDs are more commonly diagnosed among adults under 65 (Grant et al. 2004; Jackson and Burgess 2000; Sanislow and McGlashan 1998), suggesting that they disproportionately affect the individuals most likely to participate in the labor force. Evidence is mixed on whether and how prevalence rates vary by gender (Grant et al. 2004;

---

[2] As noted by Jackson and Burgess (2000), prevalence estimates for certain PDs could be conservative, due to social desirability bias; for example, an individual might be reluctant to endorse items used to assess antisocial PD, such as "I will lie or con someone if it serves my purpose."

[3] The prevalence is significantly higher when "personality disorder not otherwise specified" is included.

Table 9.2 Estimated prevalence of Axis II personality disorders among working-age, community-dwelling Americans

Personality disorder	Prevalence (entire sample)	Prevalence (women only)	Prevalence (men only)
Paranoid	4.5% (4.2%, 4.8%)	5.2% (4.8%, 5.7%)	3.7% (3.3%, 4.2%)
Schizoid	3.2% (3.0%, 3.5%)	3.4% (3.1%, 3.8%)	3.1% (2.7%, 3.5%)
Schizotypal	4.2% (3.9%, 4.5%)	3.9% (3.6%, 4.3%)	4.4% (4.0%, 4.9%)
Antisocial	3.8% (3.5%, 4.1%)	2.0% (1.7%, 2.3%)	5.8% (5.3%, 6.4%)
Narcissistic	6.2% (5.8%, 6.5%)	4.9% (4.5%, 5.3%)	7.6% (7.0%, 8.2%)
Borderline	6.1% (5.8%, 6.5%)	6.7% (6.2%, 7.2%)	5.5% (5.0%, 6.0%)
Histrionic	1.6% (1.4%, 1.8%)	1.7% (1.4%, 1.9%)	1.6% (1.3%, 1.9%)
Obsessive-compulsive	8.8% (8.4%, 9.3%)	9.0% (8.4%, 9.6%)	8.6% (8.0%, 9.3%)
Avoidant	2.5% (2.3%, 2.8%)	3.0% (2.6%, 3.3%)	2.0% (1.7%, 2.4%)
Dependent	0.4% (0.3%, 0.5%)	0.5% (0.4%, 0.7%)	0.25% (0.1%, 0.4%)

Source: 2000–2001 and 2004–2005 National Epidemiological Surveys on Alcohol and Related Conditions.

Note: The working-age population is defined as 25- to 64-year-olds. Frequencies are weighted and 95% confidence intervals are shown in parentheses. Overall, 22.5% of this population has at least one Axis II diagnosis. About 10% of the population, or almost half of individuals with any Axis II diagnosis, has more than one

Jackson and Burgess 2000; Lenzenweger 2008), but they do vary substantially across PDs. Based on data from the National Epidemiologic Survey of Alcohol and Related Conditions (NESARC), obsessive-compulsive PD is the most common, with almost 9% of working-age men and women classified with this disorder (Table 9.2).

Rates of Axis I mental and substance disorders are much higher among individuals with PDs (Franken and Hendriks 2000; Grant et al. 2005; Jackson and Burgess 2002; Lenzenweger 2006; Trull et al. 2004). About two-thirds of individuals with PDs have at least one Axis I disorder and almost one-third have three or more Axis I disorders (Lenzenweger 2006). Comorbidity is also high among PDs (Grant et al. 2005; Lenzenweger 2006); in the NESARC data, almost half of individuals with any PD have more than one. The high rates of comorbidity among Axis II disorders have been argued to be due in part to limited discriminant validity in the criteria used for assessment (Blais and Norman 1997; Blashfield and Breen 1989; Marinangeli et al. 2000). The DSM currently relies on a categorical system (yes/no) for defining whether or not an individual meets sufficient criteria for being assigned a PD diagnosis. Given the high correlation among PDs and the lack of clear differentiation among certain PDs, particularly within the same cluster (Atre-Vaidya and Hussain 1999; Blais and Norman 1997; Gunderson and Ronningstam 2001), calls have been made for future revisions of the DSM to use a dimensional rather than categorical classification for PDs (Marinangeli et al. 2000). Skodol et al. (2005) argue that PDs should be reconceptualized as hybrids of stable personality traits and intermittently expressed symptomatic behaviors. Bernstein and colleagues (2007) found that most experts believed that the DSM's categorical system of PD diagnosis should be replaced with a mixed system of categories and dimensions.

Although Axis II diagnoses have been shown to exhibit reliability comparable to that found for Axis I disorders (Zanarini et al. 2000), even PD traits that do not meet the diagnostic threshold have been shown to be associated with impaired functioning (Skodol et al. 2005). In some cases it may be more useful to consider the individual PD criteria rather than diagnosis per se, which relies on a potentially arbitrary threshold. Skodol and colleagues (2005) showed that while functional impairment was related much more strongly to PD diagnoses than to general personality functioning scores, the number of criteria met for each PD predicted functioning even better. Similarly, Cox and colleagues (2007) provide evidence supporting the current hierarchical organization of PD diagnoses, but found that dimensional models had superior performance. Westen et al. (2006) go further by proposing to replace the current criteria-based system with prototype diagnosis, which entails assessing the extent to which the patient resembles a description of a prototypical individual with the PD in question.

Given this literature, it is possible that an exclusive focus on PD diagnoses may ultimately prove less useful in understanding workforce behavior than a more comprehensive evaluation of maladaptive personality traits. In the meantime, however, this review focuses on formal Axis II diagnoses based on the current DSM methodology. DSM diagnoses may be the most useful markers for current formulation of labor policy, given their legal implications (e.g., for defining individuals whose psychiatric treatment is covered under mental health parity laws, or populations covered by the Americans with Disabilities Act).

The remainder of this chapter is organized as follows. First, a description of some of the behaviors associated with PDs that may lead to dysfunctional behavior on the job is provided. Next, the chapter reviews empirical studies of associations of PDs with work outcomes. Then a brief summary of what is currently known about treatment options and their effectiveness and cost-effectiveness is presented. The chapter concludes with a discussion of the implications of PDs for workplace policies and next steps for research in this area.

Personality Disorders in the Workplace

In contrast to Axis I disorders, which typically cause the greatest distress to the affected individuals themselves, PDs are highly problematic to the family members, friends, partners, and coworkers of the disordered individuals (Jackson and Burgess 2000; Miller et al. 2007). Although each PD is associated with different symptoms and behaviors, PDs are generally characterized by maladaptive coping mechanisms that can affect interpersonal relationships, including work relationships (The Merck Manuals Online Medical Library 2007). For example, although all individuals sometimes "project" their own unwanted feelings, desires, and thoughts onto others, personality-disordered individuals tend to do this to a greater extent. A related behavior is that they may engage in externalization of blame, in which they fault others for problems they themselves have caused (Campbell et al. 2005). Some personality-disordered individuals have difficulty praising the performance of subordinates, or are controlling and manipulative, or deceptive and vengeful, leading to interpersonal problems on the job.

Even individuals with PDs associated with abusive behavior toward others often have the capacity for great charm. Their charisma and ability to project a likeable façade when it suits their purposes often leads to divisiveness in the workplace, with coworkers having either strongly positive or strongly negative views of the disordered individual (Babiak 1995; Babiak and Hare 2006; Sankowsky 1995). Employees dealing with supervisors or coworkers with PDs often become irritated, frustrated, angry, resentful, or even depressed (The Merck Manuals Online Medical Library 2007; Trimpey and Davidson 1994). Productivity declines and turnover tends to be high (Babiak 1995; Kets de Vries and Miller 1985). The pathological effects of an individual's PD on his or her coworkers can be exacerbated by the fact that healthy individuals sometimes erroneously assume the responsibility for conflict in their relationships with disordered coworkers (Yudofsky 2005) or for failing to meet the unrealistic expectations set by disordered leaders (Judge et al. 2006; Sankowsky 1995), undermining their self-esteem and leading to adverse effects on their own mental health and workplace morale and productivity.

As demonstrated by Table 9.1, PDs vary substantially in terms of the symptoms and behaviors that characterize each disorder. Thus PDs are likely to be heterogeneous in the impact they have on workplace functioning. A study of countertransference among psychiatrists and psychologists (Betan et al. 2005) sheds some light on the problems individuals with different types of PDs might have with their interpersonal interactions in other contexts. While mental health professionals reported feeling criticized and mistreated in their interactions with individuals with Cluster A PDs (e.g., paranoid PDs), they associated parental and protective feelings with treating individuals with Cluster C PDs (e.g., depressive or avoidant PDs). Cluster B PDs (e.g., borderline and narcissistic PDs) were correlated with a wide range of negative feelings. The authors then conducted a more detailed analysis of narcissistic PD, concluding that clinicians feel anger, resentment, and dread working with patients with this disorder; feel devalued and criticized by the patient; and find themselves distracted, avoidant, and wishing to terminate the treatment (Betan et al. 2005). The suggestion that even mental health professionals trained in countertransference issues are unable to avoid experiencing negative emotions towards patients with certain PDs is indicative of the poor interpersonal functioning of such individuals (Miller et al. 2007).

The management literature has focused predominantly on Cluster B disorders, those with the greatest potential for causing serious problems in the workplace. For example, the prototypical individual with antisocial PD is deceitful, takes advantage of others, exhibits sadistic tendencies and a reckless disregard for others, and has little or no conscience or capacity for remorse (Westen et al. 2006). Numerous researchers have argued that while assignment of a formal diagnosis of antisocial PD often relies heavily on socially deviant behaviors, the same constellation of personality traits is found in mainstream society (Babiak 1995; Babiak and Hare 2006; Blackburn 1988; Board and Fritzon 2005; Lilienfeld 1998; Millon 1981), in its severe form sometimes called "psychopathy." As employees, individuals with antisocial PD are likely to break rules, be unreliable and irresponsible, be pathological liars, and sabotage projects or commit corporate fraud (Babiak 2008a, b). They use the company "grapevine" to disseminate erroneous information to denigrate the competence and loyalty of their competitors and enhance their own reputation, and they create conflicts between other employees, often by lying to each individual about what the other one said, to promote their advancement goals and distract others from noticing their own behaviors (Babiak and Hare 2006). In some cases the individual may even have the potential for violent behavior (Babiak 2008a, b).

Narcissistic PD has been described as a "white collar version" of antisocial PD (Gunderson and Ronningstam 2001), with narcissists demonstrating greater grandiosity and being more likely to engage in passive (rather than deliberate) exploitation through their inability to identify with the feelings and needs of others (Gunderson et al. 1991) and strong sense of entitlement. Whereas antisocial PD is common among incarcerated individuals (Babiak 1995; Hill et al. 2006), narcissistic PD is prevalent among executives (Kets de Vries and Miller 1985; Khoo and Burch 2007). Individuals with narcissistic PD use others to maintain their large egos (and shore up their low underlying self-esteem); to do this, they use whatever behavior may be necessary to extract admiration, attention, or even fear. As noted by Campbell and colleagues (Campbell et al. 2002), narcissists can alternately be "charming, exciting, deceptive, controlling, or nasty," depending on what is needed to achieve their goals. Judge et al. (2006) suggest that highly narcissistic individuals make particularly undesirable employees in situations in which a realistic self-assessment or team cooperation are required. They expect only positive social feedback and will demean or retaliate against anybody who disagrees with or criticizes them (Campbell et al. 2005). They often cause great distress to their coworkers, employees, and even employers, through their controlling and abusive behaviors, propensity to take credit for the achievements of others, interpersonal contempt, aggressiveness, and "petulant anger" (Brown 2002; Campbell et al. 2002, 2005; Gunderson et al. 1991). Kets de Vries and Miller (1985) argue that narcissistic leaders fail because they undertake ambitious projects without making realistic assessments of the resources available or taking the opinions of others into account; after problems arise, they do not admit to mistakes, but place the blame on others.

Individuals with borderline PD have been described as "entitled but needy" (Gunderson et al. 1991), and are widely viewed as manipulative (Gabbard 2005). In the workplace, the lack of a firm self-identity may manifest itself as passionate but ephemeral interests and an inability to commit to a career path. Individuals with borderline PD may change jobs frequently, or sabotage their chances for success by failing to meet obligations or deadlines. Betan et al. (2005) explain that individuals sensitive to rejection, such as those with borderline PD, may confirm and reinforce their internal working models of relationships by behaving in needy or angry ways or distancing themselves from others, hence soliciting the rejection they fear but expect. Manifestations of rage and hostility may serve as impediments to job success. A common phenomenon is "splitting," in which the individual alternately idealizes others as "all good" 1 min and then devalues them as "all bad" the next. This "black and white thinking" may confuse, anger, and alienate coworkers who perceive themselves to go from "doing nothing wrong" 1 min to "doing nothing right" the next. Individuals with borderline PD have a striking ability to present themselves publicly as being rational and calm even as they rage at those with whom they work closely in private settings (Kupper 2007). As with other Cluster B disorders, individuals with borderline PD have a strong need for control.

Individuals with histrionic PD are overly dramatic and superficial. They may alienate coworkers through their manipulativeness and need for attention. Sexually provocative behavior on the job, such as inappropriate attire in the workplace or seductive behavior toward coworkers or supervisors, may cause problems. Hypochondriasis often associated with histrionic PD (Demopulos et al. 1996; The Merck Manuals Online Medical Library 2007) may lead to work absenteeism.

Despite the focus on Cluster B in the industrial literature, potential workplace dysfunction is not limited to these disorders. Any PD might lead to difficulties at work, e.g., individuals with dependent PD may be unwilling or unable to make decisions on their own, or those with avoidant disorder may be reluctant to present their work to others for fear of being criticized. Yudofsky (2005) suggests that people with obsessive-compulsive PD make poor bosses because "[t]heir indecisiveness, pessimism, inflexibility, and lack of vision hamper creative development and expansion of the company, and their controlling, sour, non-nurturing, and ungrateful nature is destructive to teamwork and employee morale."

Cluster A disorders can also wreak workplace havoc. A vivid illustration is provided by the following description of a woman with paranoid PD, paraphrased from one of Yudofsky's composite case studies[4] (2005), based on patients and family members he treated as a psychiatry professor at Baylor College. The case involves a very bright young physics instructor at a university who asked a successful older female professor to be her mentor. The professor was the chair of a different department and known for being an outstanding mentor. The instructor explained that she did not have a suitable female role model in her own department. The professor agreed to help the young instructor, despite a frank conversation the professor had with the chair of the physics department, in which the chair expressed concerns about the instructor's impatience with less intelligent colleagues and competitiveness with other bright women. The professor invested a considerable amount of time helping her mentee get promoted to assistant professor and subsequently facilitating her mentee's advancement towards a tenured position.

Five years later, the senior professor had a discussion with her mentee in which the professor voiced her concern that the mentee would not attain tenure 2 years hence, because the mentee was spending too much time on committees and other professional engagements and not enough time publishing. The mentee became enraged

---

[4]Composite case studies are amalgams of actual cases, with identifying details changed to protect confidentiality. For example, the gender, occupation, geographic location, etc. of individuals described might be different, or case studies of several individuals could have been combined into a single story.

and abruptly ended the meeting. After the senior professor subsequently discussed her concerns with the chair of the physics department, the mentee cut off all ties with her mentor, accusing her of harassment and slander. One year later, when the senior professor was on the verge of being offered the job of university president, the former mentee came forward and accused the professor of having made repeated sexual advances during the course of their mentoring relationship. The mentee claimed that the senior professor had retaliated against her when the professor's advances were rejected, and she produced false evidence to support her case. The senior professor was eventually exonerated after her lawyers forced the university to produce all of the emails sent by the former mentee during the period in question. Further investigation also revealed that the former mentee had a history of making false accusations and engaging in vengeful actions in both her professional and personal relationships. Following the rejection of her sexual harassment complaint against her former mentor, the mentee filed a $10 million lawsuit against the university, which was rejected for lack of merit. She quit her job right before she came up for tenure and was not offered a job by other universities.

From this anecdote, it is clear that unchecked PDs have the potential for causing substantial disruption in the workplace. Despite their negative influence on the work life of others, however, it is not necessarily the case that individuals suffer adverse career events themselves as a result of their PD. Board and Fritzon (2005) found histrionic PD traits (e.g., superficial charm, insincerity, egocentricity, and manipulativeness) to be significantly higher among senior business managers than among forensic and psychiatric patients, suggesting that these traits may be rewarded in the workplace. Obsessive-compulsive PD is associated with perfectionism and excessive devotion to work (Board and Fritzon 2005), traits that may also be valued in many occupations. Individuals with antisocial PD typically come across as charming and intelligent and may even be perceived as more likeable than nondisordered individuals (Babiak 1995). Babiak and Hare (2006) provide numerous real-world examples of "corporate psychopaths" who achieved great success in organizations through their ability to deceive and manipulate others and were never truly punished for their misdeeds. Narcissism has also been identified as a trait of many successful, ambitious people (Kets de Vries and Miller 1985; Khoo and Burch 2007; Yudofsky 2005). Given the competitive ambition associated with narcissistic PD, it is quite conceivable that this PD may drive individuals to achieve greater, rather than lesser, career success, at least by some measures of accomplishment. This possibility is illustrated by another of Yudofsky's composite case studies.

The case involves an attractive young man from a poor farming background who managed to charm a Chicago executive into becoming his mentor. The executive assisted his mentee in obtaining a college scholarship and hired him to work in his company's financial department after graduation. The mentee went on to marry the daughter of the executive's business partner and rose rapidly through the executive ranks of the company. He repaid the generosity of his mentor and his father-in-law by blaming them for unpopular management decisions he himself made, and brought the firm to the brink of financial disaster by falsifying financial documents to make the divisions he ran appear far more profitable than they actually were. (He was later found to have a history of accepting bribes and kickbacks, and of having affairs with company employees he directly supervised.) He was ultimately terminated from his job but he was never prosecuted, and his unethical behavior had allowed him to accumulate great wealth at the expense of the company. This individual was subsequently elected to the U. S. Congress and was extremely popular. He easily won re-elections, even after strong evidence was presented to the public that he had misused government funds and accepted bribes from constituents. Ultimately, he became one of the richest men in the state. By many standard measures of work outcomes, therefore, this individual did very well for himself. This example illustrates the point that despite the harm individuals with PDs often cause to others, whether their own career success is hindered or facilitated is an empirical question. The next section describes what is known about the influence of PDs on labor market outcomes.

Studies of the Relationship Between PDs and Labor Market Outcomes

PDs and general functioning. A number of previous studies have documented associations between PDs and general functioning measures, including social and role functioning. Jackson and Burgess (2002) conclude that having any PD is significantly positively associated with several measures of disability, including role functioning, even after controlling for Axis I disorders and physical conditions. In a follow-up study, Jackson and Burgess (2004) find that among PDs, anankastic (corresponding to obsessive-compulsive) PD has the smallest association with disability. Grant et al. (2004) find that all PDs except obsessive-compulsive and histrionic are significantly associated with worse social and emotional functioning, even after adjusting for Axis I disorders and all other PDs. In contrast, Lenzenweger et al. (2006) argue that comorbid Axis I disorders largely account for the associations of individual PDs with functional impairment, although their study also finds significant associations of having any PD with instrumental role functioning. Skodol et al. (2005) suggest that psychosocial functioning improves relatively little over time among individuals with PDs.

Despite the evidence that PDs are associated with functional disability even after adjusting for Axis I disorders and that these associations largely persist over time, less is known about how PDs influence specific measures of labor market performance. Existing evidence falls into two categories: industrial or clinical case studies, primarily of high-level managers and executives with Cluster B diagnoses; and larger, quantitative studies that seek to establish generalizable associations of PDs with labor market outcomes. Each of these bodies of work is described in turn.

Industrial case studies. Although the management literature tends to focus on the role of PD traits (e.g., "toxic leadership" or "antisocial behaviors") (Giacalone and Greenberg 1997) rather than formal DSM diagnoses, a number of case studies have documented the deleterious impact individuals with PDs can have on coworkers. Goldman (2006a) describes a senior manager with borderline PD who turned around the fortunes of a fashion house, but also engaged in self-destructive behaviors and was erratic toward clients and highly emotionally abusive and derogatory towards coworkers. As a result, worker productivity and job satisfaction declined and employee turnover, absenteeism, lack of work motivation, and intent to quit increased. Over time, the situation deteriorated rapidly. Major accounts were lost and numerous complaints and formal grievances were filed against the manager. Eventually, however, a consultant hired by the company (Goldman) was able to help the manager modify his behavior sufficiently to maintain a productive employment relationship with the company (see discussion below).

Trimpey and Davidson (1994) describe a drug and alcohol rehabilitation unit director with borderline PD. She polarized her employees by engaging in "splitting behavior," in which she favored employees who socialized with her after work while devaluing those who did not. Her inconsistency in allowing her favorite patients to break rules had a deleterious effect on the functioning of the unit. She also initially idealized and attempted to seduce one of her mentors, but later filed a grievance against him when he failed to support her attempt to get him to be her supervisor. Her employers were unable to improve her performance. Trimpey and Davidson (1994) also describe a head nurse in an ICU who had obsessive-compulsive PD. This individual also caused significant workplace dysfunction, primarily through her controlling behaviors combined with her focus on minutiae, thereby frustrating her supervisees and leading to her inability to make decisions.

Sankowsky (1995) describes a paradigm for understanding how narcissistic leaders acquire power over their followers through their "symbolic status" (i.e., the tendency for followers to identify with their leaders as parental figures and seek their approval, even at the expense of their own psychological well-being). He names a number of well-known executives such as Michael Eisner as alleged examples of this phenomenon and discusses the deleterious effects their narcissism had on employees and the company itself. Based on a review of the theoretical and empirical literature on narcissistic leaders, Rosenthal and Pittinsky (2006) suggest that while narcissists make good first impressions and receive higher initial leadership ratings, over time they alienate others and are eventually doomed to poor performance due to their character flaws. They describe a field study of the computer and software industry in which companies with more narcissistic CEOs demonstrated extreme variability in performance as well as greater volatility in asset-related performance. The authors conclude that narcissistic individuals are better matches for occupations such as sales (where charisma and extraversion are important) and science (where self-absorption and grandiosity are key) than for jobs that require building trust and strong interpersonal relationships. They also suggest that narcissistic individuals may provide better leadership if their self-interest is better-aligned with the interests of the company.

Goldman (2006b) provides case studies of a surgeon whose narcissistic PD may have contributed to several failed surgeries and a physically aggressive, emotionally abusive factory manager whose antisocial PD eventually led to the organization itself taking on antisocial traits, i.e., employees mimicking his traits by being less cooperative and fighting with one another. Counseling was successful in improving the surgeon's behaviors and interpersonal relationships over time. In contrast, the manager's superiors eventually saw through the deception and forced him out of the organization. Despite the manager's abusiveness, a period of "organizational grief" was experienced, attributed to the fact that organizational systems resist change and "defend their pathologies," even at the expense of sustaining dysfunction. Goldman cautions that "just one failure to timely assess may yield dramatic interpersonal and systemic repercussions including sabotage, plunging motivation and productivity, increased turnover, and a high incidence of internal grievances, formal complaints, and litigation."

Babiak (1995) describes a case study of a young manager with "psychopathy" in an expanding electronics firm. The manager was a chronic prevaricator (including falsifying information on his résumé and plagiarizing material for reports), threw tantrums and bullied coworkers, and used company time and money to start his own business. Despite the fact that the manager routinely underperformed, he was successful because he was able to build one-on-one relationships with members of the company who were either powerful (e.g., the president and vice-president were charmed by him and protected him even in the face of evidence that he was problematic) or useful (e.g., key staff members with access to resources or information). Once the manager had gotten what he needed from a particular coworker, he would cease his relationship with the person and lead others to question the person's competence and loyalty in order to ruin his or her credibility. The manager avoided group meetings and created conflicts between coworkers so that they would not enter into discussions about him that might lead to his discovery. The manager ultimately was promoted and became even more problematic for his coworkers.

Larger-scale empirical studies. Although these case studies highlight the problems associated with PDs in the workplace and have the advantage of relying on clinical diagnoses rather than standardized assessments, they do not easily lend themselves to generalizable conclusions about the associations of PDs with labor market outcomes. A small literature has attempted to analyze data on larger samples to estimate these associations and document the workplace costs of PDs. The first few studies described here were conducted with clinical samples, so results should be interpreted with caution. Patients selected because they are in treatment for Axis I or II disorders may have different severity, prognosis, and associations with work outcomes than a community sample of individuals with PDs. The remaining studies were conducted with community samples, although each has its own unique limitations.

Using data on 88 patients with borderline PD from a clinical trial in the Netherlands, van Asselt et al. (2007) found that 42% of the total costs associated with this PD were productivity losses, primarily due to work disability. Rytsala and colleagues (2005) examined the relationships between specific classes of PDs and work disability among 269 patients diagnosed with major depressive disorder in the city of Vantaa, Finland. Severity of depression was found to be the dominant factor explaining work disability, but PDs were also independently associated with higher levels of work disability. Using data on 668 patients from four clinical sites partici- pating in the Longitudinal Personality Disorders Study, Skodol et al. (2002) show that those with schizotypal or borderline PDs were roughly one-third as likely to be employed and about three times more likely to be work-disabled as patients with major depressive disorder; however, no significant associations were found with obsessive-compulsive and avoidant PDs.

Soeteman et al. (2008) examined work productivity among 1,420 subjects with PDs who were recruited from six mental health institutes in the Netherlands. On aver- age, employed individuals with PDs had about 28 days of work absence and 20 days of reduced work efficiency annually. However, these numbers were not compared with similar measures for individuals without PDs, making interpretation difficult. Miller et al. (2007) found significant associations of narcissistic PD with a global measure of work impairment among two small convenience samples (one based entirely and the other based partly on psychiatric patients), but did not examine other PDs. Gunderson and Hourani (2003) used data from 1980 to 1992 to follow the careers of Navy enlisted personnel hospitalized for PDs, finding that they achieved lower pay grades and had fewer promotions and more demotions at the time of discharge than individuals in a comparison group. Those with PDs were less likely to complete their obligated service, less likely to be recommended for re-enlistment, and more likely to have had unauthorized absences or desertions, despite their shorter length of service. The authors concluded that PDs are an important contributor to premature discharges from the U.S. Navy and recommended screening applicants for psychopathology in order to reduce attrition rates.

Campbell and colleagues used lab experiments with human subjects to explore the relationship of narcissism with over-harvesting of renewable resources (2005), concluding that while narcissists did better than their competitors in a mock busi- ness situation, they did not do better overall. In other words, while the narcissistic individuals got a "bigger slice of the pie" as a result of their greater willingness to exploit common resources, the pie itself shrank as a result of their behavior. Thus business decisions driven exclusively by self-interest have particularly marked negative repercussions in situations involving social externalities, i.e., costs or ben- efits imposed on society that are not "internalized" by the decision-maker.

Based on a small community sample from Iowa City, Reich et al. (1989) found a marginally significant trend for a higher percentage of individuals with PDs to be unemployed longer than 6 months during the previous 5 years. Using data on full- time workers in Australia, Lim et al. (2000) found that PDs without comorbidities were associated with 0.9 monthly work-loss days and 2.4 monthly work-cutback days; PDs in combination with Axis I disorders also had a large effect. Notably, the total economic costs associated with pure PDs were higher than for affective, anxiety, or substance disorders, because of the high prevalence of PDs combined with their substantial effects on workplace productivity. Caution should be used in extrapolating these findings to the general population, however, since individuals with psychiatric disorders who are able to achieve full-time employment are likely to differ from those who cannot.

Ettner et al. (2008) used 2000–2001 NESARC data on non-institutionalized, working-age Americans to examine the relationship between labor market outcomes and seven different PDs. Even after adjusting for Axis I disorders and sociodemographic characteristics, having a PD was associated with a higher risk of chronic unemployment (and among women only, of being fired or laid off) and of reporting having had trouble with a boss or coworker in the past year. The magnitudes of the associations were large, e.g., having a PD was associated with roughly a doubling of the probability of reporting having had trouble with a boss or coworker during the past year. When looking more specifically at associations with individual classes of PDs, obsessive-compulsive, paranoid, and anti-social PDs showed the most pervasive patterns of associations with adverse work outcomes. However, borderline, narcissistic, and schizotypal PDs were not measured in this dataset.

Based on the sparse but consistent evidence, it seems clear that PDs have serious implications for workplace functioning even after accounting for comorbid Axis I disorders; although the nature and severity of these consequences, as well as the extent to which they are borne by the individual with the PD versus their coworkers and employers, vary across PDs. Just as there is heterogeneity across Axis I disorders in their implications for labor market outcomes, Axis II disorders also demonstrate substantial variability. One difference, however, is that PDs tend to be less mutable than Axis I disorders, most of which can be successfully managed, if not cured, within a relatively short time frame. The next section describes what is known about treatment of these disorders.

Treatment

The extent to which PDs are treatable is subject to debate and depends critically on the type and severity of the disorder (Gunderson and Gabbard 2000; Perry et al. 1999; Sanislow and McGlashan 1998; Stone 2006), as well as on whether the goal is amelioration of the symptoms and/or problematic behaviors versus elimination of the underlying pathology. For example, cautious optimism has been expressed in the literature regarding the prospect of successful treatment of borderline PD, although modification of the disorder may be limited to symptoms rather than elimination of internal distress (Bartak et al. 2007; Binks et al. 2006b). In contrast, many forms of antisocial PD are still considered virtually untreatable (Gabbard 2000; Gunderson and Gabbard 2000; Stone 2006).

Pharmacotherapy generally has not been shown to be effective for PDs (Binks et al. 2006a; The Merck Manuals Online Medical Library 2007) and is used primarily to treat comorbid Axis I conditions (Triebwasser and Siever 2006) or particular symptoms such as anxiety or psychotic-spectrum symptoms (Sanislow and McGlashan 1998). Instead, PDs are typically treated through psychotherapy (The Merck Manuals Online Medical Library 2007). Perry and colleagues (1999) concluded that psychotherapy treatment was associated with seven times faster recovery from PDs than what would be predicted based on the natural disease course. Other reviews also suggest that psychotherapy is effective for certain PDs (Bartak et al. 2007; Gabbard 2000). The evidence base is quite limited, however. Studies are sparse, with evidence of effectiveness limited to borderline PD, avoidant PD, and antisocial PD with concomitant depression (Bartak et al. 2007; Gabbard 2000). Long term follow-up is typically lacking. More generally, the standards applied to the research are lower than those used in evidence-based medicine (Duggan et al. 2006; Perry et al. 1999; Sanislow and McGlashan 1998), and the generalizability of the findings is limited by self-selection into, and differential attrition from, treatment (Perry et al. 1999).

Research on the cost-effectiveness of psychotherapy for PDs suffers from the same criticisms as the treatment effectiveness literature. Studies focus even more narrowly on borderline PD and the evidence is not (yet) sufficient to conclude that using psychotherapy to treat PDs represents a good investment of resources (Bartak et al. 2007; Brazier et al. 2006; Gabbard 2000). Nonetheless, it has been argued that psychotherapy has strong potential to be proven cost-effective for treating PDs, due to their high disease burden (Bartak et al. 2007; Gabbard 2000). Working against the potential benefits of psychotherapy for treating PDs are the costs, which are likely to be substantial compared to treatment of Axis I disorders (Gabbard 2000). Treatment of PDs may require long-term, intensive psychotherapy (Gabbard 2000; Perry et al. 1999), depending on the goals. Gunderson (The Merck Manuals Online Medical Library 2007) suggests that behaviors such as social isolation or temper outbursts can be changed within months through treatments such as group therapy, behavior modification, self-help groups, family therapy, or dialectical behavioral therapy (e.g., for borderline PD). However, Gunderson (The Merck Manuals Online Medical Library 2007) also notes that interpersonal problems such as distrust and manipulativeness typically require at least 1–3 years to change. The lengthy duration of treatment arises because the method used to effect changes in interpersonal relationships is for the therapist to point out the deleterious effects of the patient's defenses, beliefs, and behavior patterns until the patient finally acknowledges that they are the cause of the patient's problems (The Merck Manuals Online Medical Library 2007), a process that may take years. Some of the more severe PDs may also require psychoanalysis in order to "re-organize" the patient's attitudes, expectations, and beliefs (The Merck Manuals Online Medical Library 2007).

In addition to the direct need for services associated with Axis II disorders, PDs may have an indirect effect on treatment outcomes for Axis I disorders. PDs have been argued to complicate the treatment and worsen the prognosis of Axis I conditions (Gabbard 2000; Samuels et al. 1994; Zimmerman et al. 2005), although not all studies have found this interaction (Taner et al. 2006; van Beek and Verheul 2008). Evidence varies by diagnosis and type of therapy (Joyce et al. 2007). Craigie et al. (2007) found that worse outcomes among depressed patients were largely due to greater severity prior to treatment. A meta-analysis by Kool et al. (2005) concluded that the effectiveness of antidepressant therapy did not differ significantly for individuals with and without PDs. Joyce et al. (2007) found that PDs reduced treatment response among depressed patients randomized to interpersonal but not cognitive behavioral therapy. In contrast to the results for depression, Bottlender et al. (2006) concluded that among substance abuse treatment patients, those with PDs do tend to have worse treatment outcomes and prognosis. The high comorbidity of substance abuse and Axis II disorders such as antisocial PD therefore poses a concern for the successful treatment of addiction. Among individuals with alcohol and drug use disorders, 28.6% and 47.7%, respectively, have a PD (Grant et al. 2004); the rate is even higher among detoxified alcohol-dependent patients, with 74% having at least one PD (Bottlender et al. 2006). Furthermore, even if treatment of the addiction is successful, it may not eliminate the individual's problematic work behaviors. Based on a review of the literature, Verhuel and Van den Brink (2005) conclude that personality factors are an important etiological factor in addiction, and Verheul et al. (2000) found that remission of the substance use disorder was not significantly associated with remission of the personality pathology. The interpersonal conflict that characterizes the social interactions of many addicts may not be attenuated by substance abuse treatment if the root cause of this dysfunction is actually a comorbid PD (e.g., the "dry drunk" phenomenon).

Conclusions

Studies documenting adverse labor market effects of Axis I disorders often conclude by noting that employers and government programs have an incentive to improve insurance coverage of psychiatric treatment because they may recoup some or all of their investment in the form of increased labor market productivity and decreased turnover (Dewa et al. 2007; Ettner et al. 1997). The required investment to treat PDs, however, is likely to be greater and the outcome less certain (Gabbard 2000; Perry et al. 1999). Mental health professionals often do not even code Axis II diagnoses, since insurers assume that PDs cannot be treated and will not reimburse services for them (Gabbard 2005). Although successful treatment of Axis II disorders becomes more viable as our understanding of them grows, cost-effectiveness remains an open question, given the duration and intensity typically required. Other non-financial obstacles to treatment are clinician attitudes (Hinshelwood 1999; van Beek and Verheul 2008) and inadequate training for addressing PDs (Crawford 2007), as well as low rates of treatment-seeking among individuals not already being treated for Axis I disorders (Andrews et al. 2001), especially those with Cluster A disorders (Gabbard 2000).

Nonetheless, some PDs are treatable and treatment may even be cost-effective, depending on the time frame and perspective taken. Evidence that improved work outcomes can result from treating an employee's PD would provide greater incentive for employers to subsidize such treatment. One encouraging result is from Stevenson and Meares (1992), who found that work absenteeism declined from 4.7 months during the year prior to treatment to 1.4 months during the year after 30 patients with borderline PD received 12 months of twice-weekly psychodynamic therapy. However, the extent to which this association was due to "regression to the mean" or natural disease course is unknown.

Even in the case of treatment-refractory PDs, mental health and other support services may nonetheless be beneficial for coworkers managing their own emotional and behavioral responses to the pathology of a personality-disorder individual, especially those who may be more vulnerable to tolerating abusive behavior due to their own psychological issues, such as low self-esteem. Thus improved access to mental health care for employees is still an important response to the problem of PDs in the workforce, although the problem remains of how to get these individuals into treatment without making the company vulnerable to lawsuits related to a hostile work climate.

Greater awareness of PDs and their implications for workplace performance among Employee Assistance Program (EAP) staff, human resources personnel, and managers could also be helpful in several ways. Human resources personnel and EAP staff typically are not trained to identify and interact effectively with employees with PDs (Babiak 1995), who may require a very different approach than those with Axis I disorders such as depression and anxiety. As noted by Gunderson (The Merck Manuals Online Medical Library 2007), "Kindness and guidance alone do not change personality disorders." Setting clear expectations and limits, monitoring performance frequently, and consistently following up with consequences for inappropriate behavior may be a more effective way to manage individuals with problematic PDs than offering empathy and standard mental health counseling, which is best reserved for their victims. For example, in his case study of the factory manager prone to violence and abusive behavior toward others, Goldman (2006b) notes that EAP staff underdiagnosed severe antisocial PDs and that trivializing the manager's diagnosis by using labels such as "petty tyrant" to characterize his behavior put both coworkers and the company at risk (Goldman 2006b).

EAP staff should also be prepared to offer coworkers support and practical strategies for reducing workplace conflict with the disordered individuals. As certain Axis II disorders tend to cause the most distress to individuals interacting with the disordered employee (Campbell et al. 2002; Miller et al. 2007), the most fruitful approach might be to provide support for coworkers in circumstances in which the disordered individuals themselves may not be interested in, or capable of, change (Trimpey and Davidson 1994). A plethora of support groups and books written for the lay public is available to help people manage their interactions with personality-disordered individuals (Behary 2008; Bernstein 2001; Brown 2002; Martinez-Lewi 2008; Mason and Kreger 1998; Stern 2007), yet this information may not be accessed if the PD is not identified, at least by the EAP staff. If desirable, strategies for coping with problematic individuals can be taught to employees without reference to psychiatric diagnoses (Bernstein 2001; Stern 2007). For example, suggested approaches for coworkers interacting with employees with narcissistic or borderline traits might include firm limit-setting, compassionate assertiveness, and better communication among coworkers to avoid giving the disordered employee opportunities to foster divisiveness in the workplace (Behary 2008; Brown 2002; Martinez-Lewi 2008; Mason and Kreger 1998).

The problematic behavior of an individual with a PD can be either reinforced or attenuated by the actions of others (The Merck Manuals Online Medical Library

2007), so it may be possible to improve some contentious workplace situations even without the direct cooperation of the disordered individual. In some cases, a fundamental restructuring of the organization may be required. Gunderson notes that reducing environmental stress can ameliorate PD symptoms (The Merck Manuals Online Medical Library 2007). An example of a successful restructuring of the work environment is provided by Goldman's (2006a) case study of the executive with borderline PD, whose erratic and abusive behavior was dramatically attenuated not only through counseling but also through implementation of a co-leadership strategy and a demonstration of organizational commitment. These served to set limits on his ability to "act out" on the job, while reassuring him with regard to the instability of employment in the fashion industry. As a result, he functioned at a much higher level, with less proclivity to sabotage his own career.

The most effective means of structuring the workplace or supervising employees to minimize the dysfunctional behaviors associated with PDs depends on the specifics of the PD. Trimpey and Davidson (1994) suggest that useful tools for dealing with workers with borderline PD include work accommodation interventions such as setting clear job expectations with nonnegotiable behaviors required for continued employment, frequent feedback and performance evaluations, consistency in addressing problematic behavior, and the use of formal or informal contracting. For employees with obsessive-compulsive PD, Trimpey and Davidson (1994) recommend that supervisors should interrupt and redirect intellectualization of problems, know when to be firm and when to yield in power struggles, introduce change slowly to minimize anxiety, and foster realistic expectations about the functioning of the workplace.

Paunonen et al. (2006) argue that supervisors are deceived by the overly positive self-image projected by narcissistic employees and that only peers who are well-acquainted with each other are in the position to judge true leadership ability, defined by having high self-esteem but a low propensity to exploit others and engage in impression management. They conclude that at least in hierarchical organizations such as the military, peer assessments should play a critical role in promotion decisions. Other recommendations for minimizing the negative influence of narcissistic leaders on their organizations include distributing power more broadly, with a system of "checks and balances;" involving more employees in strategic decision making; developing a forum where others can safely express their viewpoint; using executive training; avoiding the use of self-ratings of work achievements for promotion purposes, and the use of individual performance as the sole basis for providing employment opportunities; incorporating the opinions of subordinates in regular evaluations of the leader's performance, and then basing the leader's continued employment and promotions upon those evaluations; assigning only employees with experience and strong self-esteem to work under the leader; and when necessary, transferring the leader out of harm's way (Campbell et al. 2005; Judge et al. 2006; Kets de Vries and Miller 1985; Paunonen et al. 2006; Rosenthal and Pittinsky 2006).

Particularly in the case of antisocial PD, it has been argued (Babiak and Hare 2006; Campbell et al. 2005; Goldman 2006b; Gunderson and Hourani 2003; Judge et al. 2006; Kets de Vries and Miller 1985) that managers interviewing prospective employees, evaluating current employees for promotions or assigning employees to projects might make better decisions if they conducted a formal or informal personality assessment to understand the employee's potential limitations, and in some cases financial[5] or legal risks (Kupper 2007; McDonald Jr 2002). Psychological tests should be interpreted carefully, however, in light of the likelihood that the most disturbed individuals are willing and able to "game" their responses to such tests (Babiak and Hare 2006). Moreover, an employer's ability to exclude individuals from employment is limited if they have a formal PD diagnosis, due to protections such individuals are afforded under the Americans with Disabilities Act (Goldman 2006b; U.S. Equal Employment Opportunity Commission 1997). Under certain circumstances, however, such actions can still be legally justifiable under this act, e.g., if the individual presents a "direct threat" to the health or safety of themselves or others.

Babiak and Hare (2006) advocate using selection committees, sophisticated job interview techniques, and thorough background checks to verify information on résumés as mechanisms for "weeding out" suspect job applicants with PDs that cannot be accommodated in the workplace. Due to their ability to charm and manipulate others, individuals with antisocial PD tend to interview well, coming across as self-confident and charismatic (Babiak and Hare 2006). After being offered the job, they are skilled at acquiring "patrons" in upper management who defend them, and are subsequently promoted too quickly for their true nature to be detected before the organization and other employees have suffered substantial harm (Babiak and Hare 2006). Often coworkers or supervisors who have been deceived by personality-disordered employees are either too ashamed to admit it or fear confrontation with somebody they perceive to be more powerful than themselves, so the employee is not publicly identified as problematic until serious problems occur (Babiak 1995). Thus early detection and monitoring are critical to managing serious PDs in the workforce. Some of the "red flags" identified by Babiak and Hare (2006) include disparate treatment of staff and the inability to form a team, share, tell the truth, be modest, accept blame, act predictably, react calmly, and act without aggression.

In terms of future research in this area, the impact of PDs on the job performance and satisfaction of coworkers is perhaps the strongest concern related to PDs in the workplace. Educating supervisors and coworkers about PDs may provide the greatest hope for minimizing their adverse impact and improving workplace interactions. Unfortunately, generalizable empirical evidence on this question is not available, as previous research has generally been limited to individual case studies. Given the likely "spillover" effects of PDs onto the productivity and well-being of others (Miller et al. 2007), a large-scale, quantitative examination of the associations of PDs with the job performance of coworkers would be a fruitful topic.

---

[5]Babiak and Hare (2006) cite PricewaterhouseCoopers statistics that in 2005, 45% of companies in their global survey had experienced fraud, with one-quarter committed by senior managers and executives. They also cite Hogan et al. that more executives are fired for personality problems than for incompetence.

Based on the case and research evidence provided in this chapter, work accommodation of employees with PDs is a complex and poorly understood matter. Some of them are most likely to receive accommodation on the grounds of their comorbid disorders such as depression or anxiety, and there may not be known accommodations for others. Improved understanding and identification of specific functional limitations of workers with PDs would subsequently lead to improved employer decisions on availability of work accommodations for those who are most likely to benefit from them.

References

American Psychiatric Association (2000) Diagnostic and statistical manual of mental health disorders, fourth edition, text revision. American Psychiatric Association, Washington, DC

Andrews G, Issakidis C, Carter G (2001) Shortfall in mental health service utilisation. Br J Psychiatry 179:417–425

Atre-Vaidya N, Hussain SM (1999) Borderline personality disorder and bipolar mood disorder: two distinct disorders or a continuum? J Nerv Ment Dis 187(5):313–315

Babiak P (1995) When psychopaths go to work: a case study of an industrial psychopath. Appl Psychol 44(2):171–188

Babiak P (2008a) Beware the corporate psychopath. The Meeting Professional 28(7)

Babiak P (2008b) 'Psychopath' or 'narcissist': the coach's dilemma. Worldwide Association of Business Coaches eZine 4(1)

Babiak P, Hare RD (2006) Snakes in suits: when psychopaths go to work. HarperCollins Publishers, New York

Bagge C, Nickell A, Stepp S, Durrett C, Jackson K, Trull TJ (2004) Borderline personality disorder features predict negative outcomes 2 years later. J Abnorm Psychol 113(2):279–288

Bartak A, Soeteman DI, Verheul R, Busschbach JJ (2007) Strengthening the status of psychotherapy for personality disorders: an integrated perspective on effects and costs. Can J Psychiatry 52(12):803–810

Behary WT (2008) Disarming the narcissist: surviving & thriving with the self-absorbed. New Harbringer Publications, Inc, Oakland

Bender DS, Dolan RT, Skodol AE, Sanislow CA, Dyck IR, McGlashan TH et al (2001) Treatment utilization by patients with personality disorders. Am J Psychiatry 158(2):295–302

Bernstein AJ (2001) Emotional vampires: dealing with people who drain you dry. McGraw Hill, New York

Bernstein DP, Iscan C, Maser J (2007) Opinions of personality disorder experts regarding the DSM-IV personality disorders classification system. J Pers Disord 21(5):536–551

Betan E, Heim AK, Zittel Conklin C, Westen D (2005) Countertransference phenomena and personality pathology in clinical practice: an empirical investigation. Am J Psychiatry 162(5):890–898

Binks CA, Fenton M, McCarthy L, Lee T, Adams CE, Duggan C (2006a) Pharmacological interventions for people with borderline personality disorder. Cochrane Database of Syst Rev (1), CD005653

Binks CA, Fenton M, McCarthy L, Lee T, Adams CE, Duggan C (2006b) Psychological therapies for people with borderline personality disorder. Cochrane Database of Syst Rev (1), CD005652

Blackburn R (1988) On moral judgements and personality disorders. The myth of psychopathic personality revisited. Br J Psychiatry 153:505–512

Blais MA, Norman DK (1997) A psychometric evaluation of the DSM-IV personality disorder criteria. J Pers Disord 11(2):168–176

Blashfield RK, Breen MJ (1989) Face validity of the DSM-III-R personality disorders. Am J Psychiatry 146(12):1575–1579

Board BJ, Fritzon K (2005) Disordered personalities at work. Psychol Crime Law 11(1):17–32

Bottlender M, Preuss UW, Soyka M (2006) Association of personality disorders with Type A and Type B alcoholics. Eur Arch Psychiatry Clin Neurosci 256(1):55–61

Brazier J, Tumur I, Holmes M, Ferriter M, Parry G, Dent-Brown K et al (2006) Psychological therapies including dialectical behaviour therapy for borderline personality disorder a systematic review and preliminary economic evaluation. Health Technol Assess 10(35):iii, ix–xii, 1–117

Brown NW (2002) Working with the self-absorbed: how to handle narcissistic personalities on the job. New Harbinger Publications, Oakland

Campbell WK, Bush CP, Brunell AB, Shelton J (2005) Understanding the social costs of narcissism: the case of the tragedy of the commons. Pers Soc Psychol Bull 31(10):1358–1368

Campbell WK, Foster CA, Finkel EJ (2002) Does self-love lead to love for others? A story of narcissistic game playing. J Pers Soc Psychol 83(2):340–354

Chen H, Cohen P, Crawford TN, Kasen S, Johnson JG, Berenson K (2006) Relative impact of young adult personality disorders on subsequent quality of life: findings of a community-based longitudinal study. J Pers Disord 20(5):510–523

Chen H, Cohen P, Johnson JG, Kasen S, Sneed JR, Crawford TN (2004) Adolescent personality disorders and conflict with romantic partners during the transition to adulthood. J Pers Disord 18(6):507–525

Coid J, Yang M, Tyrer P, Roberts A, Ullrich S (2006) Prevalence and correlates of personality disorder in Great Britain. Br J Psychiatry 188:423–431

Cox BJ, Sareen J, Enns MW, Clara I, Grant BF (2007) The fundamental structure of axis II personality disorders assessed in the National Epidemiologic Survey on Alcohol and Related Conditions. J Clin Psychiatry 68(12):1913–1920

Craigie MA, Saulsman LM, Lampard AM (2007) MCMI-III personality complexity and depression treatment outcome following group-based cognitive-behavioral therapy. J Clin Psychol 63(12):1153–1170

Cramer V, Torgersen S, Kringlen E (2006) Personality disorders and quality of life. A population study. Compr Psychiatry 47(3):178–184

Crawford MJ (2007) Can deficits in social problem-solving in people with personality disorder be reversed? Br J Psychiatry 190:283–284

Crawford TN, Cohen P, Johnson JG, Kasen S, First MB, Gordon K et al (2005) Self-reported personality disorder in the children in the community sample: convergent and prospective validity in late adolescence and adulthood. J Pers Disord 19(1):30–52

Daley SE, Burge D, Hammen C (2000) Borderline personality disorder symptoms as predictors of 4-year romantic relationship dysfunction in young women: addressing issues of specificity. J Abnorm Psychol 109(3):451–460

Demopulos C, Fava M, McLean NE, Alpert JE, Nierenberg AA, Rosenbaum JF (1996) Hypochondriacal concerns in depressed outpatients. Psychosom Med 58(4):314–320

Dewa CS, McDaid D, Ettner SL (2007) An international perspective on worker mental health problems: who bears the burden and how are costs addressed? Can J Psychiatry 52(6):346–356

Duggan C, Adams C, McCarthy L, Fenton M, Lee T, Binks C et al (2006) A systematic review of the effectiveness of pharmacological and psychological treatments for those with personality disorder

Ettner S, Frank R, Kessler R (1997) The impact of psychiatric disorders on labor market outcomes. Ind Labor Relat Rev 51(1):64–81

Ettner S, Maclean JC, French MT (2008) Does having a dysfunctional personality hurt your career? Axis II personality disorders and labor market outcomes

Franken IHA, Hendriks VM (2000) Early-onset of illicit substance use is associated with greater axis-II comorbidity, not with axis-I comorbidity. Drug Alcohol Depend 59(3):305–308

Gabbard GO (2000) Psychotherapy of personality disorders. J Psychother Pract Res 9(1):1–6

Gabbard GO (2005) Personality disorders come of age. Am J Psychiatry 162(5):833–835

Giacalone RA, Greenberg J (eds) (1997) Antisocial behavior in organizations. Sage Publications, Thousand Oaks

Goldman A (2006a) High toxicity leadership: borderline personality disorder and the dysfunctional organization. J Manag Psychol 21(8):733–746

Goldman A (2006b) Personality disorders in leaders: implications of the DSM IV-TR in assessing dysfunctional organizations. J Manag Psychol 21(5):392–414

Goldstein RB, Grant BF, Huang B, Smith SM, Stinson FS, Dawson DA et al (2006) Lack of remorse in antisocial personality disorder: sociodemographic correlates, symptomatic presentation, and comorbidity with Axis I and Axis II disorders in the National Epidemiologic Survey on Alcohol and Related Conditions. Compr Psychiatry 47(4):289–297

Grant BF, Hasin DS, Stinson FS, Dawson DA, Chou SP, Ruan WJ et al (2004) Prevalence, correlates, and disability of personality disorders in the United States: results from the National Epidemiologic Survey on Alcohol and Related Conditions. J Clin Psychiatry 65(7):948–958

Grant BF, Stinson FS, Hasin DS, Dawson DA, Chou SP, Ruan WJ et al (2005) Prevalence, correlates, and comorbidity of bipolar I disorder and axis I and II disorders: results from the National Epidemiologic Survey on Alcohol and Related Conditions. J Clin Psychiatry 66(10):1205–1215

Gunderson EK, Hourani LL (2003) The epidemiology of personality disorders in the U.S. Navy. Mil Med 168(7):575–582

Gunderson JG, Gabbard GO (eds) (2000) Psychotherapy for personality disorders, vol 19. American Psychiatric Press, Inc, Washington, DC

Gunderson JG, Ronningstam E (2001) Differentiating narcissistic and antisocial personality disorders. J Pers Disord 15(2):103–109

Gunderson JG, Ronningstam E, Smith LE (1991) Narcissistic personality disorder: a review of data on DSM-III-R descriptions. J Pers Disord 5(2):167–177

Hill T, Williams B, Cobe G, Lindquist K (2006) Aging inmates: challenges for healthcare and custody. A report for the California Department of Corrections and Rehabilitation. Lumetra, San Francisco

Hinshelwood RD (1999) The difficult patient. The role of 'scientific psychiatry' in understanding patients with chronic schizophrenia or severe personality disorder. Br J Psychiatry 174:187–190

Jackson HJ, Burgess PM (2000) Personality disorders in the community: a report from the Australian National Survey of Mental Health and Wellbeing. Soc Psychiatry Psychiatr Epidemiol 35(12):531–538

Jackson HJ, Burgess PM (2002) Personality disorders in the community: results from the Australian National Survey of Mental Health and Wellbeing Part II. Relationships between personality disorder, Axis I mental disorders and physical conditions with disability and health consultations. Soc Psychiatry Psychiatr Epidemiol 37(6):251–260

Jackson HJ, Burgess PM (2004) Personality disorders in the community: results from the Australian National Survey of Mental Health and Well-being Part III. Relationships between specific type of personality disorder, Axis 1 mental disorders and physical conditions with disability and health consultations. Soc Psychiatry Psychiatr Epidemiol 39(10):765–776

Johnson JG, Cohen P, Kasen S, Brook JS (2006) Personality disorder traits evident by early adulthood and risk for eating and weight problems during middle adulthood. Int J Eat Disord 39(3):184–192

Johnson JG, Cohen P, Smailes E, Kasen S, Oldham JM, Skodol AE et al (2000) Adolescent personality disorders associated with violence and criminal behavior during adolescence and early adulthood. Am J Psychiatry 157(9):1406–1412

Joyce PR, McKenzie JM, Carter JD, Rae AM, Luty SE, Frampton CM et al (2007) Temperament, character and personality disorders as predictors of response to interpersonal psychotherapy and cognitive-behavioural therapy for depression. Br J Psychiatry 190:503–508

186   S.L. Ettner

Judge TA, LePine JA, Rich BL (2006) Loving yourself abundantly: relationship of the narcissistic personality to self- and other perceptions of workplace deviance, leadership, and task and contextual performance. J Appl Psychol 91(4):762–776
Kets de Vries MFR, Miller D (1985) Narcissism and leadership: an object relations perspective. Hum Relat 38(6):583–601
Khoo HS, Burch GSJ (2007) The 'dark side' of leadership personality and transformational leadership: an exploratory study. Pers Individ Differ 44:86–97
King AR, Terrance C (2006) Relationships between personality disorder attributes and friendship qualities among college students. J Soc Pers Relat 23(1):5–20
Kool S, Schoevers R, de Maat S, Van R, Molenaar P, Vink A et al (2005) Efficacy of pharmacotherapy in depressed patients with and without personality disorders: a systematic review and meta-analysis. J Affect Disord 88(3):269–278
Kupper D (2007) In: Wolff J (ed) Borderline personality disorder and the attorney-client relationship: managing the difficult legal client to maximize positive outcomes. CEB Topics, Version 21. http://ceb.com/newsletterv21/3Lit.asp. Accessed 7 Aug 2008
Lavan H, Johnson JG (2002) The association between axis I and II psychiatric symptoms and high-risk sexual behavior during adolescence. J Pers Disord 16(1):73–94
Leible TL, Snell WE (2004) Borderline personality disorder and multiple aspects of emotional intelligence. Pers Individ Differ 37(2):393–404
Lenzenweger MF (2006) The longitudinal study of personality disorders: history, design considerations, and initial findings. J Pers Disord 20(6):645–670
Lenzenweger MF (2008) Epidemiology of personality disorders. Psychiatr Clin North Am 31(3):395–403, vi
Lenzenweger MF, Lane MC, Loranger AW, Kessler RC. DSM-IV personality disorders in the National Comorbidity Survey Replication. Biol Psychiatry. 2007 Sep 15;62(6):553-64. Epub 2007 Jan 9.
Lenzenweger MF, Loranger AW, Korfine L, Neff C (1997) Detecting personality disorders in a nonclinical population. Application of a 2-stage procedure for case identification. Arch Gen Psychiatry 54(4):345–351
Lilienfeld SO (1998) Methodological advances and developments in the assessment of psychopathy. Behav Res Ther 36(1):99–125
Lim D, Sanderson K, Andrews G (2000) Lost productivity among full-time workers with mental disorders. J Ment Health Policy Econ 3(3):139–146
Marinangeli MG, Butte G, Shinto A, Di Cisco L, Peruzzi C, Panelize E et al (2000) Patterns of comorbidity among DSM-III-R personality disorders. Psychopathology 33(2):69–74
Martinez-Lewi L (2008) Freeing yourself from the narcissist in your life. The Penguin Group, New York
Mason PT, Kreger R (1998) Stop walking on eggshells: taking your life back when someone you care about has borderline personality disorder. New Harbinger Publications, Inc, Oakland
McDonald Jr. JJ (2002) Personality disorders in employment litigation. Journal 19(4). http://www.psychiatrictimes.com/personality-disorders/article/10168/48556?pageNumber=1
McGlashan TH, Grill CM, Sanislow CA, Ravels E, Morey LC, Gunderson JG et al (2005) Two-year prevalence and stability of individual DSM-IV criteria for schizotypal, borderline, avoidant, and obsessive-compulsive personality disorders: toward a hybrid model of axis II disorders. Am J Psychiatry 162(5):883–889
The Merck Manuals Online Medical Library (2007) In: Gunderson JG (ed) http://www.merck.com/mmpe/sec15/ch201/ch201a.html. Accessed 31 July 2008
Miller JD, Campbell WK, Pilkonis PA (2007) Narcissistic personality disorder: relations with distress and functional impairment. Compr Psychiatry 48(2):170–177
Millon T (1981) Disorders of personality DSM-III: Axis II. Wiley, New York
Nestor PG (2002) Mental disorder and violence: personality dimensions and clinical features. Am J Psychiatry 159(12):1973–1978
Paris J (1998) Personality disorders in sociocultural perspective. J Pers Disord 12(4):289–301

Paris J (2003) Personality disorders over time: precursors, course and outcome. J Pers Disord 17(6):479–488

Paunonen SV, Lindquist J-E, Veradale M, Likes S, Nissinen V (2006) Narcissism and emergent leadership in military cadets. Leader Q 17:475–486

Perry JC, Banon E, Ianni F (1999) Effectiveness of psychotherapy for personality disorders. Am J Psychiatry 156(9):1312–1321

Reich J, Yates W, Nduaguba M (1989) Prevalence of DSM-III personality disorders in the community. Soc Psychiatry Psychiatr Epidemiol 24(1):12–16

Rosenthal SA, Pittinsky TL (2006) Narcissistic leadership. Leader Q 17:617–633

Rytsala HJ, Melartin TK, Leskela US, Sokero TP, Lestela-Mielonen PS, Isometsa ET (2005) Functional and work disability in major depressive disorder. J Nerv Ment Dis 193(3):189–195

Samuels J, Eaton WW, Bienvenu OJ III, Brown CH, Costa PT Jr, Nestadt G (2002) Prevalence and correlates of personality disorders in a community sample. Br J Psychiatry 180:536–542

Samuels JF, Nestadt G, Romanoski AJ, Folstein MF, McHugh PR (1994) DSM-III personality disorders in the community. Am J Psychiatry 151(7):1055–1062

Sanislow CA, McGlashan TH (1998) Treatment outcome of personality disorders. Can J Psychiatry 43(3):237–250

Sankowsky D (1995) The charismatic leader as narcissist: understanding the abuse of power. Organ Dyn 23(4):57–71

Skodol AE, Gunderson JG, McGlashan TH, Dyck IR, Stout RL, Bender DS et al (2002) Functional impairment in patients with schizotypal, borderline, avoidant, or obsessive-compulsive personality disorder. Am J Psychiatry 159(2):276–283

Skodol AE, Pagano ME, Bender DS, Shea MT, Gunderson JG, Yen S et al (2005) Stability of functional impairment in patients with schizotypal, borderline, avoidant, or obsessive-compulsive personality disorder over two years. Psychol Med 35(3):443–451

Soeteman DI, Hakkaart-van Roijen L, Verheul R, Busschbach JJ (2008) The economic burden of personality disorders in mental health care. J Clin Psychiatry 69(2):259–265

Soeteman DI, Timman R, Trijsburg RW, Verheul R, Busschbach JJ (2005) Assessment of the burden of disease among inpatients in specialized units that provide psychotherapy. Psychiatr Serv 56(9):1153–1155

Stern R (2007) The gaslight effect. Random House, Inc, New York

Stevenson J, Meares R (1992) An outcome study of psychotherapy for patients with borderline personality disorder. Am J Psychiatry 149(3):358–362

Stone MH (2006) Personaltiy-disordered patients: treatable and untreatable. American Psychiatric Publishing, Inc, Washington, DC

Taner E, Demir EY, Cosar B (2006) Comparison of the effectiveness of reboxetine versus fluoxetine in patients with atypical depression: a single-blind, randomized clinical trial. Adv Ther 23(6):974–987

Triebwasser J, Siever LJ (2006) Pharmacology of personality disorders. Psychiatric Times 23(8)

Trimpey M, Davidson S (1994) Chaos, perfectionism, and sabotage: personality disorders in the workplace. Issues Ment Health Nurs 15(1):27–36

Trull TJ, Waudby CJ, Sher KJ (2004) Alcohol, tobacco, and drug use disorders and personality disorder symptoms. Exp Clin Psychopharmacol 12(1):65–75

U.S. Equal Employment Opportunity Commission (1997) EEOC enforcement guidance on the Americans with Disabilities Act and psychiatric disabilities (No. 915.002)

van Asselt AD, Dirksen CD, Arntz A, Severens JL (2007) The cost of borderline personality disorder: societal cost of illness in BPD-patients. Eur Psychiatry 22(6):354–361

van Beek N, Verheul R (2008) Motivation for treatment in patients with personality disorders. J Pers Disord 22(1):89–100

Verheul R, Kranzler HR, Poling J, Tennen H, Ball S, Rounsaville BJ (2000) Co-occurrence of Axis I and Axis II disorders in substance abusers. Acta Psychiatr Scand 101(2):110–118

Verhuel R, Brink WVD (2005) Causal pathways between substance use disorders and personality pathology. Aust Psychol 40(2):127–136

Westen D, Shedler J, Bradley R (2006) A prototype approach to personality disorder diagnosis. Am J Psychiatry 163(5):846–856

Yudofsky SC (2005) Fatal flaws: navigating destructive relationships with people with disorders of personality and character. American Psychiatric Publishing, Inc, Washington, DC

Zaider TI, Johnson JG, Cockell SJ (2002) Psychiatric disorders associated with the onset and persistence of bulimia nervosa and binge eating disorder during adolescence. J Youth Adolesc 31(5):319–329

Zanarini MC, Skodol AE, Bender D, Dolan R, Sanislow C, Schaefer E et al (2000) The Collaborative Longitudinal Personality Disorders Study: reliability of axis I and II diagnoses. J Pers Disord 14(4):291–299

Zimmerman M, Rothschild L, Chelminski I (2005) The prevalence of DSM-IV personality disorders in psychiatric outpatients. Am J Psychiatry 162(10):1911–1918

Part III
Employment Interventions for Persons with Mental Health Disabilities

Chapter 10
Disclosure of Mental Health Disabilities in the Workplace*

Kim L. MacDonald-Wilson, Zlatka Russinova, E. Sally Rogers, Chia Huei Lin, Terri Ferguson, Shengli Dong, and Megan Kash MacDonald

Introduction

There are a number of barriers that contribute to the low employment rates of people with mental health disabilities; these barriers exist at the individual level, the programs and services level, and the systems, policy, and societal level (Anthony et al. 2002). One issue that intersects with all three is disclosure of psychiatric disability in the workplace. Individuals with mental health disabilities must weigh the personal benefits and risks of disclosing their psychiatric disability and make a number of decisions about disclosure given their particular employment circumstances. In addition, employment programs and services, and especially supported employment practitioners, must determine how to represent their services to employers, decide how to inform employers that they work with people with mental health disabilities, and plan with the individual to handle disclosure. Employers must be aware of state and federal policies regarding disability-related employment issues in the face of societal stereotypes, personal experiences with, and misunderstandings about people with mental health disabilities. And while legislation such as the Americans with Disabilities Act (ADA 1990) is in place to protect rightful access to employment, people with mental health difficulties may have little knowledge or understanding of these policies or how disclosure of disability and reasonable accommodations may allow them to enjoy full access to employment opportunities.

---

*Note about language: Throughout this paper, the term "people with mental health disabilities" is used to refer also to individuals with psychiatric disabilities or conditions. This term is meant to describe individuals who have significant mental health impairments that interfere with functioning and employment in particular and where reasonable accommodation or protection under the ADA may be indicated.

K.L. MacDonald-Wilson (✉)
Department of Counseling and Personnel Services, University of Maryland, 3214 Benjamin Building, College Park, MD, 20742, USA
e-mail: kmacdona@umd.edu

I.Z. Schultz and E.S. Rogers (eds.), *Work Accommodation and Retention in Mental Health*, DOI 10.1007/978-1-4419-0428-7_10, © Springer Science+Business Media, LLC 2011

Disclosure of disability may be a more straightforward experience for individuals with apparent disabilities, such as mobility, sensory, or physical health conditions that are visible. Such disabilities tend to be more acceptable to employers and to society in general than disabilities due to mental health conditions, substance use disorders, or cognitive disabilities (Diksa and Rogers 1996; Hernandez et al. 2000; Popovich et al. 2003; Scheid 1999). However, people with nonapparent disabilities have the option of sharing personal information about their disability. They must decide whether to disclose to an employer, and either explain experiences such as gaps in work history or work performance difficulties without referring to their disability, or choose what to say, when, how, and to whom to disclose.

Many people diagnosed with mental health conditions experience stigma, both internalized and public, and have been discriminated against in employment and other life roles based on their label (Wahl 1999). This stigma creates significant risks to disclosing in the workplace. This chapter reviews the variety of ways that disclosure of mental health disabilities in the workplace can occur and the factors involved in making disclosure decisions. Recommendations for and guidance about facilitating disclosure in the workplace is also provided.

What is Disclosure?

Disclosure is defined as the act or process of disclosing, uncovering, or revealing; bringing to light; exposure (Webster's Revised Unabridged Dictionary 1996). Disclosure, also referred to as self-disclosure, may refer to sharing information with friends, family members, doctors, insurance organizations, or employers. In this chapter, disclosure refers to revealing information about one's diagnostic label, mental health condition, or psychiatric disability to someone within the workplace (MacDonald-Wilson 2005).

While often viewed as a simple dichotomous choice (either you disclose or you do not), in practice disclosure of psychiatric disability is not a simple or straightforward process. A description of different types of disclosure that vary based on what is shared, to whom, when, and for what purposes disclosure is undertaken follows.

Full Disclosure

Full disclosure occurs when people with mental health disabilities reveal their mental health background to everyone in the workplace (Corrigan 2005; Herman 1993). Such individuals make no efforts to conceal their psychiatric disability but rather view it as part of their identity. They may share their label or diagnosis, information about their medications or other treatments, accommodations used, or the limitations they have at work due to their disability. Full disclosure may happen all at once, or may develop gradually over time on the job. As a strategy, full disclosure may be used by a job applicant to identify accepting, inclusive employers.

Selective Disclosure

Selective disclosure involves sharing information with specific or a limited number of people, or sharing specific or limited information with others. A person may choose to tell a coworker or someone in human resources about his disability, but not his supervisor, or tell his supervisor but no one else (Corrigan 2005; Ellison et al. 2003; Herman 1993). An employee may also opt to share only the more "acceptable" part of his or her disability, such as "I have a health condition" or "I take medication," instead of what might be perceived as more stigmatizing, such as a "I have schizophrenia" (Goldberg et al. 2005). Selective disclosure may be used to access protections under the ADA while minimizing risks related to stigma.

Strategically Timed Disclosure

Strategically timed disclosure is one type of selective disclosure in which employees disclose in the workplace after a period of time in which they have the opportunity to develop good working relationships and demonstrate their competence in the job (Goldberg et al. 2005). Since the ADA allows people with disabilities to disclose at any time, this strategy affords the opportunity to become known to the employer and coworkers before sharing information which might be potentially stigmatizing, particularly if the disability is not apparent and more immediate disclosure is not required. (Timing of disclosure is a dimension that will be discussed later in the chapter in the section under WHEN to disclose.)

Targeted Disclosure

With shifts in many mental health systems to a focus on recovery, many states and agencies now have jobs specifically targeted for individuals with a "lived experience" in recovery, such as peer specialists or other consumer-designated positions. Disclosure may therefore become a condition of employment or a strategy to obtain a job; in these situations individuals feel more open to disclosing their disability (Goldberg et al. 2005).

Non-Disclosure

Non-disclosure, or secrecy, reflects a choice made by individuals to keep private any information about one's psychiatric disability or treatment. Not disclosing may result in additional stress and lower self-esteem because one is hiding an aspect of one's life, but it also protects the individual from potential stigma and discrimination, and allows the person the option to "blend in" or "pass for normal" (Corrigan 2005; Goldberg et al. 2005; Herman 1993). This was evident in a recent research study (Russinova et al. unpublished manuscript) in which individuals identified fear of stigma as the predominant reason for non-disclosure in the workplace. Individuals with a mental health history chose not to disclose because of their fear of being perceived as less competent, being overlooked for a job, being fired, or having their coworker or supervisor relations adversely affected.

While non-disclosure may protect one from the negative consequences associated with stigma, it also denies the person the opportunity to gain needed accommodations and mutual support. Non-disclosure may be advisable in situations where no accommodations are anticipated, in situations where the individual has had intensely negative experiences with disclosure, or where disclosure could jeopardize work in highly sensitive positions, such as in the military, child care, or defense work. Non-disclosure may be preferred for individuals further along in their recovery who do not feel that disclosing their mental health background is necessary or relevant to their work (Goldberg et al. 2005; Russinova et al. unpublished manuscript).

Inadvertent Disclosure

While the types of disclosure described above involve the individual intentionally and actively making a choice to reveal a psychiatric disability, there are situations in which disclosure of this information is beyond the person's control. Inadvertent disclosure occurs when people who do not disclose believe that their employers or coworkers know about their disability (Goldberg et al. 2005). Inadvertent disclosure undermines the ability of employees to control the conditions of disclosure (Goldberg et al. 2005).

Forced Disclosure

Forced disclosure occurs when an individual is required by circumstances to tell an employer about the existence of a psychiatric disability or the need for accommodation.

In a survey of 350 professionals and managers with mental health conditions, about half felt forced to disclose, while only 38% disclosed when they felt ready. Disclosure occurred in response to hospitalizations, experiencing symptoms on the job they needed to explain, or upon diagnosis (Ellison et al. 2003).

Factors Influencing Disclosure Decisions

There are a variety of factors one must consider in deciding whether to disclose to an employer. The unique characteristics of individuals who have mental health disabilities, other people employed in the workplace, and the relationships between them may impact the conditions and outcomes of disclosure. Furthermore, characteristics of the job, legislation, and the work and social environment all may influence disclosure and subsequent requests for accommodations. This section of the chapter examines the research on disclosure processes and outcomes, and factors related to it.

People Factors

People factors involve characteristics of individual applicants or employees with mental health disabilities (demographic, disability-related, and past experience with disclosure), characteristics of individuals in the employment situation (supervisors, coworkers), and the interpersonal relationships between them.

Demographic characteristics. Characteristics of individuals with disabilities are described in most studies on disclosure, but few report any relationship between demographic characteristics and disclosure. Two studies have found that in supported employment programs, women are less likely to disclose than men (Banks et al. 2007; Fesko 2001b). In one other study, people with lower income levels were more likely to disclose during the application process compared to people with higher income levels (Ellison et al. 2003). No other demographic characteristics (age, race/ethnicity, or living arrangements) appeared to influence disclosure rates or status.

Disability-related characteristics. Factors such as severity, type, and visibility of disability have been found to affect disclosure. In general, the more severe the disability or the more it interferes with work performance, the more likely an individual will disclose (Conyers and Boomer 2005). An immediate medical crisis or other adverse event may necessitate disclosure (Conyers and Boomer 2005; Ellison et al. 2003; Fesko 1998). In Banks et al.'s (2007) study of people with significant mental health conditions, people with mood disorders, or individuals experiencing no symptoms at work were less likely to disclose. Similarly, professionals and managers with schizophrenia (considered a more disabling mental health condition) were more likely to disclose compared to people with other diagnoses (Ellison et al. 2003). Goldberg et al. (2005) concluded that one's stage of recovery had an impact on the job search strategy chosen (i.e., with or without assistance), which influenced the decision to disclose, at least during the early phases of the employment process. People more stabilized in their recovery, working long-term in low-wage jobs, or in competitive professional positions, tended to search for their jobs independently without the assistance of vocational rehabilitation professionals, and did not disclose in the job search process except when applying for mental health peer positions.

In general, people with more severe or disabling conditions, or those less stabilized in recovery, may have a greater need to disclose to either explain their behavior or work performance, or to request reasonable accommodations to address limitations that interfere with work. People with less severe or less apparent disabilities, or who are in later phases of recovery and who may not need accommodations, may choose not to disclose until it is needed.

Past experience with disclosure. If an individual has had negative experiences in disclosing or has perceived job discrimination, he or she may be less likely to disclose in the future, especially during the hiring process (Goldberg et al. 2005; Wahl 1999). When people with disabilities believe that disclosure will result in negative reactions from supervisors or coworkers, they are less likely to disclose (Ellison et al. 2003; Goldberg et al. 2005; Dalgin and Bellini 2008). Baldridge (2005) found that an employee's perception of work group supportiveness is an important factor in reasonable accommodation requests, such that when employees perceive the work group as more supportive, they are less likely to withhold accommodation requests.

On the other hand, higher confidence in maintaining one's professional status and a better capacity to regulate mental health symptoms at work is associated with choosing to disclose when comfortable doing so (Ellison et al. 2003). Rumrill (1999) found, in an experimental study of individuals with visual impairments, that individuals trained in disclosure were more knowledgeable, confident, and active in the disclosure and accommodation request process, thus suggesting that such training interventions might lead to more positive disclosure outcomes for individuals with mental health disabilities. Feeling that one has a choice about disclosure, optimism about the outcomes of the disclosure, and confidence and skill in how to disclose may all influence the decision to disclose.

Employer and relationship characteristics. The characteristic of employers (i.e., supervisors) most frequently associated with disclosure is supportiveness. Disclosure is more likely to occur when supervisors and coworkers are perceived as supportive (Ellison et al. 2003; Shaw et al. 2003), and is less likely to occur when supervisors are thought to be insensitive or unsupportive (Fesko 1998).

To summarize, research suggests that visibility of the disability, phase of recovery, past experiences with stigma and disclosure, confidence in disclosing, independence of the person in acquiring and maintaining a job, and supportiveness of the work environment should all be taken into account when deciding whether and when to disclose.

Job/Accommodation Factors

Nature of the job. Conflicting data exist about the effect of job level on the likelihood of disclosure. For example, Ellison et al. (2003) found that 87% of professionals and managers diagnosed with mental health conditions disclosed in the workplace, although less than a third did so during the hiring process, and more than half did so under unfavorable circumstances. Goldberg and colleagues (2005) found that people with mental health disabilities (former or current SSI or SSDI recipients) employed in professional positions tended not to disclose, except when the position was in the mental health system. People with HIV/AIDS who used accommodations at work were eight times more likely to disclose if they were in professional or managerial positions. In addition, people with visual impairments who worked part-time were less likely to disclose and request accommodations compared to full-time workers. While in general someone in a higher level and/or permanent position may be more likely to disclose, if the individual feels less secure in their position or has more concerns about the severity of their disability, they may be hesitant to disclose even when working in a professional or managerial position.

Nature of accommodations. Characteristics of needed accommodations may also be factors affecting disclosure. Individuals may be able to obtain needed accommodations without disclosing their disability status, using "natural" or "self-accommodations" (Conyers and Boomer 2005; Dalgin and Gilbride 2003). Employees who perceive that an accommodation is inexpensive or easy to provide, or that the accommodation may also help other employees, are more likely to disclose and request accommodations (Baldridge 2002). As might be expected, people who disclose are more likely to receive needed accommodations than those who do not (Banks et al. 2007).

Examination of the job tasks, level of the job, length of time in the job, and perception of job security should be explored when deciding about disclosure. Practitioners should explore the need for accommodation with the individual and determine if accommodations naturally occur in the workplace or can be easily obtained without disclosure, or if disclosure makes one eligible for a targeted position.

Work and Social Environment Factors

Knowledge of the ADA and related legislation regarding disclosure, the culture of the workplace, and prevailing societal attitudes toward people diagnosed with mental health conditions often influence disclosure of mental health disabilities in the workplace.

Knowledge and use of the Americans with Disabilities Act (ADA). People with disabilities are often unfamiliar with the ADA and related legislation that governs disclosure in the workplace. Under the ADA, an individual has the right to decide whether and when to disclose a disability to employers (EEOC 1997). If accommodations are requested, the employer has a right to request limited documentation of the disability and the need for accommodation. In Canada, there is no one federal law that governs employment of people with disabilities, but generally, the employee has the responsibility to provide information about their disability and accommodations to the employer or union (Campolieti 2004). (For additional detail on legislation related to accommodations and disclosure, see Chap. 1)

Many people with mental health disabilities are not aware of the ADA and other legislation, including guidelines about disclosure of disability in the workplace (Gioia and Brekke 2003; Granger 2000; Goldberg et al. 2005), resulting in confusion about how and when to disclose (Granger 2000). People who are more familiar with the ADA are more likely to disclose (Ellison et al. 2003). However, many people do not know about the ADA and its protections, and do not refer to it when disclosing, but still obtain accommodations requested after disclosure (Goldberg et al. 2005). In vocational rehabilitation, professionals may not address the ADA with employers when working with people with disabilities. Ironically, Granger et al. (1997) found that over half of job coaches and job developers in a national survey indicated that they never (15%) or seldom (42%) referred to the ADA when arranging job accommodations.

Workplace culture factors. The nature of the workplace is believed to influence disclosure, but limited information is available on the impact of such factors. Literature suggests that explicit organizational values and policies related to disability or diversity, and organizational flexibility are positively related to employing and accommodating employees with disabilities (Florey and Harrison 2000; Gilbride et al. 2003). Positive attitudes toward the ADA are associated with employer experience in hiring people with disabilities (Gilbride et al. 2003). In addition, employers who have a history of hiring people with disabilities have fewer concerns about the work performance of people with mental health disabilities (Diksa and Rogers 1996).

Working in certain job environments may facilitate the disclosure of a mental health disability. For example, professionals and managers who work in the mental health system have the highest rates of disclosure compared to people working in health/social services or business/technical services industries, although it is unclear whether the mental health positions were targeted for peers or consumers (Ellison et al. 2003). Peer specialist positions are reserved for people with mental health disabilities and require disclosure to be eligible for the position – in one study, every employee in the mental health system disclosed (Goldberg et al. 2005). While it is possible that employers in the mental health field are more open and supportive of disclosure, it may also be that more frequent disclosure is due to the use of disclosure for targeted positions, rather than to the openness of employers.

Attitudes toward people with disabilities. As mentioned previously, many people who have a mental health condition have felt stigmatized or report being discriminated against by employers and coworkers, which affects their willingness to disclose (Wahl 1999). These negative attitudes and experiences result in lower self-esteem or self-confidence, more likelihood of avoiding social contact, less likelihood of disclosing to others, and less likelihood of applying for a job (Wahl 1999). Furthermore, employers are less likely to interview job seekers with disabilities. The least preferred disability is a psychiatric disability, when compared to hearing impairment or mobility disabilities (Pearson et al. 2003). Employers rate people with mental health disabilities as less employable than people with physical disabilities when the disability is known (Dalgin and Bellini 2008). One interesting finding was that disclosure had no effect on actual hiring decisions or on employability ratings when comparing those who disclosed versus those who did not (Dalgin and Bellini 2008).

It is important for professionals and advocates to familiarize themselves with ADA and/or other legislation related to job accommodations and disclosure and to educate employees with mental health disabilities. Professionals can provide information about the ADA and teach individuals the interpersonal skills needed to disclose their mental health disability and request accommodations in a way that is most likely to lead to positive outcomes. It is important to explore the culture of the workplace in terms of its policies related to diversity and disability, and to determine whether it appears to be a flexible organization. It may be useful to examine attitudes and the past experiences of employers in hiring and retaining workers with mental health disabilities, and consider providing support to the employer (provided the professional has obtained the permission of the employee to get involved).

Components of Disclosure Decisions

The following section describes various components of decisions regarding disclosure, including WHY an individual may disclose, WHO discloses and TO WHOM, WHAT to say, and WHEN to say it, as well as the results or OUTCOMES of disclosing.

WHY

An individual considering disclosing a mental health disability in the workplace must consider the benefits and risks of sharing information about their mental health condition, limitations, or needed accommodations. As described above, one of the reasons cited most often by professionals and advocates for disclosure is to request job accommodations which may be used in the hiring process, on the job, or in obtaining the benefits and privileges of employment (ADA 1990; EEOC 1997, 2002). People who use vocational rehabilitation or supported employment services may be willing to have their disability disclosed so that rehabilitation professionals can assist them in developing employer contacts and job leads and/or in providing on-the-job training and support to maintain employment (Conyers and Ahrens 2003; Granger 2000; Granger et al. 1997).

Disclosure by an applicant in the job search process is helpful when the individual might otherwise not be considered for the position, for example when there is a gap in their work history or absences in previous positions (Allen and Carlson 2003; Fesko 2001a, b; Goldberg et al. 2005). As described above, disclosure may be a distinct advantage for positions which are designated in the mental health system for "consumers," or "peer specialists" (Ellison et al. 2003; Goldberg et al. 2005). Once on the job, disclosure may explain issues in productivity and other areas of work performance, or may explain the appearance of symptoms, unusual behavior, or psychiatric hospitalization (Allen and Carlson 2003; Banks et al. 2007; Ellison et al. 2003; Madaus et al. 2002) and allow the provision of supports on the job by an employment specialist or other professional (Gioia and Brekke 2003; Goldberg et al. 2005).

Other individuals may find that it is important to be honest with their employer, educate coworkers, and/or address stigma so that coworkers and supervisors have accurate information about the condition or need for accommodations, rather than fueling speculation (Allen and Carlson 2003; Ralph 2002). In addition to these practical reasons, there are some psychological benefits to disclosing. Disclosure relieves the stress of hiding information about oneself, can enhance self-esteem, and may represent emotional wellness and acceptance of one's experience (Corrigan and Matthews 2003; MacDonald-Wilson and Whitman 1995; Ralph 2002; Roberts and Macan 2006). The act of disclosing to others may also reduce a sense of isolation and facilitate interpersonal relationships by sharing personal information (Ralph 2002), and employees may gain needed support and understanding (Banks et al. 2007; Gioia and Brekke 2003; Goldberg et al. 2005; Rollins et al. 2002). See Table 10.1 for a list of reasons to disclose or not disclose.

Of course, there are also a number of reasons one might not disclose information about mental health disability in the workplace. Many people who choose not to disclose have fears that they will not be offered a job, will have an offer of employment withdrawn, will not be promoted, will be denied reasonable accommodations, will be monitored more closely on the job, or will be fired or laid off once their status is known (Dalgin and Gilbride 2003; Ellison et al. 2003; Granger 2000; Granger et al. 1997). Employees with mental health disabilities may also be afraid of negative reactions from coworkers, including being excluded from social activities, being treated differently, or being isolated and set apart from the workgroup (Allen and Carlson 2003; Dalgin and Gilbride 2003; Ellison et al. 2003; Fesko 2001a, b; Granger 2000; Madaus et al. 2002; Wahl 1999). These beliefs about stigma and discrimination are major factors that affect decisions about disclosure (Goldberg et al. 2005), and should be explored with anyone considering disclosure.

Table 10.1 Reasons for choosing to disclose or not disclose

Reasons to disclose	Reasons to not disclose
To gain the protections of the ADA	To protect my privacy
To request accommodations, access technology	To be "normal," to fit in
To explain gaps in work history, past accommodations received	To preserve self-esteem by not identifying as "disabled"
To address or explain symptoms, sudden hospitalization, or crisis issues in the workplace	Because I do not see myself as disabled, or because my condition is manageable, not disabling
To explain problems in work performance	Because there is no need for accommodation
To enlist the support of the employer	Because my job is naturally accommodating, a good job match
To increase understanding of supervisors and coworkers	Do not feel I should ask, deserve it, am eligible for accommodations – if part-time, I should not ask for accommodations
To have someone to turn to if problems arise	Do not want to be seen as asking for special treatment
To reduce fear or anxiety of coworkers	Did not know I could ask for accommodations
To make sure coworkers have accurate information instead of speculating	Fear of negative employer attitudes
To allow the involvement of a VR professional or advocate to access or maintain employment	Fear of a change in supervision
To become employed in targeted positions in the Mental Health system for "consumers" or "peers"	Fear that disclosure would lead to biased work evaluations
To serve as a role model, combat stigma, educate others	Because it is the cultural norm not to complain
To relieve the stress of keeping secrets, remembering explanations or cover stories	Fears of isolation from coworkers
To continue the process of recovery, acceptance of disability	Because of past negative experiences with disclosure in the workplace or personally
To enhance self-esteem because of choosing not to hide what others may see as a negative fact about oneself	To avoid emotionally hurtful responses
To improve psychological well-being	To avoid being more closely monitored by supervisor
To be honest, to myself and others	To avoid rejection/negative attitudes/being treated differently by coworkers or supervisors
To reduce isolation, connect with others, share personal information	To avoid harassment, gossip, social disapproval
To confirm health insurance coverage prior to accepting job	To avoid all my behavior being interpreted as due to mental illness
	To avoid discrimination, to reduce chances of not being hired, promoted, or terminated because of disability
	To avoid being thought of as less competent
	Because you need to work harder to prove your worth if they know you have a mental illness

Many people with disabilities, especially mental health disabilities, have in fact experienced discrimination in the workplace and have had negative experiences with disclosure, such as being monitored more closely, or denied a work opportunity, so that these fears may be based on past experiences, not merely future concerns (Dalgin and Gilbride 2003; Goldberg et al. 2005; Wahl 1999). Once disclosure occurs, the behavior of the employee may be interpreted through the lens of "mental illness," so that it is difficult to have a bad day, or even to have characteristics of oneself considered by others, such as accomplishments or other strengths (Gioia and Brekke 2003; Ralph 2002). Many people choose not to disclose because their coworkers or supervisors are not perceived as supportive (Allen and Carlson 2003; Baldridge 2005; Wahl 1999). Recent evidence on the impact of disclosure on employers suggests that when employers are provided with information about mental health disabilities at some point in the hiring process, they are less likely to offer interviews (Pearson et al. 2003), and rate applicants lower in employability compared to people with physical disabilities or no disability (Dalgin and Bellini 2008), a finding consistent with previous research (Berven and Driscoll 1981; Farina and Felner 1973; Diksa and Rogers 1996; Scheid 1999). Thus, the anticipation of negative outcomes from disclosure has some factual basis in the attitudes and actions of employers toward people known to have mental health conditions.

Aside from negative past experiences or anticipation of negative outcomes, many people with mental health disabilities choose not to disclose because they do not have a need for accommodations, especially if the job naturally provides the conditions that would be supportive (Conyers and Boomer 2005; Dalgin and Gilbride 2003; Granger 2000), and they otherwise wish to preserve their privacy and their self-esteem (Allen and Carlson 2003; Madaus et al. 2002; Ralph 2002). Some individuals do not view themselves as disabled even if they have a diagnostic label. If their condition can be managed at work then they would not need to disclose to an employer, nor use reasonable accommodations at work. This seems to be one of the reasons that people with mental health disabilities do not disclose (Dalgin and Gilbride 2003). Exploration of the beliefs, experiences, and reasons to disclose or not disclose psychiatric disability in the workplace is therefore needed before proceeding to decisions about how to disclose. Table 10.1 may be used as a guide to initiate exploration of decisions about whether to disclose.

WHO

A number of people may be involved in the disclosure of mental health disabilities in the workplace, including the person or persons who disclose, and the person or persons to whom disclosure is made. Under the ADA, anyone can disclose a disability or a need for an accommodation to an employer, including a rehabilitation counselor, a physician, psychiatrist, psychologist, social worker, other health professional, or even a family member or friend (EEOC 1997, 2002). A number of studies indicate that when vocational rehabilitation (VR) services are used to search for a job, implicit or explicit disclosure occurs (Goldberg et al. 2005). In focus groups with 137 people with mental health disabilities, nearly all who used VR services or job coaches had their disability disclosed (Granger 2000).

Rehabilitation or mental health professionals may share information regarding the person's mental health condition explicitly with an employer, or in other cases the employer may presume the person has a mental health condition if the employer is familiar with the agency's clientele (Banks et al. 2007). In supported employment (SE) programs, more than 80% of employees had their disability disclosed, typically by the SE staff (Banks et al. 2007). Survey and focus group data from several studies found that when individuals with mental health disabilities were working with job coaches, the applicant or employee had little role in the disclosure process, either because the job coach generally handled disclosure or the employer presumed a disability due to the employer's familiarity with the agency (Granger 2000; Granger et al. 1997). However, when individuals were not involved with VR services, the applicant or employee handled the disclosure themselves (Banks et al. 2007; Ellison et al. 2003). People not involved with VR services may be more hesitant to disclose, especially in the hiring process, and may opt to disclose only once they have been on the job for some time and are known to the employer (Ellison et al. 2003; Granger 2000).

TO WHOM

Under the ADA, applicants or employees who are disclosing disability have the option to tell only those who need to know, such as the supervisor who has the authority to adjust the job or provide other accommodations, medical personnel conducting fitness exams, first aid and safety personnel if emergency treatment may be needed, and government officials investigating compliance with the ADA (EEOC 1997). One could inform only the human resource person overseeing the hiring process and not the supervisor if no accommodation is needed once the employee is on the job. The employer is required to keep this information confidential; related documentation must be maintained separately from personnel files (EEOC 1997).

Research on disclosure suggests that most often, it is the supervisor who is told about the disability or need for accommodation (Ellison et al. 2003; Goldberg et al. 2005; Madaus et al. 2002). Professionals and managers mostly frequently informed a supervisor (80%), a coworker (73%), or both (62%), and less frequently informed human resource personnel or subordinates (Ellison et al. 2003). When interviewed in focus groups, people with mental health disabilities suggested informing the supervisor, and only rarely a trusted colleague (Granger 2000). When VR service providers are involved with the employer, the person told may be a human resource specialist, a manager or executive, or other person with whom the VR staff has contact in developing the job. Once someone is on the job, an Employee Assistance Specialist or other health professional in the workplace may be informed.

With whom the information is shared depends in part on the reasons the employee is choosing to disclose – to gain support or understanding, to explain behavior, or to request accommodations. Coworkers may learn of the disability either indirectly when observing a job coach on-site, or directly from the employee. Supervisors are often told in order to gain access to accommodations, explain work performance, or obtain other tangible supports.

Gates (2000) has developed a structured social process to facilitate the employee's return to the job after a period of absence or disability leave for mental health reasons. This process involves planning disclosure and needed accommodations, and sharing information with all important individuals in the workplace, including managers, supervisors, coworkers, and subordinates, with the intention of improving the chances for a successful return to work. She and her colleagues report success with the approach of accommodation as a social process (see Chap. 20).

WHAT

People with mental health disabilities and their advocates are often unsure about what information should be shared with an employer. Under the ADA, disclosure is required only when someone is requesting accommodations in the workplace or otherwise desires protection from discrimination under the law. Guidelines prepared by the Equal Employment Opportunity Commission (EEOC) indicate that an applicant or employee can use plain language to disclose their disability, and is not required to use any special language or technical jargon. The terms "ADA" or "reasonable accommodation" are not required, nor does the law demand that the employee submit the disclosure or accommodation request in writing (EEOC 1997, 2002). However, an employer may ask for additional, but limited documentation (EEOC 1997, 2002) to verify coverage under the ADA and the need for accommodation. Information required when documenting the disability must be specific, emphasizing the nature of the disability, the resulting functional limitations due to the disability or treatments, and any needed accommodations. Examples of such documentation are available on the Job Accommodation Network (JAN) website, a free technical assistance service (see JAN in Appendix A for a list of resources). However, for many employers, verbal disclosure may be sufficient and further documentation may not be required.

Little clear guidance exists in the literature about what to say to employers when disclosing or about the potential impact of differing approaches. In a survey of professionals and managers with mental health conditions, of those who disclosed, 64% revealed their diagnosis, 59% stated that they had a mental illness or psychiatric disability, and 51% mentioned the nature of their symptoms. Only 2% described their medications. Less often mentioned were the accommodations or job adjustments needed (31%), the difficulties in keeping the job (19%), or steps required to take care of themselves on the job (1%) (Ellison et al. 2003).

If vocational rehabilitation staff are involved in the disclosure, frequently no explicit discussion of psychiatric disability, mental health condition, or diagnoses occurs with the employer, and rarely is the ADA mentioned (Granger 2000; Granger et al. 1997). Employees with mental health disabilities report that they prefer to share information about a physical disability or limitation, and not a mental health one, if it can be avoided (Dalgin and Gilbride 2003). Gates (2000), in a qualitative analysis of case records of 12 employees with mental health conditions returning to work after disability leave, found that involvement in developing a disclosure plan and using a social and educational process in the workplace resulted in more positive return-to-work outcomes. This communication process included providing information about the impact of the disability on work, in particular the gaps in functional capacity that interfered with job requirements (Akabas and Gates 1993).

In an experimental study examining the impact of disclosure of invisible disability on employers' ratings of employability and decisions on intentions to hire job applicants, disclosure of a disability (physical, psychiatric, or none) had no significant effect on employability ratings or hiring decisions. In addition, the amount of the information shared (brief vs. detailed) made no difference in ratings or decisions. However, applicants who shared a physical disability (diabetes) were rated more highly than applicants with a mental health disability (bipolar disorder; Dalgin and Bellini 2008). Although the information that was shared included diagnosis, functional limitations, and accommodations needed, it is not clear whether different ways of presenting the information might have influenced employers differently.

WHEN

People with mental health disabilities are often unsure about the best time to disclose on the job (Granger 2000). An individual may choose to disclose a disability at any time during the application process or during any period of employment, but is not required to do so under the ADA, unless an accommodation is being requested (EEOC 2002). EEOC guidance (1997) for people with disabilities suggests that it is in an employee's best interest to disclose their disability prior to the employee's job performance decline or before problems arise on the job.

In general, when employment or vocational rehabilitation specialists are involved in the job search phase, disclosure tends to happen early in the employment process, often before or during the interview stage (Goldberg et al. 2005; Granger 2000; Granger et al. 1997). For over three-fourths of people with mental health disabilities in supported employment, disclosure occurred during the job development phase (Banks et al. 2007). People in lower-paying jobs tended to disclose earlier than those in higher-paying jobs, indicating a possible relationship between the level and status of the job and the timing of disclosure (Ellison et al. 2003).

Severity of the psychiatric disability also appears to be related to the timing of disclosure. Professionals and managers with diagnoses of schizophrenia (considered a more disabling mental health condition), part-time workers, and former SSI/SSDI recipients (receiving disability benefits due to difficulty working) tended to disclose earlier in the employment process, suggesting a need for accommodation earlier in the process and/or more difficulty concealing the impact of the mental health condition (Ellison et al. 2003). In a group of 20 young adults with schizophrenia, only 20% disclosed their disability and requested accommodations at work. Those who decided to disclose tended to be males who were using supported employment services; they tended to have the highest level of negative symptoms and the most enduring positive symptoms (Gioia and Brekke 2003), suggesting that visibility of the disability was a critical factor in disclosure.

People who are searching for jobs independently, if they choose to disclose at all, tend to disclose later in employment, and generally not in the interview stage. For example, although a large number of professionals and managers reported disclosing their mental health condition to someone on the job (87%), only one-third did so in the application process, another 16% within a year, and another 24% more than a year later (Ellison et al. 2003). Of those disclosing (87%), about half did so under unfavorable circumstances, such as the presence of symptoms or hospitalization once on the job. This forced or inadvertent disclosure tended to happen after an average of 6 months on the job. In addition, later disclosure was associated with reported difficulty in accepting one's mental health condition. Goldberg and colleagues (2005) reported that beliefs about stigma and discrimination, as well as how someone sought work (with VR assistance or not) most affected disclosure decisions.

Professionals and managers disclosing under more favorable circumstances (38%), when they actively chose to do so, tended to disclose when they felt comfortable, or after about a month on the job. Circumstances that led to comfort around disclosure included feeling that their employment was secure (32%), that disclosure would not lead to negative consequences (29%), feeling appreciated by the boss (20%), or respected by colleagues (15%).

Overall, the timing of disclosure appears to be related to the involvement of rehabilitation professionals, severity and/or visibility of the mental health condition, level of the job, and acceptance of one's condition, as well as comfort in the work environment. Exploration of these issues with individuals can facilitate informed decisions about when to disclose.

Outcomes of Disclosure

Disclosure of psychiatric disability in the workplace results in positive emotional outcomes for some employees. In a study of supported employees with mental health disabilities, employers were more likely to be supportive of employees who had disclosed than employees who had not (Banks et al. 2007). Disclosure was associated with higher levels of emotional support from coworkers and supervisors, although employees reported feeling somewhat more stress when they disclosed (Rollins et al. 2002). Perceptions of safety and support have helped some employees with schizophrenia avoid quitting their jobs (Gioia and Brekke 2003). In addition, some employees with mental health disabilities report that one of the positive effects of disclosing disability in the workplace is earning the respect and admiration of colleagues (Ralph 2002). As mentioned earlier, some practical outcomes of disclosure include receiving job accommodations (Banks et al. 2007), facilitating the hiring process (Granger 2000), having more positive return to work outcomes (Gates 2000), keeping the job, and having companies provide training in disability awareness (Banks et al. 2007).

However, disclosure may not necessarily be related to how well someone adjusts to a job (Banks et al. 2007). Some people who have disclosed at work report that their supervisor had higher expectations of them, although there is currently no evidence from employers that this actually happens (Goldberg et al. 2005). Individuals who choose not to disclose may also need support handling the stresses associated with not sharing information or accessing accommodations (Goldberg et al. 2005). About one-third of the professionals and managers report having regrets about disclosing their mental health condition in the workplace (Ellison et al. 2003). There is evidence that during the hiring process employers also rate applicants who disclose mental health disabilities lower than those who disclose physical disabilities or no disabilities (Berven and Driscoll 1981; Dalgin and Bellini 2008; Farina and Felner 1973; Jones et al. 1991). However, little is known about employer reactions toward employees who disclose mental health disabilities once on the job.

Evidence on the results of disclosure is not unequivocal. Such mixed evidence suggests that people with mental health disabilities need a structured process to weigh the potential benefits and risks of disclosure. In fact, many people describe a need for a process to assess what would be gained by disclosure before deciding what to do (Dalgin and Gilbride 2003). Others report feeling confused about how to decide the best time to disclose (Granger 2000). The next section details recommendations for practice that professionals and individuals with mental health disabilities can use to prepare for and manage disclosure in the workplace.

Recommendations for Practices on Disclosure

Findings from the literature on factors involved in decisions about disclosure and components of disclosure in the workplace suggest a number of strategies that professionals and individuals with mental health disabilities can use. There are essentially two main phases involved in disclosing: a preparation stage or "readiness to disclose" phase (Ralph 2002); and, once the decision is made, an implementation stage. The professional and the individual considering disclosure should explore these issues prior to the job search process (if in the job search phase of employment), or once the employee experiences a need to disclose on the job.

Preparing for Disclosure

Appendix A contains *Tasks for Preparing to Disclose*. These tasks have been modified from procedures originally proposed by MacDonald-Wilson (2005) and refinements from training sessions by MacDonald-Wilson (2008) and MacDonald-Wilson and Rea (2008), as well as new research findings about factors important in disclosure decisions. In summary, the professional and the individual with a mental health condition should review past experiences with and feelings about disclosing, explore the need and reasons for disclosure, examine the people factors, job/accommodation factors, and workplace culture factors that are relevant to the individual's situation, gather information about the ADA and other rights, identify any need for professional support and employer-provided accommodations, and then weigh the benefits and risks of disclosure. They should also consider the depth and timing of any disclosure.

Implementing Disclosure

Once the individual has made a decision to disclose, the next steps involve managing the disclosure, planning how to disclose (WHO should do it, WHAT to say, WHEN to say it, and TO WHOM), and then take steps to implement the planned activities related to disclosing disability. See Appendix B for the *Tasks for Disclosing*.

These tasks may vary for each person, depending on whether the purpose of disclosure is for practical reasons (e.g., to involve a professional with the employer, to gain access to a job, to request accommodations, to explain behavior), or for more personal reasons (e.g., to relieve stress, to be open and honest, to reduce isolation or connect with others, to gain support). What is disclosed may vary as well, depending on to whom one discloses – a trusted coworker, direct supervisor or manager, customer, human resources personnel, or employee assistance personnel. The tasks are not necessarily to be performed in a linear fashion. Decisions about disclosure may be recycled – within a job over time, as circumstances change, or when the individual changes jobs. Some choices about when or whether to disclose may not be available, such as when forced or inadvertent disclosure occurs, but the individual may still be able to decide what further information to share and with whom.

Professionals, employers, and individuals with mental health conditions should consider consulting the free information and technical assistance available from the Job Accommodation Network (JAN) and the ADA Disability and Business Technical Assistance Centers (see Appendix C *Resources*) at any point in the disclosure process to both prepare for disclosure and to implement disclosure. Finally, it is important for the person disclosing, whether it is the professional or individual themselves, to practice and develop the skills and confidence in communicating with the employer and in disclosing in a positive way to improve the chances for a beneficial experience with disclosure.

Conclusion

Disclosing a mental health condition in the workplace is complex and challenging, especially for people with nonapparent, but potentially stigmatized disabilities. Decisions to disclose or not require a highly individualized assessment and a weighing of the multiple factors associated with a disclosure process that will enhance the person's vocational success and overall recovery. Although many professionals and people with mental health disabilities have concerns about how best to disclose, research does not yet provide clear-cut guidelines for disclosure. However, there is a growing body of research on a number of factors related to disclosure and requesting accommodations. This chapter describes the literature on disclosure, and recommends guidelines for professionals and people who experience mental health conditions to use in disclosure of mental health disabilities to employers. The hope is that workplace disclosure will become an accepted and beneficial practice in the near future.

Appendix A

Tasks for Preparing to Disclose

1. Clarify knowledge about the ADA, accommodations, and disclosure of disability guidelines
2. Explore experiences and feelings about:

    (a) Past experiences with disclosure and/or stigma
    (b) Phase in recovery, severity and/or visibility of disability, identifying as a person with a mental health diagnosis, label, or disability, implications of recovery for disclosure
    (c) Sharing diagnosis or disability information with others, especially employers.

3. Review expectations of and needs for disclosure

    (a) Confidence and skill in disclosing and/or requesting accommodations
    (b) Confidence and skill in regulating impact of mental health condition on work
    (c) Practical and personal reasons to disclose or not disclose
    (d) Anticipated reactions of coworkers, supervisors, or others
    (e) Expected outcomes of disclosure and requests for accommodations.

4. Determine support needs for:

    (a) Involvement of VR or employment professional in hiring or once on the job
    (b) Support or connection wanted or needed from coworkers, supervisors
    (c) Explaining gaps in work history, visible signs of disability, recent hospitalization, or other unforeseen circumstances or behavior
    (d) Potential interference of disability on functioning or other issues in meeting job expectations due to disability or needed treatments.

5. Identify any potential accommodations needed and their timing

   (a) Assess job search skills and needed accommodations in hiring process:

   - Initiating contact or arranging an interview with an employer
   - Interviewing
   - Describing the disability, functional limitations, or needed accommodations
   - Providing documentation, if requested
   - Negotiating accommodations

   (b) Examine any functional limitations or issues related to meeting job expectations experienced due to the disability

   - Identify accommodations or supports
   - Determine whether accommodations/supports occur naturally in the job environment

   (c) Evaluate urgency – when accommodation is needed in employment process (in hiring, as soon as job starts, once on the job after awhile)

   (d) Explore whether accommodation will help others, if it is essential to work tasks or meeting customer needs, and other benefits to the employer and employee

6. Analyze the employer and job environment to determine potential reactions/ support

   (a) Workplace culture

   - Sensitivity to mental health issues, such as information on disabilities in newsletters, posted notices, employee education or training programs
   - Policies such as flex time, mentoring programs, telecommuting, flexible benefit plans, employee awards, or other incentives for contributing to diversity efforts.
   - Type of job that precludes disclosure of mental health disabilities, such as in child care, government high security positions, police work, etc.

   (b) Employer/supervisor knowledge of ADA and experience with people with mental health disabilities

   (c) Employee's perceptions about:

   - supportiveness of the coworkers, supervisors, customers
   - characteristics of relationships with coworkers, supervisors, and other personnel
   - likelihood of a positive reaction to disclosure or accommodation request

   (d) Employee's level of job, status in the organization, industry type, recent job performance, perception of job security, nature of job

7. Weigh the benefits and risks of disclosure. Consider:

   (a) Past experiences, and confidence and skill in disclosing
   (b) Position or status in the organization, length of time in job, job security
   (c) The need for involvement of a professional with the employer in the job search process or once on the job (i.e., visibility of condition, symptoms, or medication side effects, limited job search skills)
   (d) Urgency and need for accommodation
   (e) Personal reasons, such as:

   • identifying as a person with a disability
   • concerns about legal rights
   • need to screen out unsupportive or inflexible employers or job situations
   • relief of stress in not hiding information about self
   • interest in being a role model for other people with disabilities
   • skills in handling the experience of stigma or discrimination with coworkers, supervisors

   (f) Relationships with supervisor, coworkers, and/or customers in the workplace
   (g) Likelihood of employer and/or coworker support

8. If the decision is made NOT to disclose, prepare the following:

   (a) Decide what to say about potential issues that might lead an employer to suspect a disability, such as gaps in employment, frequent job changes, poor references, or other potentially visible signs of disability.
   (b) Plan the behind-the-scenes support that will be needed in hiring or on the job.
   (c) Research employers and jobs that provide the supports naturally, or identify self-accommodations that could be made without disclosure.
   (d) Develop alternative skills and supports to address potential functional limitations.

9. If the decision is made to disclose, plan how the disclosure will be handled – WHO, WHAT, WHEN, and TO WHOM

Adapted from: MacDonald-Wilson (2005, 2008) and MacDonald-Wilson and Rea (2008)

Appendix B

Tasks for Disclosing

1. Clarify the purpose (*Why*) of the disclosure – practical and/or personal reasons (see *Tasks for Preparing to Disclose*)
2. Choose *Who* will be handling the disclosure

   (a) Professional, if involved with employer (implicit or explicit disclosure)
   (b) Applicant/employee
   (c) Other

3. Decide *To Whom* you will disclose

   (a) Interviewer, trusted coworker, supervisor, human resources
   (b) Choose whether educating coworkers would be useful, and if so, what information would be best

4. Select *What* is best to say to whom you will disclose in describing:

   (a) Strengths and/or qualifications related to the job
   (b) Services provided to employer and/or employee (if available), benefits of disclosure and/or accommodations
   (c) Functioning in job performance affected by mental health condition or treatments
   (d) The mental health condition

      • A disability, a medical condition, an illness that is managed
      • A biochemical imbalance, neurological problem, a brain disorder
      • A mental health condition, psychiatric disorder, emotional condition
      • Recovered (In recovery) from a _____ (use preferred term)
      • Difficulty with stress, personal problems, "get a little blue"
      • Other creative expression, or not mentioned at all

   (e) Solutions to the functional issues – accommodations or other supports (presented positively)

5. Determine *When* is the best time to disclose

   (a) Before the interview (especially if accommodation needed in the interview, or if professional is involved with the employer)
   (b) During the interview (to explain gaps in work history, unusual circumstances, or visible signs of disability)
   (c) After the interview but before starting the job (when accommodations needed immediately upon starting the job)
   (d) Once on the job, before problems arise (when comfortable)
   (e) Later on the job, once the need for explanation of behavior or accommodations becomes known (last resort, may result in negative reactions of employer, potentially seen as an excuse if performance problems have arisen)

6. Research a list of the resources available to the employer if additional information about accommodations, documentation, or functional limitations is needed, such as:

   (a) Rehabilitation Counselor, Employment Specialist, Job Coach
   (b) Physician, Psychiatrist, Therapist, Counselor, Social Worker
   (c) Job Accommodation Network (JAN): 1-800-526-7234 or www.jan.wvu.edu
   (d) ADA Disability and Business Technical Assistance Centers (DBTAC): 1-800-949-4232 or www.adata.org
   (e) Other resources (see *Resources on Disclosure and Accommodations*)

7. Prepare professionals to provide documentation if requested (see *Resources,* JAN)

   (a) Nature of disability or mental health condition
   (b) Impact of disability on functioning, especially in extent of impact on major life activities (refer to *Resources*, list of functional limitations, Bazelon Center for Mental Health Law)
   (c) Specific accommodations needed for limitations in functioning

8. Prepare and practice skills of disclosing and/or requesting accommodations

   (a) Prepare a script using comfortable language
   (b) Anticipate potential questions
   (c) Prepare responses to questions using positive terms, focusing on strengths and solutions

9. Practice with trusted significant others so it feels comfortable and makes a positive impression, communicating strength, competence, and ability to handle disabilities so they will not interfere with job performance.

Example of disclosure: I am recovering from a medical condition that has been successfully treated. Currently, I am skilled in using Microsoft Office programs, answering phones, and scheduling appointments, but sometimes the medications that I take to maintain good health require using the restroom facilities every few hours. It would be helpful to have a 10-min break every 2 h, so I can manage this and do well in this job.

Adapted from: MacDonald-Wilson (2005, 2008) and MacDonald-Wilson and Rea (2008)

Appendix C

Resources on Disclosure and Accommodations

ADA Disability and Business Technical Assistance Centers (DBTACs): 800-949-4232 www.adata.org/. Ten regional centers provide information, technical assistance, and training on ADA and accommodations issues. The 800 number connects you directly to the DBTAC in your region.

Bazelon Center for Mental Health Law: www.bazelon.org. A legal advocacy organization for protection of the rights of people with mental illnesses and developmental disabilities. In addition to a variety of resources, information, and links, this site updates court decisions on the ADA and other laws.

- Civil Rights and ADA protections for people with mental disabilities: www. bazelon.org/issues/disabilityrights/index.htm
- Title I of the ADA – Prohibiting discrimination in the workplace: www.bazelon. org/issues/disabilityrights/resources/title1.htm#definition
- How the ADA Applies to people with mental health/psychiatric disabilities: www.bazelon.org/issues/disabilityrights/resources/eeocguide.htm
- List of limitations on major life activities: www.bazelon.org/issues/disabilityrights/ resources/99scotus.htm#limitations
- The Supreme Court's 1999 ADA decisions: www.bazelon.org/issues/disabilityrights/ resources/99scotus.htm
- Other ADA Court decisions: www.bazelon.org/issues/disabilityrights/incourt/ index.htm

Equal Employment Opportunity Commission (EEOC): www.eeoc.gov or 800-669-3362. The government agency overseeing and interpreting the ADA. Materials and fact sheets available.

- Enforcement guidance: Pre-employment disability-related questions and medical exams (Oct 1995): www.eeoc.gov/policy/docs/preemp.html
- Enforcement guidance on disability-related inquiries and medical examinations (July 2000): www.eeoc.gov/policy/docs/guidance-inquiries.html
- Enforcement guidance: The ADA and psychiatric disabilities (March 1997): www.eeoc.gov/policy/docs/psych.html
- Revised enforcement guidance: Reasonable accommodations and undue hardship under the ADA (Oct 2002): www.eeoc.gov/policy/docs/accommodation.html

Frequently asked questions about employees with psychiatric disabilities: Tips and resources on the ADA, job accommodations, and supervision: Booklet for employers of people with psychiatric disabilities. Kim MacDonald-Wilson, Center for Psychiatric Rehabilitation, Boston University. Free download available from www.bu.edu/cpr/resources/articles/1997/macdonald-wilson1997.pdf.

Hyman I (2008) *Self-disclosure and its impact on individuals who receive mental health services*. HHS Pub. No. (SMA)-08-4337 Rockville, MD. Center for Mental Health Services, Substance Abuse and Mental Health Services Administration, 2008. Electronic access at www.samhsa.gov. Free copies available at SAMHSA's Health Information Network at 1-877-726-4727.

Job Accommodation Network (JAN): www.jan.wvu.edu or 1-800-526-7234. A free consulting service for employers, educators, and people with disabilities, which provides individualized worksite accommodations solutions and technical assistance on the ADA and other disability legislation. Additional specific resources on the web include SOAR (Searchable Online Accommodation Resource), fact sheets and examples, and forms such as Writing an Accommodation Request Letter and Sample Medical Inquiry Form.

- Disability disclosure and interviewing techniques for persons with disabilities: www.jan.wvu.edu/corner/vol01iss13.htm
- Job accommodation process: www.jan.wvu.edu/media/JobAccommodation Process.html
- Making a written accommodation request: www.jan.wvu.edu/media/ accommrequestltr.html
- Medical inquiry in response to accommodation request and sample medical inquiry form: www.jan.wvu.edu/media/medical.htm
- Requesting and negotiating a reasonable accommodation: www.jan.wvu.edu/ corner/vol03iss04.htm
- Searchable Online Accommodation Resource (SOAR): Accommodation examples: psychiatric impairments www.jan.wvu.edu/soar/psych/psychex.html
- Reasonable accommodations for people with mental health disabilities: An online resource for employers and educators: www.bu.edu/cpr/reasaccom. Website targeted to employers and educators on accommodations and handling disclosure

Living Well with a Psychiatric Disability in Work and School – Center for Psychiatric Rehabilitation, Boston University – A website for employees and students with mental health disabilities with practical information and resources on the ADA, disclosure, and accommodations, including an interactive discussion board: www.bu.edu/cpr/jobschool. There is a related site for employers and educators at www.bu.edu/cpr/reasaccom/.

- Disclosing your disability to an employer: www.bu.edu/cpr/jobschool/ disclosing.htm
- Handling disclosure: www.bu.edu/cpr/reasaccom/employ-faq.html
- Requesting documentation: www.bu.edu/cpr/reasaccom/employ-tips. html#request

National Collaborative on Workforce and Disability for Youth (2005) *The 411 on disability disclosure workbook*. Washington, DC: Institute for Educational Leadership. Available from www.ncwd-youth.info.

References

Akabas SH, Gates LB (1993) Stress and disability management project: final report. Center for Social Policy and Practice in the workplace, Columbia University, New York

Allen S, Carlson G (2003) To conceal or disclose a disabling condition? A dilemma of employment transition. J Vocat Rehabil 19(1):19–30

Americans with Disabilities Act (ADA) (1990) Public Law No. 101-336; 42 USC Section 12211

Anthony WA, Cohen MR, Farkas MD, Gagne C (2002) Psychiatric rehabilitation, 2nd edn. Center for Psychiatric Rehabilitation, Boston University, Boston, MA

Baldridge DC (2002) The everyday ADA: the influence of requesters' assessments on decisions to ask for needed accommodation. Dissertation Abstracts International Section A: Humanities and Social Sciences 62(8-A):2807

Baldridge DC (2005) Withholding accommodation requests: the role of workgroup supportiveness and requester attributes. Academy of Management Proceedings:D1–D6

Banks BR, Novak J, Mank DM, Grossi T (2007) Disclosure of a psychiatric disability in supported employment: an exploratory study. Int J Psychosoc Rehabil 11(1):69–84

Berven NL, Driscoll JH (1981) The effects of past psychiatric disability on employer evaluation of a job applicant. J Appl Rehabil Counsel 12(1):50–55

Campolieti M (2004) The correlates of accommodations for permanently disabled workers. Ind Relat 43(3):546–72

Conyers LM, Ahrens C (2003) Using the Americans with Disabilities Act to the advantage of persons with severe and persistent mental illness: what rehabilitation counselors need to know. Work 21:57–68

Conyers L, Boomer KB (2005) Factors associated with disclosure of HIV/AIDS to employers among individuals who use job accommodations and those who do not. J Vocat Rehabil 22(3):189–198

Corrigan PW (2005) Dealing with stigma through personal disclosure. In: Corrigan PW (ed) On the stigma of mental illness: practical strategies for research and social change. American Psychological Association, Washington, DC, pp 257–280

Corrigan PW, Matthews AK (2003) Stigma and disclosure: implications for coming out of the closet. J Ment Health 12(3):235–248

Dalgin RS, Bellini J (2008) Invisible disability disclosure in an employment interview: impact on employers' hiring decisions and views of employability. Rehabil Counsel Bull 52(1):6–15

Dalgin RS, Gilbride D (2003) Perspectives of people with psychiatric disabilities on employment disclosure. Psychiatr Rehabil J 26(3):306–310

Diksa E, Rogers ES (1996) Employer concerns about hiring persons with psychiatric disability: results of the Employer Attitude Questionnaire. Rehabil Counsel Bull 40(1):31–44

Ellison ML, Russinova Z, MacDonald-Wilson KL, Lyass A (2003) Patterns and correlates of workplace disclosure among professionals and managers with psychiatric conditions. J Vocat Rehabil 18:3–13

Equal Employment Opportunity Commission (1997) Enforcement guidance on the Americans with Disabilities Act and psychiatric disabilities (EEOC Notice, Number 915.002, March 25, 1997). ADA Division, Office of Legal Counsel, Equal Employment Opportunity Commission, Washington, DC

Equal Employment Opportunity Commission (2002) Enforcement guidance: reasonable accommodations and undue hardship under the Americans with Disabilities Act. www.eeoc.gov/policy/docs/accommodation.html. Accessed 6 June 2005

Farina A, Felner RD (1973) Employment interviewer reactions to former mental patients. J Abnorm Psychol 82(2):268–272

Fesko SL (1998) Accommodation and discrimination: workplace experiences of individuals who are HIV+ and individuals with cancer. Dissertation Abstracts International: Section B: The Sciences and Engineering 59(6-B):2715

Fesko SL (2001a) Disclosure of HIV status in the workplace: considerations and strategies. Health Soc Work 26(4):235–244

Fesko SL (2001b) Workplace experiences of individuals who are HIV+ and individuals with cancer. Rehabil Counsel Bull 45(1):2–11

Florey AT, Harrison DA (2000) Responses to informal accommodation requests from employees with disabilities: multistudy evidence on willingness to comply. Acad Manage J 43(2):224–233

Frank JJ, Bellini J (2005) Barriers to the accommodation request process of the Americans with Disabilities Act. J Rehabil 71(2):28–39

Gates LB (2000) Workplace accommodation as a social process. J Occup Rehabil 10(1):85–98

Gioia D, Brekke JS (2003) Use of the Americans with Disabilities Act by young adults with schizophrenia. Psychiatr Serv 54(3):302–304

Gilbride DG, Stensrud R, Vandergoot D, Golden K (2003) Identification of the characteristics of work environments and employers open to hiring and accommodating people with disabilities. Rehabil Counsel Bull 46(3):130–139

Goldberg SG, Killeen MB, O'Day B (2005) The disclosure conundrum: how people with psychiatric disabilities navigate employment. Psychol Publ Pol Law 11(3):463–500

Granger B (2000) The role of psychiatric rehabilitation practitioners in assisting people in understanding how to best assert their ADA rights and arrange job accommodations. Psychiatr Rehabil J 23(3):215–223

Granger B, Baron R, Robinson S (1997) Findings from a national survey of job coaches and job developers about job accommodations arranged between employers and people with psychiatric disabilities. J Vocat Rehabil 9(3):235–251

Herman NJ (1993) Return to sender: reintegrative stigma-management strategies of ex-psychiatric patients. J Contemp Ethnography 22(3):295–330

Hernandez B, Keys C, Balcazar F (2000) Employer attitudes towards workers with disabilities and their ADA employment rights: a literature review. J Rehabil 66(4):4–16

Jones BJ, Gallagher BJ, Kelly JM, Massari LO (1991) A survey of Fortune 500 corporate policies concerning the psychiatrically handicapped. J Rehabil Oct/Nov/Dec:31–35

MacDonald-Wilson KL (2005) Managing disclosure of psychiatric disabilities to employers. J Appl Rehabil Counsel 36(4):11–21

MacDonald-Wilson KL (2008) Dealing with stigma and disclosure of psychiatric disability in the workplace. Webinar, Fall Webinar Series on Employment, United States Psychiatric Rehabilitation Association. (October 15, 2008)

MacDonald-Wilson K, Rea A (2008) Implementing disclosure plans in Evidence-Based Practice Supported Employment Programs. 33rd Annual United States Psychiatric Rehabilitation Association (USPRA) Conference, There's a new wind blowing: innovations in psychiatric rehabilitation. Lombard, IL (June 16, 2008)

MacDonald-Wilson K, Whitman A (1995) Encouraging disclosure of psychiatric disability. Am Rehabil 21(1):15–19

Madaus JW, Foley TE, McGuire JM, Ruban LM (2002) Employment self-disclosure of postsecondary graduates with learning disabilities: rates and rationales. J Learn Disabil 35(4):364–369

Pearson V, Ip F, Hui H, Yip N, Ho KK, Lo E (2003) To tell or not to tell: disability disclosure and job application outcomes. J Rehabil 69(4):35–38

Popovich PM, Scherbaum CA, Scherbaum KL, Polinko N (2003) The assessment of attitudes toward individuals with disabilities in the workplace. J Psychol 137(2):163–177

Ralph R (2002) The dynamics of disclosure: its impact on recovery and rehabilitation. Psychiatr Rehabil J 26(2):165–172

Roberts LL, Macan TH (2006) Disability disclosure effects on employment interview ratings of applicants with nonvisible disabilities. Rehabil Psychol 51(3):239–246

Rollins AL, Mueser KT, Bond GR, Becker DR (2002) Social relationships at work: does the employment model make a difference? Psychiatr Rehabil J 26:51–61

Rumrill PD (1999) Effects of a social competence training program on accommodation request activity, situational self-efficacy, and Americans with Disabilities Act knowledge among employed people with visual impairments and blindness. J Vocat Rehabil 12(1):25–31

Russinova Z, Kash-MacDonald M, MacDonald-Wilson KL, Wan Y, Archer M, Rogers ES (Unpublished manuscript) Decision-making processes determining disclosure of psychiatric disability at the workplace. Center for Psychiatric Rehabilitation, Boston University, Boston

Scheid T (1999) Employment of individuals with mental disabilities: business response to the ADA's challenge. Behav Sci Law 17:73–91

Shaw WS, Robertson MM, Pransky G, McLellan RK (2003) Employee perspectives on the role of supervisors to prevent workplace disability after injuries. J Occup Rehabil 13(3):129–142

Wahl OF (1999) Mental health consumers' experience of stigma. Schizophr Bull 25:467–478

Webster's Revised Unabridged Dictionary (1996) Disclosure. http://dictionary.reference.com/browse/disclosure. Accessed 20 Oct 2008

Chapter 11
Approaches to Improving Employment Outcomes for People with Serious Mental Illness

Terry Krupa

The past 20 years have witnessed the advancement of a range of innovative and promising employment initiatives for people with serious mental illness (SMI). This chapter provides a brief review of current approaches to improving employment outcomes for people with SMI. It begins with an overview of principles that provide a shared philosophical foundation for the development of these employment approaches. The chapter then offers a brief overview of three distinct approaches that have demonstrated considerable success in securing and sustaining employment for people with SMI: the Individual Placement and Support model; the community economic development approach; and the creation of affirmative employment opportunities within the mental health system.

Principles Guiding Employment Approaches

Several principles provide a foundation for the development of effective employment initiatives focusing on people with SMI. These principles encourage a shared understanding of key values, goals, and objectives, and thus improve communication and promote consensus on acceptable practices in the employment field. They also serve to explicitly address myths about the relationship between work and serious mental illness. This has been particularly important because contemporary practices represent a considerable reform of traditional approaches to psychiatric vocational rehabilitation that were largely ineffective in changing employment rates. Reforming the system of service delivery has required direct attention to inaccurate assumptions held within the field – assumptions that tended to compromise the investment of resources towards employment. For example, historically, the field was influenced by the assumption that diagnoses, particularly those with a

T. Krupa (✉)
School of Rehabilitation Therapy, Queen's University, Louise D. Acton Building,
Room 200, 31 George Street, Kingston, ON, Canada, K7L 3N6
e-mail: terry.krupa@queensu.ca

I.Z. Schultz and E.S. Rogers (eds.), *Work Accommodation and Retention in Mental Health*,
DOI 10.1007/978-1-4419-0428-7_11, © Springer Science+Business Media, LLC 2011

psychosis such as schizophrenia, were associated with a course of deteriorating function, inconsistent with employment. While largely discredited by research findings, vestiges of this idea continue to be present in the mental health system (Strauss 2008), posing a threat to the hopefulness and commitment required by both service providers and people with SMI if the field is to move ahead to achieve successful employment outcomes.

Developed and refined in response to the wisdom of accumulating evidence, these principles for employment support provide a framework for standards and accountability. Kirsh and colleagues (2005) highlight that specifying particular characteristics of effective employment approaches can facilitate quality standards while still encouraging innovation and the creation of a range of initiatives. Eight employment support principles are briefly described here.

Principle 1: Focus on Employment in the Community-Based Workforce as the Desired Outcome

Research has consistently demonstrated that, if given the supports and opportunities they require, people with serious mental illness consider participation in employment to be central to their recovery (Drake et al. 2003). Evidence suggests that in order to obtain good employment outcomes, resources and efforts should be applied directly to securing and sustaining employment (Kirsh et al. 2005; Drake and Bond 2008).

There is general agreement that employment should be characterized by the following components: involvement in the production of goods and services for the general population; integration with the general public through the workplace; and fair and equitable compensation and practice standards in the workplace. There is, however, some disagreement in the field about how these features should be operationalized in order to constitute real employment. Some of these controversies will be further developed later in this chapter, in the section that describes specific employment approaches.

Principle 2: Provide a Range of Ongoing Employment-Related Support

Finding and maintaining employment while living with a serious mental illness is complicated. A range of factors such as the episodic nature of the illness, limited work experience, the lack of work-related social networks, the lack of employer awareness, and stigma and discrimination can compromise an individual's ability to develop work-related strengths and reach full potential in work. Ensuring the availability of ongoing, non-time-limited work supports is considered critical to achieving positive employment outcomes (Drake and Bond 2008; Evans and Bond 2008).

Principle 3: Employment Practices are Client-Centered and Collaborative

Contemporary practices strive to actively engage individuals with SMI in constructing their own personal employment goals, identifying their unique employment strengths, and developing their ongoing supports and interventions. This client-centered approach is considered essential to facilitating the motivation required for sustained personal investment in work, and for addressing the many issues and challenges that might compromise employment (Kirsh et al. 2005; Marrone et al. 1998; Bond 1998).

Principle 4: Address Social Attitudes and Structures to Positively Impact Employment

While the experience of serious mental illness creates challenges to employment, the organization of work itself can create unnecessary restrictions and barriers to employment. Since work is itself socially constructed and organized, it is essential to examine how it may be organized to be disabling, exclusionary, and inequitable (Kirsh et al. 2005; McColl and Bickenbach 1998). Relevant practices might include providing education to employers to reduce stigma and discrimination, and working directly with employers to identify how work policies might be unintentionally excluding people with SMI.

Principle 5: Approach Employment from a Life-Career Perspective

Troubled by employment outcomes indicating that people with serious mental illness are largely securing entry level and low-paying jobs, the field is now adopting a life-career perspective that promotes possibilities for social-economic prosperity and advancement (Drake and Bond 2008; Kravetz et al. 2003; Salzer and Baron 2002). This life-career perspective has led to increased attention to assertively supporting the educational achievement of people with serious mental illness as a means to positively influence both employment outcomes and career progression (see for example, Mowbray et al. 2005).

Principle 6: Provide Early Intervention to Prevent Employment Marginalization

Early intervention initiatives have been developed with a view to preventing the long-term disability and social marginalization that has characterized serious mental illness. Consensus statements coming from the early intervention field have explicitly stated the expectation that within 2 years of the formal diagnosis, at least 90% of the individuals served should be participating in social roles and activities similar to their peers, and reporting satisfaction with their participation (Bertolote and McGorry 2005). Evidence for the application of evidence-based employment approaches is beginning to emerge in the early intervention field (see for example, Neuchterlain et al. 2008; Killackey et al. 2006), as are the features of initiatives that are likely to appeal to and meet the needs of young people experiencing their first episodes of mental illness (Woodside et al. 2007; Jonikas et al. 2003).

Principle 7: Integrate Vocational and Clinical Services

The integration of employment-related supports and clinical services is now considered a critical ingredient to employment success (Evans and Bond 2008). This integration increases the likelihood that symptoms and impairments of mental illness that interfere with employment will receive proper clinical treatment. It also increases the likelihood that treatments negatively influencing work performance (such as medication side effects) will be reviewed and addressed with a view to promoting employment.

There have also been concerns that clinical services are not routinely addressing employment for people with serious mental illness (Bertram and Howard 2006). Service integration is thought to be essential to increasing the likelihood that all individuals with SMI who are seen for clinical treatment will also have access to services and supports focusing on improving their chances for employment (Hanrahan et al. 2006).

Principle 8: Provider Competencies and Practice Standards are Consistent with Current Knowledge About Best Practices in the Field

Employment outcomes are believed to improve within the context of a mental health system that ensures that service providers are prepared with the attitudes, knowledge, and skills to assist people with SMI to find and keep work (McCarthy et al. 2005). Similarly, there is mounting evidence that fidelity in the application of best practices in the field leads to better employment outcomes (Bond et al. 2008a), and provides examples of how the use of training and direct consultation related to practice standards in employment can lead to better outcomes (see, for example, Lurie et al. 2007).

Approaches to Improving Employment Outcomes

Three distinct approaches to improving employment outcomes for serious mental illness are described here: the Individual Placement and Support model of supported employment, community economic development, and the creation of affirmative employment opportunities within the mental health system. These approaches have in common their focus on ensuring access to real jobs with fair compensation and employment protection, but differ with respect to how they operationalize these features and their underlying philosophies and values related to work and mental illness.

Individual Placement and Support

The Individual Placement and Support model (IPS) is a well-defined, standardized model of supported employment. It has been extensively researched and widely disseminated and is perhaps the most well-known service delivery model for improving the employment outcomes for people with serious mental illness. Bond and colleagues (2001) have stated that IPS is less a distinct model of supported employment than it is the standardization of principles for dissemination and replication.

The IPS model's structures and practices are designed with a view to enabling people with serious mental illness to secure employment in the community-based, competitive workforce. Employment is defined as work in integrated settings, where people with mental illness are competitive in securing jobs and work alongside people in the general workforce. These are "real" employment positions, belonging to the individual, rather than work positions contracted to mental health services for the purpose of enabling transition to employment.

In addition to this focus on competitive employment, IPS staff deliver proven critical ingredients of the model. A key component of the approach is rapid job search and placement coupled with direct employment support to facilitate employment success. This approach is the opposite of traditional vocational rehabilitation approaches that were characterized by lengthy assessment and pre-employment training, but ultimately unsuccessful in transitioning people to the community workforce (Bond 1998). The strength of rapid job placement likely comes from its capacity to capitalize quickly on an individual's expressed interest for work, to facilitate the management of real work demands as they emerge, and to address issues both at the level of the individual worker and the workplace context.

IPS is consistent with the employment support principles developed earlier in this chapter, with fidelity to the model requiring demonstration of collaboration with clients to identify and find jobs that match client-identified interests, preferences, strengths, and resources; the provision of indefinite formal employment supports as required; and the close integration of the employment service with the mental health treatment team (Bond et al. 2001). Recent descriptions of the model's ingredients have included providing personalized benefits counseling. This element attempts to address the employment constraints and disincentives that are posed by government income security and employment insurance plans (Drake and Bond 2008; Evans and Bond 2008).

The comprehensive and coordinated service structure of IPS is capable of delivering a broad spectrum of personalized employment interventions to neutralize challenges to employment. For example, IPS can address individual-level barriers associated with the illness, or limited work experience, by teaching coping strategies, helping the person learn to manage symptoms on the job, working with the clinical team to adjust medications interfering with performance on the job, and teaching important job related skills such as interviewing and writing resumes. IPS can address barriers within the individual's social network by engaging the emotional support and practical help of family and friends. IPS staff will collaborate with a client to address workplace challenges, perhaps by identifying and advocating for reasonable accommodations and engaging supervisor and coworker supports. Finally, IPS can address systems and societal-level barriers – for example, providing employer education to decrease the stigma related to mental illness and advocating for enabling income support policies.

The evidence in favor of IPS as a best practice is remarkable. Early studies demonstrated how traditional day treatment services could be restructured to IPS and achieve better employment outcomes (Bailey et al. 1998). A recent systematic review of randomized controlled trials conducted on the IPS model (Bond et al. 2008b) indicated that, when practiced with fidelity, the approach significantly improves employment outcomes, with an average competitive employment rate of 61% (p. 284). The approach has also been found effective in improving time to secure employment, the number of weeks worked, and duration of employment, compared to control groups. Clients of the IPS model typically worked part-time, although a significant minority (over 40%) was working more than 20 h per week.

Critiques of the IPS model of supported employment have focused on the extent to which jobs obtained are entry level and low-paying. Certainly, future initiatives will need to address the potential for advancing a career perspective by expanding the range of employment opportunities and promoting job advancement. While the approach focuses on supporting full integration within the workplace, there is a lack of research evidence at this time focusing on the nature of the social interactions experienced by people with SMI at work. Some mental health consumer advocates have argued that models of supported employment such as IPS use a service approach that depends on professional expertise. This ultimately compromises full community participation and perpetuates dependence on the mental health system and a primary personal identification with mental illness (Trainor et al. 1997).

Community Economic Development

A community economic development (CED) approach increases employment of people with serious mental illness by creating community work opportunities.

This has typically involved the development of businesses that produce goods and services for sale within the broader community. This approach to employment creation has been applied globally to alleviate the economic disadvantage and social marginalization of disenfranchised groups.

CED is grounded in the assumption that the individualism, competition, and profit orientation features of the contemporary labor market disadvantage large sectors of the population, including people with mental illness (Neufeldt and Albright 1998; Krupa 1998). Therefore businesses are constructed with a view to reducing these exclusionary conditions while still remaining competitive in their production and sale of goods and services. So, for example, the businesses may create more part-time or job-sharing opportunities, establish flexible hours of operation, and build in longer training periods and mentorship, all with a view to creating business practices that can capitalize on the strengths of the employees.

Businesses are developed to match the market gaps in the broader community with the interests, strengths, and resources of a collective of people with mental illness. So, for example, matching the interest in horticulture held by a group of mental health consumers with a local identified need for such services might be the basis for initiating the development of a landscaping business. The actual creation of a business is a multistep initiative that includes formal business planning, identifying and strengthening the required skills of the workforce, establishing business policies and practices, and securing the required resources (Krupa et al. 1999; Friesen and Viti 1994).

Several elements need to be in place to ensure that businesses are creating "real" employment. The business should demonstrate that it is consistent with regulations and standards for all business ventures. Because they are subject to comparisons to sheltered workshops, a protected model of work not associated with improved employment outcomes, it is particularly imperative for these businesses to attend to issues of community integration. Unlike IPS, which expects integration in the workplace, CED defines integration at the level of the business, with workers in the business engaging with the broader community through business organizations and networks (for example, the local Chamber of Commerce), and by marketing and delivering their products to members of the broader community. Some businesses have advanced the ideals of integration within the business itself, by creating entrepreneurial ventures that hire both people with and without mental illnesses (Warner and Mandiberg 2006; Warner 1994). Those working in the business need to receive wage compensation that is, at a minimum, fair and equitable according to community standards.

Examples of businesses using a CED approach include restaurants and catering, car washes, landscaping, cleaning, courier and delivery, contract services, and video production. Opportunities for advancement can be built in to the business design – for example, by creating supervisory positions, and by using a range of skills and expertise. Businesses will need, for example, the services of accountants, website developers, marketing specialists, administration and human resource specialists – all positions that could be filled by people with SMI.

Businesses developed from a CED approach are not considered mental health services or programs, because the intention is to create a real community business. However, it is not unusual to see psychiatric vocational rehabilitation specialists involved in partnerships with mental health consumers, business people, and others to create sustainable businesses (Krupa et al. 1999, 2003). The issue of ownership is an important defining criterion for a community business. The businesses should have ongoing job positions that "belong" to individuals hired to fill these positions. Rehabilitation agencies that develop businesses to provide employment preparation opportunities to their clients provide a valuable service, but the training and transitory nature of these initiatives limits their acceptance as real employment.

Businesses are also evaluated by the extent to which the business itself is owned and operated by the employees. Businesses fully owned and operated by people with SMI provide a vehicle for social learning (Church et al. 2000). For example, through community businesses people with SMI as a collective come to recognize their own strengths, and they gain important labor market skills and experiences while countering assumptions about incapacity held within the broader community. Warner and Mandiberg (2006) have described this sense of collective community, fundamental to the businesses, as a type of social movement.

An important outcome is the extent to which the business venture is financially viable. Like all community businesses, those developed through CED face considerable risk of failure. Some businesses have created sophisticated organizational structures, like those present in corporate structures, such as "umbrella" corporations that allow the profits raised from successful businesses to serve as seed money for new business ventures, or to support new businesses that have not yet obtained financial stability (see, for example, Krupa et al. 2003).

Research methods typically supported in evidence-based health care are not particularly applicable to businesses using CED. Instead, participatory research approaches that engage business associates directly in the study of their own business, or research approaches typically used in the business field, such as business case studies, have been applied to demonstrate both positive outcomes and to identify ongoing issues (see for example, Easterly and McCallion 2007; Krupa et al. 2003; Trainor et al. 1997).

Creating Employment Opportunities Within the Mental Health System

An important outcome of system-wide mental health reform has been the emergence of employment of people with serious mental illness within the mental health system. These positions have been created largely from the belief that a truly responsive and client-centered mental health system requires formal and legitimate structures to increase the voice and involvement of service recipients. The development of these positions is based on the idea that people with mental illness can offer their expertise on how to manage mental illness on a day-to-day basis and how to negotiate a less-than-ideal mental health system. In addition, Mowbray and colleagues (1998) suggest that these positions can augment professional expertise by taking on job duties that professionals may be unwilling to take on, and accompanying people with SMI into a broad range of community environments. For example, they may directly support socialization by accompanying an individual to recreation center programs; or meet daily living needs, such as preparing to move to new accommodations, or shopping for clothing and furnishings.

Warner (1994) has pointed out that the mental health system is itself an economic entity, producing goods and services for the treatment and care of people with mental illness, and yet historically mental health consumers have not benefited economically from this market. The affirmative development of work opportunities within this market is now viewed as an important way to improve employment outcomes for people with SMI (Hutchinson et al. 2006). Demonstrating the willingness to affirmatively hire mental health consumers and achieving positive outcomes in these efforts is also seen as important for public relations, given that the mental health field is advocating for the inclusion of people with SMI in the broader community workforce.

Most of these employment efforts have involved the creation of peer service provider positions (Hutchinson et al. 2006; Carlson et al. 2001; Chinman et al. 2001; Davidson et al. 1999). In these positions, mental health consumers are hired to provide direct support services to other mental health consumers, either in the context of multidisciplinary treatment services (Rivera et al. 2007), or within distinct peer support initiatives, such as consumer-run drop-in centers (Mowbray et al. 2005).

In addition there have been efforts to establish employment opportunities for people with serious mental illness as research associates within mental health research programs (Eastabrook et al. 2004; Henry et al. 2002; Reeve et al. 2002). While there has been a growing call for the involvement of people with mental illness in the development, design, and implementation of mental health research and evaluation, this has not widely translated into affirmative paid research positions. Advocates for these research positions have argued that people with mental illness will bring added value to research studies. They may, for example, influence the development of research studies that are particularly relevant to the needs of the population. It has also been suggested that they are able to establish trusting bonds with research participants and thus promote their commitment and genuine responses to surveys and interview questions. A study by Clark and colleagues (Clark et al. 1999) demonstrated that research participants with serious mental illness were more likely to provide extreme negative responses about their satisfaction with mental health services when they were interviewed by another person with a mental illness.

There have been challenges to creating these employment positions within mental health services. Questions emerging include the following: Should mental health consumers be hired as a peer-provider in settings where they have received treatment? How will privacy issues be managed when a peer-provider knows service recipients through other treatment or support services? Is the distinction between professional and peer-providers disempowering? What is the job classification of peer-providers who also have professional training and certification?

While these employment positions do meet the criteria for "real work" in that they are in regular work settings, concerns have been raised about the extent to which the relationship between consumer providers and professional providers and researchers is actually reflective of true integration. Strategies to improve integration are now being developed and disseminated (Gates and Akabas 2007; Macias et al. 2007). Developing essential competencies and training requirements may also serve to enhance acceptance of these positions and ultimately improve integration. Research conducted by Hutchinson and colleagues (2006) demonstrated that mental health consumers who completed a structured peer-provider training program went on to obtain mental health positions, maintained their positions for at least 1 year, and received salaries well in excess of minimum wage standards. While the merit and benefit structures inherent in the organization of mental health service provision and research should facilitate career advancements, to date there has been limited attention to this issue.

Conclusion

This chapter has provided an overview of several key principles believed to be fundamental to the design of approaches for improving the employment outcomes for people with SMI. The three approaches to promoting employment that are described in this chapter have been defined and disseminated, and have a growing body of evidence to support their effectiveness. Each of these approaches focuses on the development of "real community work opportunities," although they differ in the way they define integration in the workplace and the ways they operationalize employment support.

The success achieved by these approaches has been remarkable, bringing the full employment of persons with SMI closer to reality. In the future these, and new innovative approaches, will be faced with addressing emerging issues including: ensuring that all people with serious mental illness, regardless of gender, race and ethnicity, or geographic location have access to these approaches and employment supports; addressing the employment needs of special populations, for example those with dual serious mental illness and developmental disabilities, substance abuse, or legal involvement; and finally ensuring the dissemination and uptake of these approaches in the mental health system.

References

Bailey EL, Ricketts SK, Becker DR, Xie H, Drake RE (1998) Do long-term day treatment clients benefit from supported employment? Psychiatr Rehabil J 22(1):24–30

Bertolote J, McGorry P (2005) Early intervention and recovery for young people with early psychosis: consensus statement. Br J Psychiatry 187(48):116–119

Bertram M and, Howard L (2006) Employment status and occupational care planning for people using mental health services. Psychiatr Bull 30(2):48–51

Bond G (1998) Principles of individual placement and support model: empirical support. Psychiatr Rehabil J 22:11–23

Bond GR, Becker DR, Drake RE, Rapp CA, Meisler N, Lehman AF, Bell MD, Blyler CR (2001) Implementing supported employment as an evidence-based practice. Psychiatr Serv 52(3):313–322

Bond GR, Drake RE, Becker DR (2008) An update on randomized controlled trials of evidence-based supported employment. Psychiatr Rehabil J 31(4):280–290

Bond GR, McHugo GL, Becker DR, Rapp CA, Whitley R (2008b) Fidelity of supported employment: lessons learned from the National Evidence-Based Practice Project. Psychiatr Rehabil J 31(4):300–305

Carlson L, Rapp C, McDiarmid D (2001) Hiring consumer-providers: barriers and alternative solutions. Community Ment Health J 37:199–211

Chinman M, Weingarten R, Stayner D, Davidson L (2001) Chronicity reconsidered: improving person-environment fit through a consumer-run service. Community Ment Health J 37(3):215–229

Church K, Fontan JM, Ng R, Shragge E (2000) Social learning among people who are excluded from the labour market, part one: context and case studies. The research network to new approaches to lifelong learning. http://www.oise.utoronto.ca/depts/sese/csew/nall/res/11sociallearning.htm. Accessed 24 Aug 2008

Clark CC, Scott EA, Boydell KM, Goering P (1999) Effects of client interviewers on client-reported satisfaction with mental health services. Psychiatr Serv 50:961–963

Davidson L, Chinman M, Kloos B, Weingarten R, Stayner D, Tebes J (1999) Peer support among individuals with severe mental illness: a review of the evidence. Clin Psychol Sci Pract 6(2):165–187

Drake R, Becker DR, Bond GR (2003) Recent research on vocational rehabilitation for persons with severe mental illness. Curr Opin Psychiatry 16:451–456

Drake R, Bond GR (2008) The future of supported employment for people with severe mental illness. Psychiatr Rehabil J 31(4):367–376

Eastabrook S, Krupa T, Horgan S (2004) Creating inclusive workplaces: employing people with psychiatric disabilities in evaluation and research in community mental health. Can J Program Eval 19(3):71–88

Easterly L, McCallion P (2007) Affirmative business: examining the relevance of small business research. J Rehabil 73(1):13–21

Evans LJ, Bond GR (2008) Expert ratings on the critical ingredients of supported employment for people with severe mental illness. Psychiatr Rehabil J 31(4):318–331

Friesen M, Viti F (1994) Group hallucinations: overcoming disbelief. Consumer Survivor Business Council of Ontario and the National Network for Mental Health, Toronto

Gates L, Akabas S (2007) Developing strategies to integrate peer providers into the staff of mental health agencies. Adm Policy Ment Health Serv Res 34:293–306

Hanrahan P, Heiser W, Cooper AE, Oulvey G, Luchins DJ (2006) Limitations of system integration in providing employment services for persons with mental illness. Adm Policy Ment Health Serv Res 33(2):244–252

Henry AD, Nicholson J, Phillips S, Stier L, Clayfield J (2002) Creating job opportunities for people with psychiatric disabilities at a university-based research center. Psychiatr Rehabil J 26(2):181–191

Hutchinson DS, Anthony WA, Ashcraft L, Johnson E, Dunn EC, Lyass A, Rogers ES (2006) The personal and vocational impact of training and employing people with psychiatric disabilities as providers. Psychiatr Rehabil J 29(3):205–213

Jonikas JA, Laris A, Cook J (2003) The passage to adulthood: psychiatric rehabilitation service and transition-related needs of young adult women with emotional and psychiatric disorders. Psychiatr Rehabil J 27(2):114–120

Killackey E, Jackson H, Gleeson J, Hickie I, McGorry P (2006) Exciting career opportunity beckons! Early intervention and vocational rehabilitation in first-episode psychosis: employing cautious optimism. Aust NZ J Psychiatry 40:951–962

Kirsh B, Cockburn L, Gewurtz R (2005) Best practice in occupational therapy: program characteristics that influence vocational outcomes for people with serious mental illness. Can J Occup Ther 72(5):265–280

Kravetz S, Dellario D, Granger B, Salzer M (2003) A two-faceted work participation approach to employment and career development as applied to persons with a psychiatric disability. Psychiatr Rehabil J 26(3):278–289

Krupa T (1998) The consumer-run business: people with psychiatric disabilities as entrepreneurs. Work 11:3–10

Krupa T, McCourty K, Bonner D, Von Briesen B, Scott R (1999) Voices, Opportunities and Choices Employment Club: transforming sheltered workshops using an affirmative business approach. Can J Community Ment Health 18(2):87–98

Krupa T, Lagarde M, Carmichael K (2003) Transforming sheltered workshops into affirmative businesses: an evaluation of outcomes. Psychiatr Rehabil J 26(4):359–367

Lurie S, Kirsh B, Hodge S (2007) Can ACT lead to more work? The Ontario experience. Can J Community Ment Health 26(1):161–171

Macias C, Aronson E, Barreira PJ, Rodican CF, Gold PB (2007) Integrating peer providers into traditional service settings: the Jigsaw strategy in action. Adm Policy Ment Health Serv Res 34:494–496

Marrone J, Gandolfo C, Gold M, Hoff D (1998) Just doing it: helping people with mental illness get good jobs. J Appl Rehabil Counsel 29(1):37–48

McCarthy TP, Pelletier JR, Accordino MP (2005) Psychiatric rehabilitation education in a rehabilitation counseling graduate program: training effect on vocational rehabilitation counselor knowledge and skills. Rehabil Educ 19(4):215–224

McColl MA, Bickenbach JE (1998) Introduction to disability. W.B. Saunders Company Ltd, London

Mowbray CT, Collins ME, Bellamy CD, Megivern DA, Bybee D, Szilvagyi S (2005a) Supported education for adults with psychiatric disabilities: an innovation for social work and psychosocial rehabilitation practice. Soc Work 50(1):7–20

Mowbray CT, Holter MC, Mowbray OP, Bybee D (2005b) Consumer-run drop-in centers and clubhouses: comparisons of services and resources in a statewide sample. Psychol Serv 2(1):54–64

Mowbray CT, Moxley D, Collins ME (1998) Consumers as mental health providers: first person accounts of benefits and limitations. J BehavHealth Serv Res 25:397–411

Neuchterlain KH, Subotnik KL, Turner LR, Ventura J, Becker D, Drake R (2008) Individual placement and support for individuals with recent-onset schizophrenia: integrating supported education and supported employment. Psychiatr Rehabil J 31:340–349

Neufeldt A, Albright AL (1998) Disability and self-directed employment: business development models. Captus, Concorde,

Reeve P, Cornell S, D'Costa B, Janzen R, Ochocka J (2002) From our perspective: consumer researchers speak about their experience in a community mental health research project. Psychiatr Rehabil J 25(4):403–408

Rivera JJ, Sullivan AM, Valenti SS (2007) Adding consumer-providers to intensive case management: does it improve outcome? Psychiatr Serv 58:802–9

Salzer M, Baron R (2002) Accounting for unemployment among people with mental illness. Behav Sci Law 20:585–599

Strauss JS (2008) Is prognosis in the individual, the environment, the disease, or what? Schizophr Bull 34(2):245–246

Trainor J, Shepherd M, Boydell KM, Leff A, Crawford E (1997) Beyond the service paradigm: the impact and implications of consumer/survivor initiatives. Psychiatr Rehabil J 21(2):132–140

Warner R (1994) Recovery from schizophrenia: psychiatry and political economy. Routledge, London

Warner R, Mandiberg J (2006) An update on affirmative businesses or social firms for people with mental illness. Psychiatr Serv 57:1488

Woodside H, Krupa T, Pocock K (2007) A conceptual model to guide rehabilitation and recovery in early psychosis. Psychiatr Rehabil J 32(2):125–130

Chapter 12
Employment Interventions for Persons with Mood and Anxiety Disorders

Jason Peer and Wendy Tenhula

The occupational disability associated with mood and anxiety disorders has been receiving increasing attention from mental health and occupational researchers and practitioners, leading to the development of several work-focused interventions for these conditions. This is clearly an emerging area of research that has yet to define the critical intervention to remedy this substantial problem. This chapter describes the existing interventions with an eye toward highlighting applications to current rehabilitation and mental health practice but also the need for further development in this area. In order to provide some context, we begin first with a brief description of mood and anxiety disorders, their impact on work functioning (we refer the reader to chapters by Wald (2011) and by Lerner et al. (2011), this volume, for a more detailed discussion of these issues) and a summary of optimal treatment for these disorders. Next, we review several work-based interventions, including their key elements and available data on their effectiveness. We have organized these work-based interventions into three categories: individual level (those delivered to the individual employee), organization level (those delivered company-wide), and combined (those that combine both types). Finally, we identify common treatment foci of these interventions and how these may inform current practice.

Description of Mood and Anxiety Disorders

As outlined in the Diagnostic and Statistical Manual of Mental Disorders 4th Edition (DSM-IV-TR; APA 2000), mood disorders are defined by discrete mood episodes, and include major depressive disorder (MDD), bipolar I and bipolar II disorder (BPD I and II, respectively), dysthymia, and cyclothymia. In this chapter

W. Tenhula (✉)
VA Capitol Health Care Network Mental Illness Research Education and Clinical Center, 10 North Greene Street, Suite 6A, Baltimore, MD, 21201, USA; Department of Psychiatry, University of Maryland School of Medicine, Baltimore, MD, USA
e-mail: Jason.Peer@va.gov

I.Z. Schultz and E.S. Rogers (eds.), *Work Accommodation and Retention in Mental Health*, 233
DOI 10.1007/978-1-4419-0428-7_12, © Springer Science+Business Media, LLC 2011

we will focus primarily on interventions for major depressive episodes (MDEs). Of the mood episodes, these are most strongly associated with occupational disability (Kessler et al. 2006). They are also the most common type of mood episode and are a component of most mood disorders including MDD, BPD I and BPD II. MDD is defined by one or more MDEs, whereas BP I and BP II are defined by at least one MDE as well as a manic or hypomanic episode (i.e., a distinct period of elevated or irritable mood). Symptoms of an MDE include: depressed mood, loss of interest (anhedonia), disrupted appetite, sleep disturbance, motor retardation or agitation, loss of energy, worthlessness or excessive guilt, concentration difficulties, and suicidal thoughts or behaviors. An MDE is defined by a period of at least 2 weeks during which five or more of the preceding symptoms are present (one of these symptoms must be depressed mood or anhedonia).

The DSM-IV anxiety disorders include panic disorder, obsessive-compulsive disorder (OCD), social phobia, specific phobia, generalized anxiety disorder, and posttraumatic stress disorder (PTSD). Table 12.1 provides a summary of these disorders. While they each have distinct symptoms and clinical features, they can generally be conceptualized by three overlapping symptom domains: physiological arousal, cognitive worry or rumination, and behavioral avoidance. That is, each of these disorders to some extent is characterized by a heightened physiological response to a given situation or object; worry, or rumination associated with the situation or object; and intentional avoidance of the situation or object. For example, in the case of social phobia, an individual has a persistent fear or

Table 12.1 Summary of DSM-IV anxiety disorders (APA 2000, p. 393)

Disorder	Description
Panic disorder[a]	Recurrent, unexpected panic attacks (discrete period of intense apprehension, fear or terror accompanied by physical sensations such as shortness of breath, palpitations, chest pain)
Specific phobia	Exposure to specific feared object or situation (e.g., animals; heights) results in intense anxiety often leading to avoidance behavior
Social phobia (social anxiety disorder)	Exposure to social or performance situations results in intense anxiety often leading to avoidance behavior
Obsessive-compulsive disorder	Persistent thoughts, ideas, impulses, or images that are experienced as intrusive and distressing (obsessions) paired with repetitive behaviors (compulsions) which serve to neutralize obsessions
Posttraumatic stress disorder	Reexperiencing of a traumatic event accompanied by arousal symptoms and avoidance of stimuli associated with the trauma
Generalized anxiety disorder	Persistent and excessive anxiety and worry that lasts at least 6 months

[a] Can occur with or without agoraphobia (fear and avoidance of situations in which escape would be difficult or embarrassing)

apprehension (i.e., worry) about social situations (e.g., speaking in public). When faced with this situation the individual can experience intense physiological arousal or even a panic attack (shortness of breath, palpitations, and intense fearfulness), and he or she often intentionally avoids these situations in an effort to avoid this emotional distress. In the case of OCD, the individual experiences intrusive and persistent thoughts or ideas (e.g., being contaminated by germs) that cause marked distress. In order to avoid this distress, the individual develops behaviors (e.g., excessive hand-washing) that serve to neutralize the anxious thoughts.

While mood and anxiety disorders represent distinct diagnostic categories, they do share several commonalities. First, a critical diagnostic criterion for each of these disorders is evidence of social or occupational dysfunction. Thus, in addition to presenting with specific symptoms, an individual must also experience some related form of functional impairment in order to receive a diagnosis. A second common characteristic is that many of these disorders have an onset in early adulthood and have the potential to be long-standing with an episodic course (Kessler et al. 2008). Exacerbations in symptoms often result from psychosocial stressors. Thus these disorders can interfere both with initial career development as well as ongoing career maintenance and advancement (Waghorn and Chant 2005). Finally, it should be noted that symptoms of depression and anxiety are present in less severe forms in other disorders (e.g., adjustment disorders) which can also lead to functional disability.

Impact of Mood and Anxiety Disorders on Work Functioning

Depression has been found to impact employment status as well as job performance and productivity, resulting in significant burdens to individuals and costs to employers (Bender and Farvolden 2008). Recent 12-month prevalence estimates in a large US sample of employees indicate that 6.4% of the sample met criteria for MDD and 1.1% met criteria for BPD (Kessler et al. 2006). For each of these groups the presence of an MDE in the past year was associated with substantially impaired work performance, both in terms of number of days absent as well as reduced work productivity when on the job (i.e., *presenteeism*). Data from this survey are largely consistent with previous reviews (Mintz et al. 1992; Sanderson and Andrews 2006).

Anxiety disorders are also frequently associated with poor occupational functioning and unemployment. For example, a large Australian national survey found that, when compared to a nonaffected comparison group, individuals of working age with an anxiety disorder were over two times as likely to be unemployed, had longer periods of unemployment, and were more likely to report work restrictions (Waghorn and Chant 2005). US surveys also estimate that anxiety disorders are quite prevalent among working age individuals and that employed anxiety sufferers are more likely to experience decreased productivity and longer work absences relative to workers without an anxiety disorder (Greenberg et al. 1999).

Quantitative and qualitative studies have both highlighted the work performance impairments associated with anxiety and depression (see chapters by Wald (2011) and by Lerner et al. (2011), this volume). Individuals with anxiety disorders who are employed are more likely to report difficulties with work performance when compared to a nonaffected comparison group (Waghorn and Chant 2005). Detailed evaluations of work performance among depressed individuals have identified several specific functional limitations, such as difficulties with interpersonal, time, and workload management and physical tasks (e.g., Adler et al. 2006). These difficulties are consistent with qualitative interview and focus group studies of occupational functioning among individuals with anxiety and mood disorders. Those workers surveyed commonly reported problems with lack of energy and motivation, decreased confidence and productivity, indecision, and poor concentration, as well as social withdrawal (Haslam et al. 2005; Michalak et al. 2007). Physical symptoms such as nausea, headaches, dizziness, trembling, and insomnia are often reported to interfere with work performance; although workers are also likely to attribute these symptoms to side effects of medications prescribed for anxiety and depression (Haslam et al. 2005).

Many of these work performance limitations reflect the behavioral and cognitive symptoms associated with mood and anxiety disorders. There is fairly strong evidence that appropriate treatment of these symptoms results in subsequent improvements in work performance. An early meta-analysis of work outcomes in depression treatment trials indicated that effective treatment of symptoms by either psychotherapy or pharmacology corresponded with a reduction in work impairment, and that treatments with longer duration led to increased positive work outcomes (Mintz et al. 1992). Several more recent studies corroborate these findings in depression (Adler et al. 2006; Schoenbaum et al. 2002), bipolar disorder (Frank et al. 2008), and anxiety (Berndt et al. 1998, as cited in Waghorn et al. 2005). However, two caveats should be noted. First, while treatment appears to have a positive impact on work functioning, these functional benefits appear to occur later than initial symptom improvements. That is, there is a lag in improvements in work performance (Adler et al. 2006; Mintz et al. 1992) relative to symptom improvement. Second, despite adequate treatment of symptoms, workers may still experience residual functional limitations (Adler et al. 2006). These may be related to the initial disorder or side effects from medications commonly prescribed (e.g., Haslam et al. 2005). Nonetheless, these data suggest that the occupational impairment associated with mood and anxiety disorders can be ameliorated with appropriate treatment.

Description of Treatments for Mood and Anxiety Disorders

Depression and anxiety disorders are typically treated with medications, psychotherapy, or a combination of both. Here we provide a brief description of pharmacological and psychosocial treatments. This is not intended as an extensive review but rather a selective summary highlighting treatments that are incorporated in the work-based interventions described later in the chapter.

Pharmacological Treatments

The effectiveness of several medications for these disorders is well documented, and recommended treatment guidelines have been established (APA 2006). Common medications for the effective treatment of both anxiety and depression include selective serotonin reuptake inhibitors (SSRIs), tricyclic antidepressants, and monoamine oxidase inhibitors (MAOIs). While medication management decisions are determined by an individual and his/her prescribing physician, a few clinical points about pharmacological interventions are worth noting. First, medication response is not immediate and often occurs between 4 and 8 weeks after initiation, with substantial individual variation in response times (Nierenberg et al. 2008). Second, as discussed above, individuals taking these medications can also experience unpleasant side effects (e.g., nausea, sedation), which can result in nonadherence. Third, even when adherent, some individuals experience a suboptimal or nonresponse to medication. Thus, individuals and their prescribing physician may need to try different dosages and/or medications in order to reduce side effects while obtaining an optimal treatment response.

In the context of work-based interventions for anxiety and depression, the focus is not *which* medication is prescribed, but rather *how* it is prescribed. For example, several interventions include medication treatment algorithms or decision trees that incorporate both the workers' preference (e.g., whether or not to include medication treatment) as well as individual treatment response (e.g., adding medications for individuals who do not respond to the initial treatment). In addition, interventions also focus on ensuring access to and quality of pharmacological interventions.

Psychosocial Treatments

By far the most extensively researched psychosocial treatment for anxiety and depression is cognitive behavioral therapy (CBT). There is a well-established evidence base for the efficacy of CBT for both anxiety disorders (Barlow and Lehman 1996; Hofmann and Smits 2008) and depression (Dobson 1989; Gloaguen et al. 1998). It should be noted that other therapeutic interventions have also been found to be effective for depression (e.g., interpersonal psychotherapy; Weissman et al. 2007). However, they have not been as widely disseminated as CBT and do not figure prominently in the work-based interventions described in this chapter, so will not be described in detail.

CBT is a well-documented evidence-based practice for the treatment of depression and anxiety. The basic theory underlying CBT proposes that a person's feelings and behaviors are determined by their interpretations and attributions (i.e., thoughts) about a situation. The emotional distress associated with depression and anxiety is thought to be related to overly negative thought patterns that can be identified and changed, which thereby leads to reductions in symptoms. A key element of CBT therefore includes learning to recognize and evaluate the accuracy of negative "automatic thoughts," including the identification of specific types of cognitive errors or distortions. Examples of common cognitive distortions in depression are "all or nothing" thinking (i.e., believing that a situation is either all good or all bad, a complete success or a total failure) and mind reading (i.e., making assumptions about what other people are thinking). Examples of cognitive distortions that are quite common among persons experiencing anxiety are fortune-telling (i.e., predicting that bad things will happen) and magnification (i.e., believing things are worse than they objectively are). A more complete list of cognitive distortions that are frequently addressed in CBT can be found in *Cognitive Therapy for Depression* (Beck et al. 1987). In addition to addressing cognitive distortions, CBT typically includes behavioral strategies such as activity scheduling and behavioral activation, which help to energize and motivate persons with depression. Likewise, CBT typically involves relaxation training and exposure to feared situations for individuals who experience anxiety. Self-monitoring of thoughts, behaviors, and symptoms, as well as homework assignments are used to augment the material covered in session.

In general, the CBT therapist takes a collaborative stance, working *with* the client to help them understand the difficulties they are having (i.e., psychoeducation), set individualized goals, and develop skills that will bring about changes in their thoughts and behaviors. The therapy is present-focused and generally time-limited, although the exact duration of treatment is quite variable. CBT can be done either individually or in groups. CBT is used in many of the work-based interventions described in this chapter. Specifically, these interventions use CBT strategies to target work-related depressive and anxiety symptoms that interfere with occupational functioning (e.g., anxiety about interpersonal situations) or returning to the workforce (e.g., low energy and poor motivation). A brief clinical vignette of a work-based CBT intervention involving a worker with a long-standing anxiety disorder illustrates some of these principles.

Clinical Vignette: Use of CBT Strategies with an Older Worker with Anxiety Symptoms

Mr. B. was a 55-year-old man with a long-standing anxiety disorder who had recently begun a vocational rehabilitation (VR) program. In conjunction with the VR program, he participated in an adjunctive work-focused CBT and social skills training intervention. The CBT component of the intervention consisted of eight sessions focused on collaborative problem identification and goal setting, self-monitoring of symptoms, psychoeducation about cognitive distortions, and cognitive and behavioral symptom management strategies. During the initial problem identification and goal setting session, Mr. B. reported experiencing anxiety symptoms while at work. As part of his job duties Mr. B. had to fill specific orders with products in a small warehouse. His symptoms interfered with his ability to complete tasks carefully. Specifically, he experienced lapses of attention, feeling jittery and shaky, and feeling "rushed." Through self-monitoring, Mr. B. identified that he tended to experience these symptoms more intensely at the beginning of his shift and that over the course of the shift he became more relaxed and his performance improved. During subsequent sessions, Mr. B. and his therapist worked on identifying automatic thoughts associated with anxiety symptoms. For example, a thought Mr. B. frequently identified at the beginning of the shift was: "I will never be able to fill all these orders and if I don't fill all these orders my supervisor will be angry with me." Next Mr. B. and the therapist worked to identify the type of cognitive distortion that these thoughts represented. In this case, Mr. B. was prone to making "fortune-telling" errors, in that he automatically assumed a negative outcome, thus intensifying his anxiety symptoms. Once he was able to recognize these types of thinking errors, and how they contributed to his anxiety, Mr. B. was able to challenge their accuracy. Specifically, he and his therapist evaluated the evidence supporting the statement, "I will never be able to fill these orders." For example, Mr. B. was encouraged to ask himself whether there had ever been a time when he was unable to fill all the orders. There was not, and in fact, most days Mr. B. was able to complete the orders well before the end of his shift. During the course of subsequent sessions, Mr. B. and the therapist also worked on some relaxation training in order to provide him with coping strategies for times when his anxiety symptoms were particularly intense. At the end of the eight sessions, Mr. B. reported that he was better able to control his anxiety, particularly at the beginning of his shift, and that it had a positive impact on his work performance.

Untreated Anxiety and Depression

Despite the availability of effective treatments for anxiety and depressive disorders, they remain woefully under- or untreated in common clinical practice. Multiple surveys have documented the gap between recommended and received treatment. Wang et al. (2000) found that among adults with GAD, panic disorder, or MDD, drawn from a national sample, 54% received treatment, and of those, only 14.3% received treatment that was concordant with evidence-based practices. Young et al. (2001) found that in a sample with probable anxiety or depressive disorders, 83% saw a healthcare provider for this condition. However, of those, only 30% received care that was consistent with recommended medication or psychotherapy guidelines (e.g., appropriate number of sessions, appropriate medications, and dosage). An Australian national survey found that 40.9% of those with an identified anxiety disorder did not receive any treatment (Waghorn et al. 2005).

Recent data on workers with MDD are somewhat comparable. Slightly over 56% of the sample received treatment in the preceding year and of those only 41.7% received adequate treatment (i.e., concordant with existing treatment guidelines), indicating that only one-fourth of all individuals with MDD were obtaining adequate care (Kessler et al. 2008).

Summary of Key Points Regarding Mood and Anxiety Disorders and Their Treatment

Effective pharmacological and psychosocial interventions are available for the treatment of depression and anxiety disorders. With optimal treatment, many (although not all) individuals experience symptom relief and a corresponding improvement in work functioning. Those individuals who do not experience an optimal treatment response may require additional intervention (either medication or psychotherapy) to maximize functioning. Many mood and anxiety disorders are long-standing and are exacerbated during times of stress, which may require additional periods of more intensive treatment. Despite the availability of effective treatments, recent surveys indicate that too few individuals receive the appropriate recommended treatments for anxiety and depressive disorders.

Work-Based Interventions for Anxiety and Depression

In light of recent research highlighting the occupational disability and economic cost associated with mood and anxiety disorders, several occupational health researchers have been developing and testing work-based interventions for these conditions (Bender and Farvolden 2008). These interventions include both stand-alone treatments as well as "packages" that include multiple components delivered at both the individual level (e.g., CBT for employees) and the organizational level (e.g., depression screening programs). They also tend to differ, at least in part, due to the mental health context in which they were developed. For example, work-related interventions developed in the Netherlands have been conducted within nationalized occupational health programs (e.g., van der Klink et al. 2003), while US interventions have generally been developed for managed care companies (e.g., Wang et al. 2007). Despite these different contexts, there are some commonalities. These include the following: use of CBT strategies, use of graded activity to accelerate return-to-work, facilitation of workplace accommodations, screening and early identification of employees with an anxiety or mood disorder, and delivering and/or coordinating optimal treatment.

In order to describe this developing area, we briefly outline some recent work-focused interventions and identify their key elements. Rather than an exclusive focus on anxiety and mood disorders per se, we have chosen to include interventions that target related conditions such as work-related stress and adjustment disorders that share similar symptoms and functional limitations. While not as severe as other psychiatric disorders, the emotional distress associated with adjustment disorders can contribute to substantial occupational impairment and disability (van der Klink et al. 2003).

Based on database searches and hand searches of recent reviews (e.g., Bender and Farvolden 2008; Nieuwenhuijsen et al. 2008), we identified 11 interventions that explicitly targeted work functioning in individuals experiencing anxiety and depressive symptoms and related psychological complaints. We divided these studies into three categories of work-based intervention: individual level ($n=7$), organizational level ($n=2$), and combined ($n=2$). Individual level interventions are defined as those designed to enhance work functioning of the individual employee who is experiencing depressive or anxiety symptoms, whereas organizational level interventions are defined as those designed to reduce the impact of depression and anxiety on the workplace via screening and/or enhanced access to mental health treatment. Combined interventions represent treatment packages that include both. For each intervention, we briefly describe the context in which the intervention was delivered, the types of employees for which it is intended, its key elements, and where available, a summary of research findings on work and symptom outcomes. Table 12.2 provides a summary of the interventions and their outcomes. For the purpose of further description and synthesis, we identify three common treatment foci of these interventions that are of clinical relevance to the mental health and rehabilitation practitioner. These include a focus on returning to work, ensuring optimal treatment, and combining interventions to maximize their effectiveness.

Individual-Level Work-Based Interventions for Anxiety and Depression

Many of these interventions incorporate CBT strategies. The focus of these CBT techniques is the enhancement of the employee's coping skills and awareness of distorted thinking associated with these disorders, and the development of more adaptive behavior. The goal of these interventions is to reduce emotional distress associated with the work environment and/or with return-to-work after sick leave. Several of the individual level interventions also include an adapted *graded activity* component. Graded activity was originally developed for workers with physical limitations (e.g., musculoskeletal disorders) and has been shown to have some positive results on return-to-work times (Loisel et al. 2005), although results from more recent studies are mixed (e.g., Anema et al. 2007). In principle, a graded activity approach includes a modified or reduced schedule, modified work tasks, and workplace accommodations, allowing the worker the opportunity to remain engaged in work activities while symptomatic. Such an approach is beneficial, as it does not require the worker to be symptom-free in order to return to work, thus reducing the

Table 12.2 Summary of work-based interventions for depression and anxiety

Study	Locale/setting	Sample	Duration	Key treatment elements	Significant outcomes[a]
Individual level interventions					
Stand-alone CBT interventions					
Kidd et al. (2008)	Canada; setting not specified	Diagnosed with anxiety or depressive disorder; previous CBT	Five group sessions over 6 weeks	Vocationally focused CBT delivered by clinical psychologist	Improved sense of work competence; increased satisfaction with job supervision
Grime (2004)	UK; NHS OH department	Ten or more sick days due to anxiety, depression, or stress	8 weeks	Computerized CBT	Slight reduction in anxiety/depression symptoms
CBT plus graded activity interventions					
Van der Klink et al. (2003)(see also Brouwers et al. 2006)	Netherlands; in-house OH department of large private company	On sick leave at least 2 weeks for adjustment disorder	6 weeks (four to five sessions)	CBT with graded activity RTW focus; delivered by OH physician	Shorter time to RTW; 30% fewer sick days
Blonk et al. (2006)	Netherlands; private insurance company	Self-employed; sick leave for stress	Five to six sessions (2× per week)	CBT; graded activity; workplace intervention (consultation on accommodations); delivered by case manager	Resumed full work schedule 200 days earlier

Other individual level interventions

Study	Setting	Population/criteria	Duration	Intervention	Outcome
Schene et al. (2007)	Netherlands; academic outpatient psychiatry department	MDD with reduction in work hours of ≥50%	11 months (44 sessions)	Semistructured supportive group individual sessions focused on RTW and depression issues; delivered by OTs	RTW 90 days earlier; worked more hours
van Oostrom et al. (2007)	Netherlands	2–8 weeks of sick leave for adjustment disorder	Not specified	PW facilitation of RTW by OH case manager	RCT underway, no outcome data available
Organizational level interventions					
Godard et al. (2006)	France; large national utility companies	Screened all sick-listed employees, targeted those with anxiety or depression	Two sessions	Brief mental health screening interview; individualized health promotion; both delivered by in-house physician	Greater reduction in self reported anxiety and depressive symptoms; more employees in remission at follow-up
Charbonneau et al. (2005)	USA; large healthcare coalition	13 large companies in metropolitan Kansas City	Not specified	Multiphase intervention: depression needs assessment; educational & clinical interventions; depression community public health campaign	Needs assessment has been completed – no other outcome data available

(continued)

Table 12.2 (continued)

Study	Locale/setting	Sample	Duration	Key treatment elements	Significant outcomes[a]
Combined interventions					
Wang et al. (2007)	USA; managed behavioral healthcare company	Mental health screening in 16 large companies; targeted those with moderate depression severity	Ongoing contact over 12 months (number of sessions determined by employee preference and/or treatment algorithm)	Care manager made treatment referrals, facilitated engagement in treatment, monitored treatment progress, and when appropriate made additional recommendations based on clinical treatment algorithm; delivered CBT; provided psychoeducational materials	Worked more hours (2 weeks more per year); lower self-reported depressive symptoms at post-treatment
Vlasveld et al. (2008)	Netherlands; large OH care service	MDD with 4–12 weeks of sick leave	Maximum duration 18 weeks	Similar elements to Wang et al., with the addition of a PW component; delivered by OH physician with consultation by psychiatrist	RCT underway, no outcome data available

OH, occupational health; CBT, cognitive behavioral therapy; RTW, return to work; NHS, National Health Service; MDD, major depressive disorder; OT, occupational therapy; PW, participatory workplace; RCT, randomized controlled trial

[a] Based on results of research trial comparing intervention to comparison condition (e.g., standard care)

amount of sick leave and preventing the transition to long-term disability, as well as easing the transition back to full-time employment. Finally, several interventions were delivered by occupational case managers who also facilitated workplace accommodations for employees. In some cases these accommodations were part of the graded activity components, in other cases the intervention was more specifically focused on facilitating workplace accommodations (van Oostrom et al. 2007).

We have divided the individual level interventions into three subcategories: stand-alone CBT interventions, CBT plus graded activity interventions, and other individual level work-based interventions. Stand-alone CBT interventions are defined as those that utilize CBT techniques to enhance management of anxiety and depression symptoms in a work context. CBT plus graded activity interventions are embedded within occupational case management and use a treatment protocol with CBT strategies to accelerate the employees' return-to-work. The "other" subcategory includes related problem-solving interventions, as well as a case management protocol designed to facilitate workplace accommodations for individuals with anxiety and depressive symptoms.

Stand-Alone CBT Interventions

Kidd et al. (2008) developed and tested a brief vocationally focused CBT intervention for people with mood and anxiety disorders. The intervention consisted of five group sessions. Four initial sessions were held weekly and focused on CBT strategies for anxiety reduction, job task completion, interpersonal interactions, response to feedback, and other potential work-related stressors. A fifth session was conducted 2 weeks later that focused on problem solving for areas in which participants were still having difficulty. The intervention was delivered by a clinical psychologist and vocational specialist. Results of a small pre-post pilot trial of this intervention ($N=16$) indicated that participants who were employed ($n=8$) experienced an increase in self-reported competence in their job performance and satisfaction with job supervision. Those participants who were unemployed ($n=8$) experienced an increase in self-reported expectancy for vocational success. Data on work outcomes were not collected as part of this trial. Also of note, participants in the study were required to have had a prior course of CBT.

Investigating a slightly different angle, Grime (2004) conducted a feasibility study of a computerized CBT intervention, "Beating the Blues," with employees who had ten or more stress-related sick days in a 6-month time period. The computerized CBT intervention consisted of eight sessions delivered in an educational format, and explored cognitive distortions associated with anxiety and depressive symptoms. The intervention also introduced behaviorally oriented problem-solving strategies. It was implemented at employees' work sites during scheduled work hours. Results of a randomized controlled trial comparing computerized CBT to conventional care indicated a slight advantage for the computerized CBT with regard to reduction in symptoms. No data on work productivity or absenteeism were provided. Additional qualitative data gathered were also informative and highlighted some challenges to implementing this intervention. Several participants declined to participate in the intervention because they preferred face-to-face contact with a therapist. Of those who did participate, some indicated that attending sessions during work time made them uncomfortable (e.g., they felt they had limited privacy), while others indicated that the sessions were quite helpful. The study concluded that offering the computerized CBT in a more flexible format (i.e., online) may afford more access and privacy.

CBT Plus Graded Activity Interventions

Van der Klink et al. (2003) developed and tested a CBT and graded activity intervention designed to accelerate employees' return-to-work who are on sick leave because of an adjustment disorder (e.g., milder symptoms of anxiety and/or depression). The intervention was designed to be delivered by occupational physicians who were part of an in-house occupational health department in a large company of over 100,000 employees in the Netherlands. It used a three-stage CBT intervention focused on problem-solving and behavioral activation. During the first stage, work-related problems or stressors were identified and participants were encouraged to engage in more nondemanding daily activities. During the second phase, problem-solving strategies were developed for the identified stressors. The third phase focused on implementation of these strategies and increasing the participants' activity level to include more demanding work-related tasks. The intervention was relatively brief and delivered over four to five sessions during the first 6 weeks of sick leave. Compared to a group ($n=83$) receiving routine occupational physician care (e.g., nonspecific counseling, psychoeducation about stress), the intervention group ($n=109$) had a significantly more rapid return-to-work and a 30% reduction in sick days during the year following the intervention. There were no significant differences between groups on psychological measures of stress-related symptoms or coping, as both groups significantly improved on these measures. It should be noted that a second study using this intervention with a similar sample failed to replicate the more rapid return-to-work findings (Brouwers et al. 2006). However, in contrast to van der Klink et al., the intervention in this study was delivered by social workers in a primary care setting.

Blonk et al. (2006) also developed and tested an enhanced work-focused CBT plus graded activity intervention to be delivered by an occupational case manager involved in return-to-work. It was designed for individuals who were self-employed and experiencing psychological complaints (stress, anxiety, or depression) resulting in an inability to work. This intervention was conducted in the Netherlands and used labor experts (employed by the insurance company) who functioned as case managers for employees on sick leave or disability. They focused on assisting the employee in returning to work through workplace interventions and accommodations. It was proposed that provision of CBT-related stress management in tandem with workplace intervention and graded activity would enhance employees' ability to return to work. The CBT portion of this intervention consisted of five to six individual sessions focused on stress management, including psychoeducation, relaxation exercises, self-help books, and homework assignments. Hour-long sessions were held at the workplace or home of the employee twice a week. The workplace intervention component consisted of consultation on ways to reduce job demands and increase effectiveness (e.g., prioritization of tasks, planning/organizing tasks, conflict management), with the goal of increasing work activity to a partial or full resumption of work. This enhanced CBT intervention ($n=40$) was compared with a standard CBT intervention ($n=40$) and a no-treatment control group ($n=42$) in a randomized controlled trial. Participants were self-employed and experiencing symptoms of adjustment disorder that resulted in work difficulty and obtaining temporary disability benefits. The interventions began on average 2–3 weeks after disability leave was initiated. The intervention phase of the study (which included a pre- and post-assessment) lasted 4 months, and there was a follow-up assessment at 10 months. Results indicated that the enhanced CBT intervention resulted in a more rapid return to partial and/or full work capacity when compared to the standard CBT protocol and the no-treatment control condition. Participants receiving the enhanced CBT intervention resumed a partial work schedule 17 days earlier than the standard CBT group and 30 days earlier than the control group. The enhanced CBT intervention group resumed a full work schedule approximately 200 days earlier than the comparison groups. All three groups demonstrated a reduction in psychological complaints (e.g., stress, anxiety, depressive symptoms) over the 10-month study regardless of the type of intervention received.

Other Individual Level Interventions

In one of the more intensive interventions to date that specifically targeted employees with more severe depression (i.e., MDD), Schene et al. (2007) designed and tested a multiphase manualized intervention focused on work reintegration. The 11-month (44-session) intervention was delivered through an academic outpatient psychiatry clinic in the Netherlands. It consisted of a five-session diagnostic phase that included personal occupational history, functional assessment, collateral contact with the employee's occupational physician, and development of a return-to-work plan. This was followed by the active intervention phase, which consisted of 24 weekly group sessions (i.e., a semistructured supportive group therapy format) followed by 12 weekly individual sessions. These sessions focused on exploring themes (e.g., work problems, perfectionism, conflicts), as well as planning for and returning to work. Three individual sessions were conducted during a 20-week follow-up phase. This adjunctive occupational therapy (OT) intervention was tested in a randomized controlled trial, comparing OT plus treatment as usual (TAU) to TAU alone. TAU consisted of high-quality outpatient pharmacological treatment for depression.

Sixty-two participants who had experienced a reduction in work hours of at least 50% because of depression were randomized (OT plus TAU $n=30$; TAU alone $n=32$). Outcomes included depressive symptoms, time until return-to-work, hours worked, work stress, and cost effectiveness. Outcomes were assessed at post-treatment and at 2.5-year follow-up. Results indicated that participants receiving the OT intervention returned to work, on average, 92 days sooner than those in the TAU alone condition and worked more hours during the first 18 months of the study period. With regard to depression outcomes, both groups demonstrated a significant decrease in depressive symptoms and a significant reduction in the number of participants who met the criteria for MDD, although there was little indication of the superiority of one treatment over the other. A cost-effectiveness analysis indicated that the OT intervention was more costly as a result of the additional group and individual therapy sessions. However, because participants in the OT plus TAU condition worked more hours, there was a net benefit (earnings subtracted from cost of treatment). That is, the effect of the treatment (more rapid return-to-work) essentially paid for the cost of the intervention. Finally, there was no indication of an increase in stress as a result of return-to-work, and the data did not suggest that working exacerbated depressive symptoms.

Van Oostrom et al. (2007) developed an intervention that also employed an occupational case manager, but in contrast to the other individual level interventions, this case manager functioned as a facilitator between the employee and employer. Specifically, they adapted a *participatory workplace* (PW) intervention for stress-related mental disorders. The PW intervention was originally developed for workers with low back pain and demonstrated a reduction in sick leave and subsequent accelerated return-to-work (Anema et al. 2007). Although the original study protocol focused on low back pain, stress- and mental health-related concerns were frequently identified barriers to return-to-work among these employees, which was the impetus for the present study. The key element of this intervention consisted of work with an occupational health case manager who functioned as a return-to-work facilitator. Through semistructured interviews with both the employee and employer, barriers to return-to-work were identified based on a task analysis. Next, the case manager, employee, and employer participated in a brainstorming session to identify possible solutions or accommodations for identified barriers and the necessary actions to implement them. The rationale for this intervention is that reduction in sick days and more rapid return-to-work has positive benefits for employee mental health and employer productivity and efficiency. No outcome data are available presently for this intervention.

Organizational Level Work-Based Interventions for Anxiety and Depression

Godard et al. (2006) developed and tested a work-based health promotion intervention for anxiety and depressive disorders. The study was conducted in France with sick-listed employees ($N=9,743$) in two large national utility companies with in-house health insurance. Sick-listed employees are required to consult an in-house physician who authorizes sick leave and provides treatment recommendations (but not treatment itself). The intervention was comprised of two key elements. The first was a brief screening interview conducted by the physician that assessed for common depressive and anxiety disorders. Second, the results of the screening interview were discussed with employees, and information about specific disorders and their treatment was provided. Employees were strongly encouraged to consult with their treating physician. They tested the intervention in a quasi-experimental design in which company-based medical centers were allocated to deliver the intervention ($n=8$ active sites) or standard care ($n=13$ control sites). This allowed for the intervention to be delivered country-wide and balanced across local settings (e.g., metropolitan vs. rural). Employees completed self-report symptom questionnaires at baseline, 6 weeks, and 6 months; the screening interview was re-administered at 1 year. Results of the initial screening identified 10.6% ($n=912$) of the sample with an anxiety or depression diagnosis. Of those, 367 received the subsequent health promotion intervention and demonstrated a significant reduction in self-reported symptoms at 6 weeks and 6 months when compared to the group receiving standard care ($n=476$). This group also had a significantly higher portion of employees with remission from their initial diagnosed disorder (51% vs. 40.5%). Employees' subjective assessment of the intervention was largely favorable, as a large proportion reported that the health promotion information and consultation with physician were useful. No data on return-to-work or absenteeism were provided.

Charbonneau et al. (2005) described a multidisciplinary healthcare coalition's development of a work-based multiphase intervention for depression that is currently underway. This intervention includes three phases. Phase 1 consisted of a needs assessment of employees' and employers' knowledge of and attitudes towards depression and its treatment. Phase 2 will consist of education and clinical interventions designed to enhance access to treatment for depression. Phase 3 will consist of a community public health campaign designed to increase depression awareness and reduce stigma. Results are available for the first phase of the intervention, a needs assessment survey. An anonymous survey was completed by employees ($N=6,399$) of 13 companies surrounding Kansas City (a 16% response rate). Respondents were knowledgeable about the signs and symptoms of depression and demonstrated positive attitudes about working with colleagues suffering from depression. Respondents also reported being willing to seek help for depression, with herbal remedies, self-help, primary care, medications, and counseling being the most frequently endorsed treatment options. Respondents less frequently endorsed psychiatric or company programs as treatment options for depression. Also of interest, survey results indicated employees were reticent to discuss depression concerns with their supervisor, and 49% of respondents felt there were few employee depression resources available (even though all employers offered a standard package of mental health benefits). However, 50% of respondents indicated they would access help through an employee assistance program if available. These results offer some options with regard to points of intervention. Most notably, the authors identify improving access to depression care and worksite mental health resources as key areas for future work.

Combined Work-Based Interventions for Anxiety and Depression

There is increasing recognition that a multitude of factors can affect the impact of mood and anxiety on work functioning. These include, for example, access to appropriate treatment, the work environment, job satisfaction, job stress, related psychological factors such as perceived social support, burnout, and stigma and associated discrimination (Bender and Farvolden 2008; Sanderson and Andrews 2006). Increasingly, multicomponent interventions have been developed that attempt to address some of these factors at the individual as well as the organizational level.

In the most comprehensive treatment package to date, Wang et al. (2007) developed and tested a multiphase enhanced depression screening and treatment program that was implemented in a large managed behavioral healthcare company in the USA. Phase 1 began with a health risk appraisal survey that included screening questions for depression. Employees whose screen indicated the possible presence of depression were invited to participate in a telephone survey to assess depression in more detail. Those with at least "moderate" depression severity as measured by this telephone survey were invited to participate in phase 2 of the study. Phase 2 consisted of a randomized controlled trial comparing the intervention ($n=304$) to usual care ($n=300$). A critical component of the intervention was the "care manager," who performed multiple functions, including assessment of need for treatment, facilitation of engagement in recommended treatments, and ongoing monitoring of adherence to treatment and clinical progress. Recommended treatments included in-person psychotherapy, pharmacotherapy, or a combination of both, for which the care manager made the necessary authorization and referrals. An innovative feature of this intervention was its flexible treatment algorithm. Specifically, it accommodated individual employee treatment preferences, a range of severity of depressive symptoms, and the individual's responsiveness to treatment. For example, employees who declined treatment recommendations were provided psychoeducation, self-help materials, and ongoing telephone contact to monitor their symptoms by the care manager. Those employees who continued to experience significant depressive symptoms after 2 months were offered an eight-session CBT intervention delivered by the care manager. For employees who participated in the in-person treatment recommendations, and who continued to experience significant depressive symptoms after 2 months, the addition of a second mode of treatment was recommended (e.g., if receiving psychotherapy, adding pharmacotherapy was recommended). Following active treatment, booster sessions to monitor treatment progress were conducted approximately every 4–8 weeks throughout the duration of the 12-month study period.

The results of the study indicate that participants in the intervention group worked significantly more hours, at a rate equivalent to 2 more weeks per year, relative to the comparison group. Participants in the intervention group also experienced lower levels of depressive symptoms at 6-month and 12-month assessments, had significantly more treatment contacts, and were more likely to obtain specialized mental health care, relative to the comparison group. An additional outcome of interest was that there was no difference in intervention effects between those with mild, moderate, or severe depression. This suggests that the intervention accommodated and effectively treated a range of severity of depressive symptoms.

Vlasveld et al. (2008) have developed a similar multicomponent intervention using a care manager for employees with MDD within the Netherlands occupational health setting. This intervention is designed for employees who have been on sick leave for MDD between 4 and 12 weeks. This time frame was selected in order to intervene early to reduce the likelihood of the employee transitioning to long-term disability. The treatment elements are quite similar to Wang et al. (2007), and include provision of self-help materials and an occupational physician (OP) care manager who makes treatment recommendations (e.g., medications), monitors treatment progress and treatment adherence, and provides a brief manualized CBT intervention. The intervention also included a PW element similar to van Oostrom et al. (2007), whereby the OP care manager acts as a facilitator between employee and employer to develop workplace accommodations to accelerate the employee's return to work. A treatment plan is developed collaboratively between the employee and OP care manager based on treatment recommendations and employee preference. For example, the employee may opt to begin pharmacotherapy immediately, whereby the OP care manager would initiate a medication treatment algorithm. The elements of the intervention are intended to be delivered concurrently (i.e., CBT, self-help, and PW) and monitored every 2 weeks. If depressive symptoms do not respond to treatment, the protocol allows for the addition of more CBT sessions and/or medication or medication changes. No outcome data are available for this intervention and a randomized clinical trial is currently underway.

Summary of Work-Based Interventions, Implications for Clinical Practice, and Barriers to Implementation

While there is some variation in these interventions, there are arguably at least three common clinical foci that emerge: a focus on returning to work, a focus on ensuring optimal treatment, and a focus on combining interventions to maximize effectiveness. In this section we discuss these three themes with an eye toward potential implications for clinical practice as well as potential barriers to implementation.

Focus on Return to Work

Several of the individual level interventions specifically focused on facilitating a return to work. The content of these intervention sessions explicitly addressed the management of symptoms of depression and anxiety in order to facilitate reintegration to the work place. Specifically, these interventions included common CBT and related problem-solving strategies. In addition to these strategies, two specific elements were utilized to enhance progress towards this goal: graded activity and facilitation of workplace accommodations.

Graded Activity

The rationale for use of graded activity is compelling for several reasons. First, it focuses on preventing the long-term occupational disability that can be associated with anxiety and depression by following a specific time course of work activity that is independent of the course of symptoms (van der Klink et al. 2003). The goal is to foster confidence and promote a sense of success, such that individuals can engage in productive work activity even though they may still be experiencing symptoms. Indeed, in naturalistic longitudinal studies of depressed workers, partial and full work resumption had a positive impact on the course of depressive symptoms (Brenninkmeijer et al. 2008). Clinically, this approach is also relevant, as prolonged absence from work can perpetuate negative consequences of depression (isolation, poor self-esteem), and/or heighten the anxiety associated with returning to work (Bilsker et al. 2006). This approach is also quite consistent with the behavioral component of CBT interventions. In the case of depression, increasing activity levels, particularly activities from which one derives a sense of pleasure or mastery, represents a fundamental, effective behavioral intervention for depressive symptoms. Second, in the case of anxiety, exposure to stress and/or feared situations (in the context of therapy) offers the worker opportunity to utilize adaptive coping strategies and correct cognitive distortions associated with these situations (Blonk et al. 2006). In either case, graded activity paired with CBT provides the worker with psychoeducation and cognitive coping strategies to deal with anxiety and depression at work (i.e., CBT), while simultaneously providing in vivo opportunities to practice skills (i.e., graded activity). In contrast, stand-alone CBT cannot necessarily ensure in vivo opportunities for practice.

Data suggest that in at least two studies, CBT plus graded activity yielded more rapid return-to-work than standard care (van der Klink et al. 2003) or standard CBT (Blonk et al. 2006). A third study, while not explicitly utilizing a graded activity protocol, specifically targeted planning for return-to-work in group- and individual-based psychotherapy sessions and also yielded accelerated return-to-work times (Schene et al. 2007). In all three studies, there was no evidence of an increase in symptoms upon return-to-work, suggesting that the stress of reintegration to the workplace was adequately managed by the intervention design. In part, this may have been the result of coordination between the interventionist and employer in the facilitation of return-to-work in these interventions.

It should be noted that an additional study (Brouwers et al. 2006) that utilized a similar intervention as van der Klink failed to yield a positive benefit with regard to return-to-work times. These authors suggest that this may be the result of delivering the intervention at a more distant site from the workplace, or less rigorous fidelity to the graded activity component of the intervention. Interventions with favorable return-to-work outcomes were delivered by occupational case managers or physicians more closely linked to the workplace. Such an approach is innovative in that it seeks to maximize the effectiveness of these occupational personnel (who already have considerable knowledge of the workplace and job tasks) to address the individual employee's specific difficulties in returning to work. Determining by whom and in which context an intervention should be delivered warrants further research.

Clinical implications. There appears to be some functional and clinical benefit from the resumption of work activities among workers with depression and anxiety. While this is a clear focus of vocational rehabilitation practitioners, it may be less so among mental health practitioners. The inclusion of work-related treatment goals among clinicians and, when possible, coordination with vocational rehabilitation practitioners, may have the potential to reduce long-term disability. Indeed, the impetus for several of these interventions was the recognition that the occupational consequences of depression and anxiety are often overlooked in standard care.

Barriers to implementation. Several caveats are warranted regarding the graded activity interventions. First, these interventions have largely been tested among individuals with less severe symptoms (e.g., adjustment disorder). Of the interventions discussed, only one with a return-to-work focus was tested with individuals diagnosed with MDD. This intervention was more intensive and of longer duration than those interventions developed for stress-related mental disorders or adjustment disorders (e.g., 44 sessions vs. 5 sessions). While the graded activity data for depression are promising and the rationale for the intervention is clinically intuitive, no graded activity studies with more severe anxiety disorders have been conducted. Implementation of graded activity for more severe disorders will likely require a longer duration intervention and, given the recurring nature of many of these disorders (e.g., MDD), maintenance or booster sessions may be necessary. More research is needed in order to identify clear clinical guidelines for such an intervention. For example, with regard to more severe disorders, what degree of symptom resolution should be attained before return-to-work is initiated? What is the optimal duration of the intervention, and what if any maintenance treatment should be provided?

A second implementation limitation is the requirement of coordination of the graded activity schedule with the employer. This represents a substantial hurdle for both the employee, who may be required to disclose their mental health problem to obtain this accommodation; and the employer, who will, for example, have to reorganize other aspects of the workplace in order to accommodate this employee. Indeed, one of the graded activity studies was conducted with self-employed individuals for the specific purpose of obviating some of these challenges (Blonk et al. 2006). Further, it should be noted that graded activity interventions have primarily been conducted in the Netherlands, where there is comprehensive occupational health coverage. This country has also implemented preventive measures and legislation designed to reduce short- and long-term disability claims, including occupational health treatment guidelines (Buijs et al. 2007). It is unclear as to how graded activity interventions would be implemented in other healthcare contexts (e.g., US managed care).

In sum, a graded activity approach is promising, and its combination with CBT has a particularly appealing clinical rationale. However, its effectiveness with more severe disorders, and clinical guidelines for its implementation, need to be more fully developed. Furthermore, not all individuals may be willing to disclose a mental health problem in order to obtain accommodations for a modified return-to-work schedule. As discussed below, the issue of disclosure is a well-recognized challenge to employment and mental health interventions (refer to the chapter by MacDonald-Wilson et al. (2011), this volume, for further discussion of disclosure).

Facilitation of Workplace Accommodations

In addition to graded activity return-to-work components, two individual-based interventions explicitly focused on making accommodations in the workplace for individuals with anxiety and depressive disorders. The Blonk et al. (2006) enhanced CBT intervention used case managers who were specialists in occupational health, as well as work efficiency and processes, with expertise in workplace accommodations. Van Oostrom et al. (2007) adapted a PW to enhance return-to-work times. This intervention used a case manager to deliver a structured three-session protocol designed to facilitate communication between employees and employers about workplace adaptations to accommodate anxiety and depression symptom-related limitations. A PW component has recently also been incorporated into multifaceted intervention for workers with MDD (Vlasveld et al. 2008). While PW interventions have been found to be effective in decreasing return-to-work times in workers with physical disabilities or injuries (Anema et al. 2007), no data on their effectiveness for individuals with anxiety or depressive disorders are available, although trials are currently underway.

Clinical implications. Workplace accommodations have been a significant component of vocational rehabilitation for individuals with psychiatric disabilities for decades (Granger et al. 1997). As outlined by Granger et al. (1997), examples of common categories of accommodations for individuals with psychiatric disabilities include: flexible scheduling (e.g., changes in work schedule to accommodate medical appointments); job description modification (e.g., removing tasks from an individual's job responsibilities or changing job task requirements); communication facilitation (e.g., adding alternative ways or changing the way information is communicated by a supervisor, such as having an on-site job coach to assist in relaying information); and physical space accommodations (e.g., having access to a work area with reduced noise to minimize distractions).

Many accommodations can assist individuals experiencing symptoms of anxiety and depression to function better at work. While anxiety and depression are diagnostically distinct, they share several overlapping functional limitations. These include cognitive limitations (distraction, poor concentration), interpersonal limitations (social withdrawal or avoidance), motivational limitations (low energy sedation related to medication side effects), as well as periodic symptom exacerbations (i.e., an episodic course of disorder). Some examples of accommodations relevant for these common limitations that may impact job performance are shown in Table 12.3.

Surveys and qualitative interviews offer additional insight into accommodation strategies for anxiety and depression. Waghorn and Chant (2005) found that almost one-fourth of individuals with an anxiety disorder in a large population survey indicated the need for a "support person" (i.e., a job coach) in order to be able to work. This is consistent with findings from surveys of other individuals with psychiatric disabilities, who indicate that a job coach is among the most frequent accommodations (Granger et al. 1997). Individuals with mood disorders frequently develop effective self-management strategies to cope with symptoms of depression that parallel some of the listed accommodations. These include: reducing stress, workload, and/or stimulation when experiencing symptom exacerbations; using the work routine to maintain structure; drawing on support from trusted coworkers; and seeking assistance from their mental health clinicians (Michalak et al. 2007).

There is information on reasonable accommodations for psychiatric disabilities that is publicly available. The Job Accommodations Network (JAN) is one of the most comprehensive resources for accommodations. Sponsored by the US Department of Labor's Office of Disability Employment Policy, JAN offers consultation to employees with disabilities, employers, and rehabilitation professionals on reasonable accommodations as well as extensive web-based information (http://www.jan.wvu.edu/links/about.htm). The US Equal Employment Opportunity Commission offers information on the Americans with Disability Act (ADA) with regard to psychiatric disabilities, including examples of reasonable accommodations (http://www.eeoc.gov/policy/docs/psych.html; see also the Center (2011) chapter, this volume, for additional information on the ADA).

Barriers to implementation. While accommodations have been used to help individuals with mental health disabilities maintain employment, there are still many barriers to their implementation. Some accommodations are informal arrangements (see Granger et al. 1997). However, in formal cases, reasonable accommodations, as defined by the ADA, require an employee to disclose a disabling condition to their employer in order to obtain this accommodation. By far the most commonly cited barrier to accommodations is that of stigma associated with disclosure of a disability (Conyers and Ahrens 2003). Individuals with anxiety and mood disorders commonly report feeling stigmatized at work as a result of their mental health condition (e.g., Haslam et al. 2005; Michalak et al. 2007). This leads to reluctance to disclose the condition to a supervisor and, in turn, receive disability-related accommodations. Further evidence for this reluctance can be seen in the

Table 12.3 Examples of workplace accommodations for common functional limitations associated with mood and anxiety disorders[a]

Functional limitation	Accommodation
Cognitive(*e.g., poor concentration, disorganization, and/or poor time management, resulting in reduced work quality or efficiency*)	– Relocate workspace to quieter area to reduce distraction – Provide divider, partition, or soundproofing at workspace – Modify job to include less complex tasks or restructure tasks into smaller steps with more frequent supervisor feedback – Provide written instructions/checklists for assigned job tasks
Interpersonal(*e.g., social withdrawal, social anxiety and avoidance, and/or irritability resulting in reduced interpersonal effectiveness*)	– Increase job structure and supervision – Provide job coach support (either on site or available via telephone) – Provision of a coworker buddy – Modify job tasks to limit public contact
Motivational(*e.g., low energy, motor retardation, and/or sedation resulting in slower work pace*)	– Provide regular supportive feedback – Adjust work schedule to times when individual can function best – Increase breaks – Modify job to include periodic tasks with more physical activity
Symptom exacerbations(*e.g., increased stress, reoccurrence of acute symptoms, resulting fluctuations in clinical presentation and functioning*)	– Permit the use of accrued paid leave or unpaid leave for treatment-related activities – Permit the use of flexible work schedule – Allow for a reduced work schedule (fewer hours per shift) – Modify job to include less demanding job tasks

[a] Adapted from Conyers and Ahrens (2003), EEOC (1997), and Granger et al. (1997)

intervention studies reviewed above. Some individuals participating in an on-site computerized CBT intervention reported significant concerns about their coworkers' perceptions when participating in this worksite-based intervention, even though it was computerized and provided in a private location (Grime 2004). Many respondents to a needs assessment survey were reticent to discuss their mental health condition with a supervisor, and only 50% indicated they would access treatment through an employee assistance program if available (Charbonneau et al. 2005). The issue of stigma is a prominent target of one of the organizational level interventions, which includes a community public health campaign designed to increase depression awareness and decrease stigma (Charbonneau et al. 2005).

In sum, facilitating workplace accommodations for workers with anxiety and depression appears to be worthwhile from a rehabilitation perspective. However, this needs to be accomplished on a case-by-case basis depending on worksite, employer, and the employee's decision to disclose their disability. While use of occupational specialists and the PW interventions to facilitate return-to-work are promising, data from pending trials are needed before clear clinical guidelines can be established. Even then, the existing stigma toward mental illness presents a substantial barrier to implementation of these types of interventions. Refer to the chapter by Baldwin and Marcus (2011), this volume, for further discussion of stigma.

Focus on Ensuring Optimal Treatment of Anxiety and Mood Disorders

Ensuring optimal treatment is a critical clinical issue for workers with mood and anxiety disorders for several reasons. First, well-established evidence-based pharmacological and psychotherapeutic interventions are available for the treatment of depression and anxiety. Second, several studies indicate that adequate treatment of anxiety and depression is associated with improved work functioning and vocational outcomes. Third, in spite of the availability of evidence-based treatments that can impact work outcomes, several survey studies indicate anxiety and depression are often under- or untreated. In large part, these findings provide the impetus for several of the interventions. Notably, interventions developed by Wang et al. (2007) and Vlasveld et al. (2008) both target MDD and utilize an active case management model. Identified depressed employees are followed by a care manager who provides psychoeducation and treatment recommendations, and monitors treatment adherence and progress. Both interventions use a treatment algorithm in order to ensure an optimal treatment response, such that among individuals who continue to experience symptoms, additional treatment options are provided (e.g., adding psychotherapy to pharmacotherapy). This active case management model focuses on ensuring that each employee receives the best treatment, but also seeks to maximally engage employees in their treatment. Results of the research of Wang et al. are promising and have demonstrated clinical and vocational, as well as some economic benefit (i.e., reduced absenteeism). The latter finding increases the chance that this intervention may be more widely adopted by employers and managed care companies interested in reducing the economic burden associated with depression in the workplace.

A related commonality across interventions regarding optimal treatment was the utilization of CBT strategies. The evidence base for CBT for depression and anxiety is strong. Thus it is logical to incorporate these strategies into work-based interventions that focus on specific situations, such as returning to work after a period of sick leave. CBT interventions have additional advantages in that they are often manualized, thus making them more easily implemented in a variety of settings and, provided they are adequately trained, by a variety of interventionists. Several interventions have successfully used occupational case managers (Blonk et al. 2006), occupational health physicians (van der Klink et al. 2003; Vlasveld et al. 2008), and masters level mental health clinicians (Wang et al. 2007) to deliver CBT. CBT treatment manuals and accompanying client workbooks as well as CBT clinical reference materials are widely available.

Clinical implications. For both mental health clinicians and rehabilitation practitioners there are some key clinical implications. First, for the mental health clinician there is a need to ensure that the worker is receiving the best treatment available for the identified symptoms. Not surprisingly, this notion largely parallels the active case management model discussed above. This model may involve delivery of CBT or related evidence-based interventions for anxiety and depression, monitoring of treatment progress, and/or referral for pharmacological treatment. Similarly, rehabilitation counselors should be equipped with knowledge about appropriate treatments for anxiety and depression in order to make appropriate mental health referrals where possible. Second, as discussed above in the context of graded activity interventions, work- related symptom complaints or difficulties should routinely be assessed. This is likely more commonplace in the clinical practice of rehabilitation counselors, but assessment by the mental health clinician can facilitate the tailoring of mental health treatment in order to reduce the occupational impairment associated with these disorders (e.g., developing symptom coping strategies for work; developing behavioral goals to reduce absenteeism). Third, inherent in ensuring optimal treatment is the recognition of the potential for relapse following a period of symptom remission. Given the episodic course of mood disorders and many anxiety disorders, mental health and rehabilitation counselors alike should work with clients to develop relapse prevention strategies. These may include the use of ongoing maintenance treatment, either with medication or regular booster psychotherapy sessions, as well as regular assessment for breakthrough symptoms.

Barriers to implementation. Several barriers may interfere with ensuring optimal treatment for workers with anxiety and mood disorders. One limitation to utilization of CBT strategies for work-based interventions is that it requires training and experience to be delivered effectively. Many of the work-based CBT interventions were brief and time-limited and used a manualized protocol. However, in these studies, implementation of CBT was supervised by clinicians with expertise in this area, in order to ensure the quality of the intervention. For less experienced clinicians, the availability of CBT supervision may not be easily accessible, which raises the risk that implementation may be suboptimal, thereby reducing CBT's effectiveness.

A larger, more global barrier to ensuring optimal treatment includes the availability of and access to such treatment. Survey research findings indicate that access to appropriate treatment is influenced by several factors, including the quality of health insurance coverage (Kessler et al. 2008; Wang et al. 2000; Young et al. 2001). Further, data indicate that anxiety and depressive disorders are frequently treated in primary care settings rather than mental health settings (which are more likely to deliver optimal treatment, Kessler et al. 2008) and that anxiety and depression symptoms are underdiagnosed in primary care settings (Godard et al. 2006). Finally, survey data from workers indicate some reluctance to accessing more specialized mental health care due to concerns about stigma (e.g., Charbonneau et al. 2005). Thus, multiple factors at both the organizational and individual level, many of which are beyond the control of practitioners, can impede efforts to ensure that workers with anxiety and depression obtain optimal treatment. Such barriers require more comprehensive approaches, which we briefly discuss below.

In summary, delivery of optimal treatment for depression and anxiety is a critical feature of many work-based interventions. Outcomes of these interventions echo findings that effective treatment results in both symptom remission and a reduction in occupational impairment. Use of evidence-based practices, such as CBT, in a work-focused manner shows promise for reducing symptom-related distress in order to enhance work functioning. However, while effective treatment for workers with mood and anxiety disorders is a worthy goal with substantial benefits to employees and employers alike, many factors can present substantial barriers to achieving this goal.

Focus on Combined Interventions to Maximize Effective Treatment of Mood and Anxiety Disorders

Recognition of some of the barriers to delivery of effective treatments for workers with mood and anxiety disorders has informed the development of several work-based treatment programs that combine organizational and individual level interventions. We have highlighted some of these in the previous section, but will briefly discuss some of the organizational level components. Several interventions have utilized screening measures to identify workers who may potentially be experiencing anxiety and depressive symptoms. Godard et al. (2006) found that a simple screening procedure combined with a brief psychoeducational health promotion intervention resulted in significant reductions in anxiety and depressive symptoms. In recognition of the poor identification of depression in the workplace, Wang et al. (2007) used a brief self-report screen to identify workers who might benefit from their work-based intervention for depression. In addition, several interventions under development worked to incorporate stakeholders in designing the intervention in order to enhance implementation effectiveness. For example, van Oostrom et al. (2007) utilized focus group interviews with employees and employers in the development of their PW intervention to enhance return-to-work times for employees with stress-related mental disorders. Charbonneau et al. (2005) have organized a comprehensive community initiative that is currently underway. This intervention seeks to target some of the more systems level barriers such as stigma through utilization of public awareness campaigns, employment-based education programs, and employer benefits policies. As discussed above, the first phase of this community initiative consisted of a needs assessment, the results of which will inform the latter phases of this program.

Conclusion

This chapter highlights some recent developments in work-based interventions for anxiety and depression. These interventions included either one or more of the following: work-focused CBT, graded activity, workplace accommodations and supports, appropriate screening, and optimal pharmacological and psychosocial treatment. Increasingly, there appears to be a shift toward combined or multicomponent treatment packages. In part, this development can be traced to the recognition that several other situational factors (e.g., stigma, access to appropriate treatment) have the potential to impact the treatment of psychiatric disorders in a work context. These combined interventions are just beginning to address these issues, and more innovation and development are needed to tackle the complex nature and severity of these mental health problems as they relate to the workplace. Nonetheless, there are some promising findings that have the potential to inform clinical practice.

Acknowledgements Many thanks to Christine Calmes Ph.D. for her thoughtful comments on earlier drafts of this paper.

References

Adler DA, McLaughlin TJ, Rogers WH, Chang H, Lapitsky L, Lerner D (2006) Job performance deficits due to depression. Am J Psychiatry 163:1569–1576

American Psychiatric Association (2000) Diagnostic and statistical manual of mental disorders, 4th edn, TR. Author, Washington, DC

American Psychiatric Association (2006) Practice guidelines for the treatment of psychiatric disorders: compendium. Author, Washington, DC

Anema JR, Steenstra IA, Bongers PM, de Vet HCW, Knol DL, Loisel P et al (2007) Multidisciplinary rehabilitation for subacute low back pain: Graded activity or workplace intervention or both? A randomized clinical trial. Spine 32:291–298

Baldwin M, Marcus S (2011) Stigma, discrimination, and employment outcomes among persons with mental health disabilities. In: Schultz IZ, Sally Rogers E (eds) Handbook of work accommodation and retention in mental health. Springer, New York

Barlow DH, Lehman CL (1996) Advances in the psychosocial treatment of anxiety disorders: Implications for national health care. Arch Gen Psychiatry 53:727–735

Beck AT, Rush AJ, Shaw BF, Emery G (1987) Cognitive therapy for depression. Guilford, New York

Bender A, Farvolden P (2008) Depression and the workplace: A progress report. Curr Psychiatry Rep 10:73–79

Berndt ER, Finkelstein SN, Greenberg PE, Howland RH, Keith, A (1998) Workplace performance effects from chronic depression and its treatment. J Health Econ 17:511–535

Bilsker D, Wiseman S, Gilbert M (2006) Managing depression-related occupational disability: A pragmatic approach. Can J Psychiatry 51:76–83

Blonk RWB, Brenninkmeijer V, Lagerveld SE, Houtman ILD (2006) Return to work: A comparison of two cognitive behavioural interventions in cases or work-related psychological complaints among the self-employed. Work Stress 20:129–144

Brenninkmeijer V, Houtman I, Blonk R (2008) Depressed and absent from work: Predicting depressive symptomatology among employees. Occup Med 58:295–301

Brouwers EPM, Tiemens BG, Terluin B, Verhaak PFM (2006) Effectiveness of an intervention to reduce sickness absence in patients with emotional distress or minor mental disorders: A randomized controlled effectiveness trial. Gen Hosp Psychiatry 28:223–229

Buijs PC, van Dijk FJH, Evers M, van der Klink JJL, Anema H (2007) Managing work-related psychological complaints by general practitioners, in coordination with occupational physicians: A pilot study developing and testing a guideline. Ind Health 45:37–43

Center C (2011) Law and job accommodation in mental health disability. In: Schultz IZ, Sally Rogers E (eds) Handbook of work accommodation and retention in mental health. Springer, New York

Charbonneau A, Bruning W, Titus-Howard T, Ellerbeck E, Whittle J, Hall S et al (2005) The community initiative on depression: Report from a multiphase work site depression intervention. J Occup Environ Med 47:60–67

Conyers LM, Ahrens C (2003) Using the Americans with Disabilities Act to the advantage of persons with severe and persistent mental illness: What rehabilitation counselors need to know. Work 21:57–68

Dobson KS (1989) A meta-analysis of the efficacy of cognitive therapy for depression. J Consult Clin Psychol 57:414–419

Equal Employment Opportunity Commission (1997) Enforcement guidance on the Americans with Disabilities Act and psychiatric disabilities. Retrieved January 12, 2009, from http://www.eeoc.gov/policy/docs/psych.html

Frank E, Soreca I, Swartz HA, Fagiolini AM, Mallinger AG, Thase ME et al (2008) The role of interpersonal and social rhythm therapy in improving occupational functioning in patients with bipolar I disorder. Am J Psychiatry 165:1559–1565

Gloaguen V, Cottraux J, Cucherat M, Blackburn I (1998) A meta-analysis of the effects of cognitive therapy in depressed patients. J Affect Disord 49:59–72

Godard C, Chevalier A, Lecrubier Y, Lahon G (2006) APRAND Programme: An intervention to prevent relapses of anxiety and depressive disorders. First results of a medical health promotion intervention in a population of employees. Eur Psychiatry 21:451–459

Granger B, Baron R, Robinson S (1997) Findings from a national survey of job coaches and job developers about job accommodations arranged between employers and people with psychiatric disabilities. J Vocat Rehabil 9:235–251

Greenberg PE, Sisitsky T, Kessler RC, Finkelstein SN, Berndt ER, Davidson JRT et al (1999) The economic burden of anxiety disorders in the 1990s. J Clin Psychiatry 60:427–435

Grime PR (2004) Computerized cognitive behavioural therapy at work: a randomized controlled trial in employees with recent stress-related absenteeism. Occup Med 54:353–359

Haslam C, Atkinson S, Brown SS, Haslam RA (2005) Anxiety and depression in the workplace: effects on the individual and organization (a focus group investigation). J Affect Disord 88:209–215

Hofmann SG, Smits JAJ (2008) Cognitive-behavioral therapy for adult anxiety disorders: a meta-analysis of randomized placebo-controlled trials. J Clin Psychiatry 69:621–632

Kessler RC, Akiskal HS, Ames M, Birnbaum H, Greenberg P, Hirschfield R et al (2006) Prevalence and effects of mood disorders on work performance in a nationally representative sample of U.S. workers. Am J Psychiatry 163:1561–1568

Kessler RC, Merkangas KR, Wang PS (2008) The prevalence and correlates of workplace depression in the national comorbidity survey replication. J Occup Environ Med 50:381–390

Kidd SA, Boyd GM, Bieling P, Pike S, Kazarian-Keith D (2008) Effect of a vocationally-focused brief cognitive behavioral intervention on employment-related outcomes for individuals with mood and anxiety disorders. Cogn Behav Ther 37:247–251

Lerner D, Adler D, Hermann R, Rogers W, Chang H, Thomas P, Greenhill A, Katherine Perch (2011) Depression and work performance: The Work and Health Initiative Study. In: Schultz IZ, Sally Rogers E (eds) Handbook of work accommodation and retention in mental health. Springer, New York

Loisel P, Buchbinder R, Hazard R, Keller R, Scheel I, van Tulder M et al (2005) Prevention of work disability due to musculoskeletal disorders: The challenge of implementing evidence. J Occup Rehabil 15:507–524

MacDonald-Wilson KL, Russinova Z, Rogers ES, Lin CH, Ferguson T, Dong S, Kash MacDonald M (2011) Disclosure of mental health disabilities in the workplace. In: Schultz IZ, Sally Rogers E (eds) Handbook of work accommodation and retention in mental health. Springer, New York

Michalak EE, Yatham LN, Maxwell V, Hale S, Lam RW (2007) The impact of bipolar disorder upon work functioning: A qualitative analysis. Bipolar Disord 9:126–143

Mintz J, Mintz LI, Arruda MJ, Hwang SS (1992) Treatments of depression and the functional capacity to work. Arch Gen Psychiatry 49:761–768

Nierenberg AA, Ostacher MJ, Huffman JC, Ametrano RM, Fava M, Perlis RH (2008) A brief review of antidepressant efficacy, effectiveness, indications and usage for major depressive disorder. J Occup Environ Med 50:428–436

Nieuwenhuijsen K, Bultmann U, Neumeyer-Gromen A, Verhoen AC, Verbeek J, van der Feltz-Cornelis CM (2008). Interventions to improve occupational health in depressed people (Review). Cochrane Database of Syst Rev, Issue 2. Art No.: DCOO637. doi:10.1002/14651858. CD006237.pub2

Sanderson K, Andrews G (2006) Common mental disorders in the workforce: recent findings from descriptive and social epidemiology. Can J Psychiatry 51:63–75

Schene AH, Koeter MWJ, Kikkert MJ, Swinkels JA, McCrone P (2007) Adjuvant occupational therapy for work-related major depression works: Randomized controlled trial including economic evaluation. Psychol Med 37:351–362

Schoenbaum M, Unutzer J, McCaffrey D, Duan N, Sherbourne C, Wells KB (2002) The effects of primary care depression treatment on patients clinical status and employment. Health Serv Res 37:1145–1158

van der Klink JJL, Blonk RWB, Schene AH, van Dijk FJH (2003) Reducing long term sickness absence by an activation intervention in adjustment disorders: A cluster randomized controlled design. Occup Environ Med 60:429–437

van Oostrom S, Anema JR, Terluin B, Venema A, de Vet HCW, van Mechelen W (2007) Development of a workplace intervention for sick listed employees with stress-related mental disorders: Intervention mapping as a useful tool. BMC Health Serv Res 7:127

Vlasveld MC, Anema JR, Beekman ATF, van Mechelen W, Hoedeman R, van Marwijk HWJ et al (2008) Multidisciplinary collaborative care for depressive disorder in the occupational health setting: Design of a randomized controlled trial and cost-effectiveness study. BMC Health Serv Res 8:99

Waghorn G, Chant D (2005) Employment restrictions among persons with ICD-10 anxiety disorders: Characteristics from a population survey. J Anxiety Disord 19:642–657

Waghorn G, Chant D, White P, Whiteford H (2005) Disability, employment and work performance among people with ICD-10 anxiety disorders. Aust NZ J Psychiatry 39:55–66

Wald J (2011) Anxiety disorders and work performance: In: Schultz IZ, Sally Rogers E (eds) Handbook of work accommodation and retention in mental health. Springer, New York

Wang PS, Berglund P, Kessler RC (2000) Recent care of common mental disorders in the United States prevalence and conformance with evidence-based recommendations. J Gen Intern Med 15:284–292

Wang PS, Simon GE, Avorn J, Azocar F, Ludman EJ, McCulloch J et al (2007) Telephone screening, outreach, and care management for depressed workers and impact on clinical and work productivity outcomes: a randomized controlled trial. J Am Med Assoc 298:1401–1411

Weissman MM, Markowitz JC, Klerman GL (2007) Clinician's quick guide to interpersonal psychotherapy. Oxford University Press, New York

Young AS, Klap R, Sherourne CD, Wells KB (2001) The quality of care for depressive and anxiety disorders in the United States. Arch Gen Psychiatry 58:55–61

Chapter 13
Employment Interventions for Persons with Mild Cognitive Disorders

Robert T. Fraser

Introduction

For purposes of this chapter, we will focus our review on people with mild traumatic brain injury (TBI) and the employment interventions for this population. There are many similarities between mild cognitive impairment (MCI) and mild TBI. MCI has been reported to be between 2.8 and 5.3% of the population (Larrieu et al. 2002) and typically presents as issues with memory, inefficient attention and concentration, and concerns with cognitive fatigue. Londos et al. (2008) indicate that these issues relate more to dementia in the elderly, but in a general cohort, issues such as vascular changes and depression can affect areas of cognitive functioning. Unfortunately, the annual rate of progression to dementia in MCI is in the 12–15% range (Palmer et al. 2003). People with other disabilities such as multiple sclerosis can also have cognitive limitations including fluctuations in memory capacity, word fluency, and speed of information processing (Clemmons et al. 2004). To understand clinical and vocational rehabilitation issues associated with MCI, we will utilize the mild brain injury paradigm because the cognitive issues at 6 months to a year post-injury are relatively stable. With MCI, rehabilitation and job coach personnel are generally dealing with a relatively stable condition rather than progressive cognitive deterioration, which may lead to supported or sheltered employment settings.

Perspectives on Mild Traumatic Brain Injury

On an annual basis, 1.5–2 million individuals in the United States experience a TBI, with an estimated 25% of the individuals requiring hospitalization (Langlois et al. 2003). Approximately 90% of all traumatic brain injuries are estimated to be mild (Kraus et al. 1996) and the preponderance of these will often be unreported (Kay 1993).

R.T. Fraser (✉)
Harborview Medical Center, Box 359745325 Ninth Avenue, Seattle, WA, 98104, USA
e-mail: rfraser@u.washington.edu

I.Z. Schultz and E.S. Rogers (eds.), *Work Accommodation and Retention in Mental Health*, 263
DOI 10.1007/978-1-4419-0428-7_13, © Springer Science+Business Media, LLC 2011

The American Congress of Rehabilitation Medicine proposed the following definition of mild TBI in 1993:

"A mild traumatic brain injury is a traumatically induced physiological disruption of functioning as manifested by at least one of the following:

1. *any period of loss of consciousness;*
2. *any loss of memory for events immediately before or after the accident;*
3. *any alteration in mental state at the time of the accident;*
4. *focal neurological deficit that may or may not be transient but where the severity does not exceed the following: (1) loss of consciousness approximately 30 minutes or less, (2) an initial Glasgow Coma Scale of 13 to 15 after 30 minutes, and (3) posttraumatic amnesia of no more than 24 hours in duration."*

It must be noted that with TBI, there are a number of other uncommon somatic and emotional complaints that include headache, pain, visual disturbance, dizziness, fatigue, depression, and irritability, in addition to cognitive concerns such as attention and memory. As noted by Comper et al. (2005), many individuals will report only one or two symptoms. In a number of cases, concern is not for the cognitive issues, which are generally transitory, but rather the treatment of headache, pain, balance issues, or mood disturbance. Most of these problems will resolve within 3–6 months post-injury and many individuals will receive no formal treatment (Duff et al. 2002). There is, however, a subgroup of approximately 5–15% of individuals who continue to have persistent dysfunction often involving somatic or psychosocial issues.

Some individuals with a mild TBI experience persistent cognitive concerns which often relate, in actuality, to a complicated mild TBI with intracranial lesions, which presents a much greater challenge. Kashluba et al. (2008) have shown that at one year post-injury, those with complicated mild TBI have similar impairments in the areas of memory, speed of information processing, verbal fluency, and abstractive capabilities to those with a moderate TBI. Individuals with complicated mild TBI will thus more likely need employment support or accommodations when returning to work (see Chap. 8, Guilmette and Giuliano (2010) on moderate to severe TBI, in this book). Fluctuating accounts of success in the return-to-work among those with mild TBI are probably due to this lack of differentiation between mild and complicated mild TBI, the latter involving more chronic residual brain impairment. Nevertheless, a number of specific predictors of work status in mild TBI cases were identified: premorbid intelligence, premorbid internalizing psychiatric difficulties, the interactions of race by region of current residence, education by period of altered consciousness, and race by region of current residence by period of altered consciousness. Presence of altered consciousness was not predictive of work status but interacted with other variables to negatively affect work status (Vanderploeg et al. 2003). Empirically derived risk factors for disability are important considerations in early intervention and rehabilitation in mild TBI, although more research is still needed in the field.

Review of Treatment for Uncomplicated Mild Traumatic Brain Injury

There have been two recent systematic reviews of treatment for chiefly uncomplicated mild TBI by Comper et al. (2005) and Snell et al. (2009). These reviews suggest a number of considerations and steps prior to vocational rehabilitation.

Education and Reassurance

First and foremost, education and reassurance regarding expected recovery is needed. Information about the transitory nature of a number of symptoms including memory and concentration concerns, and somatic problems such as balance, dizziness and fatigue as well as advice about managing these symptoms is needed. There should be a discussion of expected resumption of normal pre-injury activities with the person who experienced the injury.

Scope and Intensity of Rehabilitation

There appears to be little benefit, across all individuals with mild TBI, from intensive multidisciplinary treatment. This is not necessarily true in relation to complicated mild TBI and uncomplicated mild TBI associated with comorbid conditions such as chronic pain, depression, or posttraumatic stress disorder (PTSD). Both Ghaffar et al. (2006) and Elgmark-Andersson (2007) had negligible findings with their randomized intervention studies. Although routine treatment for all mild TBI patients was of limited benefit, Ghaffar et al. did find some significant benefit for a subgroup of participants, namely those with mild brain injury history and a mental health disorder.

Cognitive Rehabilitation

There appears to be no established benefit related to intensive cognitive rehabilitation interventions focused on attention, memory, speed of information processing, etc., as reviewed by Comper (2005) and Snell et al. (2009). Although studies in this area have had a number of internal validity problems, the intensity, resources, and expense of these neuropsychological rehabilitation interventions do not appear to be justifiable (Snell et al. 2009).

Preexisting and Concurrent Mental Health Vulnerabilities

There can be value in identifying subgroups with mild TBI for whom specific treatment approaches might be of particular use, such as those with psychiatric and trauma histories. This would include, but not be limited to, preexisting anxiety disorders, PTSD as a function of the injury, and to those reporting an excessive number of symptoms in the early period following injury, as indicated by Elgmark-Andersson et al. (2007).

Focus on Most Prominent Symptoms

Those with mild TBI and other presenting symptoms may receive a postconcussive syndrome diagnosis which can involve multiple cognitive, emotional, and somatic concerns. However, the focus should remain on treating the most prominent symptoms, such as headache or depression, which create the majority of dysfunction. In many cases, a focus on the most prominent symptoms, such as impairing headaches or acute PTSD symptoms, can lead to considerable return of functionality. Pharmacology, including antidepressants and anxiety medications, can also result in significant improvement.

Psychological Assessment

For those who have difficulty returning to work and appear to have no identifiable persistent work-related cognitive issues, a detailed psychological assessment can be extremely important. Such an evaluation can reveal preexisting dysfunctional personality styles and learning disabilities, significant trauma history, poor coping, mental health disorders such as depression or anxiety, substance abuse, and chronic pain and concurrent major stressors that need to be recognized and addressed in the rehabilitation process. These factors, as opposed to direct injury effects, can result in an increasingly complex clinical picture (Snell et al. 2009). Shames et al. (2007) and Friedland and Dawson (2001) emphasize the comorbidity between mild TBI and PTSD, particularly involving motor vehicle accidents in which there is a high incidence of PTSD. Friedland and Dawson (2001) emphasize that much of the cognitive symptomology in disability can be more related to the PTSD than brain injury sequelae per se.

Contextual Factors

There are many difficult to disentangle psychosocial factors that affect disability following mild TBI (Kay et al. 1992). Understanding the individual's life situation at the time of the injury can be a helpful diagnostic starting point and is consistent with a comprehensive, biopsychosocial perspective to cognitive and vocational rehabilitation. Personal and life stressors at the time of the injury, such as an impending divorce or a business that is about to fail, may interact with life disruption created by postconcussive symptoms and medico–legal factors such as involvement of workers' compensation or automotive insurance company litigation. Litigation has been implicated in poorer functional outcomes in mild TBI (Binder and Rohling 1996; Cassidy et al. 2004; Greiffenstein et al. 1994), although the interpretations of these findings vary. At the time of intake into a vocational rehabilitation program, it is important to identify whether a litigation or work-related disability claim is currently being pursued or will be pursued at a later date, and to carefully consider psychosocial or medical-legal factors in intervention planning.

The Importance of Comprehensive Assessment

For those individuals with mild TBI and persisting vocational difficulties months after the injury, comprehensive multidisciplinary assessment is very important. Some individuals, particularly those from rural areas, will not have access to neurologists, physiatrists, or rehabilitation medicine specialists focusing on the physical, cognitive, or sensory concerns related to the accident. Many individuals have not been to an emergency room, a hospital, or any center where medical diagnostic expertise is available. Following a comprehensive assessment by a physiatrist, for example, an individual may be referred to physical therapy or to a balance clinic for purposes of improving physical stability or perhaps to a neuroophthalmologist for assessment of visual concerns. In sum, appropriate assessment can resolve a number of the issues which get clouded within a general postconcussive syndrome diagnosis.

Those with persistent cognitive issues will definitely benefit from neuropsychological assessment. A neuropsychologist can be extremely helpful in identifying and differentiating cognitive versus emotional/personality or motivational concerns. Considerations include identification of existing cognitive limitations and assets, problems in emotional functioning, and clarification of current stressors and supports, including individual or family functioning dynamics which may interface with the cognitive concerns, prognosis as to recovery, and potential substance abuse issues.

The neuropsychologist, in the case of mild TBI, often dictates the scope and intensity of the desired treatment (Uomoto 2000). Neuropsychologists will frequently refer individuals to a speech and language pathologist who is experienced in brain injury for purposes of remediating attention and concentration, providing memory aids, improving communication, or assisting with organizational strategies. An occupational therapist, an assistive technologist, or a rehabilitation professional certified in accommodation strategies can be valuable in making recommendations relative to procedural changes in work activity, workstation modifications, or the use of assistive technology in order to complete work proficiently.

Additional Accommodation Resources

If resources relating to accommodation such as speech and language pathology expertise or assistive technology expertise are not available in one's area of residence, additional resources are available at a regional and national level. These include the national Job Accommodation Network (JAN) at West Virginia University, and ABLEDATA.

The JAN is a service provided by the U.S. Department of Labor's Office of Disability Employment Policy (ODEP) and housed at West Virginia University. This service, which is accessible by both internet and telephone, is the most comprehensive resource for job accommodations available nationally. Additionally, it has been expanded to provide information on the Americans with Disabilities Act (ADA). Consultants available through this site have specialized training in areas ranging from rehabilitation counseling to engineering. Specific team collaboration is available at the site in relation to different categories of disability: physical, cognitive, emotional, and so forth. JAN has recently developed their Searchable Online Accommodation Resource (SOAR), which includes not only accommodations, but also searchable resources for assistive technology.[1] Another helpful resource providing objective information on assistive technology and rehabilitation equipment available from domestic and international sources is ABLEDATA at www.abledata.com.

Of particular benefit for ADA issues and accommodations referral are the regional Disability and Business Technical Assistance Centers (DBTACs), which are supported by the National Institute of Disability and Rehabilitation Research (NIDRR). There are ten of these centers nationally and, in addition to phone consultation, specialists from these centers are often available to actually go out to employment sites and consult with or negotiate for accommodations between employers and qualified workers with disabilities. These regional centers will provide technical assistance, training, and resources on all aspects of the ADA.[2]

Levels of Intervention in Mild Brain Injury

It is important to note that "one size does not fit all" regarding employment support, and this is particularly true with regard to mild TBI. While some individuals with complicated mild TBI involving brain lesions may need more intensive intervention (one-to-one employment support/job coaching, etc.), the majority of those with mild brain injury may need relatively minimal, but targeted intervention. The amount and sophistication of the intervention is often related to the complexity of required job activities. In other words, even though an individual can have memory impairments, if the job is relatively routine and task-sequenced, no intervention may be necessary.

---

[1] The Job Accommodation Network can be contacted toll free at 1-800-526-7234, or at www.jan.wvu.edu.

[2] www.adata.org, or 1–800–949–4232.

On the other hand, with increasing complexity, even mild decrements related to memory or speed of information processing can negatively affect job functioning.

Minimal Intervention

Some clients with MCI can be helped simply by advisement relative to scheduling, procedures, workstation, or other environmental modifications that will enable them to be proficient on the job. Some individuals are resourceful themselves, while others profit from basic guidance by a rehabilitation counselor or employment specialist. Input can also be provided from several of the resources, such as JAN or DBTACs, as previously described. It can be helpful, however, for people to have an accommodation letter from a rehabilitation provider in the event that they do encounter job performance issues. An example of this type of letter is provided in the Appendix.

More Intensive Jobsite Support

More intensive forms of on-site placement support have been described by Wehman et al. (2000) at Virginia Commonwealth University. These include the individual one-to-one supported employment specialist model, and natural supports in the workplace, such as the employer or supervisor-as-training-mentor or coworker assistance. In some cases, these have involved on-the-job training (OJT) contracts with state vocational rehabilitation agencies or some other OJT funding agency. In some instances, coworker assistance has been provided on a gratis basis within a company. More intensive coworker-as-trainer intervention has been described by Curl et al. (1987), Curl and Hall (1990). In the case of mild brain impairment, a coworker might be used as a mentor on a nonpaid or paid basis. Job coaches might also be used, but typically they will be involved in only the most critical tasks. They might also be involved on a short-term basis in order to establish a template or protocol for procedural job strategies for the worker with mild brain injury, assistance in modifying the work station, or implementing the use of assistive technology or equipment by the worker (see review chapter by Guilmette and Guiliano, in this book, on moderate to severe TBI).

Cognitive Accommodation Framework

Below are accommodations frequently utilized for people with mild TBI. They are presented by cognitive category, and in each case, for purposes of simplicity, it is noted whether a procedural, physical workstation, or assistive technology/equipment is recommended in this category. These options are brief examples and are not meant to be all-inclusive. Many of the concerns for people with mild brain impairment relate to difficulties with attention/concentration, impaired memory, processing speed, and decreased verbal fluency. These same issues are very common among individuals with other neurological conditions, such as multiple sclerosis and epilepsy. A complete list of these accommodations across all levels of brain injury severity can be found at the national JAN website. That website also provides a fact sheet for universal design and assistive technology options in the workplace.

Attention/Concentration

Procedural: Consider steps to reduce distractions in work area; begin work early to reduce distractions; provide a private office or work space away from the center of activity and office traffic; increase natural lighting or provide full-spectrum lighting. Schedule uninterrupted work time to include no phone contact; divide large assignments into smaller tasks and steps; use routine and repetition as much as possible.

Workstation Modification: The workstation should be well organized with consideration given to using work carrels as necessary. Provide adequate work space on desktop and consider using a flat-panel computer monitor (LCD) with an adjustable mounting arm. Tool and material sequencing may be color-coded or outlined on workspace. Reduce the need to travel to other workplaces for task completion by providing appropriate equipment or materials at the employee's own workstation. Give consideration to environmental factors such as heat from windows or vents, cold drafts from doors or air conditioning units, or poor lighting, which may also lead to difficulties concentrating.

Assistive Technology: Consider the use of "white noise" in the background to mask distractions. Use headphones with noise cancellation, which limits auditory distractions and still provides access to computer, telephone, or hand-held devices. Occasionally an "invisible clock" or pager can be used to remind individuals to stay on task. Other equipment that can help with concentration might include ear plugs or flash cards with a list of procedures.

Organization/Problem Solving

Procedural: A job can be restructured in order to include not only the essential functions, but critical functions in the most efficient priority order. Consider color-coding files or items in order of use. A manager or coworker can be assigned as a mentor to assist with essential functions. During meetings, a coworker may be assigned to make duplicate notes of meeting highlights, timelines, or responsibilities. A job coach or a paid coworker can also be assigned to assist in mentoring. These sessions can be conducted over lunchtime, on breaks, or before the work day begins. Role playing possible scenarios or problems that may be encountered assists with problem-solving skills, increases retention, and builds confidence. In general, it is extremely helpful to do the most complex work early in the morning and the less complex work later in the day. It can also be helpful to set aside time at the work day's end to plan for the next morning and organize one's desk for an optimal next-day start.

Assistive Technology: Make daily to-do lists using Outlook calendar programs or a personal digital assistant (PDA), noting meetings and deadlines. Encourage email reminders from oneself or others relative to tasks. Color code email inbox folders to correspond with area of importance, e.g., reminders, to-do, correspondence, etc. Commercially available web-based communication products such as jott.com provide numerous options for quickly and easily leaving organizational messages or reminders via cell phone. Divide larger assignments into smaller tasks or steps. Use paper, cardboard, or Plexiglas to cover unnecessary buttons, data, or visual stimuli that are not necessary to the task at hand.

Memory

Procedural: Utilize written checklists and environmental cues, including color-coding or bulletin boards. Use yellow sticky notes to indicate work activity being conducted when interrupted. Have instructions provided in both verbal and written format, or use pictures, diagrams, or video. Allow the employee to tape record meetings and provide a written or email summary of each meeting. Post instructions or pictures over all frequently used equipment. It is also helpful to use relevant visual or auditory prompts, if that type of memory is better. Use easy-to-remember acronyms and categorized information. Verbally rehearse information with file cards or hand-held devices.

Assistive Technology: Train employees to email/phone mail reminders from oneself or others. Use Microsoft Outlook or other available digital options providing calendar functions. Use personal digital assistants, such as a wrist watch or hand-held device, that provide text or alarms. Consider faxing instructions the day of or the day before activities are scheduled to begin. Use digital cameras, computer software, and color printers to provide powerful visuals or memory aides. New hand-held devices allow use of video clips for training, as well as ongoing memory aids. These videos provide quick reminders of tools and materials used, as well as sequencing of tasks. Small hand-held video recorders allow for quick edits or review, and include auditory supports. Combining use of a headset and the recording device frequently improves concentration by masking external auditory distractions.

Current cell phones (smart phones) provide a wide variety of options for many of the previously mentioned solutions.

Speed of Information Processing

Procedural: Work coming into the work station should be provided in a structured and simplified format. Employees should use priority sequencing based upon essential functions of the job. Employees might use attentional control, asking that a supervisor or coworker "slow down" the verbal speed at which information is being conveyed or provide information in written format. Employees should develop task confidence in use of electronic devices needed for training or practice, and identify appropriate accommodations for mathematics, reading, or spelling using job-specific software or locally developed software programs.

Assistive Technology: Consider dictating previously outlined reports using voice-activated software. Consider using jigs or "cheat sheets" to lessen decision making. Cut down on available options through use of masking or key guards. Use task-specific computer software, calculators and PDAs. When appropriate, use text-to-speech applications in order to improve reading comprehension or word prediction applications for spelling deficits. Consider use of and proper placement of adjustable document holders for improved data retrieval.

Cognitive Fatigue

Procedural: Consider eliminating any unnecessary physical exertion, and schedule periodic rest breaks (this may include napping in the car or within one's office at lunch). Utilize a flexible work schedule, which could involve coming in early to do more complex work activity and leaving early to avoid traffic. Complete less complex work activity at home. Allow employee to work from home on a scheduled basis, e.g., a mid-week day. Employees who are post-brain-insult frequently benefit from initially working on a part-time basis and gradually increasing worksite time. Determine stress inducers and identify appropriate reducing agents or reasonable accommodation. Some cognitive fatigue can be reduced by attention to physical wellness, including nutritious meals, sufficient sleep, and appropriate exercise.

Assistive Technology: Use nonglare monitor screens to combat eye fatigue. Consider augmenting visual reading activities with text-to-speech applications to decrease focused reading time on one's computer. Explore use of adaptive, modified, ergonomic, or electric devices that increase efficiency and reduce effort during task completion.

Mild Brain Injury: Accommodation and Employment Intervention Examples

Case of Joe

Deficits: Memory, organizational difficulties

Joe was hired as a customer service representative for a company owning numerous parking lots in the Seattle area. He has difficulty remembering which procedures to follow when customers encounter difficulties. There was a large manual detailing these procedures, but it was unwieldy for Joe and difficult for him to remember critical information.

In interacting with customers, Joe seems to lack the ability to sequence and prioritize their concerns. An additional issue arose from the fact that the supervisor, although a caring person, was fairly brusque and provided information in a quick, staccato format. Joe missed much of this information, because of both the speed of delivery and his memory problems.

In this case, the employment specialist who placed Joe contacted the supervisor for purposes of intervention. The employment specialist reviewed the entire manual and "earmarked" the critical information necessary to the essential functions of Joe's job. This included flagging the important pages and highlighting the critical information with a colored pen. The employment specialist also provided Joe with a phone script and sequence of activities for problems called in by the parking lot leasors. Finally, the employment specialist modeled to the supervisor a "tell, show, watch, and provide feedback" sequence, providing instruction which was very helpful to Joe, and which also improved the supervisor's capacity in general for providing instruction to employees. Joe profited greatly from the employment specialist's intervention. In fact, the intervention actually "saved" his job. This intervention was targeted to performance deficits related to the essential functions of the job and involved approximately 32 h of intervention over 2 weeks. There was some follow-up post-intervention, but it was very minimal, involving brief tutorial phone calls over another few weeks.

Case of Brandy

Cognitive Deficits: Speed of information processing/divided attention, abstractive capacities

In a number of cases, individuals with MCI can improvise their own adaptations, which actually work quite well. Brandy is an individual who initially worked as a call center operator with a vacation timeshare sales company. She was promoted to a higher level customer service representative, and later call center supervisor. She had some significant difficulties with divided attention/speed of information processing and low average abstractive abilities as a result of impairment related to childhood seizures.

Despite these difficulties, Brandy had made a number of personal adaptations that were quite successful. Although call center specialists would frequently turn to her for help throughout the day, Brandy was excellent at prioritizing their concerns and dealing with only one issue at a time. Her employees became accustomed to the fact that she could not multitask, but that she was quite effective once contact was made with them. She had disclosed her disability to them and, partially as a function of her pleasant personality, had solicited their cooperation. As problems arose and were sent to her desk, Brandy used multiple colored folders in order to catalog them specifically by the type of issue that needed her attention. The other call center supervisor needed no such system but this was imperative for Brandy in order to stay organized. Despite company procedures, Brandy would not use email to provide advice to her employees because this took too much time for her. She would see each employee individually at the beginning of the day in order to provide specific instructions and repeat this type of activity as needed throughout the day, which took much less time than producing emails. Finally, Brandy explained her slowed learning curve (abstractive deficits) to the owners and how she needed more time in order to learn new policies and procedures. Once grasped, she recalled material quite well. Again, as a function of her directness and pleasing personality, the owners seemed more than happy to make these accommodations.

Brandy had only come to a rehabilitation counselor in order to secure a letter documenting her concerns, as supported by her neuropsychological test results, because she was leaving the company in order to move East. She wanted to have a letter reviewing her cognitive strengths and deficits in functional terms specifically in order to seek job accommodations if issues arose in a new job and she was unsure about the response she might receive from a new company. Given her past success and demonstrated skill in self-determination, it seemed highly likely that she would be successful in job relocation.

Conclusion

In the case of MCI, accurate assessment and diagnostic specificity are crucial. Without adequate assessment, the type and intensity of employment interventions can be misdirected. This can certainly be the case for those with mild TBI and mental health disorders for whom substantive efforts in cognitive remediation or accommodation can be made mistakenly, while the rehabilitationists should be dealing with the client's emotional and personality issues. For those with persisting cognitive difficulties, we can access vocational rehabilitation professionals, occupational therapists, neuropsychologists, and rehabilitation psychologists, as well as regional/national consultation resources as previously reviewed that can assist individuals with MCI to function vocationally.

Acknowledgement Support of the University of Washington TBI Model Systems grant H113A980023 is acknowledged from the National Institute of Disability and Rehabilitation Research (NIDRR), U.S. Department of Education.

Appendix: Disclosure Support Letter

To whom it may concern:

Re: PF

Ms. PF has been a vocational client of the University of Washington's Neurological Vocational Services for some time. As part of this employment program, we provided a complete vocational assessment and thorough cognitive testing. Ms. PF has excellent abstractive and language skills, in the upper 10% of the population. She has excellent verbal fluency and general fund of information. Her interests tend toward office support activities, and she is particularly project-oriented and well suited to data entry and development projects.

Ms. PF, however, will have some difficulty with multitasking and may require more time to learn a procedure in contrast to the average worker. She simply does an outstanding job if her focus is one project at a time, despite how large the project. Once learned, Ms. PF will perform a job function very well and on a consistent basis. It is hoped that this information is useful to a supervisor, because during initial training the disparity between her learning curve and her actual comprehension abilities may not be well understood. Once learned, she can easily make key decisions. Again, she has strong problem-solving abilities, but can become less efficient if she is spread across multiple duties/projects simultaneously.

Kind regards,
Robert T. Fraser, Ph.D., CRC

References

Binder LM, Rohling ML (1996) Money matters: a meta-analytic review of the effects of financial incentives on recovery after closed-head injury. Am J Psychiatry 153:7–10

Cassidy JD, Carroll LJ, Peloso PM, Borg J, von Holst H, Holm L et al (2004) Incidence, risk factors and prevention of mild traumatic brain injury: results of the WHO collaborating centre task force on mild traumatic brain injury. J Rehabil Med 36:28–60

Clemmons DC, Fraser RT, Getter A, Johnson KL (2004) An abbreviated neuropsychological battery in multiple sclerosis (MS) vocational rehabilitation. Rehabil Psychol 49(2):100–105

Comper P, Bisschop SM, Carnide N, Tricco A (2005) A systematic review of treatments for mild traumatic brain injury. Brain Inj 19(11):863–880

Curl RM, Hall SM (1990) Put that person to work! A co-worker training manual for the co-worker transition model. Outreach, Development and Dissemination Division, Developmental Center for Handicapped Persons, Utah State University, Logan, UT

Curl RM, McConaughy EK, Pawley JM, Salzberg CL (1987) Put that person to work! A co-worker training manual for the co-worker transition model. Outreach, Development and Dissemination Division, Developmental Center for Handicapped Persons, Utah State University, Logan, UT

Duff MC, Proctor A, Haley K (2002) Mild traumatic brain injury (MTBI): assessment and treatment procedures used by speech-language pathologists (SLPs). Brain Inj 16(9):773–787

Elgmark-Andersson E, Emanuelson I, Bjorklund R, Stalhammar DA (2007) Mild traumatic brain injuries: the impact of early intervention on late sequelae. A randomized controlled trial. Acta Neurochir 149:151–159, discussion, 160

Friedland JF, Dawson DR (2001) Function after motor vehicle accidents: a prospective study of mild head injury and posttraumatic stress. J Nerv Ment Dis 189(7):426–434

Ghaffar O, McCullagh S, Ouchterlony D, Feinstein A (2006) Randomized treatment trial in mild traumatic brain injury. J Psychosom Res 61:153–160

Greiffenstein MF, Baker WJ, Gola T (1994) Validation of malingered amnesia measures with a large clinical sample. Psychol Assess 6:218–224

Guilmette TJ, Giuliano AJ (2010). Brain injury and work performance. In: Schultz IZ, Rogers ES (eds) Handbook of work accommodation and retention in mental health. Springer, New York, pp 141–161

Kashluba S, Hanks RA, Casey JE, Millis SR (2008) Neuropsychologic and functional outcome after complicated mild traumatic brain injury. Arch Phys Med Rehabil 89:901–911

Kay T (1993) Neuropsychological treatment of mild traumatic brain injury. J Head Trauma Rehabil 8:74–85

Kay T, Newman B, Cavallo M, Ezrachi D, Resnick M (1992) Toward a neuropsychological model of functional disability after mild traumatic brain injury. Neuropsychology 6:371–384

Kraus J, McArthur J, Silberman T et al (1996) Epidemiology of brain injury. In: Narayan RK, Wilberger JE Jr, Povlishock JT (eds) Neurotrauma. McGraw-Hill, New York, pp 13–30

Langlois JA, Kegler SR, Butler JA, Gotsch KE, Johnson RL, Reichard WW, Webb KW, Coronado VG, Selassie AW, Thurman DJ (2003) Traumatic brain injury-related hospital discharges. Results from a 14-state surveillance system, 1997. MMWR Surveill Summ 52(4):1–20

Larrieu S, Letenneur L, Orgogozo JM et al (2002) Incidence and outcome of mild cognitive impairment in a population-based prospective cohort. Neurology 59:1594–1599

Londos E, Boschian K, Lindén A, Persson C, Minthon L, Lexell J (2008) Effects of a goal-oriented rehabilitation program in mild cognitive impairment: a pilot study. Am J Alzheimers Dis Other Demen 23(2):177–183

Palmer K, Fratigilioni L, Winblad B (2003) What is mild cognitive impairment? Variations in definitions and evolution of nondemented persons with cognitive impairment. Acta Neurol Scand Suppl 179:14–20

Shames J, Treger I, Ring H, Giaquinto S (2007) Return to work following traumatic brain injury: trends and challenges. Disabil Rehabil 29(17):1387–1395

Snell DL, Surgenor LJ, Hay-Smith E, Jean C, Siegert RJ (2009) A systematic review of psychological treatments for mild traumatic brain injury: an update on the evidence. J Clin Exp Neuropsychol 31(1):20–38, Psychology Press, iFirst, 1–19

Uomoto J (2000) Application of the neuropsychological evaluation. In: Fraser RT, Clemmons DC (eds) Traumatic brain injury rehabilitation: practical vocational, neuropsychological, and psychotherapy interventions. CRC, Boca Raton, FL

Vanderploeg RD, Curtiss G, Duchnick JJ, Luis CA (2003) Demographic, medical, and psychiatric factors in work and marital status after mild head injury. J Head Trauma Rehabil 18(2):148–163

Wehman P, Bricout J, Targett P (2000) Supported employment for persons with traumatic brain injury: a guide for implementation. In: Fraser R, Clemmons D (eds) Traumatic brain injury rehabilitation: practical vocational, neuropsychological, and psychotherapy interventions. CRC, Boca Raton, FL, pp 201–240

Chapter 14
Return to Work After Traumatic Brain Injury: A Supported Employment Approach

Pamela Targett and Paul Wehman

Introduction

Traumatic brain injury (TBI) is one of the most prevalent disabilities in the United States. Of the estimated 1.4 million who sustain a TBI annually, about 1.1 million are treated and released from the emergency room (Langlois et al. 2006). Another 235,000 are hospitalized, and 80,000 to 90,000 experience permanent disabilities (Langlois et al. 2006) TBI is three times more common in men than in women, with the most common causes for injury being motor vehicle accidents, falls, and violence (Greenwald et al. 2003).

TBIresults in a variety of difficulties with cognitive, physical, sensory, and psychosocial functioning (Horn and Sherer 1999). As a result, individuals often experience difficulty returning to work or becoming employed post-injury and maintaining employment for extended periods of time (Asikainen et al. 1998; Curl et al. 1996; Keyser-Marcus et al. 2002a; Kreutzer and Morton 1988; McMordie and Barker 1988). Some of the difficulties a person may face after a TBI are listed in Table 14.1 (see review of neuropsychological and work performance issues in moderate to severe TBIs in Chap. 8, Guilmette and Giuliano 2010).

Young males have the highest risk of sustaining TBI (Abrams et al. 1993). Because survival rates post-injury have dramatically improved due to medical advances, individuals often experience normal life expectancy. Research indicates that aside from economic factors, employment is an important psychological predictor related to well-being, quality of life, social integration, and recovery of individuals with TBI (O'Neil et al. 1998; Abrams et al. 1993; Wehman et al. 2005).

Despite the importance of employment, studies have consistently reported that individuals with TBI experience difficulties with returning to work (Brooks et al. 1987; Targett and Yasuda 2006; West et al. 2007; Keyser-Marcus et al. 2002b). For

P. Targett (✉)
VCU RRTC, Virginia Commonwealth University, 1314 W Main Street, Richmond, VA, 23284, USA
e-mail: psherron@vcu.edu

I.Z. Schultz and E.S. Rogers (eds.), *Work Accommodation and Retention in Mental Health*, 277
DOI 10.1007/978-1-4419-0428-7_14, © Springer Science+Business Media, LLC 2011

Table 14.1 Examples of difficulties with posttraumatic brain injury

Cognitive

Poor short-term memory

Distractibility

Difficulty solving problems

Problems with reasoning, planning, and understanding

Slowed thinking

Problems with attention and concentration

Difficulties with speech and language

Physical and sensory

Unsteady gait

Vision problems

Inability to regulate body temperature

Difficulties with fine motor skills

Hearing problems

Dizziness

Emotional or behavioral

Depression

Irritability

Apathy

A tendency to be confrontational

Quick-tempered

Frustration

example, data from the TBI Model Systems dataset indicate that while 59% of those who sustained TBI were employed at the time of injury, only 23% were competitively employed at 1 year post-injury (TBI Model Systems National Data Center 2001).

Improving employment rates after a TBI is critical for the vocational rehabilitation field and to meet increasing demands to assist individuals with TBI with return-to-work post-injury. Effective models of vocational service delivery are needed and are being developed.

Supported Employment

Supported employment, introduced in the 1980s (Wehman 1981), is an approach to helping individuals with the most severe disabilities to gain and maintain employment. It is an approach that is frequently found in successful rehabilitation programs serving individuals with TBI (Shames et al. 2007). In the late 1980s, as practical strategies to assist individuals with TBI were evolving, Wehman et al. pioneered the development of a supported employment approach for individuals with TBI. The work was based on supported employment demonstration projects with other disability populations (Moon et al 1985; Wehman et al. 1993b; Inge et al. 1996; Drake et al. 1999).

A driving force behind the advent of supported employment as a service option was the fact that there were thousands of people with severe mental disabilities who were viewed as incapable of work (Wehman 1981). Many of these individuals spent their time in day programs, adult activity centers, and sheltered workshops, and most lived their adult lives at home with their parents or other family members or in institutions. In the mid to late 1970s research began to investigate ways to assist severely disabled people with going to work in their communities (Mank et al. 1986; Parent and Everson 1986). This was fueled by the concept of employability; the notion that everyone possesses the potential to work. The concept of employability means that all individuals, including those with the most severe disabilities, should have the opportunity to participate in real work for real pay in his or her community. With the introduction of supported employment as a vocational rehabilitation service option, working for subminimal wages in a sheltered workshop setting (a place where only people with disabilities worked) was no longer the primary option.

Today the numbers are growing; currently about 140,000 people are working while using supported employment services (Rizzolo et al. 2004; Wehman et al. 1998). Some of the guiding values and principles of supported employment are provided in Table 14.2.

By 1986, supported employment had become a vocational rehabilitation service option by way of the 1986 Amendments to the Rehabilitation Act (Federal

Table 14.2 Supported employment guiding principles

Individuals with disabilities do not have to "get ready" to work. Instead, a job search aims at locating or developing a job that will capitalize on a person's current skills and abilities and the orchestration of workplace supports to promote success on the job
The consumer of services (the individual with the disability) should be involved in service planning
Services should be delivered on an individualized (not large groups or aggregate formats) basis to each person
Support should be provided or facilitated in a flexible manner, changing in type, level, and degree of intensity as needed to best meet the needs and desires of the person with the disability
All individuals regardless of the severity of their disability should have the opportunity to work in the community
The consumer of services should make the final decision about whether or not to accept a job offer and when to end employment
People with disabilities should be paid prevailing wages similar to those of others without disabilities who are performing similar work
On-the-job skills training is most effective when it takes place once employed in a real work place, not in settings offering simulated environments or using work samples
The business that hires the person with the disability is also a customer of the supported employment provider and should receive services specifically tailored to its preferences and needs and in a way that enhances but does not supplant its existing organizational practices

Table 14.3 Supported employment services

Personalized assessments to help determine vocational abilities, interests, and potential on-the-
job support needs and accommodations using a functional approach

Tailored help with locating and applying for jobs in the community

Individualized negotiations with interested employers aimed at modifying existing work options
(i.e., restructuring essential job duties or implementing accommodations such as schedule or
environmental changes, etc.) or creating a new work opportunity via job carving or creation

Individualized assistance with participating in and completing the employer's new employee
orientation and training program

Additional one-to-one on job skills training after the completion of the employer's new
employee skills training program

Customized assistance with identifying an array of workplace supports or accommodations to
enhance job and job-related performance

Provision and facilitation of workplace supports needed to assist the new employee with
learning the job and job-related duties

Personalized assistance to help the individual problem-solving issues, either at or away from
work, that could prevent separation from employment

Register 1987). It is important to note that supported employment is not a suitable
option for every one with a disability who desires to work. It is usually reserved for
an individual with a significant disability who requires one-to-one assistance from
a vocational rehabilitation professional, often called an employment specialist or
job coach.

In supported employment services, the specialist first must get to know the job
seeker, their vocational history, job goals, functional limitations, strengths, and
abilities. Potential work opportunities can be developed for the person once there is
a clear understanding of the individual's job goals and strengths. Once the job
seeker is hired, the specialist should be available to assist the person with learning
to perform the job to meet the expectations and standards set forth by the employer.
Another key feature of supported employment is ongoing, long-term follow-up and
support throughout the person's employment. Once the employee is proficient on
the job, the specialist fades his or her presence away from the job site, but still
remains in contact with the employee and employer to ensure things continue to go
well at work. As indicated, the specialist returns to the job site to provide or facilitate additional on-the-job assistance and support. Table 14.3 describes the key elements of supported employment as a vocational service option.

The following is a brief review of the supported employment and TBI literature.

Traumatic Brain Injury and Supported Employment

Evidence suggests that persons with TBI who use supported employment services
can obtain and successfully maintain employment. In 1990, Wehman et al. reported
on 51 individuals who received supported employment services. Findings indicated
that 77% of those individuals became successfully employed in competitive

employment in their communities. The average hours worked per week were 31 and average earning was $4.61, which was slightly above the minimum wage at the time. Notably, individuals were employed in positions, and received wages and work hours, that were comparable to their pre-injury employment histories. Job retention rates were as follows: 37% remained employed for 1–6 months, 39% worked 7 months to 1 year, 7% worked 13–18 months, and 12% worked for more than 2 years.

Subsequently, Wehman used clinical ratings by direct service staff (i.e., employment specialists) to identify individuals with TBI who were perceived as the least or most difficult to serve. They found that individuals rated "most difficult" tended to be younger, had problems with visual and fine motor skills, and lacked numerous other work-related skills (Wehman et al. 1993a).

In terms of the cost of providing supported employment service, Wehman and his colleagues found that 54% of employment specialists' intervention/service time was spent providing on-the-job site skills training and advocacy to the employee with TBI (Wehman et al. 1994). The mean cost was $10,198 for the first year of services. Once employed, individuals reached stabilization on their job after 18 weeks of intervention, on average. Once stabilized, weekly intervention time dropped to an average of 2.24 h per week, which resulted in a modest cost of $71 per week. Wehman et al. (2003) conducted an analysis of long-term follow-up costs and program efficiency among 59 individuals with TBI. Data were collected from 1985 to 1999. The average length of employment for subjects was 42.58 months. Average gross earning was $26,129, while the cost of services averaged $10,349.

Another study (Sale et al. 1991) examined reasons behind the job separations of 38 supported employment placements. By using employment specialists as key informants, five categories of reasons for termination from employment emerged. The five categories, presented in order of highest to lowest frequency of problems were: (1) issues related to the employment setting, (2) interpersonal relationship problems, (3) mental health problems and criminal activities, (4) other problems, and (5) economic layoffs. The study found that 68% of the separations occurred in the first 6 months of employment. Notably, excluding economic layoffs, more than three-fourths, or 75%, of the separations resulted from two or more problems. The authors concluded that separation was most often a process rather than an event. In other words, termination did not generally occur as a result of a single event; rather, a series of incidents led to separation (Sale et al. 1991).

Wehman et al. conducted an analysis of long-term follow-up costs and program efficiency among 59 individuals with TBI. Data were collected from 1985 to 1999. Results of this preliminary investigation provided additional support for the conclusion that supported employment is cost-effective for individuals with disabilities, including TBI, and that the costs of these services decreases over time. With regard to cost efficiency, participant income was an average of $17,515 greater than the cost of supported employment services received over the 14-year time period. The researchers noted that, while encouraging, these results were only preliminary.

The Agency for Health Care Policy and Research awarded a contract to Oregon Health Sciences University for a review of published reports and a compilation of an evidence-based report on supported employment. Two panels of experts worked with a research team to identify key questions, with one being: "Does the application of supported employment enhance outcomes for people with TBI?" An extensive search of the relevant literature yielded 93 articles, after an expert review. A three-level system was used to rate the studies. Well-designed randomized controlled trials (RCTs) were rated as Class I. Studies rated as Class II were RCTs with design flaws, well done, prospective, quasi-experimental or longitudinal studies, and case control studies. Case reports, uncontrolled case series, and expert or consensus opinions were generally rated Class III. The review found Class II evidence that supported employment can improve the employment outcomes of TBI survivors. However, the findings have not been replicated in other settings by other centers, so the generalization of these programs remains untested and more research on supported employment for individuals with TBI is needed.

Individualized Approach to Supported Employment

The remainder of this chapter will examine how Wehman et al. have implemented supported employment services to assist individuals with TBI in returning to work. As mentioned earlier, supported employment can be helpful to anyone with a disability, but is particularly useful for people with more severe disabilities. Using this strategy the type and intensity of services and supports are adapted to each individual's needs, and under some circumstances involve extensive intervention. The time needed to perform supported employment services varies depending upon the needs of the service recipient and the work context and demands.

As previously mentioned, individuals with TBI accessing supported employment generally require intensive one-to-one assistance from an employment specialist to help them identify vocational strengths and potential support needs, conduct a job search, and negotiate employment details (Sherron and Groah 1990). Once employed, the new hire receives on- and off-the-job support (Killam et al. 1990). Since supported employment offers individualized and intensive support services in obtaining employment, the person with TBI does not have to "get ready" to work, as we commonly assumed necessary in the older "train-and-place" models. Instead the specialist provides assistance with helping the person identify his or her abilities and support needs, contacting employers to discuss hiring needs, identifying and negotiating work opportunities and on-the-job supports and accommodations, and providing or facilitating on- and off-the-job workplace support. The type, level, and intensity of support the person will need on the job does not remain static, but changes over time. Most individuals with TBI who use supported employment services will benefit from an array or combination of workplace supports. Over time and based on the employee's performance, the specialist will begin to fade away from being at the job site (West and Hughes 1990), but unlike with other vocational support services, the employment specialist maintains contact with both the employee and employer intermittently and provides additional assistance if deemed necessary to insure ongoing success at work.

Specific activities involved in implementation of supported employment at Virginia Commonwealth University (VCU) include referral and intake, functional vocational assessment, job development, on- and off-the-job support, long-term support, and job retention services. These services are described below.

Referral and Intake

Any person with a moderate to severe TBI who wants to return to work may be referred for services at VCU or similarly funded supported employment programs. Referrals may come from physicians, nurses, social workers, therapists, vocational rehabilitation counselors, the person with TBI, family members, or others. Usually, the individual is not employed at the time of referral; however, there may be exceptions. Upon confirmation of funding sources, the employment specialist will schedule a time to meet with the person in a setting of his or her choice. Often this is a home visit. At the time of referral the level of involvement of a large team of rehabilitation staff will vary, depending upon the amount of time since the person's injury and their current medical and disability status.

Sometimes the person is employed at the time of referral. In such an instance, it is important that someone like the employment specialist, a designated team member, and/or family member serve as a liaison with the employer to keep them abreast of the individual's progress in rehabilitation and return-to-work status. However, this is often not the case, as the person with a severe injury may be out of work for an extended time with initial hospitalization, followed by both inpatient and outpatient therapies. In some instances, return to previous work may be feasible. The individual may not be able to return to the same position, but there may be some other type of work within the organization that suits the individual's current strengths and abilities. Occasionally a person may be able to return to work gradually while involved in day rehabilitation. If designated to do so, the employment specialist will work closely with the rehabilitation team and help them gather information needed about the position from the employer to assist them with making return-to-work recommendations (including job accommodations such as schedule adjustments or provision of distraction-free environment and memory aids). The employment specialist will also communicate with the employer about the individual's progress and return-to-work plans. Experience suggests however, that most individuals with TBI enter the supported employment program unemployed. Additionally, many of those referred have no, limited, or unstable work histories.

Functional Vocational Assessment

Prior to looking for work, the employment specialist must get to know the person through an individualized functional assessment (Sherron and Groah 1990; Targett et al. 1998; West et al. 2007). The employment specialist will meet the person, recommend assessment activities, and affirm the direction of the job search. The plan is based upon information gathered from interviews with the person and/or those who know him or her best, and a review of relevant records. This lays the foundation that guides the job search, the jobsite training, and long-term support. This also is a time when rapport should develop between the job seeker and the employment specialist.

As previously mentioned, the person with TBI either will be looking to return to pre-injury employment (i.e., either in his or her preemployment occupation, with or without accommodation, or in a new position), or will be seeking new employment. Our focus here will be on the approach used to assist those who are seeking new employment with a new community-based employer after the TBI.

A functional assessment can provide valuable career development and work accommodation information to individuals with disabilities, and particularly to those who have more significant barriers to work. All too often, persons with severe disabilities are considered either too disabled to work or not ready for employment. One advantage of a functional approach to assessment is that it focuses on the person's current strengths and abilities and the types of supports that will enable a person to succeed in the workplace. Since the assessment is customized, the design will depend upon a number of factors, including the person's perceived barriers to work, previous work history, and so forth. The amount of time needed to conduct the assessment will vary depending upon what activities and components are utilized (Targett et al. 1998; West et al. 2007).

At a bare minimum the specialist usually visits the person in their home and spends some time with him or her in the community. If the individual does not have the option to return to the pre-injury employer, but desires information on whether or not he or she can use a specific skill set in the workplace, a Residual Skills Assessment may be warranted. If the person is unemployed, and/or has been absent from the workplace for an extended period of time, then a Career Exploration Assessment may be conducted. Brief descriptions of these approaches follow.

Residual Skills Vocational Assessment

In order to make an informed decision about whether or not to return to a previous occupation, the person with TBI may need information about his or her current skills, abilities, and possible accommodations. In this instance, a customized vocational assessment specifically designed to explore current level of functioning and support needs for a previous line of work may prove useful, and may instill the confidence the person needs to determine whether or not to pursue a career in a specific field (Targett et al. 1998; West et al. 2007).

Depending on the nature and complexity of the occupation in question, a consultant or "expert" who is actively engaged in the desired occupation may be required to assist with designing and implementing an assessment. The employment specialist usually takes responsibility for locating and securing the consultant's service, coordinating the assessment, assisting with the assessment design and tools, and writing a report. The specialist actively participates in the assessment process by making observations and collecting information about the individual's abilities and support needs. The goal of this assessment is to assess the individuals skills and abilities vis-à-vis a known skill set, and to have this assessment be implemented by a person with first-hand knowledge of the job skills and demands for a specific occupation.

Career Exploration Assessment

Individuals with TBI may have no, limited, or an unstable work history. A Career Exploration Assessment can help the person establish a desirable career path. Using this approach, the individual's current abilities, interests, and support needs will be examined (Targett et al. 1998; West et al. 2007). This information will also provide an initial direction for the job search, or in some instances identify the type of education or training that would be necessary to pursue a specific occupation.

A Career Exploration Assessment can be particularly helpful when a person does not know what he or she can do and/or what types of supports may enable him or her to pursue employment. The assessment provides an opportunity to participate in one or a combination of the following activities: community-based situational assessment, job shadowing, and informational interviews.

Situational Assessments

Situational assessments provide the person with the opportunity to perform job tasks in real work environments in the community (Sherron and Groah 1990; Targett et al. 1998; West et al. 2007). Usually, situational assessments are conducted across three or four different types of jobs which are representative of the local labor market. First, the specialist develops sites for the assessment to take place. This involves contacting community employers to locate sites, and developing tools to evaluate the person as they perform work and work-related duties in that setting. The specialist then accompanies the person to the employment sites, observes, and collects information, while the individual performs various work tasks. This takes place across multiple environments and should provide insight into the individual's work preferences, interests, skills, abilities, learning style, and support needs.

Job Shadowing

Job shadowing typically involves observing or "shadowing" a worker who is performing a job. The employment specialist makes the necessary arrangements with area businesses for a job shadowing, then accompanies the person to the jobsite.

Observing an employee performing his or her job duties provides an opportunity to survey the individual about his or her likes and dislikes related to various work conditions and helps further clarify vocational interests and preferences. It also provides an opportunity for the employment specialist to observe the person's social and communication, physical, cognitive, and other abilities in a community-based business environment.

Informational Interviews

Informational interviews involve arranging opportunities to meet with an employer and to interview personnel to learn what they can do on the job (Targett et al. 1998; West et al. 2007). Generally, this activity involves meeting with a business representative to learn about jobs across an organization. This is followed by a tour of the workplace, which provides opportunities for the job seeker to observe employees performing various types of jobs in a particular industry. The specialist makes observations and takes notes about the individual's reaction to a variety of work-related factors. This information is used to further discussions about a direction for the person's job search.

Interacting in the Community

Sometimes instead of participating in an organized activity such as those mentioned above, the specialist and job seeker simply spend time together in the community (Sherron and Groah 1990; Targett et al. 1998; West et al. 2007). Much can be learned by visiting a person in his or her home or accompanying the person on a trip to the mall or going out for coffee. This also gives both parties an opportunity to get to know one another. Other activities may relate to researching local businesses on the internet, or riding around the community to identify potential places for employment.

After vocational assessment activities are completed, the information learned is used to develop a vocational profile (Sherron and Groah 1990; Targett et al. 1998; West et al. 2007). This profile includes a variety of information related to work history, vocational preferences and expectations, vocational strengths and abilities, and potential needs for vocational support. After the assessment, the employment specialist will meet with the individual and other relevant parties (for example, family members, the state vocational rehabilitation counselor, etc.) to coordinate and facilitate the next steps to assist the person with achieving employment. Once the person has considered important aspects of employment and what he or she has to offer the business, strategies for returning to work and presenting his/her unique skills and talents to the business community are developed. Then the job search begins.

Job Development

Conducting a successful job search is not a simple process. Because of the continuously changing and competitive nature of the job market, there are no guarantees that any particular job search strategy will work. Therefore, a diverse approach is recommended (Sherron and Groah 1990; Targett 2006). For example, the employment specialist may brainstorm a list of prospective businesses where the individual might want to work, survey the individual's personal network, identify job leads and referral services, and contact former employers and businesses that advertise employment opportunities. Personal business networks are also developed and used. In this procedure, the job seeker lists names of individuals to contact about employment opportunities. The network may include family, friends, family members of friends, church members, support group members, people they do business with, etc. For each person reached, the names of new contacts are requested. This leads to the development of a network referral chain.

Job leads and referral services are also used to identify jobs. These services include job banks, telephone job information lines, and professional networks. If the individual has worked in the past, whether paid or volunteer, former supervisors and coworkers may be contacted for leads. Sometimes current employment leads are used to locate work opportunities. Sources of information include newspapers, government listings (federal, state, city, and county), help wanted signs or announcements posted on public bulletin boards, etc. However, the focus is usually on tapping into the "hidden job market," jobs that do not currently exist or jobs not being advertised. Therefore, a major thrust of job development efforts is to become known or acquainted with an employer. The specialist's initial contact focuses on offering services and building rapport with the employer without immediately asking for something in return.

During the initial interview with a prospective employer, the specialist will be seeking to obtain key pieces of information about the company and, if applicable, specific job openings.

After obtaining information concerning an employer's hiring needs, when possible, the specialist may observe someone performing various jobs. This observation helps the employment specialist understand essential job functions, critical skills needed for each job duty, the approximate time spent engaged in each job task, and so forth.

These activities provide insight into the employer's expectations as well. Later, if applicable, this information is presented to the job seeker and serves as a first step toward assisting him or her with making a decision about whether or not to interview for a particular position.

Job Restructuring

Some individuals with TBI will not be qualified or will not be able to perform the essential functions of existing jobs. Under these circumstances, the employment specialist may attempt to negotiate or "restructure" a job with a willing employer.

Job restructuring involves redistribution of job tasks that cannot be performed by the employee who has TBI. These tasks are given to a coworker in exchange for a task that the job seeker can do. In other instances, a new job may be created or carved out.

Job restructuring or carving are excellent strategies to use with a job seeker who may not be necessarily "qualified" or able to perform all aspects of a current job opening (Griffin and Targett 2006). Job restructuring ideas may not be obvious and therefore require creativity on the part of the employment specialist and employer. The specialist must spend time working with the business to determine the potential for restructuring by exploring their needs undiscovered needs of the employer. Some of the best ideas for job restructuring have resulted from informal observations and conversations with employers about their needs and how they do business. A proposal is then needed to make recommendations on specific tasks the job seeker could do, the advantages that will accrue to the business, recommended hours of work, and the next steps in the negotiations.

Involvement of Individuals with TBI

Individuals with TBI may actively participate, or simply stay informed, during the job development or job search process. Once a job option is developed, then job seekers are supported to engage in preemployment activities. For example, this may include one-to-one assistance with completing an employment application, participating in an interview, or arranging for or providing accommodation if needed to complete any required preemployment testing or questionnaires.

Prior to the interview, and again after employment is offered, the pros and cons of the particular employment situation are reviewed with the job seeker. This information is based upon a comparison of the vocational profile and data collected when meeting with the employer. This analysis helps determine whether or not the employment offer is acceptable to the person, and what types of workplace support and accommodations may be necessary.

At the point in time when an employer makes an offer of employment, the job seeker must make a decision about whether or not to accept it. The job seeker, with the support of the employment specialist, if needed, makes the final decision about whether or not to accept or reject an employer's offer.

On- and Off-the-Job Supports

Once employed, the employment specialist provides support on and sometimes off the job site (Killiam et al. 1990; West et al. 2007). Supports are provided to assist the new hire with learning how to do the job effectively, negotiating accommodations, communicating with job site personnel, and handling issues outside of work such as accessing transportation systems, depositing paychecks, and so forth. The employment specialist is prepared to take on the role as a facilitator of independent, self-directed learning and/or a job skills training specialist. Examples follow of workplace supports that may be used.

Support Prior to Starting Work

Before starting employment it is imperative that the specialist notify significant others (if applicable) and ensure that arrangements are made for transportation or transportation training; and if appropriate, tax credit information is completed and signed by the employer. Other activities that may need to be completed include referral to Social Security for reporting procedures, assistance with the purchase of uniforms or other work-related items, and so forth. The employment specialist may also need to discuss any concerns or fears that the individual has about starting job, and provide information or assistance that can allay these fears.

Additional on the Job Site Skills Training and Data Collection

Most employers offer new workers an employee training program. The employment specialist should participate in this training along side the new hire. Then once the training is completed as needed additional training takes place. It is not unusual for the new employee with TBI to need additional skills training. Because of difficulties with learning and physical challenges, it often takes the person longer to learn the job than the time the employer typically offers to employees. This is when the employment specialist can provide additional training (Killiam et al. 1990; West et al. 2007). Systematic instruction is one method that has been used to teach job duties and job-related tasks in a consistent manner. The use of systematic instruction also provides a method of data collection regarding the person's performance level. Using this information to monitor progress during training will provide structure and minimize confusion for the trainee as well as the employment specialist. Production training may also be required. This involves determining production standards for the job and the individual's current rate of performance. A comparison reveals if there is a need to increase productivity and if so the training focuses on ways to enhance it.

Data related to skill may also be needed and if so training takes place. Acquisition and productivity is essential for helping the specialist make decisions about the effectiveness of their instruction. It is also used to design a fading schedule in which the specialist removes himself gradually from the job site.

Job Completion Guarantee

Depending upon the type of job and the job context, the employment specialist may actually perform some of the work while the new employee is solidifying his or her work skills. This insurance to complete the needed work can keep the employer satisfied and provide the additional time necessary for the employee to learn the job.

Compensatory Memory Strategies

Job accommodations including the implementation of compensatory memory strategies may prove useful (Wehman et al. 2000; West et al. 2007). Adjustments to either the job tasks or work environment can be cognitively-oriented. These strategies promote learning and can provide ongoing reminders of what to do and how to do it. Some strategies are external and involve the use of objects like check-lists, decision trees, pictorial lists, alarm watches, day planners, and the like. For example, a cashier may follow written steps to remember how to balance a cash drawer or process a credit card purchase; a groundskeeper may use a color coded map that outlines the route to follow and what tasks to perform on certain days of the week; a hotel worker may be taught to drape different types of linens over bins to highlight where items go during the sorting process; or a food service worker may follow pictures to help him determine how to prepare a particular type of sandwich. Internal or mental strategies may also prove useful for some individuals and might involve teaching a person to state aloud the steps needed to complete a task as it is performed. Helping the employee devise a rhyme or other mnemonic to remember something he or she often forgets might be useful (like "don't be late, set a date") to remind the person to write in changes in sequence of duties in his or her day planner.

It is important to note that rarely will a person be able to use these strategies without instruction. The compensatory strategy must be taught. So instead of teaching steps in a job skill sequence, the specialist may instead be engaged in teaching the employee to use the strategy to follow the steps to complete the task on hand. Again, it is important to collect data to determine the effectiveness of these strategies to help determine when to fade out the instructions. Development and implementation of strategies usually occurs when an individual encounters difficulty in learning a new task, shows great variability in task performance, or is unable to meet the required production standard. It is crucial that the employee perceive the usefulness of any strategy that is also recommended, and therefore he or she should be involved in selecting strategies.

Some new employees require less assistance with learning how to do the job, but more support related to interpersonal relations – specifically, communicating and working effectively with job site personnel. The specialist is available to help in this area as well.

Sometimes going to work and learning a new task can be very challenging. The inability to see immediate success can lead to frustration. Frustration appears in many ways, such as verbal outbursts, or even threats to quit a job. If this arises, the specialist is there to assist by teaching techniques to keep the outward signs of frustration at a minimum and manage these feelings in a way that is appropriate for the workplace.

Compensatory memory strategies may also be used to build awareness of behaviors that need to be ameliorated (rather than acquired or used) on the job. For example, a phone operator has a post-it note on his computer that says "Operators who ask customers out on dates will be terminated" to remind him or her not to solicit for dates. Or a person who gets very upset with coworkers who interrupt his routine is taught to put up a "do not disturb" sign or post times of days when interruptions are acceptable.

Eventually, when the employee is performing the job to the employer's standard, the employment specialist can begin to fade from the job site. This is a gradual process, whereby the specialist spends less and less time on the job site until he or she is no longer present on a regular basis.

Long-Term Support and Job Retention Services

Supported employment, unlike time-limited vocational support services, offers ongoing support for as long as the individual is employed (West and Hughes 1990; Wehman et al. 2000; West et al. 2007). This feature distinguishes it from other vocational service options. It is important for the employee to learn to perform the job well, and just as crucial to keep it.

A multitude of factors can influence the need for support, such as a change in business management or coworkers, job functions, or the person's transportation or residential situation. Therefore, in order to proactively help problem-solve issues that if left unresolved could lead to termination, the employment specialist must keep abreast of how things are going both at and away from work. Follow-up services must be available throughout the duration of employment. During this phase, the employment specialist is no longer at the work site on a regular basis, but should continue to have ongoing employee and employer contact and be available to provide additional support services as needed.

To ensure that employment is stable throughout the person's employment, the specialist gathers information on how the employee is performing at work. This may relate to the employee's job performance, the employer's perceptions of performance, effectiveness of work accommodations, relations with job site personnel, job satisfaction, and factors outside of work that may affect job performance. Information is collected by making observations and communicating with knowledgeable work site personnel, and forms the basis for additional assistance as needed either by the employer or the employee with TBI.

Conclusion

Supported employment can help individuals with TBI to gain and maintain employment. Professionals and others directly or indirectly involved with service delivery should note that no two individuals are alike, and that there is tremendous variability in occupations and work environments. All of these factors create the need for an independent and customized approach to supported employment service. The type, level, and intensity of support will also depend on the person's abilities and the particular circumstances on hand. Therefore, the amount of time needed to provide services will also depend upon the situation. Data should be collected to document and determine the need to increase or decrease employment services.

This chapter provided an overview of a supported employment program for individuals with TBI based out of VCU. Although research has shown the promise of supported employment for individuals with TBI, to date the findings have not been replicated in other settings by other centers, so the generalization of these programs remains untested.

Although promoted as an innovative and promising approach to assist individuals with TBI with employment, supported employment services continue to be underutilized. A review of the biennial RSA 911 data from 1993 until 2001 indicated the vocational rehabilitation system is serving increasing numbers of individuals with TBI. However, only 10% of those rehabilitated received supported employment services. There continues to be a need for evidence-based, prospective studies that compare the efficacy of different supported employment interventions.

References

Abrams D, Barker LT, Haffey W, Nelson H (1993) The economics of return to work for survivors or traumatic brain injury: vocational services are worth the investment. J Head Trauma Rehabil 8:59–76

Asikainen L, Kaste M, Sarna S (1998) Predicting late outcome for patients with traumatic brain injury referred to a rehabilitation programme: a study of 508 Finnish patients 5 years or more after injury. Brain Inj 12(2):95–107

Brooks N, McKinlay W, Symington C, Beattie A, Campsie L (1987) Return to work within the first seven years of severe head injury. Brain Inj 1:5–19

Curl RM, Fraser RT, Cook RG, Clemmons D (1996) Traumatic brain injury vocational rehabilitation: preliminary findings for the coworkers as trainer project. J Head Trauma Rehabil 11(1):75–85

Drake RE, McHugo GJ, Bebout RR, Becker DR, Harris M, Bond GR (1999) A randomized clinical trial of supported employment for inner-city patients with severe mental illness. Arch Gen Psychiatry 56:627–633

Greenwald B, Burnett D, Miller M (2003) Congenital and acquired brain injury: epidemiology and pathophysiology. Arch Phys Med Rehabil 84(Suppl 1):S3–S7

Griffin C, Targett P (2006) Job carving and customized employment. In: Wehman P (ed) Life beyond the classroom: transition strategies for young people with disabilities. Paul H. Brookes, Baltimore, MD, pp 289–308

Guilmette TJ, Giuliano AJ (2010) Brain injury and work performance. In: Schultz I, Sally Rogers E (eds) Handbook of work accommodation and retention in mental health. Springer, New York, pp 141–161

Horn IJ, Sherer M (1999) Rehabilitation of traumatic brain injury. In: Grabois M, Garrison SJ, Hart KA et al (eds) Physical medicine and rehabilitation: the complete approach. Blackwell Science, Cambridge, MA, pp 1281–1304

Inge KJ, Wehman P, Kregel J, Targett PS (1996) Vocational rehabilitation for persons with spinal cord injury and other severe physical disabilities. Am Rehabil 22(4):2–12

Keyser-Marcus L, Bricout J, Wehman P, Campbell B, Cifu D et al (2002a) Acute predictors of return to employment after traumatic brain injury: a longitudinal follow-up. Arch Phys Med Rehabil 83:635–641

Keyser-Marcus L, Briel L, Targett P, Yasuda S, Johnson S et al (2002b) Enhancing the schooling for students with traumatic brain injury. Teach Except Child 34(4):62–67

Killiam S, Diambra J, Fry R (1990) Supported employment phase II: job-site training and compensatory strategies. In: Wehman P, Kreutzer J (eds) Vocational rehabilitation for persons with traumatic brain injury. Aspen, Rockville, MD

Kreutzer J, Morton MV (1988) Traumatic brain injury: supported employment and compensatory strategies for enhancing vocational outcomes. In: Wehman P, Moon S (eds) Vocational rehabilitation and supported employment. Paul H. Brookes, Baltimore, MD, pp 291–311

Langlois JA, Rutland-Brown W, Thomas KE (2006) Traumatic brain injury in the United States: emergency department visits, hospitalizations, and deaths. Centers for Disease Control and Prevention, National Center for Injury Prevention and Control, Atlanta, GA

Mank D, Rhodes L, Bellamy GT (1986) Four supported employment alternatives. In: Kiernan W, Stark J (eds) Pathways to employment for adults with developmental disabilities. Paul H. Brookes, Baltimore, MD, pp 139–153

McMordie WR, Barker SL (1988) The financial trauma of head injury. Brain Inj 2:357–364

Moon MS, Goodall P, Barcus M, Brooke V (1985) The supported work model of competitive employment for citizens with severe handicaps: a guide for job trainers. Published guide. Virginia Commonwealth University Rehabilitation, Research, and Training Center on Supported Employment, Richmond, VA

O'Neil J, Hibbard MR, Brown M et al (1998) The effect of employment on quality of life and community integration after traumatic brain injury. J Head Trauma Rehabil 13(4):68–79

Parent W, Everson J (1986) Competencies of disabled workers in industry: a review of the business literature. J Rehabil 52(4):16–23

Rizzolo MC, Hemp R, Braddock D, Pomeranz-Essley A (2004) The state of the states in developmental disabilities. American Association on Mental Retardation, Washington, DC

Sale P, West M, Sherron P, Wehman P (1991) Exploratory analysis of job separation from supported employment for persons with traumatic brain injury. J Head Trauma Rehabil 6:1–11

Shames J, Treger J, Ring H, Giaquinto S (2007) Return to work following traumatic brain injury: trends and challenges. Disabil Rehabil 29(17):1387–1395

Sherron P, Groah C (1990) Supported employment phase I: job placement. In: Wehman P, Kreutzer J (eds) Vocational rehabilitation for persons with traumatic brain injury. Aspen, Rockville, MD

Targett P (2006) Finding jobs for young people with disabilities. In: Wehman P (ed) Life beyond the classroom: transition strategies for young people with disabilities. Paul H. Brookes, Baltimore, MD, pp 255–288

Targett P, Feguson SS, McLaughlin J (1998) Consumer involvement in vocational evaluation. In: Wehman P, Kregel J (eds) More than a job: securing satisfying careers for people with disabilities. Paul H. Brookes, Baltimore, MD, pp 95–118

Targett P, Yasuda S (2006) Applications for youth with traumatic brain injury. In: Wehman P (ed) Life beyond the classroom, 4th edn. Paul H. Brookes, Baltimore, MD

The state supported employment service program final regulations. Federal Register (1987) vol 52, pp 30546–30552

The traumatic brain injury model systems national data center. Brain injury facts and figures (2001) vol 7, pp 8

Wehman P (1981) Competitive employment: new horizons for severely disabled individuals. Paul H. Brookes, Baltimore, MD

Wehman P, Bricout J, Targett P (2000) Supported employment for persons with traumatic brain injury: a guide for implementation. In: Fraser R, Clemmons D (eds) Traumatic brain injury rehabilitation: practical vocational, neuropsychological, and psychotherapy interventions. CRC, USA, pp 201–240

Wehman P, Kregel J, Keyser-Marcus L, Targett P, Campbell B, West M et al (2003) Supported employment for persons with traumatic brain injury: a preliminary investigation of long-term follow up costs and program efficacy. Arch Phys Med Rehabil 84(2):192–196

Wehman P, Kregel J, Sherron P, Nguyen S, Kreutzer J, Fry R et al (1993a) Critical factors associated with the successful supported employment placement of patients with severe traumatic brain injury. Brain Inj 7(1):31–44

Wehman P, Kregel J, West M, Cifu D (1994) Return to work for patients with traumatic brain injury: analysis of costs. Am J Phys Med Rehabil 73:280–282

Wehman P, Revell G, Kregel J (1998) Supported employment: a decade of rapid growth and impact. Am Rehabil 24(1):31–43

Wehman P, Sherron P, Kregel J, Kreutzer J, Tran S, Cifu D (1993b) Return to work for patients following severe traumatic brain injury: supported employment outcomes after five years. Am J Phys Med Rehabil 72(6):355–363

Wehman P, Targett P, West M, Kregel J (2005) Productive work and employment for persons with traumatic brain injury: what have we learned after 20 years? J Head Trauma Rehabil 20:115–127

West M, Hughes T (1990) Supported employment phase III: job retention techniques. In: Wehman P, Kreutzer J (eds) Vocational rehabilitation for persons with traumatic brain injury. Aspen, Rockville, MD

West M, Targett P, Yasuda S, Wehman P (2007) Return to work following TBI. In: Zasler N, Katz D, Zafonte R (eds) Brain injury medicine: principles and practice. Demos Medical Publishing LLC, New York, pp 1131–1150

Chapter 15
Company-Level Interventions in Mental Health

Karen A. Gallie, Izabela Z. Schultz, and Alanna Winter

The effects of mental health are not just mental.
What is good for individual mental health is good
for firm performance.

– Prof. E. Kevin Kelloway, Saint Mary's University, Halifax, Canada
(Standing Senate Committee on Social Affairs, Science and Technology 2004)

According to North American statistics, the prevalence of mental disorders annually among adults ranges from 12% (Statistics Canada) to 26% (Kessler et al. 2005). In contrast to rates of mental disorders among adults, which remain stable over time, rates of emotional disturbance and conduct problems among adolescents, our future work force, demonstrated alarming increases between 1974 and 1999 (Collishaw et al. 2004).

It has been estimated that in Canada mental health disabilities comprise close to 25% of all diseases and injuries, with 13% of that number attributable to depression alone (Dewa et al. 2004). Canadian workers with mental health disabilities, as compared to the rest of the working population, have a higher number of days where they are either unproductive or unable to function at full capacity. Annual productivity losses related to mental health disorders in Canada total about (CAD) $17 billion (Lim et al. 2008). Members of Canada's Global Business and Economic Roundtable on Addiction and Mental Health (2005) believe that the direct and indirect costs of mental health in the workplace equate to 14% of the net annual profits of all companies, sixteen billion dollars, or roughly 3% of Canada's Gross Domestic Product (2004). Similarly, proportionately high costs are reported for Australia, the United Kingdom, and the United States (Standing Senate Committee 2004; Stewart et al. 2003).

The World Bank (1993) reported that disorders due to mental health and neuropsychological problems entailed a higher economic burden than cancer, and more than 15% of the total burden of all diseases. Workplace absenteeism related to

K.A. Gallie (✉)
Director/CQU Academic and Past-President, OHS Educator Chapter,
Safety Institute of Australia
e-mail: abcsolaris@gmail.com

I.Z. Schultz and E.S. Rogers (eds.), *Work Accommodation and Retention in Mental Health*,   295
DOI 10.1007/978-1-4419-0428-7_15, © Springer Science+Business Media, LLC 2011

mental health problems accounts for about 7% of total payroll, one of the main contributors to overall absences (Watson Wyatt Worldwide 2000). Fifty percent of those who miss work because of mental or emotional problems will take either 13 or more days off, or will never return to their jobs (Perez and Wilkerson 1998). While these current costs are staggering, it is their projected increases and the associated exponentially rising healthcare costs that cause great concern for governments, policy makers, employers, insurers, and other stakeholders. Review the chapter by Dewa and McDaid (2010), in this book, for the comprehensive epidemiological and economic rationale for investing in the mental health of the labor force worldwide.

Employers thus have a vested economic and social interest in the mental health condition of their workers and would benefit from developing and instituting programs and policies within the workplace designed to detect, intervene with, and accommodate individuals with mental health disabilities. However, there are limited examples of such programs in existence. This chapter provides a review of documented programs and policies that exist internationally and nationally. In addition, practice and research recommendations and future directions will be explored, focusing on how employers can best address mental health disability issues within the workplace.

Mental Health in the Workplace: Need for Employer-Based Approaches

The more enlightened employers ... are less focused on costs and more focused on keeping a healthy workforce, on their human capital, on productivity...

– Dr. Carli, Medical Director for Partnership Health (APA 2004)

The existing literature suggests that there is a variety of approaches to workplace health, largely predicated on the laws and social context of the countries in which employers operate. For example, in the United States, Americans with Disabilities Act (ADA) (1990) ensures broad discrimination protections for workers (see Center 2010). The ADA requires that all American employers with 15 or more workers make reasonable accommodations for their employees with disabilities. In conjunction, U.S. parity laws require health insurance issuers to adopt the same annual and lifetime dollar limits for mental health as for general medical benefits, encouraging employers to focus on mental health in their workplaces (U.S. Department of Labor 2007). Canada does not have similar national legislation. Provincially, Ontario introduced the Ontarians with Disability Act, Bill 118, in 2001 (amended to Accessibility for Ontarians with Disabilities Act, in 2005), although other provinces have yet to follow their lead. This legislation was initiated by the Ontarians with Disability Act Committee, described as a voluntary coalition of individuals and community organizations wishing to secure the passage of a new law which would achieve a barrier-free society for persons with disabilities.

The European Union has not yet followed the United States in terms of legislation to protect the rights of workers with disabilities. This may be a result of the view among European national systems that return-to-work (RTW) and rehabilitation activities are the responsibility of the state (McAnaney and Wynne 2004). Until quite recently, European "decision makers have been slow to recognize the importance of promoting mental health within the workplace" (McDaid et al. 2005, p. 365). As a result, a review of the international literature yields limited, but growing, information on interventions and performance outcomes with respect to protecting the rights of employees.

In the United States (where the bulk of this literature is based), there appears to be a shift from a medical to a social model of workplace health, where the workplace environment is viewed as impeding or facilitating the ability of individuals to be healthy and engaged in productive long-term employment (McAnaney and Wynne 2004; Unger 2002). Larger workplaces have employed interventions that both prevent and accommodate disability in the workplace (APA 2005). However, most of these initiatives focus on prevention activities in the absence of good outcome indicators (Demyttenaere et al. 2004; Stein 2005). Depression has become the focus for many workplace initiatives, as it dominates health-related lost labor time costs (Greenberg et al. 1993; Stewart et al. 2003). There is an increasing recognition, however, that anxiety and drug and alcohol addictions also require workplace attention (APA 2005). Chronic workplace stress is also being more widely recognized as a precursor to mental health problems (Grebner et al. 2005).

A growing number of employers understand that mental health conditions can develop into long-term disorders and disability (World Health Organization 2000). There appears to be an increasing awareness that the circumstances that lead to work disability may develop and progress while an individual is employed (McAnaney and Wynne 2004; WHO 2000). Deterioration is typically hastened when a person becomes unemployed (WHO 2000). To date, however, most investigations have focused on the employed rather than those wishing to re-enter the workplace (Goldner et al. 2004). Because of their rising healthcare costs and liabilities, employers are becoming aware of the need for workplace activities designed to facilitate and retain fully engaged, productive employees (APA 2005).

"There is a growing awareness that disability is not so much an impairment of the individual as a product of the environment in which he or she lives."

– International Labour Organization (ILO 1998)

An employer-driven workplace integration model may increase the amount and focus of stakeholder involvement in addressing workplace disability (McAnaney and Wynne 2004). Stakeholders typically include Human Resources/Training and Occupational Health and Safety personnel, supervisors, unions, and workers (APA 2005). Workplace activities and interventions provided by these stakeholders are dependent on the committed leadership of the business executives (HERO 2005). In practical terms, it is often the company's Medical, Occupational Health and Safety, or Human Resources director that identifies worksite issues and then gains support from the company CEO to facilitate the necessary internal workplace interventions (Lawson 2004).

Given that most worksites are small to mid-size, they often do not have the necessary on-site expertise or resources to effectively address workers' mental health issues (Highet et al. 2005). As a result, employers often engage the assistance of outside professionals, such as clinical psychologists or occupational therapists from healthcare companies and/or universities (APA 2005). Thus, proactive workplaces seem to have embraced an employer-specialist model, in which interventions occur at the collective worksite level (Prevention/Education) as well as on an individualized worker level (Employee Assistance Program (EAP) Assessment/Treatment/Accommodation/RTW). Concurrently, employer-driven think tanks are emerging to promote the sharing of ideas and intervention activities, as exemplified in the U.S. Business Group on Health (Washington, 2000), the U.K. Institute for Health and Productivity Management (IHPM) (2005), and the Canadian Global Business and Economic Roundtable on Addiction and Mental Health (2005).

Regardless of the specific clinical diagnosis of the employee, intervention strategies tend to share similar characteristics (Krupa 2007; Krupa (2010a, b); Peer and Tenhula (2010)) and often include EAPs.[1] These programs are believed to enhance access to focused psychological interventions for a range of mental health disabilities (Bender and Kennedy 2004; Henderson et al. 2003). EAPs can be an effective means of treating people for mental health issues, as suggested by respondents at the Canadian Standing Senate Committee of Mental Health, Mental Illness and Addiction, who reported that the vast majority of people getting treatment in an EAP required no further assistance (2004).

However, systematic outcome research on EAPs has not kept pace with the growing proliferation and popularity of these programs. Various methodological concerns with respect to an existing evidentiary base for the effectiveness of EAPs have been raised by researchers. They include the following: (1) an excessive reliance on self-report of client satisfaction, which does not necessarily translate into behavioral outcomes, such as increased productivity or earlier RTW, and into economic outcomes for the employer; (2) a shortage of randomized controlled trials, (3) a lack of the investigation of the interrelationships among various outcome measures, (4) an insufficient link between EAP approaches and the reality of the workplace, (5) and lastly, much of the research was done on selective mental health conditions such as addictions. Claims related to the effectiveness of EAPs in producing increased rates of RTW, reduced recurrence of absences related to illness, and lower employer bottom line costs still require research substantiation (Arthur 2000). Also, a more nuanced understanding of what aspects of EAPs work better for whom and when, as well as what does not, is much needed to develop improved, clinically effective and cost-efficient programs and expand EAP interventions into the work environment. While benefiting from the inclusion of EAPs in their workplace, employers should not base their strategies on these programs alone. They are necessary, yet a more comprehensive approach is needed to target

---

[1] Between 60% and 80% of Canadians employed in large to medium-sized companies have access to some form of EAPs, primary portals to mental health and addiction treatment (Standing Senate Committee 2004).

mental health policies in the workplace, utilize multiple interventions, and provide relevant education.

Initiatives in the United States

The more organizationally advanced workplaces (typically large multinational corporations) often include Health Promotion, Employee Education, and Occupational Health and Safety components. There appears to be a growing acceptance of public health concepts in the workplace, which assumes a connectedness between social and workplace circumstances (Stewart-Brown 2005; WHO 2000). This approach is exemplified by the Johnson and Johnson Corporation, which relies on management support with scientifically based resources and a business plan to promote a multidisciplinary approach with a focus on good "corporate health" (APA 2005).

> "To be a sustainable organization, we must be a caring organization. Care for employees is also a hallmark of our corporation...we develop policies that are designed to attract and retain a talented workforce. We have implemented a rigorous workplace safety program, based on a goal of zero injuries and illnesses, with the belief that every accident is preventable. Our employee health and wellness programs are recognized as among the best in the world ..." (Johnson & Johnson 2003, p. 8).

Johnson & Johnson is the world's largest manufacturer of healthcare products and employs 110,600 employees in 57 countries (Johnson & Johnson 2003). In 2003 it was recognized as the most "caring" corporation (Caring Institute 2003), and was ranked by Fortune Magazine as the seventh most admired company in America (Harrington 2004). Johnson & Johnson is viewed by many as a corporate leader in the arena of workplace health, wellness, and safety. It is a founding member of the American National Partnership for Workplace Mental Health (APA 2005) and sits on the UK's Institute for Health and Productivity Management Advisory Council (IHPM 2005).

Johnson & Johnson has achieved its reputation through the integration of a series of initiatives that are often found as individual components in medium to smaller companies. In relation to mental health issues in the workforce, it integrates preventive and defensive efforts with underpinnings in the domains of Public Health, Occupational Health and Safety (OHS), and EAP, Disability Management, and RTW programs. Integration of these components provides a powerful model for achieving overall corporate/employee health and well-being.

As with many worksites, Johnson & Johnson instituted its model by first training managers and human resources professionals in how to recognize and support employees who need help before serious performance or health issues developed (Johnson & Johnson 2003). This initiative follows research showing that supervisor support imparts protective qualities for the mental health of workers (Bromet et al. 1992; Stansfeld et al. 1999; Weinberg and Creed 2000).

Utilizing internal and external workplace advisors and consultants, Johnson & Johnson then developed a Mental Health Interactive Screening tool in 2002. Workplace Response™ is an anonymous telephone and online screening tool for symptoms of depression, bipolar disorder, generalized anxiety, posttraumatic stress disorder, substance abuse, and eating disorders. This screening tool is linked to the EAP and to trained professionals such as social workers, counselors, and psychologists. In its first year, over 5,000 employees and family members made use of the service (Johnson & Johnson 2003).

> *"I had suspected that I might be suffering from depression, but felt too self-conscious to reach out for help. I saw the Mental Health Interactive Screening announcement through an email, took the questionnaire, and got results immediately that told me I should follow up with a mental health professional. For whatever reason, those results helped motivate me to make a call. I'm in treatment now and can't believe the relief I feel knowing that I don't have to continue feeling the way I had been for so long" (Johnson & Johnson 2003, p. 27).*

In the spring of 2003, Johnson & Johnson launched a related effort with its "Healthy People" program, providing employees with access to a variety of wellness programs (including 10,000 steps, a 6-week walking program) and lifestyle counselors. This program recognizes the role of exercise in both the treatment and promotion of mental health (Jones and O'Beney 2004). Research has established exercise as an effective treatment for clinical and subclinical depression, anxiety, and possibly schizophrenia (Brown et al. 2005; Dunn et al. 2005; Faulkner and Biddle 1999; Faulkner and Sparks 1999; Martinsen 1987). There is also limited evidence that exercise may be an adjunctive treatment for somatoform disorder and psychosis (Chamove 1986; Faulkner and Biddle 1999). Other mental health benefits of regular exercise include mood elevation and improved self-esteem, especially in conjunction with psychological therapy (Jones and O'Beney 2004; Skrinar 2003).

With the Healthy People program in place, Johnson & Johnson realized a 20% decline in employee inactivity from 2003 to 2005. In addition, absenteeism declined and there were concurrent improvements in work performance and employee self-reports of emotional well-being (APA 2005). Recipients of EAP, when asked to complete a mental health status questionnaire, improved on average approximately two points on a 10-point scale (5.6 at intake versus 7.5 at follow-up). Johnson & Johnson interpreted these findings, in conjunction with other data, as indicating that EAP interventions not only improve employees' emotional well-being, but also improve their work performance (APA 2005). From 1995 to 1999 (the time in which primary and secondary prevention activities of Work-Life programs, OHS, EAP, and Disability Management were integrated), the company estimated that it saved 42.5 million dollars and reduced each employee's medical costs by $224.00/year (APA 2005). Key Performance Indicators from Johnson & Johnson's 2003 annual report show a related decline in general OHS statistics, suggesting that this integrated model has merit for improving "corporate health." Lost workday case rates declined (from 0.31 to 0.13 cases per 100 employees); severity rate declined (from 7.88 to 3.04 lost workdays per 100 employees); and serious injury/illness rate also declined (0.07 to 0.06 serious illness/illness cases per 100 employees). No direct statistics pertaining to mental health outcome have been publicly reported.

Currently, Johnson & Johnson is complementing these aforementioned activities with the development of a stress management/resilience program. Included in this program are health screening tools, resilience training, healthy eating programs, fitness and wellness promotion, disease management, and stress management techniques, all in an effort to "harness our organizational power – especially the strength of our people and partnerships – to make those improvements and reach our ultimate goal: Healthy People, Healthy Planet, Healthy Futures" (Johnson & Johnson 2003, p. 9).

Recognizing that, as a major global corporation, they "have the capacity to positively influence the behavior of their contractual partners," Johnson & Johnson shares information about its programs with suppliers who, in turn, are expected to follow its lead. Johnson & Johnson also partners with other organizations to encourage its best practices (APA 2005).

Another internationally recognized company that has reported findings on its programs is Rolls-Royce, a company with 35,000 employees serving 150 countries in its civil and defense aerospace, marine, and energy divisions (Rolls-Royce 2004). As a U.K.-based engineering company, it has taken a Risk Management/ Audit approach rather than a Health Promotion model. Thus, there is the assumption that Occupational Health and Safety processes used to identify, assess, control, monitor, and review physical elements are equally applicable to mental health issues. This initiative is spearheaded by Rolls-Royce's Chief Medical Officer, in partnership with the Director of Human Resources and Chief Executive and Chairman of the Health, Safety, and Environment Committee.

Similar to Johnson & Johnson, Rolls-Royce introduced its mental health initiatives by focusing its managers on stress[2] in the workplace with presentations to promote "awareness" in 1992. Rolls-Royce produced a booklet titled "A Manager's Guide to Mental Health at Work" (2000). This guide provides information and clarification about the differences between occupational and nonoccupational stress, differentiating between corporate and individual responsibilities, behaviors, and appropriate actions. In contrast to Johnson & Johnson, Rolls-Royce makes clear distinctions between achieving "organizational health" versus "individual mental health" conditions. Employee mental health is viewed as being influenced by both work and non-work factors such as "personality, preference, and individual susceptibility – the latter being influenced by physical health and fitness, lifestyle, ability and facility to relax, and external support" (Lawson 2004, p. 3). Managers are told that both the employer and employee hold responsibilities in facilitating workplace health and in creating stress. This responsibility includes employee decisions to "create their own stress," such as when opting for excessive overtime. However, the manager's guide opens managerial actions and omissions in relation to workplace health to scrutiny as well.

---

[2] Bourbonnais et al. (2001) found that individuals experiencing work-related stress were twice as likely to have a mental illness.

When Rolls-Royce managers suspect that there is or may be a "stress problem" in their department, they are charged with conducting a risk assessment. Line managers are guided on the "need to be sensitive to ... employees and ... show they are committed to taking action and providing support where needed" (Lawson 2004, p. 9). The company expects managers to oversee work activities to minimize stress. For example, including employees in decision-making, ensuring that employees have the skills, knowledge, and confidence required for the jobs they are given, establishing reasonable production or performance goals, and assisting employees in using time effectively. Limited statements are made about alcohol and drug abuse, mental illness, and posttraumatic stress disorder; and managers are encouraged to confer with the Occupational and Health Department. Rolls-Royce sends a clear message to its managers that, together with their employees, they must strive to prevent and resolve problems (Lawson 2004, p. 15).

In 2002, Rolls-Royce moved to the second stage of its strategy to address workplace stress by developing an "audit tool" for its general employees. This tool comprises 48 questions containing items pertaining to relationships at work, career development, work scheduling, social isolation; and long-distance commuting. Using an internet format for data collection, employees receive immediate feedback and links to relevant internet sites, including PowerPoint presentations on stress. Department managers are provided with summaries of their employees' online responses and receive assistance from the Occupational Health department to address stress-related issues. To date, interventions have focused on improving communications, sorting out bottlenecks, and replanning work (Lawson 2004). Seven work style maxims are promoted by Rolls-Royce including such things as "Beware of the long hours," "Be a holiday spacer and time pacer," "Take regular lunch and workstation breaks," and "Walk and Talk" (instead of sending emails).

Rolls-Royce reports that stress-related absences have dropped as a result of these initiatives. Career uncertainty and job overload remain their two main areas of employee stress. Underway is what Rolls-Royce sees as the next logical step to reducing stress in their workplace – a new course for managers regarding care and rehabilitation of absent employees with "stress problems."

Canadian Initiatives

Unfortunately, there appear to be few best practice initiatives in Canada regarding workplace health (Bill Wilkerson, CEO Canadian Global Business and Economic Roundtable on Addiction and Mental Health, personal communication, April 6, 2006). Wilkerson believes we do a fair job of supporting physical injuries but that employees with mental health disabilities typically disappear once on disability leave (Wilkerson, as per Galt, March 2006). Organizations like Wilkerson's Roundtable and the Canadian Mental Health Association (CMHA) are left to advocate for workplace programs and policies designed to aid those with mental health disabilities. In the CMHA publication *Working Well: An Employer's Guide to Hiring and Retaining People with Mental Illness* (Vandergang 2002), it is estimated that less than half of all Canadian companies have developed and implemented any type of wellness or disability management program. The reason for this seems to be the length of time required before employers see benefits in employees' behavior, or a return on their investment.

Despite this, *Working Well* (2005) has documented the efforts of some Canadian companies to address the needs of employees with mental health disabilities. For instance, the Canadian telecommunications company Telus is stated to have seen a $717,000 savings in productivity in one year, alone as a result of implementing a wellness program and health centers that allow employees to attend a nurse-led RTW program. Weyerhauser's Chemainus Sawmill, a relatively small workplace of 170 employees, was provided as another example of a company with a successful disability management program. They set up a committee of three workers to meet monthly with rehabilitation advisers and workers with disabilities (including mental health) in support of RTW plans.

Another successful disability management program mentioned in *Working Well* was created in 1999 by the Canadian Imperial Bank of Commerce (Vandergang 2002). They experienced excellent results in the first year, with a 30% reduction in disability-related absences and a savings of $6.5 million. Liberty Health Insurance is also cited as an example of a leading edge employer. This company began providing early intervention for "high risk" cases by getting in touch with employees within 10 days after they go onto Short-Term Disability. Their approach has resulted in the prevention of movement to Long-term Disability in 90% of these cases. The final company mentioned by the CMHA publication is CP Rail (Vandergang 2002). Senior management, working with union representatives, made RTW a priority by including accommodations for workers with disabilities in their collective agreement. As seems apparent from the examples documented in *Working Well*, companies choosing to make the investment in a wellness or disability management program that includes mental health considerations do experience considerable rewards.

One final example of a Canadian company, provided by Wilkerson (2006), as leading the way in this area is Ontario Power Generation Inc. (OPG). Under the direction of its wellness division, it obtained executive support in partnership with workers and the union for making Mental Health its wellness priority for 2005. Its three pillars are Prevention, Organizational Effectiveness, and Disability Management. Prevention at OPG consists of a wellness website, lunch and learn sessions, and supervisor training and videos. OPG's Organizational Effectiveness revolves around creating a high energy environment where stress is positive and motivating. Its Disability Management consists of mental health case management initiatives based on early intervention and treatment. Key features include: monthly reports made to identify cases requiring follow-up and additional attention; use of nurse care coordinators to contact employees after 5 days of leave to discuss RTW plans; and rewarding managers when staff do well (Galt 2006).

Policy and Program Recommendations for Employers

The Canadian Global Business and Economic Roundtable on Addiction and Mental Health (2005) encourages employers to embrace new strategic and operational approaches to mental health in the workplace by proposing a 12-step business plan comprised of the following:

Step One: Brief the CEO on mental illness and addiction
Step Two: Early detection of mental illness and addiction
Step Three: Reform EAP and health plans
Step Four: Establish a mentally healthy workplace
Step Five: Reduce the overflow of email and voicemail messages
Step Six: Develop flexible return-to-work policies
Step Seven: Educate managers and supervisors on connections between mental illness and physical illness
Step Eight: Reduce emotional work hazards
Step Nine: Promote work/life balance policies
Step Ten: Encourage people to seek the necessary professional help
Step Eleven: Monitor health status of the organization through specific targets
Step Twelve: Eliminate sources of workplace stress.

While these 12 steps provide a good guide, both macrolevel support and detailed "how to" guidelines for employers are also required. The Roundtable has recently announced their success in attracting greater involvement from Canadian stakeholders such as university and government research and policy centers, as well as a greater focus on the need for earlier detection of mental health issues (e.g., awareness, prevention, and cure; W. Wilkerson, personal communication, April 6, 2006).

Australia, in many ways, "leads the world in mental health promotion, mental illness prevention, early intervention initiatives, and stigma reduction" (Standing Senate Committee 2004, p. 7). "Beyondblue" is a bipartisan initiative of the Australian State and Territory governments vested with the goal of raising community awareness about depression and reducing stigma associated with mental health problems. Beyondblue works with its societal partners to bring together expertise on depression and other forms of mental health disabilities. It does this in several ways, including its workplace program, in which trainers provide direct consultation to workplaces and assistance with training. Beyondblue's CEO reports receptivity to this program from Australian public sector worksites. Unfortunately, no data is currently available to determine the effectiveness of these workplace interventions (N. Highet, personal communication, October 25, 2005).

Another Beyondblue initiative is to establish greater community awareness of depression and related mental health problems. To achieve this, Beyondblue sponsored a series of television advertisements highlighting Australians with mental health disabilities and concurrently promoting the schedule of a national caravan that travels the country providing information. A subsequent telephone survey of 3,200 Australians found that 65% of respondents were now aware of someone with depression. Sixty-nine percent of these respondents recalled viewing something about depression in the media within the last year. The conclusion reached by Beyondblue is that active dissemination of information has increased community recognition of depression and the impact it has on individuals, families, and organizations, and has, in turn, raised related community services access (Highet et al. 2005). Furthermore, Beyondblue initiatives may have led to improved community attitudes and increased positive health behaviors, such as pursuing prevention and treatment when required (Highet et al. 2005).

Future Directions

A review of the literature suggests that employer-driven workplace activities addressing effective mental health prevention are in the early stages of development and implementation. There is a paucity of published documentation on these types of initiatives and little systematic research to document their effectiveness. Government activities, such as those seen in the U.S. with the ADA and the efforts of Federal and State/Territory Governments in Australia, have helped to raise societal awareness, and thus workplace acknowledgement regarding the importance of this issue. As such, it will be interesting to follow the resulting effects that the Accessibility for Ontarians with Disabilities Act will have on Ontario's workplaces. As addressed in earlier sections of this chapter, activities at the macro and micro levels, while important, must also be complemented by investigations of effectiveness.

An efficacious future model of workplace mental health prevention strategies would likely be multisystem, multilevel, multidimensional, and integrative. It would combine evidence-informed occupational and clinical approaches embedded in a social model of work, where the focus is on ability/wellness instead of disability and the traditional medical model. The levels of intervention that emerge from current literature reviews appear as follows:

1. Society-wide public awareness campaigns using a social marketing approach such as public advertising that targets attitudes towards mental health with an associated societal behavioral change. The anticipated outcome of this intervention would be the gradual dissipation over time of both the stigma surrounding mental health and the unhelpful separation of "health" and "mental health" in society and its institutions.
2. Legislative and public policy support for mental health parity in health and insurance/compensation systems and in the workplace, together with fair employment practices for persons with mental health conditions.
3. Prevention and educational strategies embedded in a health promotion model in the workplace integrated with individual, worker-level strategies, involving assessment, clinical and occupational interventions, RTW, and job accommodation. This approach must be situated within a Disability Prevention and Management Model.
4. The often fragmented employer resources such as occupational health and safety units, disability management services, EAPs, and corporate training departments

must be synchronized and integrated accordingly for the common goal, as some of the best employer examples currently indicate.

5. At the organizational level of the work unit, training of management and supervisory staff in mental health risk audit, early detection and intervention with employees with mental health performance difficulties at all stages of disability, and setting up the best organizational conditions for a mentally healthy work unit, are of critical importance.

Last but not least, we must recognize that even with improved practises in mental health in the workplace, real progress will be hampered by a lack of outcomes based research. There is a substantial need for methodologically rigorous investigations into the effectiveness of the various components of mental health in the workplace strategy, the interrelationships among its various aspects and levels, and the differential outcomes of the interventions. The research community at this stage is poorly prepared to meet this challenge. The type of research that is needed defies traditional research disciplines and paradigm boundaries. It should be transdisciplinary and focus on multisystem interactions. The methodology for such research is just emerging but has not yet been applied to the mental health field (Gallie and Jessup 1998; Schultz and Gatchel 2005). Moreover, grant support for applied research in this field is lagging behind other disciplines, mostly because of the traditional boundaries between medical and social science research, together with the high costs, complexity, and labor intensiveness of such investigations.

Without further well-designed and innovative research, employers will continue to be left to their best intentions, resources, and ideas. Evidence-supported practices are in short supply and generalization from one work setting to another is a major challenge. Likewise, the transfer of knowledge from other areas of research that are more advanced than mental health disabilities in the workplace, such as musculoskeletal pain and disability, has not yet been initiated.

In the meantime, employers who plan new organizational initiatives to improve mental health are best advised to add occupational health researchers to their program development teams to ensure that evaluation and research activities are included. The ability to generalize the findings to other settings and the knowledge of what works for whom and when (and under what circumstances) would then be possible.

Acknowledgment This chapter is based on the research completed for a Social Development Canada–funded project, "Towards Evidence-Informed Best Practice Guidelines for Job Accommodations for Persons with Mental Health Disabilities."

References

Accessibility for Ontarians with Disabilities Act, S.O. (2005) Chapter 11. Legislative Assembly of Ontario. http://www.e-laws.gov.on.ca/DBLaws/Statutes/English/05a11_e.htm. Accessed 18 April 2006
Americans with Disabilities Act (1990) U.S. Department of Justice. http://www.usdoj.gov/crt/ada. Accessed 18 Aug 2004

American Psychiatric Association (2004) Ford Motor Company achieves positive ROI by integrating Med/Psych services. Ment Health Works 1:4–5

American Psychiatric Association (2005) Johnson & Johnson integrates health, wellness, and safety. Ment Health Works 1:3–5

Arthur AR (2000) Employee assistance programmes: the emperor's new clothes of stress management? Br J Guid Counc 28(4):549–559

Bender A, Kennedy S (2004) Mental health and mental illness in the workplace: diagnostic and treatment issues. Healthc Pap 5(2):54–68

Bourbonnais R, Larocque B, Brisson C, Vézina M (2001) Contraintes psychosociales du travail. Institut de la Statistique du Québec, Québec, pp 267–277

Bromet EJ, Dew MA, Parkinson DK (1992) Effects of occupational stress on the physical and psychological health of women in microelectronics plant. Soc Sci Med 34:1377–1383

Brown WJ, Ford JH, Burton NW, Marshall AL, Dobson AJ (2005) Prospective study of physical activity and depressive symptoms in middle-aged women. Am J Prev Med 29(4):265–272

Caring Institute (2003) Johnson & Johnson named 2003 most caring corporation in America. Caring 22(12):48–53

Center C (2010) Law and job accommodation in mental health disability. In: Schultz IZ, Rogers ES (eds) Handbook of work accommodation and retention in mental health. Springer, New York

Chamove AS (1986) Positive short-term effects of activity on behavior in chronic schizophrenic patients. Br J Clin Psychol 25:125–133

Collishaw S, Maughan B, Goodman R, Pickles A (2004) Time trends in adolescent mental health. J Child Psychol Psychiatry 45(8):1350–1362

Demyttenaere K, Bruffaerts R, Posada-Villa J, Gasquet I (2004) Prevalence, severity, and unmet need for treatment of mental disorders in the World Health Organization – World Mental Health Surveys. J Am Med Assoc 291(21):2581–2590

Dewa CS, Lesage A, Goering P, Caveen M (2004) Nature and prevalence of mental illness in the workplace. Healthc Pap 5(2):12–25

Dewa CS, McDaid D (2010) Investing in the mental health of the labour force: epidemiological and economic impact of mental health disabilities in the workplace. In: Schultz IZ, Rogers ES (eds) Handbook of work accommodation and retention in mental health. Springer, New York

Dunn AL, Trivedi MH, Kampert JB, Clark CG, Chambliss HO (2005) Exercise treatment for depression: efficacy and dose response. Am J Prev Med 28(1):1–8

Faulkner G, Biddle S (1999) Exercise as an adjunct treatment for schizophrenia: a review of the literature. J Ment Health 8(5):441–457

Faulkner G, Sparks A (1999) Exercise as therapy for schizophrenia: an ethnographic study. J Sport Exerc Psychol 21:52–69

Gallie KA, Jessup BA (1998) Accident repetition – breaking the chain. Occup Health Safety Canada 14(3):18–20

Galt V (2006) Out of the shadows: mental health at work. http://www.workopolis.com. Accessed 7 April 2006

Global Business and Economic Roundtable on Addiction and Mental Health (2005) http://www.mentalhealthroundtable.ca. Accessed 12 Sept 2005

Goldner E, Bilsker D, Gilbert M, Myette L, Corbière M, Dewa CS (2004) Disability management, return to work and treatment. Healthc Pap 5(2):59–70

Grebner S, Semmer NK, Elfering A (2005) Working conditions and three types of well-being: a longitudinal study with self-report and rating data. J Occup Health Psychol 10(1):31–43

Greenberg PE, Stiglin LE, Finkelstein SN, Berndt ER (1993) The economic burden of depression in 1990. J Clin Psychiatry 54:405–418

Harrington A (2004) America's most admired companies recent bad press hasn't dimmed the business world's affection for Wal-Mart. Fortune Magazine, 149, 5. http://money.cnn.com/magazines/fortune/fortune_archive/2004/03/08/363675/index.htm. Accessed 23 April 2007

Health Enhancement Research Organization – HERO (2005) Forum for Optimal Employee Health. http://www.the-hero.org/. Accessed 8 Nov 2005

Henderson M, Hotopf M, Wessely S (2003) Workplace counseling (editorial). Occup Environ Med 60:899–900

Highet N (2005) Personal communication on 25 Oct 2005

Highet NJ, Luscombe GM, Davenport TA, Burns JM, Hickie IB (2005) Positive relationships between public awareness activity and recognition of the impacts of depression in Australia. Journal of Psychiatry 40:55058

Institute for Health and Productivity Management (2005) http://www.ihpm.org. Accessed 1 Nov 2005

International Labour Organization (1998) World employment report, 1998–99. Employability and the global economy – how training matters. ILO, Geneva, p 35

Johnson & Johnson (2003) Sustainability Report. http://www.jnj.com/community. Accessed 1 Nov 2005

Jones M, O'Beney C (2004) Promoting mental health through physical activity: examples from practice. J Ment Health Promot 3(1):39–47

Kessler RC, Chiu WT, Demler O, Merikangas KR, Walters EE (2005) Prevalence, severity, and comorbidity of 12-month DSM-IV disorders in the National Comorbidity Survey Replication. Arch Gen Psychiatry 62(6):617–627

Krupa T (2007) Employment interventions to facilitate employment for individuals with psychological, psychiatric and neuropsychological disabilities. Canadian Journal of Psychiatry, 52(6):339–45

Krupa T (2010a) Approaches to improving employment outcomes for people with serious mental illness. In: Schultz IZ, Rogers ES (eds) Handbook of work accommodation and retention in mental health. Springer, New York

Krupa T (2010b) Employment and serious mental health disabilities. In: Schultz IZ, Rogers ES (eds) Handbook of work accommodation and retention in mental health. Springer, New York

Lawson I (2004) Auditing and assessing work stress at Rolls-Royce. IRS Employment Review, 812

Lim KL, Jacobs P, Ohinmaa A, Schopflocher D, Dewa CS (2008) A new population-based measure of the economic burden of mental illness in Canada. Chron Dis Can 28(3):92–98

Martinsen EW (1987) The role of aerobic exercise in the treatment of depression. Stress Med 3:93–100

McAnaney D, Wynne R (2004) Employment and disability: back to work strategies. European Foundation for the Improvement of Living and Working Conditions

McDaid D, Curran C, Knapp M (2005) Promoting mental well-being in the workplace: a European policy perspective. Int Rev Psychiatry 17(5):365–373

Peer J, Tenhula W (2010) Employment interventions for mood and anxiety disorders. In: Schultz IZ, Rogers ES (eds) Handbook of work accommodation and retention in mental health. Springer, New York

Perez E, Wilkerson B (1998) Mindsets. Homewood Centre for Organizational Health at Riverslea, Oakwood, ON

Rolls-Royce (2000) A manager's guide to mental health at work. Rolls-Royce plc, Derby, UK

Rolls-Royce (2004) Annual report. http://www.rolls-royce.com. Accessed 4 Nov 2004

Schultz IZ, Gatchel RJ (2005) Research and practice directions in risk for disability prediction and early intervention. In: Schultz IZ, Gatchel RJ (eds) Handbook of complex occupational disability claims. Early risk identification, intervention, and prevention, 1st edn. Springer Science + Business Media, Inc., New York, pp 1–566

Skrinar G (2003) Mental illness. In: Durstine J, Morre G (eds) ACSM's exercise management for persons with chronic diseases and disabilities, 2nd edn. Human Kinetics, Champaign, IL

Standing Senate Committee on Social Affairs, Science and Technology (2004) Mental health, mental disorder and addiction: overview of policies and programs in Canada, report 1. Government of Canada Printing Office, Ottawa, ON, pp 1–247

Stansfeld SA, Fuhrer R, Head J, Shipley MJ (1999) Work characteristics predict psychiatric disorder: prospective results from the Whitehall II study. Occup Environ Med 56:302–307

Stein MB (2005) Sweating away the blues – can exercise treat depression? Am J Prev Med 28(1):140–141

Stewart WF, Ricci JA, Chee E, Hahn SR, Morganstein D (2003) Cost of lost productive work time among US workers with depression. J Am Med Assoc 289(23):3135–3144

Stewart-Brown S (2005) Interpersonal relationships and the origins of mental health. J Public Ment Health 4(10):24–46

Unger DD (2002) Employers' attitudes toward persons with disabilities in the workforce: myths or realities. Focus Autism Other Dev Disabil 17(1):2–10

U.S. Department of Labor (2007) Mental Health Parity Act [MHPA]: Employee Benefits Security Administration. http://www.dol.gov/ebsa/newsroom/fsmhparity.html. Accessed 28 April 2007

Vandergang AJ (2002) Working well: an employer's guide to hiring and retaining people with mental illness. Canadian Mental Health Association (CMHA) – National Office, Toronto

Washington Business Group on Health (2000) Large employer experiences and best practices in design, administration, and evaluation of mental health and substance abuse benefits

Watson Wyatt Worldwide (2000) Staying at work 2000/2001 – the dollars and sense of effective disability management. Watson Wyatt Worldwide, Vancouver

Weinberg A, Creed F (2000) Stress and psychiatric disorder in healthcare professionals and hospital staff. Lancet 355:533–537

Wilkerson W (2006) Personal communication on 6 April 2006

World Bank (1993) World development report. Investing in health – world development indicators. Oxford University Press, New York

World Health Organization (2000) Mental health and work: impact, issues and good practices. World Health Organization, Geneva, Switzerland

Chapter 16
Service Integration in Supported Employment

Jane K. Burke-Miller, Judith A. Cook, and Lisa A. Razzano

Introduction

Research literature exploring the importance of service integration to positive employment outcomes is relatively sparse. One study in this area was conducted by Drake and colleagues (2003), and focused on the Individual Placement and Support (IPS) model of supported employment (SE). Investigators analyzed quantitative and qualitative data from three supported employment studies and identified four consistent advantages of services integration, including: (1) more effective engagement and retention of consumers; (2) better communication between employment specialists and mental health clinicians; (3) promoting understanding and a focus on employment among clinicians; and (4) incorporation of clinical information into vocational plans (Bond 2004; Drake et al. 2003). Another advantage of services integration identified in prior studies is that it necessitates a multidisciplinary "team" approach that can result in an individualized application of a variety of evidence-based practices from the different fields that make up the service delivery team (Liberman et al. 2001).

Previous research suggested that service integration could occur in a variety of different ways within diverse types of organizational settings and model programs. For example, individual staff could provide both clinical and vocational services, using what is known as a "generalist" model of psychosocial rehabilitation (Dincin 1975). Alternatively, employment staff could coordinate employment services with those of other providers working together on multidisciplinary teams, using a "vocational specialist" model. These teams typically consist of psychiatrists, case managers, and vocational staff, along with other providers such as peers, addictions counselors, and benefits counselors. There was also diversity among team approaches, since team members could have responsibility for either an individual caseload or a shared caseload. Advantages of this type of integration in supported

J.K. Burke-Miller (✉)

Department of Psychiatry, Center on Mental Health Services Research and Policy, University of Illinois at Chicago, 1601 W. Taylor Street MC – 912, Chicago, IL, 60612, USA

e-mail: jburke@psych.uic.edu

I.Z. Schultz and E.S. Rogers (eds.), *Work Accommodation and Retention in Mental Health*, 311
DOI 10.1007/978-1-4419-0428-7_16, © Springer Science+Business Media, LLC 2011

employment were thought to include an enhanced ability to engage and retain clients, more efficient and effective communication between different types of providers, enlistment of clinical staff in support of clients' vocational attainment, and incorporation of clinical issues into the vocational rehabilitation process (Drake et al. 2003).

The Employment Intervention Demonstration Program

The Employment Intervention Demonstration Program (EIDP) was a multisite, federally funded study designed to identify innovative service delivery models that combined vocational rehabilitation with other types of services and supports (Cook et al. 2002). Extensive information on over 1,650 participants with psychiatric disabilities at eight sites nationwide was collected by the EIDP from February 1996 through May 2000. The EIDP demonstrated that compared to nonintegrated services, supported employment featuring integrated vocational and mental health services resulted in superior employment outcomes, even in the face of demographic, clinical, and economic barriers to employment (Burke-Miller et al. 2006; Cook et al. 2006; Razzano et al. 2005). While the overall success rate of EIDP participants was noteworthy (Cook et al. 2005a, b), the statistics alone do not convey the great range and variety of service integration that contributed to positive employment outcomes. This chapter highlights the different ways in which supported employment services were combined with medical, clinical, case management, addiction, and other behavioral health services to enhance employment outcomes. In this analysis, we use a series of case examples to illustrate the ways in which service integration occurred and how it led to vocational success.

In the EIDP, integrated services were delivered in a variety of organizational settings, including outpatient clinics, community mental health centers, and rehabilitation programs. A number of different models were used, including IPS (Drake et al. 1999), assertive community treatment (ACT) (Russert and Frey 1991), Integrated Services (Cook and Rosenberg 1994), and other models of SE that were specially tailored for psychiatric populations. The operational definition of services integration used in the study involved four components: (1) vocational and mental health services (such as medication management, case management, and individual therapy) were provided by the same agency; (2) vocational and mental health services were delivered at the same program location; (3) vocational information and mental health information about the client were combined in a single case record; and (4) regularly scheduled team meetings occurred at least three times per week to coordinate treatment planning and service delivery, while enhancing staff communication and coordination. An example of a matrix for assessing programs' level of services integration on these components is shown in Table 16.1. In this example, Program "A" has each of the four components required to operationalize integrated vocational and mental health services, and therefore has an integrated services score of 100%. Other programs are missing at least one of the four components,

Table 16.1 Level of vocational and mental health services integration matrix

Program	Vocational and mental health services are… (each of the four integration components is worth 25% of the total percent score)				
	… at the same agency	… at the same location	…in a single case record	…discussed in three or more scheduled team meetings/week	Percent services integration
A	Y	Y	Y	Y	100%
B	Y	Y	N	Y	75%
C	Y	N	Y	Y	75%
D	Y	N	Y	N	50%
E	N	N	Y	N	25%
F	N	N	N	N	0%

and therefore have scores ranging from 75% down to 0%. In the EIDP, the information necessary for completing the matrix was gathered from program directors using a manual developed specifically for that purpose (Burke and Cook 2000).

Published EIDP analyses showed that those EIDP supported employment models with high levels of services integration were more effective than models with low levels of integration in producing positive outcomes (Cook et al. 2005 b). Study participants who received the integrated services were over twice as likely to achieve competitive employment, and almost one and one-half times as likely to work 40 or more hours per month, even adjusting for demographic, clinical, and work history employment predictors. The advantage of those receiving integrated services over those who did not increased over the course of the 2-year study period. In addition, higher cumulative amounts of vocational services were associated with better employment outcomes, whereas higher cumulative amounts of clinical services were associated with poorer work outcomes. The study's results confirm the importance of provider communication and coordination of mental health and rehabilitation services in working toward vocational goals.

The following brief stories contributed by the sites involved in this study describe some of the ways services can be integrated, and highlight how each program model flexed to accomplish integration by responding to participant needs. Common to all of these experiences was the attention paid to specific client needs and preferences, and ongoing support both in and out of the workplace. The individuals described all live with serious and persistent mental illness, but represent great diversity, not only in geographic region, but in age (31–53 years), education (grade school through graduate school), ethnicity (African American, American Indian, Hispanic/Latino, and White), and diagnosis (schizophrenia spectrum and mood disorders). Their experiences and jobs reflect their individuality, and are also diverse in nature, hours, and pay, although none were sheltered work or were paid less than minimum wage. Taken together, these case studies illustrate the myriad ways in which integration can be achieved and the common elements that need to be present may interact.

The Importance of Service Integration to Engagement

In Arizona, study participants received a full array of case management and supported employment services from an integrated treatment team composed of psychiatrists, case managers, rehabilitation counselors, employment specialists, job developers, peer specialists, and benefits specialists, all located in the same office. One participant at this site had been avoiding engaging in services with vocational staff for several months after she joined the EIDP. After several attempts to meet with her, the employment specialist solicited the help of case management staff to aid with her engagement into the program. Together, the team eventually discovered that she was dealing with several serious family issues, including an abusive relationship with her partner and her mother's Alzheimer's disease. With the team's help, she began to receive supportive counseling for her feelings about her mother's chronic condition and how she could deal with her sense of helplessness. She was also assisted in terminating the abusive relationship and relocating to a new residence. Eventually, she decided to begin training to become a certified nursing assistant, which she completed successfully and followed up with a job in her new field. This participant reported being very happy to be at work and able to help others in ways that were personally meaningful.

Combining Cognitive Skills Training, Family Therapy, and Vocational Rehabilitation

The SE model tested in Maine involved working with a consortium of employers committed to promoting employment for mental health consumers. Clinical services were delivered using a family-aided assertive community treatment (FACT) model. This story is about a participant who had not worked competitively for over 10 years because of cognitive deficits, including difficulty with information processing and executive functions such as memory and organization. The team first provided him with cognitive skills training while simultaneously evaluating his job skills and career interests. They found that his strengths and weaknesses called for a careful job match with a supportive supervisor, a low-stress environment, and opportunities to use his exemplary proofreading skills. Once his cognitive skills improved, employment staff arranged, through the employer consortium, an opportunity for him to "job shadow" a proofreader at a small area newspaper. Next, he had a chance to work part-time for pay, at a situational assessment placement doing proofreading, again arranged with a consortium employer. This assessment opportunity turned into a permanent position with flexible hours. Employment staff provided him with on-site support, which was gradually reduced as he demonstrated increased confidence and decreased anxiety. Program staff continued regular phone contact as well as monthly in-person contact with him and his relatives through multifamily group therapy. He continued to work at the same job, and he and his family noted that he felt less depressed and anxious, and more outgoing.

Service Integration for Working Mothers

The Connecticut program used the IPS SE model. IPS focuses on rapid placement with continued follow-along support, and seeks to find employment opportunities that are consistent with consumers' preferences, skills, and abilities. In this particular instance, the study participant had only an eighth grade education, had never held a paying job, and had experienced six psychiatric hospitalizations in 4 years prior to EIDP participation. She was raising three children and coping with family members who were having difficulties with alcohol. The team recognized the need to carefully coordinate medication management services in order to help her remain out of the hospital, which she did for the rest of her time in the program. At the same time, she was helped to enroll in Al-Anon and received supportive counseling for dealing with her relatives' alcohol problems. Next, employment staff helped her to get a job as a greeter at a child care center that was part of a larger chain. Shortly thereafter, another location in the chain hired her into a position closer to her home and promoted her to teacher's assistant, with an increase in salary. The team continued to work with this consumer, providing vocational support, psychiatric treatment, individual counseling, and case management. She continued to work in this job for over a year and a half, leaving only because she moved away from the area.

Coordination of Symptom Management with Vocational Support

The Maryland site also used the IPS model as part of a community-based comprehensive services program. One of their study participants had no previous competitive work history, was unsure about his employment goals, and often experienced delusions and paranoia. Employment staff continued to work through his psychiatrist and case manager to let him know they were standing by to help if he should decide that work was a goal. After 7 months of program involvement, he approached his therapist to express a renewed interest in work. The participant, his therapist, and vocational staff then met to develop a plan of action that involved small steps at the client's preferred pace. Initially, he restricted the job search to only certain days and hours during the week. However, after working with vocational staff and increasing his comfort level with vocational services, he started to apply for part-time positions regardless of the hours or days of the week. As they got to know him better, employment staff noticed that this participant was an avid sports fan. With some encouragement and accompaniment by vocational staff, he attended a job fair at Camden Yards (the home of the Baltimore Orioles) and was hired on the spot as a member of the janitorial crew. He worked there through the entire baseball season and discovered, to his surprise, that he was able to rearrange his daily activities without assistance. Other staff on the team began to notice the absence of delusional and paranoid thinking as he grew in self-confidence after he began work.

Integrating Substance Abuse Treatment with Employment Services

One of the programs studied in Massachusetts was the International Center for Clubhouse Development (ICCD), involving a planned community of staff and consumers who work together daily to provide and receive services such as meals, companionship, skills training, and paid work. One of their participants had been unemployed continuously for over 7 years when she enrolled in the EIDP, and had been recently hospitalized. Shortly after becoming a member of the Clubhouse, she got a job working at a retail business, but left after a few months because of difficulties with substance abuse. Through careful coordination between her psychiatrist, case manager, and peer supporters, she entered addictions treatment, managed to stop using drugs, and moved back into a stable living environment. Once involved in a support group at the Clubhouse, she started working a part-time job. With assistance from the Clubhouse, she continued to stay drug free even while dealing with the additional stresses involved in a new job, and confronting a major personal crisis. She worked at this job for 9 months and then was hired permanently by the same company. As of the end of EIDP data collection, she was still working in the position, with good relationships and good work performance. She remained an active member of the Clubhouse and she had not been hospitalized since becoming a participant in the EIDP.

Importance of Peer Support to the Career Journey

The Pennsylvania site evaluated a model of long-term employment training and supports delivered by vocational staff working together with mental health consumer peers. The story at this site is about a woman who had been treated by psychiatrists on and off throughout her life, including being hospitalized for a suicide attempt. She had a variety of work experiences, most recently as an HIV/AIDS educator and consultant. Unfortunately, she had also experienced prejudice and stigma based on race and disability in the course of her work history. She had a lot of insecurity issues, and felt that she had received little emotional support from her service providers. With support from her peer specialists, she became the Executive Director of her own education program on issues of sexually transmitted diseases, sexual identity, and teen pregnancy awareness. She also became employed as a radio personality with her own call-in radio show on HIV/AIDS issues. She noted that the support she received from her peer groups has been instrumental in helping her realize that a career was possible for her, and that she should strive to get her message out to others.

"It Takes a Village" to Coordinate Recovery

The South Carolina site studied the vocational effects of a program that combines ACT and IPS. The story from this site presents an outstanding example of workplace accommodation and personal development of disability insight, increase in self-efficacy and effective problem-solving, and success of a client through her full partnering with a clinical treatment/rehabilitation team. The ACT-IPS team helped a participant who was experiencing frequent hospitalizations to find a job as a personal caretaker. The team worked out an arrangement with her employer such that when she required time-off for hospitalization or to deal with the stress of partial relapse, she was quickly given that time; whenever she was ready to return to work, she was taken back immediately. With the constant support of the ACT-IPS team, her numerous hospitalizations became progressively briefer and less frequent. Toward the end of the EIDP, she presented on her own to the local ER for crisis intervention and a desire to be hospitalized. Her case manager of the ACT portion of the team met her there, and persuaded her that she ought to return home, with the guarantee of twice-daily visits from the team and daily monitoring of medication adherence. On this occasion her recovery was so rapid that in a couple days, she was back at work.

Service Integration Combining Professional Treatment with Natural Supports

The program located in Texas combined "rapid entry" supported employment with social network enhancement. Services were designed to help participants move from support networks characterized primarily by professional support to more balanced networks that were larger, more diverse, and more reciprocal. The participant profiled here had been employed intermittently for 10 years prior to joining the EIDP. After 3 months in the program, she was hired as manager of an apartment complex that contained 40 affordable housing units for low-income individuals in recovery. Her duties included those traditionally associated with apartment management as well as peer counseling. At times, however, she worked so hard that it seemed she was bound for a "meltdown." With support from employment staff and her peers, she learned to turn her beeper off at night and on the weekends. After 2 years, she was still at this salaried position with full benefits. She continued to refer her building's residents to the vocational program, and volunteered to speak to groups of job seekers about her experiences. After hearing her speak, a consumer who had been negative, frustrated, and withdrawn experienced a change of heart, saying, "if she [can do this], then maybe there is hope for me too."

Summary and Conclusions

As these case examples illustrate, the integration of services requires extraordinary effort from service delivery staff who often "go that extra mile" and "think outside the box" when designing individually tailored vocational services and providing unobtrusive job supports. Also required is the development of a high degree of trust with clients, since insight into the myriad problems they face is often needed for effective service coordination. The importance of peer services also is evident, as is the coordination of natural supports provided through family and friends. In understanding service coordination, it is important to remember that there are important personal stories behind the models that are studied, and these stories point to the need to address each client's specific situation along with his or her strengths as well as weaknesses, using available support and service delivery networks.

The successful integration of vocational and mental health services is also a step toward more comprehensive integration of additional services, including all aspects of psychosocial rehabilitation such as housing, education, social relations, and economic security, with physical health and other services (Kopelowicz and Liberman 2003). An example of an expanded matrix for assessing programs' level of services integration on these components is shown in Table 16.2. Criterion components could be as specific as those shown in Table 16.1, or could simply be the presence or absence of a type of expanded service (e.g., education), or that the expanded service meets all or some of the best practice standards for that service. The number and nature of the components incorporated in the matrix would be determined by the user's purpose. In the example in Table 16.2, there are ten components, meaning each would be worth 10% of the total integration score. More elaborate matrices in which differing components are awarded different weights according to importance can also be constructed.

In addition to integrating services at the individual client level, one recent proposal (Cook 2006) calls for a "multisystemic approach" to supported employment that would encompass federal, state, and local systems whose combined efforts are needed in order to help people become and remain employed (see also chapter 24 by Schultz et al. 2010, this volume, for a discussion of this topic).

These systems include vocational rehabilitation, mental health, social security (SSI, SSDI), health care (Medicaid, Medicare), housing, education, legal aid, criminal justice, asset accumulation, and other social services, as well as the business and mental health advocacy communities. Such an approach is necessary in order to blend the different political interests, mandates, and funding streams that are needed in order to create programs in which services can be combined, administered, and evaluated by a single entity.

In the meantime, service providers must become more aware of the critical importance of service integration in vocational rehabilitation of individuals with mental health disabilities. Employment service professionals can educate themselves about different ways of combining and integrating services and supports not

Table 16.2 Expanded level of services integration matrix

Mental health, case management, housing, education, social relations, economic security, physical health, and other services are... (each of the ten integration components is worth 10% of the total percent score)

Program	Component 1	Component 2	Component 3	Component 4	Component 5	Component 6	Component 7	Component 8	Component 9	Component 10	Percent services integration
A	Y	Y	Y	Y	Y	Y	Y	Y	Y	Y	100%
B	Y	Y	Y	Y	Y	Y	Y	Y	Y	N	90%
C	Y	Y	Y	Y	Y	Y	Y	N	N	N	70%
D	Y	Y	Y	Y	Y	N	N	N	Y	Y	50%
E	N	N	N	N	N	N	N	N	Y	Y	20%
F	N	N	N	N	N	N	N	N	N	N	0%

only in multidisciplinary teams, but also through staff teleconferencing, interagency cooperative agreements, cross-training, job sharing between vocational and mental health providers, and involvement of natural and peer supports. The "art" of service coordination involves working closely with clients to identify their full range of needs, being aware of a wide array of available community resources, and being able to coordinate and integrate service delivery in creative, thoughtful, and effective ways. While this is undeniably a daunting challenge, the outcomes are worth the additional effort.

Acknowledgement This chapter is part of the Employment Intervention Demonstration Program (EIDP), a multisite collaboration among eight research demonstration sites, a Coordinating Center, and the Center for Mental Health Services (CMHS), Substance Abuse and Mental Health Services Administration (SAMHSA). This research was funded by Cooperative Agreement No. SM51820 from CMHS/SAMHSA. The views expressed herein are those of the authors and do not necessarily reflect the policy or position of any federal agency.

References

Bond GR (2004) Supported employment: evidence for an evidence-based practice. Psychiatr Rehabil J 27(4):345–359

Burke JK, Cook JA (2000) Employment Intervention Demonstration Program Cross-Site Program Measure and Documentation Manual. University of Illinois Center for Mental Health Services Research, Chicago, IL. http://www.psych.uic.edu/eidp/eidpdocs.htm

Burke-Miller JK, Cook JA, Grey DG, Razzano LA, Blyler CR, Leff HS, Gold PB, Goldberg RW, Mueser KT, Cook WL, Hoppe SK, Stewart M, Blankertz L, Dudek K, Taylor AL, Carey MA (2006) Demographic characteristics and employment among people with severe mental illness in a multisite study. Community Ment Health J 42(2):143–159

Cook JA (2006) Employment barriers for persons with psychiatric disabilities: a report for the President's New Freedom Commission. Psychiatr Serv 57(10):1–15

Cook JA, Rosenberg H (1994) Predicting community employment among persons with psychiatric disability: A logistic regression analysis. J Rehabil Admin 18:6–22

Cook JA, Carey MA, Razzano LA, Burke J, Blyler C (2002) The pioneer: The Employment Intervention Demonstration Program. New Dir Eval 94:31–44

Cook JA, Leff HS, Blyler CR, Gold PB, Goldberg RW, Mueser KT, Toprac MG, McFarlane WR, Shafer MS, Blankertz LE, Dudek K, Razzano LA, Grey DD, Burke-Miller J (2005a) Results of a multisite randomized trial of supported employment interventions for individuals with severe mental illness. Arch Gen Psychiatry 62:505–512

Cook JA, Lehman AF, Drake RE, McFarlane WR, Gold PB, Leff HS, Blyler CR, Toprac MG, Razzano LA, Burke-Miller JK, Blankertz LE, Shafer MS, Pickett-Schenk SA, Grey DD (2005b) Integration of psychiatric and vocational services: A multi-site randomized, controlled trial of supported employment. Am J Psychiatry 162(10):1948–1956

Cook JA, Mulkern V, Grey DD, Burke-Miller J, Blyler C, Razzano LA, Onken SJ, Balser RM, Gold PB, Shafer MS, Kaufmann CL, Donegan K, Chow CM, Steigman PA (2006) Effects of unemployment rate on vocational outcomes in a randomized trial of supported employment for individuals with psychiatric disabilities. J Vocat Rehabil 25(2):71–84

Dincin J (1975) Psychiatric rehabilitation. Schizophr Bull 1(13):131–147

Drake RE, McHugo GJ, Bebout RR, Becker DR, Harris M, Bond GR, Quimby E (1999) A randomized clinical trial of supported employment for inner-city patients with severe mental illness. Arch Gen Psychiatry 56:627–633

Drake RE, Becker DR, Bond GR, Mueser KT (2003) A process analysis of integrated and non-integrated approaches to supported employment. J Vocat Rehabil 18:51–58

Kopelowicz A, Liberman RP (2003) Integrating treatment with rehabilitation for persons with major mental illnesses. Psychiatr Serv 54(11):1491–1498

Liberman RP, Hilty DM, Drake RE, Tsang HWH (2001) Requirements for multidisciplinary teamwork in psychiatric rehabilitation. Psychiatr Serv 52(10):1331–1342

Razzano LA, Cook JA, Burke-Miller JK, Mueser KT, Pickett-Schenk SA, Grey DD, Goldberg RW, Blyler CR, Gold PB, Leff HS, Lehman AF, Shafer MS, Blankertz LE, McFarlane WR, Toprac MG, Carey MA (2005) Clinical factors associated with employment among people with severe mental illness: Findings from the Employment Intervention Demonstration Program. J Nerv Ment Dis 193:705–713

Russert MG, Frey JL (1991) The PACT vocational model: A step into the future. Psychosoc Rehabil J 14:7–18

Schultz IZ, Krupa T, Rogers ES (2010) Towards best practices in accommodating and retaining persons with mental health disabilities at work: Answered and unanswered questions: In: Schultz IZ, Rogers ES (eds) Work accommodations and retention in mental health. Springer, New York, pp 445–466

Part IV
Barriers and Facilitators to Job Accommodations in the Workplace

Chapter 17
Employer Attitudes Towards Accommodations in Mental Health Disability

Izabela Z. Schultz, Ruth A. Milner, Douglas B. Hanson, and Alanna Winter

Introduction

It has been almost a decade since the World Health Organization brought to light the impact of mental health disorders worldwide (WHO, 2001). The report states that mental health disorders are one of the ten leading causes of disability and account for 10.5% of the global burden of all illness (Standing Senate committe, 2004). Moreover, depression alone constitutes the second highest burden of disease worldwide (Murray and Lopez 1996). The costs associated with mental health disabilities are rising and are predicted to continue in this fashion. The aggregate yearly cost of mental health disabilities constitutes about 2.5% of the Gross National Product (GNP) (WHO 2001). Furthermore, estimated costs to the U.S. economy reached $47.4 billion by 1992 (Rice et al. 1992). Clearly, mental health disabilities are prevalent and costly (as described in the chapter by Dewa and McDaid 2010, this volume which provides a comprehensive analysis of the epidemiological and economic impact of mental health disorders in the workplace).

The impact of mental health disabilities in the workplace is most keenly felt in productivity losses. The most frequently identified workplace mental health disabilities are those arising from dysthymia, major depression, posttraumatic stress disorder, panic disorder, and social phobia (Alonso et al. 2004). The presence of a mental health disability has been found to reduce productivity in employment by approximately 11% (Ettner et al. 1997). Specifically, this number translates into an average of 1 day of work absence and 3 days of reductions in work per month for an individual with a mental health disability. For mood disorders the number of days reduced per month was six, while for anxiety disorders it was four and a half (Lim et al. 2000).

I.Z. Schultz (✉)
Vocational Rehabilitation Counselling Program, University of British Columbia,
Scarfe Library 297, 2125 Main Mall, Vancouver, BC, V6T 1Z4, Canada
e-mail: ischultz@telus.net

I.Z. Schultz and E.S. Rogers (eds.), *Work Accommodation and Retention in Mental Health*, 325
DOI 10.1007/978-1-4419-0428-7_17, © Springer Science+Business Media, LLC 2011

There is consensus in the literature that one of the key barriers to employment of persons with disabilities is social stigma in the workplace and associated discrimination. (see the chapter by Baldwin and Marcus 2010, this volume and the chapter by Wang 2010, this volume). The presence of social stigma among employers, coworkers, and customers has consistently been postulated to be more disabling than the primary condition, especially in serious mental illnesses (e.g., Jenkins and Carpenter-Song 2009). Stigma has been identified as affecting opportunities for securing employment, career advancement, and tenure (Kusznir 2001), and more generally as the most salient barrier to work (WHO 2001). The social stigma surrounding mental health disability has been largely based on workplace myths that over the years have served to perpetuate employers' discriminatory attitudes (Harnois and Gabriel 2000). Myths include perceptions that mental illness is synonymous with mental retardation, is not amenable to treatment, and may cause workers to become disruptive or harmful to others. Research on employer attitudes demonstrated that 90% of employers would hire a person with a physical disability but only 20% would hire a person with mental illness (McFarlin et al. 1991).

More positive attitudes among employers towards employing individuals with mental health disorders have been found to be associated with the following factors: previous positive experience with a person with a mental health disability, being in the government sector, being a female employer, possessing a higher level of education, and receiving on-site training from a supportive agency (Graffam et al. 2002). Akabas (1994) indicated that the cultural characteristics of a successful company are the same factors that are necessary for successful accommodation of workers with mental health disabilities (see also the chapter by Gates and Akabas 2010, this volume; and the chapter by Kirsh and Gewurtz 2010, this volume). Gilbert and Myette (2003) emphasized the role of employer and coworker discrimination in preventing workers with mental health disabilities from seeking appropriate diagnoses and treatment and in avoiding disclosure of their condition. Other findings suggest that employers discriminated the most towards individuals with emotional and cognitive disorders, as compared to other disabilities (Greenwood and Johnson 1987).

Recent literature has emphasized the importance of having employers take responsibility for establishing healthy work environments. This attitude involves being informed as well as promoting antidiscriminatory attitudes and practices towards workers with mental health disorders (Akabas 1994; Bender and Kennedy 2004; Gallie et al. 2010, this volume; Gates and Akabas 2010, this volume; Goldner et al. 2004; Kirsh and Gewurtz 2010, this volume).

A review of the literature completed by Akabas (1994) revealed that many employers believed persons with mental health disabilities are unproductive in the workplace and are of limited employability. Consequently, employers do not consider accommodations to be effective options for retaining persons with mental health disabilities in the workplace. Akabas (1994), in response, advocated for studies which would investigate the relationship between employers who offer positive, flexible, innovative, supportive, collaborative, and accommodating work environments for all employees and the employment outcomes of persons with mental health disorders.

Though Akabas' recommendation has since been followed in many studies on employment interventions with persons with mental health disabilities (Krupa et al. 2005), there continues to be limited research in the area of job accommodation. The authors of this chapter undertook a study to explore the relationship between employer attitudes toward workers with mental health disabilities and their knowledge and use of appropriate job accommodations. We anticipated that our findings would: (1) inform future conceptual and practical models of combating stigma of mental health disorders through attitudinal and organizational change in the workplace, and (2) yield recommendations for best workplace practices and research in this field.

Our study entailed a survey of different employers from across Canada. The sample of employers was randomly selected from InfoCan, a commercially available database of more than 900,000 businesses. The inclusion criteria were: a minimum number of 100 employees from across Canada, and from the ten industrial groups of the National Occupational Classification System. Employers surveyed were from the following industries: Agriculture, Forestry, and Fishing; Manufacturing; Mineral Industries; Construction Industries; Wholesale Trade; Retail Trade; Transportation, Communication, and Utilities; Finance, Insurance, and Real Estate; Service Industries; and Public Administration.

Employers agreeing to participate provided the names of the managerial and labor representatives from the company's Occupational Health and Safety Committee. These individuals were then contacted to complete our telephone survey. The survey we employed was chosen after an extensive review of instruments which had been developed to investigate these attitudes. It measures employer attitudes towards the work personality, work performance, and symptomatology demonstrated by employees with mental health disabilities (Diksa and Rogers 1996).

Employer Perspectives on Job Accommodations for Persons with Mental Health Disabilities

The relevant findings of our survey study are detailed below.

Employer Profiles

Table 17.1 shows that of the 83 employers surveyed across English-speaking Canada, various industry categories were represented.

Most employers (59%) had 100–499 employees, with 11% of employers having 500–999 employees and 30% of employers having more than 1,000 employees.

Table 17.1 Industry category

Industry category	Frequency	Percent
Agriculture	4	4.8
Mining	5	6.0
Construction	4	4.8
Manufacturing	8	9.6
Transportation	15	18.1
Wholesale	9	10.8
Retail	6	7.2
Finance	4	4.8
Service	15	18.1
Public administration	13	15.7
Total	83	100.0

A total of 68% of surveyed employers were for-profit companies, 18% were not-for-profit, and 14% were government agencies. When questioned about hiring policies, we determined that almost 40% of employers had a policy regarding the hiring of persons with disability and a similar number had hired a person with a mental health disability in the past.

Employer Attitudes

We asked employers to respond to a list of concerns about hiring individuals with mental health disabilities. Notably, mental health disabilities were defined broadly, ranging from serious mental illness, through depression and anxiety disorders, to brain injury and learning disorders. Table 17.2 shows the outcomes of these queries, and suggests that in the area of work personality, employers are most concerned about how individuals with mental health disabilities adjust to the work environment, and how reliable they are. About half of the employers raised concerns of some type about various aspects of work personality, except for taking pride in work, which was of concern only to 36% of employers.

Similarly, employers were queried about their concerns regarding the work performance of employees with mental health disabilities. A review of Table 17.3 suggests the highest concerns reside in the following areas: being able to tolerate the working conditions, being able to perform job tasks safely, being able to perform job tasks, being able to produce an acceptable quality of work, and possessing adequate problem-solving skills. The lowest level of concern was around possession of academic skills.

Table 17.4 shows employer concerns around symptoms exhibited by persons with a mental health disability. The most significant concerns were seen in the area

Table 17.2 Employer concerns about work personality

Employer concern	Frequency	Percentage
Work personality		
Adjusting to the work environment	57	68.7
Being reliable	53	63.9
Benefiting from the standard amount of supervision	48	57.8
Responding to criticism	48	57.8
Communicating with others	46	55.4
Seeking assistance to perform job better	44	53.0
Showing up for scheduled shifts	43	51.8
Keeping the job	43	51.8
Being on time	39	47.0
Being flexible	39	47.0
Respecting authority	37	44.6
Getting along with coworkers and supervisors	37	44.6
Taking pride in work	30	36.1

Table 17.3 Employer concerns about work performance

Employer concern	Frequency	Percentage
Work performance		
Being able to perform job tasks safely	62	74.7
Being able to tolerate the working conditions	62	74.7
Being able to perform job tasks	55	66.3
Being able to produce an acceptable quality of work	55	66.3
Possessing adequate problem-solving skills	52	62.7
Being able to produce an acceptable quantity of work	52	62.7
Possibility of incurring unknown costs	44	53.0
Ability to discipline or fire the person once hired	44	53.0
Being able to be accepted by the public or customers	44	53.0
Being able to be accepted by coworkers	40	48.8
Being able to advance in organization	39	47.0
Possessing adequate academic skills	21	25.3

of maintaining emotional stability, exhibiting bizarre behaviors, being able to tolerate work pressure and stress, and concerns about the potential to become violent. Generally, employer concerns in this category were much more prevalent than with respect to work personality or work performance.

Taken together, it is clear that employers continue to have significant concerns about hiring individuals with mental health disabilities. They are concerned about personality issues such as the ability to be conscientious and flexible with regard to work, about having the individual work safely and effectively in the job, and about the ability to refrain from unusual behavior in the workplace.

Table 17.4 Employer concerns about symptomatology

Employer concern	Frequency	Percentage
Symptomatology		
Having the ability to maintain emotional stability	69	83.1
Bizarre behaviors	67	80.7
Ability to tolerate work pressure and stress	66	79.5
Becoming violent	63	75.9
Ability to leave personal problems outside work	56	67.5
Withdrawing into own world	56	67.5
Showing poor judgment	54	65.1
Ability to pay attention to detail	51	61.4
Having poor memory	43	51.8
Lacking enthusiasm	41	49.4
Having poor grooming skills	40	48.2
Lacking initiative	39	47.0
Needing time off for medical appointments	36	43.9
Having poor physical coordination	31	37.3

Employer Experience with Job Accommodations for Persons with Mental Health Disabilities

Next, we queried employers about their experiences with various types of accommodations that could be used to ensure good work performance among employees with mental health disabilities. Table 17.5 summarizes the data on employers' experience with these various types of job accommodations.

Table 17.5 suggests that employers are most aware of the usefulness of open communication, positive reinforcement and praise, written instructions, and the need for additional time to learn job responsibilities as accommodations that are potentially useful to improve job performance and retention. Apart from direct work performance, employers are most aware that having flexible leave for health issues and time for telephone calls for healthcare and support can be useful on the job to support individuals with mental health disabilities.

As can be seen in Table 17.5, accommodations in the area of stamina, concentration, memory, organization, work with supervisors, interactions with coworkers, handling stress, and attendance issues were much less utilized by employers. Overall, the least utilized job accommodations in terms of the work schedule itself included working from home and job sharing. The use of white noise, allowing the employee to play music, providing space enclosures, reducing distractions, and providing uninterrupted work time all can be useful accommodations to boost concentration, yet they are not widely known or used, according to the employer respondents. Allowing the employee to tape record meetings and use software organizers can be helpful for organization and to boost memory, respectively, but again are not widely known or utilized.

Table 17.5 Employer experience with job accommodations

Job accommodation	Moderate–great deal of experience frequency	Percentage
Stamina		
Flexible scheduling	60	73.2
Allow longer work breaks	55	67.1
Additional time to learn new responsibilities	67	81.7[a]
Self-paced work load	45	54.9
Backup coverage when employee needs breaks	52	63.4
Time off for counseling	63	76.8
Supported employment and job coaches	45	54.9
Allow employee to work from home	20	24.4[b]
Part-time work schedules	51	62.2
Concentration		
Reduce distractions in work area	39	47.6[b]
Provide space enclosures	30	36.6[b]
Allow for white noise	20	24.4[b]
Allow employee to play music	38	46.3[b]
Increase natural light	38	46.3[b]
Allow employee to work from home	18	22.0[b]
Plan for uninterrupted work time	33	40.2[b]
Allow for frequent breaks	51	62.2
Divide assignments into smaller tasks	58	70.7
Restructure job	42	51.2
Organization		
Make daily to-do lists	48	58.5
Use calendars to mark meetings and deadlines	45	54.9
Remind employee of important deadlines	62	75.6
Use electronic organizers	42	51.9
Divide large assignments into smaller tasks	59	72.0

(continued)

Table 17.5 (continued)

Job accommodation	Moderate–great deal of experience frequency	Percentage
Memory		
Provide day planners	56	58.3
Allow employee to tape record meetings	25	30.5[b]
Provide typewritten minutes	53	64.6
Provide written instructions	69	84.1[a]
Allow additional training time	68	82.9[a]
Provide written checklist	61	74.4
Allow software organizers	38	46.3[b]
Allow email	54	65.9
Provide cues	47	57.3
Post instructions on equipment	52	63.4
Provide positive praise and reinforcement	72	87.8[a]
Provide written job instructions	67	81.7
Effective work with supervisors		
Develop written work agreements	57	69.5
Allow for open communications with managers	76	92.7[a]
Establish written long-term and short-term goals	62	75.6
Develop strategies to deal with problems	56	68.3
Develop procedure to evaluate accommodations	53	64.6
Provide extra supervision	55	67.1
Educate all employees on right to accommodation	58	70.7
Interaction with coworkers		
Provide sensitivity training to coworkers and supervisors	48	58.5
Do not mandate employees to attend social functions	58	70.7
Employees to move non-work-related conversations out	53	64.6
Pair individual with a mentor	29	35.4

Difficulty handling stress

Provide praise and positive reinforcement	72	87.8[a]
Refer to counseling and employee assistance program	60	73.2
Allow telephone calls to doctors and others for support	67	81.7
Allow presence of support animal	25	30.5
Allow employee to take breaks as needed	59	72.0
Provide flexible leave for health problems	67	81.7[a]

Attendance issues

Provide self-paced work load and flexible hours	47	57.3
Allow employee to work from home	23	28.0[b]
Provide part-time work schedule	49	59.8
Allow employee to make up time	53	64.6
Employ job sharing	36	43.9[b]
Recognize that change is difficult	53	64.6
Maintain open channels of communication	60	73.2
Provide weekly or monthly meetings	60	73.2
Provide retraining for alternate duties	56	68.3

[a] Indicates the most widely used accommodations
[b] Indicates the least used accommodations

Prediction of Attitudes and Job Accommodation by Employer

Regression analyses were performed using eight categories of job accommodations: stamina, concentration, organization, memory, employee–supervisor relationship, employee–coworker relationship, stress, attendance, and three aggregate scores for the domains of concern, i.e. work performance, work personality, and symptomatology. We analyzed the extent to which industry type, size of industry, and profit vs. non-profit status were predictive of these three dimensions of employer attitudes. We also examined whether the presence of company policy on hiring persons with disabilities and whether or not a company had prior experience with an employee with a mental health disability were predictive of attitudes and job accommodations.

Results of these analyses suggest that having a company policy on hiring persons with disabilities and whether or not the company had prior experience in having a person with a mental health disability were more predictive of attitudes toward hiring and providing job accommodations than company size. Company size was a predictor of concerns about work performance and accommodations related to employee attendance and employee–supervisor relationship, while having a company policy was broadly predictive of concerns about work performance, work personality, and attitudes towards accommodations. Having hired an individual with mental health disability was a predictor of attitudes towards aspects of symptomatology and, broadly, toward job accommodations.

The variables associated with the use of employee stamina-related job accommodations were industry category and work personality. Accommodations related to employee concentration, employee memory, employee–supervisor relationships, and employee stress were best predicted by industry category and size of company. Accommodations related to employee organization were best predicted by work performance. Accommodations related to employee attendance were predicted by work personality, size of company, and industry category.

Discussion

Consistent with the literature on negative attitudes that employers hold towards persons with mental health disabilities (Baldwin and Marcus 2010, this volume; Gilbert and Myette 2003; Greenwood and Johnson 1987; Harnois and Gabriel 2000; Jenkins and Carpenter-Song 2009; Kusznir 2001; McFarlin et al. 1991), our study indicated that over 50% of Canadian employers (in midsize to large companies) have important concerns about workers with mental health disabilities. The most significant concerns included: employees' capacity to remain mentally stable, exhibiting bizarre behaviors, the ability to tolerate work pressure and stress, becoming violent in the workplace, and being able to tolerate working conditions.

These findings confirm the presence of social stigma, with at least half of all the employers in this study expressing significant concerns. Our analyses of the most prominent employer concerns suggest that persons with mental health disabilities are still perceived by many in the workplace as unstable, bizarre, violent, and unable to tolerate working conditions and work stress. This portrayal is consistent with society's fear of persons with mental health disabilities, which is often exacerbated by media accounts sensationalizing mental illness in the workplace (e.g., reports of rare but deadly violence in the workplace). Concerns about bizarre and violent behavior are in fact at odds with the epidemiology of mental health disorders in the workplace.

The most prevalent disorder in the workplace is now depression (Gnam 2005; Goldberg and Steury 2001; Olsheski et al. 2002; Stewart et al. 2003; Zuberbier and Schultz 2007), and its clinical symptoms are not at all consistent with the presence of bizarre behaviors, violence, or complete inability to manage stress. Employers' perceptions of mental health disabilities in the workplace continue to be based on myths that in turn perpetuate negative attitudes and emphasize disruptiveness and harmfulness to others (Harnois and Gabriel 2000). On a positive note, few employers (25.3%) worry about insufficient academic training of employees with mental health disabilities, which may signify dissolution of the myth that persons with mental health disabilities have mental retardation or are cognitively impaired. It is the social behavior of persons with mental health disabilities that continues to be a concern to employers.

Study findings on employers' experience with various types of job accommodations for persons with mental health disabilities demonstrated that the most commonly known and used accommodations include open communication with managers, praise and reinforcement, written instructions, extra training time, and allowing phone calls for support. However, there exists a large segment of job accommodations that most employers have no experience with. Between 53 and 74% of employers had no experience with work environment changes, such as working from home, job sharing, use of white noise, space enclosures, uninterrupted work time, reduced distraction, or the use of natural light and music. Among organizational and memory aids, many employers had little experience with the use of electronic organizers and the potential to tape record meetings to aid with memory concerns.

It appears that employers have more experience with no- or low-cost communication/interaction-based management strategies for dealing with employees with mental health disabilities, but little experience in structural changes to work environment and organization. The latter are more costly, are complex to implement, and may require increased organizational support from various levels. This finding has clear implications for the direction of future management training and organization of the workplace to accommodate persons with mental health disabilities. The findings of the study reported here with respect to employers' experience with job accommodations are consistent with the observation by Akabas (1994) that the workplace conditions known to assist with mental health disabilities are the same conditions that make the workplace culture healthy and productive for all employees: innovation, flexibility, support, collaboration, and accommodation. In order to be most effective in retaining employees with mental health disabilities, the accommodating workplace incorporates the appropriate structure and organization of work, not just a more positive interaction between a manager and a person with a mental health disability (Kirsh and Gewurtz 2010, this volume).

There has been no direct empirical research to date on predictors of the use of job accommodations for persons with mental health disabilities. Our study showed that the following factors are important: industry category, size of company, and employer perceptions of worker personality (for accommodations related to stamina and attendance); and worker performance (for accommodations related to organization). Notably, employer perception of worker symptomatology did not appear to be relevant in predicting the use of job accommodations.

At the same time, the data indicate that having a company policy on, and having direct experience with, hiring a person with a mental health disability were more significant predictors of both attitudes and practices regarding job accommodations than the size of the company. Interestingly, employer attitudes were predictive of the use of accommodations in the employee stamina, employee organization, and employee attendance domains only. Industry category and company size were generally stronger predictors than attitudes. Company size was a predictor of attitudes towards work performance; presence of a company policy on hiring persons with disabilities was a predictor of both attitudes towards work performance and work personality; and experience with hiring a person with a mental health disability was a predictor of attitudes towards symptomatology.

While enhancing employers' attitudes towards persons with mental health disabilities is an important goal, changes in attitudes alone are not likely to improve employer practices in job accommodation. Structural and organizational changes to the workplace will be necessary to form the context for accommodating employees with mental health disabilities. The size and type of employer, the presence of company policies on hiring persons with disabilities, and the employer's experience in hiring a person with a mental health disability, in conjunction with knowledge of the wide range of accommodations available for mental health disabilities, are much more directly related to job accommodation practices than attitudes alone. Even with positive attitudes, most employers are lacking practical tools, from both the organizational and management perspectives, to provide the most effective accommodations for persons with disabilities.

Conclusions and Recommendations

The Canadian employers surveyed in the current study remain adversely affected by myths about mental health disorders, with about 50% of them holding negative attitudes towards workers with mental health disabilities. Of particular concern is the strength of the myth that persons with mental health disorders display bizarre behavior, can be violent, and cannot cope with work conditions. This perception is at odds with epidemiological evidence showing that the most common disorder in the workplace is depression, which is not generally associated with the symptomatology that employers are concerned about. Attitudes of those employers who have a policy on hiring persons with disabilities, those who have experience in hiring person(s) with mental health conditions, and those with a larger number of employees are more positive than their counterparts. Likewise, those employers who have policies on, and have hired, persons with mental health disabilities, have more experience with job accommodations for these workers, and in general may be more comfortable with providing reasonable accommodations.

Employers are generally familiar with communication- and interaction-oriented, management-based job accommodations such as open communication, positive reinforcement, and additional training. They are much less familiar, however, with the structural, environmental, and organizational underpinnings of job accommodation, such as the use of work-at-home, part-time work, or job sharing. These accommodations do require multilevel support within the workplace and are more complicated to implement, given the likely demand for improved working conditions among nondisabled employees who struggle with workplace stress as well. Having positive attitudes towards persons with mental health disabilities in the workplace does not automatically imply improved practices in job accommodation.

We know from this survey and similar research that larger companies often have more established policies and procedures in the human resources area, may provide more management training, and may have more flexibility in implementing accommodations. The existence of written policies on hiring persons with disabilities may be indicative of the maturity of the organizational culture, senior management commitment to issues of diversity and accommodation, and access to improved organizational resources to assist with implementation of best practices. Past experience in hiring person(s) with mental health disability provides valuable employer experience in handling the accommodations, may desensitize fears regarding bizarre symptoms and violent behavior, and may raise awareness of the needs of persons with mental health disabilities. Taken together, research suggests that increased employer awareness of job accommodations for persons with mental health disabilities may extend to organizational, environmental, and structural changes in the workplace that are needed to provided optimal accommodations. Furthermore, employer training in how to match, implement, and maintain different job accommodations to different workers with varied impairments and work-related issues, is sorely needed.

Based upon the study we conducted and existing literature in this area, we offer the following recommendations to improve work accommodation and retention outcomes for individuals with mental health disabilities:

1. Awareness raising, education, and fear reduction for employers and coworkers regarding employees with mental health disabilities;
2. Awareness, education, and training regarding accommodations for mental health disabilities, including possibilities for:

- Structural and organizational changes to be implemented to ensure accommodations regarding flexibility, self-pacing in schedule and location, and reduction of unnecessary and adverse environmental and social demands;
- Increased knowledge of specific accommodations for stress management and cognitively based difficulties;

3. Increased research and training focus on accommodations for persons with mental health disabilities in the context of the "blue-collar" work environment, and for smaller companies. Notably, we experienced difficulties locating suitable survey respondents in these companies, and as a result they were omitted in our survey;
4. Development of written company policies regarding employment and job accommodations for persons with mental health disabilities;
5. Offering volunteer or subsidized employment for persons with mental health disabilities to desensitize employers' fear regarding mental disorders and provide them with valuable management and job accommodation experience.

Much of the responsibility for safe and sustainable transitions to the work environment rests with mental health and medical professionals, vocational rehabilitation counselors, and other rehabilitation professionals, as well as disability managers and employment counselors. The anachronistic divide between physical health and mental health, and between mental health treatment and the workplace, must be addressed to ensure an integrated, coordinated, and transdisciplinary approach to job accommodation for persons with mental health disabilities (see chapter 19, Schultz et al. 2010, this volume, for a fuller discussion on this topic). Employers do not have, and are not expected to have, knowledge and training in matching job accommodations to an individual's functional limitations; they require ongoing and interactive support from mental health and rehabilitation professionals who can assist in this regard.

Last but not least, individuals with mental health disabilities need to be equipped with the self-knowledge and skills that will allow them to advocate for and assertively obtain effective job accommodation solutions in the workplace without fear of social stigma.

Future Research

Future research is needed on assessing the functional capacities of persons with mental health disabilities that can lead to the development of appropriate and effective job accommodations. Further, research on interactions among the various factors that can change employers' attitudes towards persons with mental health disability, and enhance knowledge and application of proactive job accommodations practices for employees with mental health disorders, is also necessary. Evaluation of the effectiveness of specific job accommodations in the workplace would advance the field by providing empirical support for various accommodation strategies. Moreover, research on multisystem interactions among the healthcare and rehabilitation systems, the insurance system, and the workplace, in service of the retention and productivity of persons with mental health disabilities, constitutes a major future direction in the still fragmented field of work accommodation.

Acknowledgement This chapter was informed by the research report for a project funded by Social Development Canada, "Towards Evidence-Informed Best Practice Guidelines for Job Accommodations for Persons with Mental Health Disabilities."

References

Akabas SH (1994) Workplace responsiveness: Key employer characteristics in support of job maintenance for people with mental illness. Psychosoc Rehabil J 17(3):91–101

Alonso J, Angermeyer MC, Bernert S, Bruffaerts R, Brugha TS, Bryson H et al (2004) Prevalence of mental disorders in Europe: Results from the European Study of the Epidemiology of Mental Disorders (ESEMeD) project. Acta Psychiatr Scand 109(Suppl 420):21–27

Baldwin ML, Marcus SC (2010) Stigma, discrimination, and employment outcomes among persons with mental health disabilities. In: Schultz IZ, Rogers ES (eds) Work accommodation and retention in mental health. Springer, New York, pp 53–69

Bender A, Kennedy S (2004) Mental health and mental illness in the workplace: Diagnostic and treatment issues. Healthc Pap 5(2):54–68

Dewa CS, McDaid D (2010) Investing in the mental health of the labour force: Epidemiological and economic impact of mental health disabilities in the workplace. In: Schultz IZ, Rogers ES (eds) Work accommodation and retention in mental health. Springer, New York, pp 33–52

Diksa E, Rogers SE (1996) Employer concerns about hiring persons with psychiatric disability: Results of the Employer Attitude Questionnaire. Rehabil Couns Bull 40(1):31–43

Ettner SL, Frank RG, Kessler RC (1997) The impact of psychiatric disorders on labour market outcomes. Ind Labour Relat Rev 51(1):64–82

Gallie KA, Schultz IZ, Winter A (2010) Company-level interventions in mental health. In: Schultz IZ, Rogers ES (eds) Work accommodation and retention in mental health. Springer, New York, pp 295–310

Gates LB, Akabas SH (2010) Inclusion of people with mental health disabilities into the workplace: Accommodation as a social process. In: Schultz IZ, Rogers ES (eds) Work accommodation and retention in mental health. Springer, New York, pp 375–392

Gilbert M, Myette L (2003) Depression in the workplace: Shared costs and responsibilities. Rehabil Rev 24(7):13–15

Gnam WH (2005) The prediction of occupational disability related to depressive and anxiety disorders. In: Schultz IZ, Gatchel RJ (eds) Handbook of complex occupational disability claims. Early risk identification, intervention, and prevention. Springer, New York, pp 371–386

Goldberg RJ, Steury S (2001) Depression in the workplace: Costs and barriers to treatment. Psychiatr Serv 52(12):1639–1643

Goldner E, Bilsker D, Gilbert M, Myette L, Corbière M, Dewa CS (2004) Disability management, return to work and treatment. Healthc Pap 5(2):76–90

Graffam J, Shinkfield A, Smith K, Polzin U (2002) Factors that influence employer decisions in hiring and retaining an employee with a disability. J Vocat Rehabil 17(3):175–181

Greenwood R, Johnson VA (1987) Employer perspectives on workers with disabilities. J Rehabil 53:37–45

Harnois G, Gabriel P (2000) Mental health and work: Impact, issues and good practices. Joint publication of the WHO and the International Labour Organization [ILO]. WHO/ILO, Geneva, Switzerland

Jenkins JH, Carpenter-Song EA (2009) Awareness of stigma among persons with schizophrenia: marking the contexts of lived experience. J Nerv Ment Dis 197(7):520–529

Kirsh B, Gewurtz R (2010) Organizational culture and work issues for individuals with mental health disabilities. In: Schultz IZ, Rogers ES (eds) Work accommodation and retention in mental health. Springer, New York, pp 393–408

Krupa T, Kirsh B, Gerwurtz R, Cockburn L (2005) Improving the employment prospects of people with serious mental illness: Five challenges for a national mental health strategy. [Special electronic supplement on mental health reform for the 21st century in partnership with the School of Policy Studies and with the Centre for Health Services and Policy Research of Queen's University, Kingston, ON]. Canadian Public Policy, 31(Suppl.):60–63. Retrieved March 11, 2010, from http://economics.ca/cgi/jab?journal=cpp&view=v31s1/CPPv31s1p059.pdf

Kusznir A (2001) Understanding return to work challenges with depression. Paper presented at the second Annual National Symposium of the National Institute of Disability Management and Research, Ottawa, Ontario

Lim D, Sanderson K, Andrews G (2000) Lost productivity among full-time workers with mental disorders. J Ment Health Policy Econ 3:139–146

McFarlin DB, Song J, Sonntag M (1991) Integrating the disabled into the work force: A survey of Fortune 500 company attitudes and practices. Employee Responsib Rights J 4(2):107–123

Murray CJL, Lopez AD (1996) The global burden of disease. WHO. Harvard University Press, Cambridge, MA

Olsheski JA, Rosenthal DA, Hamilton M (2002) Disability management and psychosocial rehabilitation: Considerations for integration. Work 198:63–70

Rice DP, Kelman S, Miller LS (1992) The economic burden of mental disorder. Hosp Community Psychiatry 43(2):1227–1232

Schultz IZ, Rogers ES, Krupa T (2010) Towards best practices in accommodating and retaining persons with mental health disabilities at work: Answered and unanswered questions. In: Schultz IZ, Rogers ES (eds) Work accommodation and retention in mental health. Springer, New York, pp 353–372

Standing Senate Committee on Social Affairs, Science and Technology (2004) Mental health, mental disorder and addiction: Overview of policies and programs in Canada (Report 1 of 3). Government Printing Office, Ottawa, ON

Stewart WF, Ricci JA, Chee E, Hahn SR, Morganstein D (2003) Cost of lost productive work time among U.S. workers with depression. J Am Med Assoc 289(23):3135–3144

Wang JL (2010) Mental health literacy and stigma associated with depression in the working population. In: Schultz IZ, Rogers ES (eds) Work accommodation and retention in mental health. Springer, New York, pp 341–352

World Health Organization [WHO] (2001) The World Health Report 2001. Mental health: new understanding, new hope. WHO, Geneva, Switzerland

Zuberbier OA, Schultz IZ (2007) Mental disorders and the Canadian workplace: Epidemiological estimates of rates of disorders in Canada. Unpublished manuscript

Chapter 18
Mental Health Literacy and Stigma Associated with Depression in the Working Population

Jian Li Wang

Introduction

The societal and economic burden of depression is evident in North America (Kessler et al. 2003; Statistics Canada 2002) and worldwide (Üstün et al. 2004). Based upon the first Canadian national mental health survey data, released in 2004, over 1,120,000 Canadians aged 15 years or older have had a unipolar depressive disorder in the past 12 months (Statistics Canada 2002). However, the rate of mental health service utilization in people with major depression has not significantly increased in the past 10 years (Wang et al. 2005). In the United States, the National Comorbidity Survey–Replication showed that the lifetime prevalence of major depressive disorder was 16.2% (32.6–35.1 million US adults) and for 12 months was 6.6% (13.1–14.2 million US adults) (Kessler et al. 2003). Workers in the US with depression reported significantly more total health-related lost productive time than those without depression (5.6 h/week vs. an expected 1.5 h/week) (Stewart et al. 2003). Major depression accounts for 48% of the lost productive time among those with depression, again with a majority of the cost explained by reduced performance while at work (Stewart et al. 2003). The high prevalence of major depression has significant public health implications, given its impact on the individuals' functioning and quality of life and its economic burden on society (Statistics Canada 2002; Dyck et al. 1998; Parslow and Jorm 2002). Comprehensive strategies including primary and secondary prevention and effective self-management are needed (see Chap. 6 by Lerner et al. 2010). However, these efforts are impeded by lack of sufficient knowledge about depression, negative attitudes towards evidence-based interventions, and stigma (Jorm et al. 2003).

Mental health literacy is "knowledge and beliefs about mental disorders which aid their recognition, management, or prevention" (Jorm 2000). It consists of several

J.L. Wang (✉)
Departments of Psychiatry and of Community Health Sciences, Faculty of Medicine,
University of Calgary, Room 4D69, Teaching, Research & Wellness Building.
3280 Hospital Dr. NW, Calgary, T2N 4Z6, Canada
e-mail: jlwang@ucalgary.ca

I.Z. Schultz and E.S. Rogers (eds.), *Work Accommodation and Retention in Mental Health*,   341
DOI 10.1007/978-1-4419-0428-7_18, © Springer Science+Business Media, LLC 2011

components, including: (a) the ability to recognize specific disorders; (b) knowledge and beliefs about risk factors and causes; (c) knowledge and beliefs about professional help and self-help interventions; (d) attitudes which facilitate recognition and appropriate help-seeking; and (e) knowledge of how to seek mental health information (Jorm 2000). This theoretical model has been widely used in studies examining mental health literacy in various populations. The measurement of mental health literacy involves eliciting people's responses to a vignette of a person (John or Mary) suffering from a mental health problem such as major depression, according to the Diagnostic and Statistic Manual for Mental Disorders – Fourth Edition (DSM-IV) criteria (American Psychiatric Association, 1994). Respondents are asked two open-ended questions: what would you say, if anything, is wrong with John or Mary? How do you think John or Mary could best be helped? The rest of the assessment contains questions about one's knowledge of risk factors for mental health problems, attitudes towards various interventions, and beliefs associated with stigma.

There have been many studies examining mental health literacy in various populations. In terms of the knowledge about depression in the general population, the first Australian survey conducted in 1995 (Jorm et al. 1997a, b) showed that about 39% of individuals in the community could correctly identify depression described in the case vignette. A higher percentage (51.4%) was found in South Australia (Goldney et al. 2001). Eight years later, the percentage increased to 67.3%, according to the second survey carried out in 2003–2004 (Jorm et al. 2006). In Germany, the percentage was 28% (Angermeyer and Matschinger 1999). In Switzerland, the depression vignette was correctly recognized by 39.8% (Lauber et al. 2003), while in a study conducted during 2006 in Alberta, Canada, the percentage was relatively high, at 75.6% (Wang et al. 2007a, b).

Previous studies indicated that mental health specialists and pharmacological treatments were less likely than general practitioners and self-management to be perceived by the general public as helpful for depression. The ratings of general practitioners, counselors, close friends, or family as helpful for managing depressive symptoms were higher than those of psychiatrists and clinical psychologists in the Australian population (Jorm et al. 1997a, b). Despite trial evidence supporting their effectiveness, antidepressants were recognized as helpful by only 29% and as harmful by 42% in the Australian population (Jorm et al. 1997a, b). On the other hand, vitamins, minerals, tonics, or herbal products were considered helpful by 57% and as harmful by 3% (Jorm et al. 1997a, b). Regarding nonpharmacological treatments, nonstandard interventions (physical activity, stress management, meditation, reading about people with similar problems) were regarded as helpful and not harmful by over 80% of the public (Jorm et al. 1997a, b). Psychotherapy was seen as helpful by 34% and harmful by 13% (Jorm et al. 1997a, b). It was similar in Germany, where relaxation and talk with friends/family was highly recommended (88.8%), while drug treatment was recommended by only 28.5% (Angermeyer and Matschinger 1999). Only 10% of respondents advised turning to a psychiatrist or psychologist for help in the first place (Angermeyer et al. 1999). Negative attitudes towards mental health specialists and pharmacological treatments might largely be based on general health beliefs, which are shaped by media and personal experience, rather than by scientific evidence. Through social networking, the public's opinion may shape the attitudes of those in direct contact with individuals experiencing depression.

Psychosocial factors are more likely than biological factors to be considered as causes for depression. Studies conducted in Germany (Angermeyer and Matschinger 1999, 2003a, b; Dietrich et al. 2004), Mongolia, and Russia (Dietrich et al. 2004) show that most people in the community perceive that psychosocial factors, particularly work stress and difficulties in partner or family relationships (over 70%), are the causes of depression. About 45% consider that depression is a brain disease or due to heredity (Angermeyer and Matschinger 1999, 2003a, b; Dietrich et al. 2004). The popular view of psychosocial factors as potential causes for depression is consistent with findings from epidemiologic studies (Wang 2004, 2005; Kendler et al. 1999, 2000, 2001a, b, 2004; Wade and Kendler 2000; Sanathara et al. 2003; Foley et al. 2001). Nevertheless, depression is increasingly recognized as the result of interaction between biological and psychosocial factors (Kendler et al. 1999, 2000, 2001a, b, 2004; Wade and Kendler 2000; Sanathara et al. 2003; Foley et al. 2001). Holding certain views about the causes of depression may alter the ways of seeking help and treatment (Jorm 2000).

The experience of depression may affect people's knowledge and attitudes toward depression; however, findings from previous studies have not been consistent in this regard. In the first Australian national survey on mental health literacy, having experienced depression in oneself or others made no difference in relation to beliefs about treatment (Jorm et al. 1997a, b). In a survey conducted in New South Wales, Australia, the authors reported that people who had sought help for depression were less likely to believe in the helpfulness of lifestyle interventions and more likely to be believe in medical interventions including antidepressants (Jorm et al. 2000). Those who had a history of depression but who had not sought help were more likely to rate counseling as helpful; those with current depressive symptoms were less likely to rate telephone counseling, family, and friends as helpful (Jorm et al. 2000). In another study conducted in South Australia, those who met the criteria for major depression were less likely to recommend seeing a doctor, but were more likely to rate antidepressants as helpful, than those who did not have major depression (Goldney et al. 2001). However, those with major depression were two times more likely than those without depression to report that they did not know how depression could be helped (Goldney et al. 2001). They did not differ in other elements of mental health literacy, including recognition of depression (55.6% vs. 48.6%) (Goldney et al. 2001).

In the United States, Link and his colleagues reported 33% of the participants believed that persons with major depression were violent (Link et al. 1999). The results of this study indicated that the public's attitudes toward depression in terms of unpredictability and dangerousness did not depend on their levels of depression literacy or whether they had personal contact with someone with depression. Recognition of depression may have a positive effect on changing the view that depression is a sign of personal weakness. Personal contact appeared to have a positive effect on such a view only in women (Link et al. 1999).

Stigma toward persons with mental illness exists in the general population worldwide (Phelan et al. 2000; Stuart and Arboleda-Flôrez 2001). Although depressive disorders are prevalent and impose considerable burden on society, there is a paucity of studies investigating stigma against depression. The general public appears to increasingly view depression as a health problem, and has become more comfortable than before about interacting with individuals with depression. In a national survey by Mental Health America in 2006, 72% of the Americans surveyed considered depression as a health problem, compared to 38% ten years ago (News & Notes 2007). 63% of participants felt comfortable interacting with individuals with depression, which was higher than the percentages of being comfortable interacting with those with alcohol and drug problems (43%) and schizophrenia (45%). However, response rates were much less than the percentages of feeling comfortable in having a relationship with individuals who have cancer or diabetes (98%; News & Notes 2007).

Previous research has mainly focused on stigma associated with schizophrenia. Some researchers have measured stigma related to both schizophrenia and depression (Angermeyer et al. 2004; Lauber et al. 2005), using social distance as a measure of stigma. The findings about demographic correlates for social distance from people with mental illness have not been consistent across studies. These inconsistencies were discussed in detail in a recent review (Angermeyer and Dietrich 2006). Dietrich and colleagues' study indicated that endorsing biogenetic explanations appeared to decrease the likelihood of social acceptance of people with schizophrenia and major depression (Dietrich et al. 2006). Endorsing biological factors as the cause of major depression was associated with a greater desire for social distance from people with major depression (Dietrich et al. 2004). Recognizing major depression depicted in a case vignette had no effect on public attitudes towards people with major depression (Angermeyer and Matschinger 2003a, b). In studies investigating stigma associated with schizophrenia, Lauber and colleagues found that recommending pharmacological therapies was related to greater social distance (Lauber et al. 2005).

Many strategies have been used to reduce stigma against mental illness, including education programs that raise awareness about mental illness, improving mental health literacy, and promoting contact with persons with mental illness (Rusch et al. 2005). However, these stigma-reducing approaches are not uniformly effective. Gaebel and Baumann (2003) reported that, in Germany, a film portraying the experience of a young man with schizophrenia actually strengthened negative and stereotyping attitudes of the audience and increased social distance. Lauber and colleagues suggested that one unintended consequence of the initiatives aiming at improving mental health literacy might be greater social distance from persons with mental illness (Lauber et al. 2005). Dietrich and colleagues also noted that improving mental health literacy could possibly increase stigma against depression (Dietrich et al. 2006). On the other hand, studies conducted in Australia have shown that mental health literacy interventions have a small impact on reducing social distance and the stigma associated with depression (Christensen et al. 2004; Kitchener and Jorm 2004; Griffiths et al. 2004).

Despite the population-based studies on mental health literacy and stigma towards mental illness, there is a paucity of epidemiological studies on this topic in working populations. Thus, knowledge about depression and the attitudes towards depression in the working population are relatively unknown, and the relationship between mental health literacy and stigmatizing attitudes towards depression in workers is not clear. In the next section, the author describes a depression literacy study conducted in the year of 2006 in Alberta, Canada. The description focuses on levels of depression literacy and stigma against depression in the Albertan working population. Additionally, the relationships among mental health literacy, personal contact with depression, and stigma against depression are described.

A Population-Based Study of Depression Literacy

The Alberta Depression Literacy Study was carried out between January and June, 2006. The target population was household residents in Alberta, Canada, who were between the ages of 18 and 74 years old. Alberta is one of the ten provinces of Canada. At the time of the survey, Alberta had a population of 3 million people. Participants in this study were selected and recruited using random digit dialing methods, and the interviews were conducted using the method of computer-assisted telephone interview. For this study, 3,047 interviews were completed. The response rate was 75.2%. The study used the questionnaire for depression literacy developed for the Australian studies (Jorm et al. 1997a, b, 2006). The major depression module of the International Neuropsychiatric Interview (M.I.N.I.; Sheehan et al. 1997, 1998) was used to determine lifetime major depression. (Detailed information about sampling procedures, weighting methods, and measurements used in this study can be found in related publications (Wang et al. 2007a, b). For the purpose of this chapter, the results related to individuals who were working at the time of the interview ($n=2,162$) will be presented.

The mean age of the working participants was 43.5 years. Of the participants, 56.6% were men and 43.4% were women. Additional detail can be found in the publication of this project (Wang et al. 2007a, b). In this study, the measurement of depression literacy started with a vignette of a person (John or Mary) suffering from major depression. Respondents were asked two open-ended questions, the first being: What would you say, if anything, is wrong with John or Mary? Among participants who were working at the time of interview, 75.8% correctly attributed depression to the provided description. 7% reported "Don't know." Women (87.4%) were more likely to correctly identify depression than men (66.9%); men (9.4%) were more likely to report "Don't know" than women (3.8%). As expected, workers who had a university education (80.9%) and those who completed a high school education (76.2%) were more likely to recognize depression than those who had education less than high school (69.7%). Health professionals (85.7%) were more likely to recognize depression than others (74.8%). However, the difference was only about 10%.

The other open-ended question following the case vignette was: "How do you think John or Mary could best be helped?" Participants were allowed to provide more than one answer to this question. It was found that general practitioners or family doctors (45%) were most highly rated, followed by counselors (18.3%), and medication (9.8%). Of the participants, 7.9% reported "Don't know" about the best help and 35.3% stated "Others" as best help. Only 7.9% of the working population endorsed psychiatrists as the best help.

Over 61% of the participants thought that antidepressants were helpful for depression. However, over 62% considered vitamins and minerals to be of assistance. A total of 24.5% of workers endorsed sleeping pills as being helpful, followed by antipsychotic drugs (18.7%), antibiotics (11.1%), and pain relievers (11%).

With respect to potential causes of the problem described in the case vignette, the majority of the respondents (over 90%) endorsed psychological factors (day-to-day problems, child abuse, traumatic events, death of close friend) as potential causal factors for having depression. However, 42.6% of the working population reported that "weakness of character" was the cause of having depression.

In the Alberta Depression Literacy Study, the nine-item Personal Stigma sub-scale of the Depression Stigma Scale (Griffiths et al. 2004) was used. The subscale had a good internal reliability in this study ($\alpha=0.715$). The questions of the nine-item scale can be found in a published paper based on the data (Wang et al. 2007b). The percentages of specific stigmatizing attitudes were estimated overall and by job status (managers, middle level supervisors, and ordinary workers). Over 45% of the participants considered that people with a problem like John or Mary's are unpredictable. Employees at the three levels did not differ in this regard. However, differences between the levels of employees were found in three stigmatizing attitudes. Compared to others, managers were more likely to endorse "I would not vote for a politician if I knew they suffered a problem like John or Mary's" (45.0% vs. 38.9% for supervisors, 36.2% for ordinary workers); "I would not employ someone if I knew they had a problem like John or Mary's" (30.0% vs. 15.7% for supervisors, 18.9% for ordinary workers); and "If I had a problem like John or Mary's I would not tell anyone" (17.2% vs. 9.6% for supervisors, 12.0% for ordinary workers).

We classified stigmatizing attitudes into four groups based on the number of stigmatizing attitudes endorsed (0=no stigma, 1=mild stigma, 2–3=moderate stigma, 4–9=severe stigma). The percentage of participants in each group was 26.3%, 26.1%, 33.0%, and 14.6%, respectively. We found that more managers than mid-level supervisors held severely stigmatizing attitudes (16.5% vs. 10.9%, $p<0.05$). Participants with major depression were less likely to endorse severely stigmatizing attitudes than those without major depression (10.8% vs. 23.7%, $p<0.001$).

In separate analyses, we investigated relationships among mental health literacy, personal contact with depression, and stigma against depression. (More detail about this study can be found in another manuscript (Wang and Lai 2008). We found that of the nine personal stigma questions, differences were found in six of them in relation to the levels of depression literacy (i.e., case recognition of depression and agreement with health professionals about treatment), and in five of them in relation to whether a family or friend had depression. Enhancing case recognition, agreement with health professionals about treatment, and contact with a person with depression appeared to affect men and women differently in terms of stigmatizing attitudes. In men who had a family member or friend with depression, enhanced case recognition might have reduced stigma towards hiring a person with depression. In women who did not have a family member or friend with depression, the ability to recognize depression seemed to change their negative view that persons with depression could "snap out of it." There was no evidence that improving depression literacy would increase stigma against depression. Correct recognition of individuals with depression was negatively associated with stigma against depression. Thus, case recognition may have a positive effect on changing the views that depression is a sign of personal weakness. Personal contact appeared to have a positive effect on such a view only in women (Wang and Lai 2008).

Conclusions

Depression is prevalent in the working population and can have a significant impact on workers' productivity. Depression is also a stigmatizing condition. For employees with depression, there may be profound consequences of stigma, including diminished employability, limited career advancement, and poor quality of working life. Because of the fear of stigmatization, employees with depression are often not willing to seek professional help, or they may underutilize professional assistance. Stigma against depression in the workplace represents a major and potent target for intervention and prevention.

In our study of depression literacy, we found that the majority of Albertans could correctly recognize depression – more than in other similar studies conducted elsewhere. In spite of this, there is room for improvement in the general population's understanding of depression. There were misconceptions about the causes of depression and the helpfulness of various interventions for it, which should be a target of future interventions. Stigma against depression was found to be prevalent in the working population. Over 45% of workers considered that depression was unpredictable. Furthermore, a marked proportion of top-level managers held severely stigmatizing attitudes against depression, which was consistent with the study by Schultz et al. (2010) on attitudes of Canadian employers (see Chap. 17). Therefore, it is critical to change the attitudes of managers so as to promote mentally healthy workplaces and stigma reduction in the working population. Case recognition, agreement with health professionals about treatment, and contact with a person with depression may have positive effects on reducing stigma. However, these effects may vary somewhat based on gender. Future studies are needed to understand the mechanisms of stigma and mental health literacy, and to develop and test problem-specific intervention strategies using rigorous study designs.

It is also important for employers to understand depression and the resulting stigma. In any given year, 5% of their employees may suffer from an episode of clinical depression. Though depression can have significant impacts on workers' functioning and productivity, and the company's bottom-line, it is treatable. To restore productivity of workers with depression, offering professional services through an Employee Assistance Program and treatment by general practitioners and mental health specialists, in combination with work accommodation and other workplace interventions, is critical (see Chap. 6, Lerner et al. 2010). More importantly, employers should advocate for a work climate promoting open communication where issues of depression can be discussed and dealt with appropriately. Advancing mental health literacy through education can also be helpful; employees can be aware of early signs of depression and seek professional help before the symptoms rise to the level of a clinical disorder. Education can also help employees to understand that depression, like physical illness, is a medical condition that can be effectively treated.

Mental health service providers should understand that various work environment factors may trigger, maintain, or exacerbate depression; and that employees with and without depression have different levels of knowledge about depression and hold different attitudes towards treatments. Practitioners must seek to understand clients' understanding and explanations of depression to gain insight about ways to build an effective treatment alliance. Service providers should work closely with managers, employees, and other mental health professionals using a multiprong early intervention approach for depression, and also plan sustained return-to-work procedures for employees who require extended leaves for their disability. Organizational culture and the views of managers and top executives can positively or negatively affect the efforts of education about depression, early intervention, and case management by service providers. Thus both organizational culture and beliefs held by senior management should be considered in interventions in the workplace (see Chap. 21 by Kirsh and Gewurtz 2010 on organizational culture and Chap. 23 by Harder et al. 2010 on disability management in mental health).

References

American Psychiatric Association (1994) Diagnostic and statistic manual for mental disorders (4th edn) (DSM-IV). American Psychiatric Association, Washington, DC

Angermeyer MC, Matschinger H (1999) Lay beliefs about mental disorders: a comparison between the western and the eastern parts of Germany. Soc Psychiatry Psychiatr Epidemiol 34:275–281

Angermeyer MC, Matschinger H, Riedel-Heller SG (1999) Whom to ask for help in case of a mental disorder? Preferences of the lay public. Soc Psychiatry Psychiatr Epidemiol 34:202–210

Angermeyer MC, Matschinger H (2003a) Public beliefs about schizophrenia and depression: similarities and differences. Soc Psychiatry Psychiatr Epidemiol 38:526–534

Angermeyer MC, Matschinger H (2003b) The stigma of mental illness: effects of labeling on public attitudes towards people with mental disorder. Acta Psychiatr Scand 108:304–309

Angermeyer MC, Matschinger H, Corrigan PW (2004) Familiarity with mental illness and social distance from people with schizophrenia and major depression: testing a model using data from a representative population survey. Schizophr Res 69:175–182

Angermeyer MC, Dietrich S (2006) Public beliefs about and attitudes towards people with mental illness: a review of population studies. Acta Psychiatr Scand 113:163–79

Christensen H, Griffiths KM, Jorm AF (2004) Delivering interventions for depression by using the internet: randomized controlled trial. Br Med J 328:265

Dietrich S, Beck M, Bujantugs B, Kenzine D, Matschinger H, Angermeyer MC (2004) The relationship between public causal beliefs and social distance toward mentally ill people. Aust NZ J Psychiatry 38(5):348–354

Dietrich S, Matschinger H, Angermeyer MC (2006) The relationship between biogenetic causal explanations and social distance toward people with mental disorders: results from a population survey in Germany. Int J Soc Psychiatry 52:166–74

Dyck RJ, Bland RC, Newman SC, Orn H (1998) Suicide attempts and psychiatric disorders in Edmonton. Acta Psychiatr Scand Suppl 338:64–71

Foley DL, Neale MC, Kendler KS (2001) Genetic and environmental risk factors for depression assessed by subject-rated symptom check list versus structured clinical interview. Psychol Med 31(8):1413–1423

Gaebel W, Baumann AE (2003) Interventions to reduce the stigma associated with severe mental illness: experiences from the Open the Doors program in Germany. Can J Psychiatry 48:657–662

Goldney RD, Fisher LJ, Wilson DH (2001) Mental health literacy: an impediment to the optimum treatment of major depression in the community. J Affect Disord 64(2–3):277–284

Griffiths KM, Christensen H, Jorm AF, Evans K, Groves C (2004) Effect of web-based depression literacy and cognitive-behavioral therapy interventions on stigmatizing attitudes to depression: randomized controlled trial. Br J Psychiatry 185:342–349

Harder H, Hawley J, Stewart AM (2010) Disability management approach to job accommodation for mental health disability. In: Schultz IZ, Sally Rogers E (eds) Handbook of work accommodation and retention in mental health. Springer, New York

Jorm AF, Griffiths KM, Christensen H, Korten AE, Parslow RA, Rodgers B (2003) Providing information about the effectiveness of treatment options to depressed people in the community: a randomized controlled trial of effects on mental health literacy, help-seeking and symptoms. Psychol Med 33(6):1071–1079

Jorm AF (2000) Public knowledge and beliefs about mental disorders. Mental health literacy. Br J Psychiatry 177(5):396–401

Jorm AF, Korten AE, Jacomb PA, Christensen H, Rodgers B, Pollitt P (1997a) "Mental health literacy": a survey of the public's ability to recognise mental disorders and their beliefs about the effectiveness of treatment. Med J Aust 166(4):182–186

Jorm AF, Korten AE, Rodgers B, Pollitt P, Jacomb PA, Christensen H, Jiao Z (1997b) Belief systems of the general public concerning the appropriate treatments for mental disorders. Soc Psychiatry Psychiatr Epidemiol 32(8):468–473

Jorm AF, Christensen H, Medway J, Korten AE, Jacomb PA, Rodgers B (2000) Public belief systems about the helpfulness of interventions for depression: associations with history of depression and professional help-seeking. Soc Psychiatry Psychiatr Epidemiol 35(5):211–219

Jorm AF, Christensen H, Griffiths KM (2006) The public's ability to recognize mental disorders and their beliefs about treatment: changes in Australia over 8 years. Aust NZ J Psychiatry 40(1):36–41

Kendler KS, Karkowski LM, Prescott CA (1999) Causal relationship between stressful life events and the onset of major depression. Am J Psychiatry 156(6):837–841

Kendler KS, Thornton LM, Gardner CO (2000) Stressful life events and previous episodes in the etiology of major depression in women: an evaluation of the "kindling" hypothesis. Am J Psychiatry 157(8):1243–1251

Kendler KS, Thornton LM, Prescott CA (2001a) Gender differences in the rates of exposure to stressful life events and sensitivity to their depressogenic effects. Am J Psychiatry 158(4):587–593

Kendler KS, Thornton LM, Gardner CO (2001b) Genetic risk, number of previous depressive episodes, and stressful life events in predicting onset of major depression. Am J Psychiatry 158(4):582–586

Kendler KS, Kuhn J, Prescott CA (2004) The interrelationship of neuroticism, sex, and stressful life events in the prediction of episodes of major depression. Am J Psychiatry 161(4):631–636

Kessler RC, Berglund P, Demler O, Jin R, Koretz D, Merikangas KR et al (2003) The epidemiology of major depressive disorder: results from the National Comorbidity Survey Replication (NCS-R). J Am Med Assoc 289(23):3095–3105

Kitchener BA, Jorm AF (2004) Mental health first aid training in a workplace setting: a randomized controlled trial [ISRCTN13249129]. BMC Psychiatry 4:23

Kirsh B, Gewurtz R (2010) Organizational culture and work issues for individuals with mental health disabilities. In: Schultz IZ, Sally Rogers E (eds) Handbook of work accommodation and retention in mental health. Springer, New York

Lauber C, Nordt C, Falcato L, Rossler W (2003) Do people recognize mental illness? Factors influencing mental health literacy. Eur Arch Psychiatry Clin Neurosci 253(5):248–251

Lauber C, Carlos N, Wulf R (2005) Lay beliefs about treatments for people with mental illness and their implications for antistigma strategies. Can J Psychiatry 50:745–752

Lerner D, Adler D, Hermann R, Rogers W, Chang H, Thomas P, Greenhill A, Perch K (2010) Depression and work performance: the work and health initiative study. In: Schultz IZ, Sally Rogers E (eds) Handbook of work accommodation and retention in mental health. Springer, New York

Link BG, Phelan JC, Bresnahan M, Stueve A, Pescosolido BA (1999) Public conceptions of mental illness: labels, causes, dangerousness, and social distance. Am J Public Health 89:1328–1333

News & Notes (2007) MHA survey shows progress in public knowledge about mental disorders, but stigma remains a problem. Psychiatr Serv 58(7):1020

Parslow RA, Jorm AF (2002) Improving Australians' depression literacy. Med J Aust 177(Suppl):S117–S121

Phelan JC, Link BG, Stueve A, Pescosolido BA (2000) Public conceptions of mental illness in 1950 and 1996: what is mental illness and is it to be feared? J Health Soc Behav 41:188–207

Rusch N, Angermeyer MC, Corrigan PW (2005) Mental illness stigma: concepts, consequences, and initiatives to reduce stigma. Eur Psychiatry 20(8):529–539

Sanathara VA, Gardner CO, Prescott CA, Kendler KS (2003) Interpersonal dependence and major depression: etiological inter-relationship and gender differences. Psychol Med 33(5):927–931

Schultz IZ, Milner R, Hanson D, Winter A (2010) Employer attitudes towards accommodations in mental health disability. In: Schultz IZ, Sally Rogers E (eds) Handbook of work accommodation and retention in mental health. Springer, New York

Sheehan DV, Lecrubier Y, Harnett-Sheehan K et al (1997) The validity of the Mini-International Neuropsychiatric Interview (MINI) according to the SCID-P and its reliability. Eur Psychiatry 12:232–241

Sheehan D, Lecrubier Y, Sheeahan K, Amorim P, Janavs J, Weiller E, Hergueta T, Baker R, Bunbar G (1998) The mini-international neuropsychiatric interview (M.I.N.I.): the development and validation of a structured diagnostic psychiatric interview for DSM-IV and ICD-10. J Clin Psychiatry 59(suppl 20):22–33

Statistics Canada (2002) Canadian Community Health Survey Mental Health and Well-being. Statistics Canada Catalogue no. 82–617-XIE. Ottawa. Version updated September 2004. Ottawa. http://www.statcan.ca/english/freepub/82–617-XIE/tables.htm. Accessed 19 Oct 2009

Stewart WF, Ricci JA, Chee E, Hahn SR, Morganstein D (2003) Cost of lost productive work time among US workers with depression. J Am Med Assoc 289(23):3135–44

Stuart H, Arboleda-Flôrez J (2001) Community attitudes toward people with schizophrenia. Can J Psychiatry 46:245–52

Üstün TB, Ayuso-Mateos JL, Chatterji S, Mathers C, Murray CJ (2004) Global burden of depressive disorders in the year 2000. Br J Psychiatry 184:386–392

Wade TD, Kendler KS (2000) Absence of interactions between social support and stressful life events in the prediction of major depression and depressive symptomatology in women. Psychol Med 30(4):965–974

Wang JL (2004) Perceived work stress and major depressive episode(s) in a population of employed Canadians over 18 years old. J Nerv Ment Dis 192(2):160–163

Wang JL (2005) Work stress as a risk factor for major depressive episode(s). Psychol Med 35(6):865–871

Wang JL, Patten SB, Williams JV, Currie S, Beck CA, Maxwell CJ et al (2005) Help-seeking behaviors of individuals with mood disorders. Can J Psychiatry 50(10):652–659

Wang J, Adair C, Fick G, Lai D, Evans B, Perry BW, Jorm A, Addington D (2007a) Depression literacy in Alberta: findings from a general population sample. Can J Psychiatry 52(7):442–449

Wang JL, Fick G, Adair C, Lai D (2007b) Gender specific correlates of stigma toward depression in a Canadian general population sample. J Affect Disord 103(1–3):91–97

Wang JL, Lai D (2008) The relationship between mental health literacy, personal contacts and personal stigma against depression. J Affect Disord 110(1–2):191–196

Chapter 19
Systemic Barriers and Facilitators to Job Accommodations in Mental Health: Experts' Consensus

Izabela Z. Schultz, Danielle Duplassie, Douglas B. Hanson, and Alanna Winter

Introduction

According to the World Health Organization (WHO 2001a), mental health disorders comprise 40% of the top ten leading individual causes of disability worldwide and account for 10.5% of the total global burden of disease. It is estimated that 450 million people worldwide are affected by at least one mental health disorder. Moreover, mental health disorders are projected to account for 15% of disability internationally by the year 2020, and depression is projected to account for the second highest burden of disease, second only to cardiovascular disease (Scott and Dickey 2003; WHO 2001b). Further research indicates that the economic costs of mental health disabilities are immense and are incurred at multiple levels, as reviewed in the chapter by Dewa and McDaid (2010, this volume).

Further, mental health disabilities are becoming increasingly common in the workplace. According to the Canadian Community Health Survey, in 2002 approximately one in every five individuals over the age of 15 experienced psychological symptoms over the course of the year, including depression, anxiety disorders, and alcohol/drug dependence (CCHS; Statistics Canada 2003). About 10% of the working population has at least one mental health disorder (Dewa et al. 2007). The annual societal costs of mental health disorders in North America is estimated at $83.1 billion (Greenberg et al. 2003) and in the European Union at EU$320 billion (Andlin-Sobocki et al. 2005). Between 30% and 60% of the costs are losses associated with reduced productivity. Absenteeism related to mental health problems accounts for about 7% of the total payroll (Watson Wyatt Worldwide 2000). Stephens and Joubert (2001) suggested that the costs to Canadian employers in lost productivity arising from presenteeism (i.e., less than optimal job performance) and absenteeism due to mental health problems are in

I.Z. Schultz (✉)
Vocational Rehabilitation Counselling Program, University of British Columbia, Scarfe Library 297, 2125 Main Mall, Vancouver, BC, V6T 1Z4, USA
e-mail: ischultz@telus.net

I.Z. Schultz and E.S. Rogers (eds.), *Work Accommodation and Retention in Mental Health*, DOI 10.1007/978-1-4419-0428-7_19, © Springer Science+Business Media, LLC 2011

the billions of dollars each year. Given this estimate, there are economic, social, and humanitarian reasons to implement workplace initiatives which will reduce the incidence of mental health distress in the workplace as well as facilitate the rehabilitation of those employees affected by mental health disabilities (Dewa and McDaid 2010, this volume).

However, Goldner and colleagues (2004) reported that there is limited research to guide disability management practices for mental health disabilities. Furthermore, despite emerging consensus on the importance of the integration of clinical treatment and vocational rehabilitation, the current literature is still fragmented, focused on selected diagnoses, and weak at identifying functional limitations cross-diagnostically. The field lacks an overarching conceptual model and satisfactory evidentiary basis for best practices to support job accommodation and retention of individuals with mental health disabilities.

In this chapter we explore the complexity and dynamic of multisystem barriers and facilitators related to best practices in job accommodation and retention in mental health. The chapter is based on the results of a qualitative study that aimed to obtain the perspective of experts in the field in order to improve the work outcomes of individuals with mental health disabilities. In interviewing both senior researchers in the field and leading professionals who provide front-line treatment and rehabilitation, and who initiate job accommodations for employees with mental health impairments, we sought practical information and guidance regarding policies, procedures, and best practices in job accommodation and retention in mental health.

The objective of the study was to explore the following: (1) the current and recommended practices regarding accommodations for individuals with mental health disabilities; (2) barriers to accommodations for persons with mental health disabilities; and (3) solutions to remove these barriers. The goal in exploring these areas was to obtain a better understanding of the barriers to workplace accommodations and to aid in the provision of realistic and concrete solutions to overcome them. By obtaining information in these areas, stakeholders, employers, managers, and service providers should be better positioned to provide support and interventions for persons with mental health disabilities. Well-matched and well-implemented accommodations in the workplace should provide increased support for employees with mental health disabilities, ultimately decrease costs for employers, and at the same time enhance the social and occupational inclusion of persons with mental health disabilities.

Interviewing the Experts: Methods

To gain the perspective of experts, a qualitative semistructured interview was utilized. The interview focused on the following aspects of job accommodations: current practices, recommended practices, barriers, and solutions to overcome these barriers.

Prior to conducting these semistructured interviews, a literature search was conducted to identify and categorize job accommodations for individuals with mental health disabilities. We relied on the Job Accommodation Network (JAN 2007), an organization to which employers can submit examples of successful accommodations from their work site. These accommodations are then categorized and made available to other employers, rehabilitation professionals, and the general public. Commencing with this taxonomy, examples of accommodations by prominent authors in the field were added (Franche et al. 2004; Gates 2000; MacDonald-Wilson et al. 2002; Sahi and Kleiner 2001). The result was four broad categories of accommodations for individuals with mental health disabilities. The categories included changes in the work environment, and accommodations needed for cognitive, social, or emotional limitations. An interview protocol was developed, refined, and pilot-tested, and the enhanced taxonomy of accommodations was made available to the participants as a reference during the semi-structured interviews.

A sample of 41 individual experts with recognized training and experience in the rehabilitation and accommodation of individuals with mental health disabilities was interviewed. The participants were selected from a variety of backgrounds to ensure participant triangulation (Creswell 1998). The 41 participants included:

- 20 female and 21 male respondents;
- 23 Canadians, 17 Americans, and 1 Australian;
- 14 psychological rehabilitation counselors, 19 psychologists, 3 psychiatrists, 2 physicians, 1 nurse manager, 1 sociologist, and 1 human resources consultant.

Of these respondents, 24 had academic positions at universities and 23 were practitioners. Primary positions held by the participants included: president/CEO, VP/Medical Director, medical/department directors; university professors/instructors; senior rehabilitation counselors; senior human resources consultants; managers, occupational physicians, occupational therapists, social workers, and private practice psychologists. Primary places of employment included universities, international disability insurance companies and financial institutions, national human resource consulting firms, international services organizations for individuals with disabilities, provincial, regional, and national disability services organizations, municipal employee health organizations, workers compensation boards, hospitals, veterans administration agencies, and other primary industries.

Data were coded into categories which were then examined for themes both within and across categories. The results are presented below. Only aggregated data is reported, with no specific comments attributable to any specific individual.

Removing the Barriers to Accommodating Persons with Mental Health Disabilities at Work: Study Results

The major categories resulting from the qualitative analysis of the interviews were guided by the research questions. Synthesis of these categories identified major themes relevant to the removal of barriers and the implementation of reasonable accommodations for individuals with mental health disabilities.

The process of coding resulted in five broad categories and four to seven related subcategories. The subcategories are reported hierarchically on the basis of the frequency of related data.

Current Practice Barriers

Under this theme respondents identified a variety of factors that served as barriers to accommodations for persons with mental health disabilities, including the following subcategories:

Limitations

The most referenced subcategory of barriers consisted of personal limitations of people with mental health disabilities (e.g., low self-esteem, inability to multitask), treatment limitations, funding limitations, limitations of insurance company policies/procedures, and employer programs/services. Financial resources were identified as being inadequate for governments and companies to fund effective vocational rehabilitation programs in mental health. According to respondents, this situation has resulted in inadequate treatment, interventions, and follow-up, as well as insufficient training for employers, coworkers, and service providers. Participants reported that the traditional medical model is not an effective paradigm for the treatment and rehabilitation of individuals with mental health disabilities. A more holistic biopsychosocial model, conducted in a collaborative process involving the employee, employer, physician, vocational rehabilitation counselor (VRC), and other service providers, is recommended (see Schultz et al. 2010, this volume).

Some disability insurance company practices were reported as being adversarial and cumbersome, and were purported to cause conflict, confrontation, and anguish, resulting in an increase in the worker's symptoms. The net effect of this adversarial stance can be to deter the return-to-work process which can lead to higher claims costs per case. Participants also noted that some employers' receptiveness to accommodations was dependent on how skilled and valuable the employee was.

Lack of Education/Knowledge/Awareness/Research

This was the second most referenced category of barriers to accommodation. Participants primarily emphasized deficiencies in the education of individuals with mental health disabilities, their employers, coworkers, and even service providers. Key issues for employers were their lack of understanding and knowledge about symptoms and the resulting functional implications and accommodation needs of workers with mental health disabilities. According to respondents, employers did not understand impairments and disability related to mental health problems and tended to be apprehensive and fearful about hiring individuals with a mental health disability. Service providers (e.g., physicians, psychiatrists, psychologists, employee assistance providers, and VRCs) were perceived as lacking training in psychological aspects of vocational rehabilitation and meaningful accommodations for individuals with various mental health problems. To a lesser degree, participants were concerned about the lack of research to inform practice in the area of job accommodation and retention.

Attitudes/Fear/Stigma

This subcategory describes workplace attitudes, primarily related to the impact of stigma, which were also identified as a substantial factor hindering the accommodation process. Coworker attitudes were referenced as a main barrier to the success of accommodations. Coworkers were reported to be resentful of the perceived preferential treatment given to the worker with a mental health disability, resulting in reluctance to participate in assisting the individual with accommodation needs. Coworkers were said to avoid contact with colleagues who experience mental health problems, and employer representatives were reported to be reluctant to engage in the accommodation process with employees diagnosed with a mental health disability. Stigma was seen as related to employers' and coworkers' lack of knowledge about mental health disability, its causes, and its implications in the workplace. It was noted that many employers had negative attitudes towards, and were fearful of, individuals with mental health disabilities, and were more likely to understand the employment and accommodation needs of employees with physical conditions or disabilities. Respondents often reported that the attitude among business people towards mental health disabilities was that it represented a "weakness of character." Feelings of guilt and shame that arise from such stigmatizing attitudes were thought to often exacerbate the employee's mental health condition, resulting in increased symptoms and a decrease in job functioning.

Lack of an Employee/Client-Centered Approach and Employees' Self-Preserving, Counter-Productive Responses

This was the fourth most referenced subcategory. Participants reported that persons with mental health disabilities are often "short-changed" because they do not receive what they need in order for accommodations to be successful. Issues reported as needing attention included employee support, relationships at work, communication, flexibility, control/autonomy, specialized consideration, and advocacy. Issues delineating the employees' responses included confidentiality/disclosure and workplace avoidance.

The workplace was reported to be adversarial when the individual with a mental health disability was not contacted by or supported by the employer during absences from work due to their mental health disability. The lack of support during a work absence may promote disconnection from the workplace and complicate return-to-work or increase the chance of a relapse of symptoms. Participants reported that "most illness in the workplace [is] related to the lack of meaningful relationships with supervisors and coworkers." Management, and the corporate culture it cultivates, was thought to contribute significantly to promoting these relationships. Respondents also mentioned a lack of communication and collaboration between the treatment and work components of the recovery or return-to-work process for employees with mental health disabilities. Unions were reported to usurp human rights legislation relating to the "duty to accommodate" by invoking the "seniority rights" clause in contracts which may create barriers for workers with mental health disabilities with low seniority to access certain positions. On the other hand, employers will often resort to a claim of "undue hardship" to avoid providing suitable accommodations. Participants identified that the main causes of health care costs were the workers feeling a lack of control, autonomy, and uniqueness in the performance of their duties. In some work situations, trust between the employee with the mental health disability and their VRC may be an issue if the latter is employed by the employer or insurer. The absence of a trusting relationship with the counselor can compromise the success of any job accommodations process.

Employees' resistance to disclose confidential information regarding their mental health problems was seen as a considerable barrier to vocational rehabilitation, return-to-work and accommodation. Employees often do not disclose for fear of the stigma from employers and coworkers. In an adversarial workplace, the employee may "fight to prove he/she is sick … in order to avoid the workplace [which] is the source of depression."

Employer Characteristics and Expectations

This subcategory of barriers focuses on an understanding of contextual workplace factors that impact accommodations for employees with mental health disabilities. These include the substantial influence of top management over corporate culture, and both attitudes and practices towards workers with mental health disabilities. According to our respondents, changing corporate culture is difficult and might be compared to changing a personality: the company must desire a change, otherwise efforts to intervene will not be successful. The main barriers to accommodating employees are the employers' expectations, in addition to their receptiveness to the provision of accommodations. If their prior experience of persons with mental health disabilities was not positive, they are not likely to be receptive to accommodating such employees. Participants noted that companies are resource-lean and the pressure on managers is for productivity. Employers often expect the individual returning from sick leave to be fully recovered and perform the essential job duties without difficulty, when in fact support and accommodation may continue to be needed to facilitate optimal work performance.

Current Practice for Processes/Facilitators/Solutions to Remove Barriers

This category was the least referenced by the participants and describes the current processes that take place to facilitate accommodations and remove barriers. The key issues that participants' identified as necessary for resolving the barriers to accommodations were primarily related to employee support, increased knowledge by all stakeholders, the need for efficient assessments, and funding issues.

Help from Others

In this subcategory participants discussed the idea that vocational rehabilitation in mental health should use a "systemic approach" involving collaboration, involvement, and commitment among the (VRC), physician or psychiatrist, employer, client, coworkers, and union (if applicable). This approach offers the opportunity to repair relationship issues with supervisors and coworkers if they have been damaged as the person's job performance suffered because of their disability. Using such an approach, physicians would be informed of the "patient's" graduated return-to-work (GRTW) opportunities, and the VRC maintains informed communication between all stakeholders. All of these efforts should be initiated for the benefit of the employee.

Flexibility in the accommodation of employees with mental health disabilities is the primary message in planning the necessary, individually tailored return-to-work process. This approach is required as a result of the recurring nature of mental health problems, as well as all employees' "unique style of working."

Participants emphasized that some employers are "proactive" with respect to providing accommodations for individuals with mental health disabilities. It is often the larger employers who have the human and financial resources and disability management organizational structure to initiate reasonable accommodations. Coworkers were also identified as key to the success of accommodation initiatives. They often require additional time to support the individual, help coordinate job tasks, and provide regular, concrete feedback. In using the natural supports available through coworkers, it is important to resolve any previous relationship issues between the individual and coworkers, and to work collaboratively with coworkers to ensure that accommodations are acceptable and realistic to them.

Education/Knowledge/Awareness/Research

Education and awareness for all levels of management, union representatives, and coworkers were seen as ongoing requirements for effective disability management programs. Educational themes outlined by participants included the following: awareness/knowledge of the needs of individuals with a mental health disability;

the impact of one's symptoms on work function; the benefits of rehabilitation interventions and/or accommodations; the challenges faced by individuals as they transition back to the workforce; and the environmental conditions that are required in order to make accommodations successful. When available, unions can both support and participate in the education process.

Participants also indicated that employers (especially large employers) are becoming more aware of the magnitude of the impact of mental health problems on their workforce. These employers have acknowledged that, "it will be the workplace where the solutions are developed." As stated above, the most receptive employers and coworkers were those who have had a previous, positive experience with someone with a mental health disability.

The participants recommended that a "much more expansive knowledge base [is needed] regarding the efficacy of specific interventions and accommodations." On the other side of the equation, the VRC "must be aware of the social behaviors and norms inherent in the corporation's culture" to facilitate the best possible accommodations and workplace environment for the worker with a disability.

Meeting the Basic Requirements of a Job Description

This subcategory addressed the process involved in assisting persons with mental health disabilities to meet the requirements of a job description. Participants reported the need for the person with the disability to take responsibility by initiating services, seeking out assistance when needed, and/or working diligently to meet the requirements of the job. Comments about the need for skills training, work hardening (use of progressively complex simulated and real workplace tasks in an effort to strengthen the cognitive, social, and coping skills of the worker with mental health disorder), and modifications to job descriptions and/or environments were made by the interviewees as important in the accommodations process.

Assessment

Participants recommended that assessments be conducted in order to establish suitable modified work and/or reasonable accommodations for the employee with a mental health disability. They noted that the assessment process should be collaborative, involving the employee, employer, and union representative, if applicable. One suggestion was to utilize the workplace to assess functioning in relation to the essential job duties.

Conditions for Current Practices to Be Successful

This subcategory addressed the specific and practical components required for current practices to be successful. Participants discussed the importance of good management, recognition of the impact of mental health disabilities in the workplace, taking the feelings of the individual with a mental health disability into account, the need to focus on abilities vs. disabilities, the need for confidentiality, the need for funding, and knowledge of the legal obligations of companies to provide accommodations. Without these fundamental conditions, the process of developing and implementing effective accommodations may be difficult.

Current Practice in Accommodations

This category addressed the accommodations that are currently being facilitated in the workplace. Participants reported the following accommodations: changing job tasks/work requirements; providing social support in the workplace, changes to the work environment; providing social skills training for the employee with the disability; and providing memory/physical/technological enhancements for the employee. Participants suggested that insurers and employers deal with the impact of mental health disabilities in the workplace as this "problem has not disappeared." Employers who have been proactive in the provision of accommodations appear to practice good management policies and practices for their employees. Both presenteeism and absenteeism require the employers' attention when considering accommodations. Participants discussed the need for the return-to-work process to plan for success, as opposed to failure, and to include early intervention to reduce the likelihood of chronic problems. A GRTW process should be implemented on a transitional basis, involving a process of progressive psychological work hardening. The goal of all accommodations is to return the employee to full regular duties, if possible.

Changing Job Tasks/Work Requirements/Work Environment

Participants noted that temporary changes in job requirements are effective accommodations for individuals experiencing mental health disabilities. Reducing the hours of work, eliminating some tasks from the job, moving the individual to a private office, allowing the employee to temporarily work from home or relocating him/her to a different department are just a few of the effective reasonable accommodations identified. The use of dividers to reduce noise, headphones to reduce distractions, and the use of a private office to reduce stimuli and interactions were all mentioned as potentially useful accommodations.

Workplace Social Support

Interviewees suggested that prior to initiating a GRTW, the VRC should collaborate with the individual's supervisor and relevant coworkers in order to address and resolve any relationship difficulties as noted above. This situation may arise as mental health symptoms increase and are manifest in the workplace. The VRC should offer counseling to the individual regarding issues of guilt, shame, and/or fear in relation to potential stigma. The supervisor and coworkers should be informed of the effect of these issues during the return-to-work process. Job modifications and accommodations were thought to be best if arrived at by consensus and if perceived as realistic for all parties.

Social Skills Training

Individuals with mental health disabilities often develop social limitations due to low self-esteem, fear of the stigma from others regarding their condition, agitation and irritability, and/or difficulties with decision-making or fears and phobias. Participants indicated that these individuals can benefit from social skills training, facilitated by a qualified counselor. Some of the specific training opportunities that were identified by interviewees included conflict resolution, the development of an awareness of workplace cultural expectations, stress management, and assertiveness training.

Accommodations for Memory Deficits/Technological Enhancements

Mental health disabilities can affect the individuals' ability to concentrate and make decisions. Participants noted several of the compensatory strategies that are used to accommodate such problems. Some of these include note-taking, the use of diagrams as visual instructions, reminder booklets, the use of color-coding systems, and the use of pictures and maps as tools to help employees know where things belong. Assistive technologies might include the use of timers, voice organizers, or rehabilitation computer programs that instruct employees on how to perform tasks.

Recommended Practice for Removal of Barriers to Work Accommodations

This category included the need for substantial enhancements in overall education, knowledge, awareness, and understanding of the needs and abilities of individuals with mental health disabilities. Study participants discussed the need to increase appropriate support from resources, both internally and externally. Concern was raised regarding the need for objective assessments in terms of an individual's functioning; the psychosocial contextual requirements of a job; and the evaluation of outcomes pertaining to interventions and accommodations. A collaborative, biopsychosocial approach to treatment and vocational rehabilitation comprised the paradigm shift that was believed to be essential to optimize intervention and accommodation outcomes. Participants also discussed the idea that all models of effective mental health disability management must begin with the commitment, support, and involvement of top management.

Education/Knowledge/Awareness/Understanding

This subcategory drew the most input from participants. Participants discussed the importance of education to promote increased knowledge, awareness, and understanding to all involved in the rehabilitation process – the individual with a mental health disability, physicians, VRCs, occupational therapists, psychologists, employers, coworkers, and the community at large. By improving awareness of the incidence, causes, symptoms, and successful interventions for mental health disabilities, participants believed that greater acceptance of mental health disorders would occur, having an impact on the stigma associated with them.

Participants reported a need for greater education regarding the following: the magnitude and incidence of mental health disorders in the workplace, the signs and symptoms of various disorders, the resources and accommodations available to employees with mental health disabilities, communication skills, social skills, empathic communication skills, the need for support from top management, the benefits of early intervention, and the various models of mental health disability management (e.g., the benefits of a biopsychosocial model).

Although education is required in these important areas, participants indicated that it is difficult to engage companies that are unaware of having problems with employees with mental health problems. The lack of awareness by these companies does not seem to be resolved through training. One participant reported that, "education and knowledge does not change [corporate] minds." Participants suggested that education should be directed toward suitable companies who are "receptive to change ... diversity ... [and] interested in the process of accommodation." It was believed by some that the reluctant companies would soon "follow suit." In essence, "no one wants to be first, but one is always interested in comparing oneself to peers."

Help from Others

This subcategory focused on the issues of collaboration, communication, individually tailored support, employer support, service provider support, and institutional support. The main issue emphasized by the participants was the need for a collaborative effort by the employer, employee, coworkers, union, physicians, and service providers in order to prepare the individual for a successful work placement. Participants discussed how imperative it is for top management to endorse and actively participate in the development of a corporate culture that supports the accommodation of all employees with disabilities. Top management should establish partnerships with experts in mental health disability management and rehabilitation as a way to facilitate the necessary changes within the corporate culture. A collaborative corporate culture is typified by "flattening the corporate hierarchy," "cultivating communication between all workers and management so everyone has a say and [is] listened to," and "ongoing communication which ensures that employer/employee values are expressed and concerns are reviewed."

In essence, this subcategory represents the need for support and assistance by all who are affected and/or involved in the provision of accommodations for persons with mental health disabilities – top management, employers, supervisors, coworkers, rehabilitation counselors, physicians, and others. Participants discussed the importance for all to be involved and to demonstrate a supportive attitude and desire to work for the common good.

Assessments

Although current practices include the use of assessments, this subcategory suggests that more assessments need to be administered on an ongoing basis to better determine a person's readiness for employment, in addition to the accommodations that would be necessary for the individual with a mental health disability. Types of assessments that were recommended in this subcategory included psychological, neuropsychological, vocational, medical, functional capacity, and situational work assessments. Participants stressed the importance of these assessments to focus on abilities and individual needs, as opposed to disabilities and the use of a cookie-cutter approach to accommodations.

Implications for Best Practices in Work Accommodation and Retention in Mental Health: Discussion

Job accommodations are recognized in legislation, research, and practice models as one of the most critical strategies for promoting the employment retention of individuals with mental health disabilities. Yet there is little empirical information to guide the design and implementation of effective job accommodations for mental health problems in practice (Goldner et al. 2004; Schultz et al. 2010, this volume). In an effort to address this paucity, the current study focused on the exploration of best practices in job accommodations for persons with mental health disabilities. Using a broader, system-based contextualization of job accommodations, experts' input was solicited for an improved understanding of the barriers to job accommodation and larger, systemic solutions to the encountered problems. This approach underscored the need for a broader ecological, or person-environment model of mental health disability in the workplace (WHO 2001a), which can be translated into the interactive model of job accommodation for persons with mental health disabilities (Fig. 19.1).

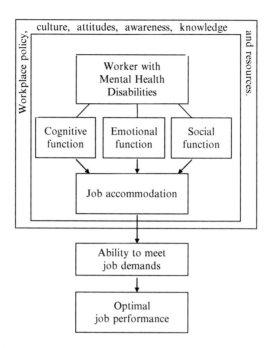

Fig. 19.1 Optimizing work performance of workers with mental health disability (mental health disabilities) through job accommodation

Barriers

Multiple limitations to the utilization of job accommodations were reported. They included limitations at the *microsystem level*, or the functional impairment of persons with disabilities themselves (Baril and Berthelette 2000; Bronfenbrenner 1979). Limitations of clinical treatment and rehabilitation systems were pointed out; as were those of insurance companies, the financial systems, and the employers. These constitute a *mesosystem*, i.e., characteristics of primary institutional stakeholders of mental health disability (Baril and Berthelette 2000; Bronfenbrenner 1979). Importantly, *macrosystem* limitations seen in the ongoing reliance on the medical model were reiterated through the expert interviews, together with the need for implementation of an integrated biopsychosocial model of mental health disability (Baril and Berthelette 2000; Bronfenbrenner 1979; Schultz 2005, 2008; Schultz and Stewart 2008).

A significant cluster of barriers to effective job accommodation practices included the lack of knowledge, education, awareness, and research on mental health disability in the workplace. This is consistent with concerns identified in the literature (Goldner et al. 2004) that are seen as arising from a lack of knowledge, leading to the stigma, fears, and negative attitudes towards persons with mental health disabilities held by employers and coworkers. These fears and lack of knowledge were highlighted as significant barriers to job accommodation. Social stigma was perceived by the experts as the leading cause for the absence of meaningful relationships between workers and their superiors and coworkers. The literature has been consistent in indicating that the social stigma associated with mental health conditions is often more disabling than the primary condition itself and should be considered as the most salient barrier to work on (Kusznir 2001; WHO 2001a; Baldwin and Marcus 2010, this volume). Harnois and Gabriel (2000) indicated that the stigma is frequently based on myths perpetuated in the workplace, including the notion that mental health disabilities are synonymous with mental retardation, are not amenable to treatment, cause lack of productivity, and may cause workers to become disruptive and harmful to others. As evidence of this stigma, researchers found that 90% of employers would hire a person with a physical disability but only 20% would hire a person with mental illness (McFarlin et al. 1991).

The lack of a worker-centered approach in the workplace was pointed out by the experts as another mesosystem-based barrier to job accommodation. Client-centered practice is based on the assumption that employees with mental health disabilities should be actively involved in identifying their unique employment problems, strengths, and needs in the design, implementation, and ongoing evaluation of work-related interventions (Yasuda et al. 2001).

Experts cautioned that the lack of an employee-centered approach for workers with mental health disabilities is likely to result in their unwillingness to disclose disability in the workplace. Yet, disclosure of mental health disabilities is critical for the implementation of job accommodations (Carling 1993; Gioia and Brekke 2003; Marrone et al. 1998; MacDonald-Wilson et al. 2010, this volume).

Optimal Workplace Interventions

The current study has clearly demonstrated that optimal interventions to mental health disabilities in the workplace need to involve multisystem and multiplayer collaboration in a flexible, proactive, collaborative, and awareness-raising manner. Optimal approaches should involve a team that includes the worker, coworkers, union, employer, physician, psychologist, and vocational rehabilitation consultant. This is consistent with a biopsychosocial model of occupational disability, which postulates the integration of vocational and clinical services and emphasizes the worker's strengths and abilities rather than psychopathology and diagnosis (Goldner et al. 2004; Loisel et al. 2001; Schultz et al. 2007; Schultz 2008; Schultz and Stewart 2008).

At the workplace (*mesosystem*) level, development of a corporate culture that supports accommodations, with senior management commitment, collaborative communication between workers and management, and partnerships with experts in occupational aspects of mental health in the workplace, was highly recommended. This is consistent with best practices in disability management (Harder et al. , this volume; Harder and Scott 2005). Interestingly, Akabas (1994) indicated that the cultural characteristics of a successful company are the same factors necessary for successful accommodation of workers with mental health disabilities: positive, flexible, innovative, collaborative, and inclusive of all employees (review chapters by Gates and Akabas 2010, this volume; Kirsch and Gewurtz 2010, this volume; Gallie et al. 2010, this volume).

Goldner and colleagues (2004) recommended the use of professionals who have expertise in the vocational rehabilitation of workers with mental health disabilities. Results of the current study stipulate the importance of a coordinated continuum of intervention involving primary prevention: education, awareness-raising, and skill training in the workplace, and management of workplace risk factors for mental health problems; secondary prevention: early identification of mental health issues in the workplace and early workplace intervention to prevent chronic disability; and tertiary prevention: improvement of work functioning of persons with mental health disabilities, including job accommodation. Consistent with the findings of the current study, Fabian et al. (1993) reported that the most salient accommodation for employees with mental health disabilities was the training and orientation of superiors and coworkers.

The study demonstrated that coworkers are central to the success of accommodations for persons with mental health disabilities; workplace interventions need to be realistic and acceptable to them. DeLisio et al. (1986) suggested that the disabled worker's relationships with family and coworkers were salient factors in supporting, motivating, and facilitating the individual's rehabilitation and successful return-to-work.

Coworkers may need extra time to mentor the worker with a disability and any relationship difficulties should be addressed and resolved prior to the commencement of a GRTW process.

Accommodations

An appropriate situational and functionally oriented ("ecological") assessment of the employee in the work environment was recommended by the experts as a key prerequisite in the development of suitable job accommodations. This is consistent with the literature (MacDonald-Wilson et al. 2003; Spirito-Dalgin and Gilbride 2003). The categories of job accommodations for persons with mental health disabilities that were identified by most experts were as follows: changes to the work environment, change of job tasks/work requirements, workplace social support, skill training, and memory enhancements. These recommendations were also supported by the existing literature (Bell et al. 2003; Coles 1996; Granger et al. 1997; Hirsh et al. 1996; Holzberg 2001; MacDonald-Wilson et al. 2002; MacDonald-Wilson 1994; Mancuso 1993; Means et al. 1997; Mpofu et al. 1999).

Environmental modifications were mentioned by our experts and also appear in the literature, including the use of an enclosed office; provision of partitions, room dividers, or other enhancements of soundproofing and visual barriers between workplaces; offering reserved parking; blocking noise through volume control of telephones; and allowing for personal space and access to a quiet rest area (Coles 1996; Granger et al. 1997; Holzberg 2001; Mancuso 1993; MacDonald-Wilson 1994; Means et al. 1997; Weed and Brown 1993).

Workplace social support, a supportive work culture, employer policies, and disability management practices that encourage respect, trust, and communication are all important and mentioned in this study and by other researchers (Goldner et al. 2004). These were salient determinants of successful return-to-work programs. Provencher et al. (2002) suggested that accommodations incorporate components that reinforce the individual's "self-definition, empowerment, connection to others, meaning of work, vocational future, and meaning of recovery" (p. 135). Supportive coworkers as natural supports (Storey and Certo 1996) can be instrumental in encouraging, prompting, and instructing the employee with mental health disability, and may also provide a supportive function if the employee experiences a recurrence of symptoms (Scott and Dickey 2003). Changes in the work environment might also include relationship accommodations and acknowledgement of the social impact of the accommodation (Gates 2000). This approach has been referred to in the literature as relationship accommodation.

Social skills training may be helpful and can include conflict resolution, awareness and development of workplace cultural expectations, stress, and anger management. Coping skills training may enhance the individual's management of emotions in response to difficult situations at work and the development of a repertoire of effective behavioral skills (Smith et al. 1996; Wallace et al. 1999). Such training often involves time and stress management, energy conservation, assertiveness and communication skills training, anger/frustration management, and problem solving (Bell et al. 2003; Bozzer et al. 1999; Hesslinger et al. 2002; Parente et al. 1991; Storey 2002; Wallace et al. 1999; Weiss and Murray 2003). Accommodations for cognitive difficulties postulated by the experts included assistive devices and compensatory strategies of different degrees of technical complexity.

Conclusions

Multisystem barriers to the development, implementation, and evaluation of job accommodations for persons with mental health disabilities have been identified in this study. These barriers included all levels, from *macrosystem* (societal attitudes, public policy, and institutional challenges), to *mesosystem* (organizational characteristics of the workplace that make it more or less accommodating), and *microsystem* (the person with the mental health disability).

All of these systems interact in a way that promotes or discourages accommodating workplace practices. The most significant barriers to the implementation of proactive accommodations for persons with mental health problems seem to be social in nature, consistent with the conceptualization of accommodation as a social process (Gates 2000; Gates and Akabas 2010, this volume). A lack of awareness and understanding of mental health problems, coupled with negative societal and employer attitudes, stigma, and fear continue to perpetuate in societal institutions which are work disability stakeholders, such as healthcare institutions, insurance companies, and employers. These attitudes translate into an unsupportive work environment laden with relationship difficulties among the worker with mental health disability, supervisors, and coworkers.

A negative social climate in the workplace can breed inflexibility instead of flexibility, reactivity instead of proactivity, and rule conformity instead of innovation and creativity. Under these conditions, often associated with policy and financial constraints, the development and application of job accommodations for persons with mental health disabilities is difficult to introduce. Scarce research on job accommodations and poor knowledge transfer to employers complicate the job accommodation process further.

Overall, the study results underscore the importance of social factors in the job accommodations process for mental health disabilities: societal attitudes and stigma; social and institutional policy; multisystem collaboration among the disability stakeholders and the worker; relationships among the employee, coworkers, unions, and management; social climate in the workplace; prominence of relationship-based job accommodations; and the need for social skills training for persons with mental health disabilities.

A paradigm shift away from the traditional psychopathology-based medical model towards an integrated biopsychosocial model combining clinical and occupational interventions and focusing on function, not on diagnosis, is a necessary requirement for successful accommodation of mental health disability in the workplace. New professional training approaches are therefore needed in clinical and vocational rehabilitation, disability management, human resources, management, and other professions to meet the challenges of mental health disability in the workplace.

Last but not least, enhanced interdisciplinary, qualitative, and quantitative research paradigms relating to the integration of persons with mental health disabilities in the workplace, and thus providing much needed evidence-informed practices, are also critically needed.

Acknowledgement This chapter was based on a study completed for a project funded by Social Development Canada: "Towards Evidence-Informed Best Practice Guidelines for Job Accommodations for Persons with Mental Health Disabilities."

References

Akabas SH (1994) Workplace responsiveness: Key employer characteristics in support of job maintenance for people with mental illness. Psychosoc Rehabil J 17(3):91–101
Andlin-Sobocki P, Jonsson B, Wittchen HU, Olesen J (2005) Cost of disorders of the brain in Europe. Eur J Neurol 12(Suppl 1):1–27

Baldwin ML, Marcus SC (2010) Stigma, discrimination, and employment outcomes among persons with mental health disabilities. In: Schultz IZ, Rogers ES (eds) Handbook of work accommodation and retention in mental health. Springer, New York

Baril R, Berthelette D (2000). Les composantes et les déterminants organisationnels des interventions de maintien du lien d'emploi en enterprise. Institut de recherche en santé et en sécurité du travail du Québec, Montréal; Rapport No. R-238

Bell MD, Lysaker PH, Bryson G (2003) A behavioural intervention to improve work performance in schizophrenia: Work behaviour inventory feedback. J Vocat Rehabil 18(1):43–50

Bozzer M, Samson D, Anson J (1999) An evaluation of a community-based vocational rehabilitation program for adults with psychiatric disabilities. Can J Community Ment Health 18(1):165–179

Bronfenbrenner U (1979) The ecology of human development: Experiments by nature and design. Harvard University Press, Cambridge, MA

Carling PJ (1993) Reasonable accommodations in the workplace for individuals with psychiatric disabilities. J Consult Psychol J 45(2):46–62

Coles A (1996) Reasonable accommodations in the workplace for persons with affective mood disorders: A case study. J Appl Rehabil Couns 27(4):40–44

Creswell JW (1998) Qualitative inquiry and research design: Choosing among five traditions. SAGE Publications, Thousand Oaks

DeLisio G, Maremmani I, Perugi G, Cassano GB, Deltito J, Akiskal HS (1986) Impairment of work and leisure in depressed outpatients. A preliminary communication. J Affect Disord 10(2):79–84

Dewa CS, McDaid D (2010) Investing in the mental health of the labour force: Epidemiological and economic impact of mental health disabilities in the workplace. In: Schultz IZ, Rogers ES (eds) Handbook of work accommodation and retention in mental health. Springer, New York

Dewa CS, Lin E, Kooehoorn M, Goldner E (2007) Association of chronic work stress, psychiatric disorders, and chronic physical conditions with disability among workers. Psychiatr Serv 58(5):652–658

Fabian ES, Waterworth A, Ripke B (1993) Reasonable accommodations for workers with serious mental illness: Type, frequency, and associated outcomes. Psychosoc Rehabil J 17(2):163–172

Franche R-L, Cullen K, Clarke J, MacEachen E, Frank J, Sinclair S et al (2004) Workplace-based return-to-work interventions: A systematic review of the quantitative and qualitative literature. Institute for Work and Health, Toronto, ON

Gallie KA, Schultz IZ, Winter A (2010) Company-level interventions in mental health. In: Schultz IZ, Rogers ES (eds) Handbook of work accommodation and retention in mental health. Springer, New York

Gates LB (2000) Workplace accommodation as a social process. J Occup Rehabil 10(1):85–98

Gates LB, Akabas SH (2010) Inclusion of people with mental health disabilities into the workplace: Accommodation as a social process. In: Schultz IZ, Rogers ES (eds) Handbook of work accommodation and retention in mental health. Springer, New York

Gioia D, Brekke J (2003) Knowledge and use of workplace accommodations and protections by young adults with schizophrenia: A mixed method study. Psychiatr Rehabil J 27:59–68

Goldner E, Bilsker D, Gilbert M, Myette L, Corbière M, Dewa CS (2004) Disability management, return to work and treatment. Healthc Pap 5(2):59–70

Granger B, Baron R, Robinson S (1997) Findings from a national survey of job coaches and job developers about job accommodations arranged between employers and people with psychiatric disabilities. J Vocat Rehabil 9:235–251

Greenberg PE, Kessler R, Birnbaum HG, Leong SA, Lowe SW, Bergland PA, Corey-Lisle PK (2003) The economic burden of depression in the United States: How did it change between 1990 and 2000? J Clin Psychiatry 64(12):1465–1475

Harder HG, Scott LR (2005) Comprehensive disability management. Elsevier, London

Harder H, Hawley J, Stewart A (2010) Disability management approach to job accommodation for mental health disability. In: Schultz IZ, Rogers ES (eds) Handbook of work accommodation and retention in mental health. Springer, New York

Harnois G, Gabriel P (2000) Mental health work: Impact, issues and good practices. Joint publication of the World Health Organization [WHO] and the International Labour Organization [ILO], pp 1–66. WHO/ILO, Geneva, Switzerland

Hesslinger B, Tebartz van Elst L, Nyberg E, Dykierek P, Richter H, Berner M et al (2002) Psychotherapy of attention deficit hyperactivity disorder in adults: A pilot study using a structured skills training program. Eur Arch Psychiatry Clin Neurosci 252(4):177–184

Hirsh A, Duckworth K, Hendricks D, Dowler D (1996) Accommodating workers with traumatic brain injury: Issues related to TBI and ADA. J Vocat Rehabil 7:217–226

Holzberg E (2001) The best practice for gaining and maintaining employment for individuals with traumatic brain injury. Work 16:245–258

Job Accommodation Network [JAN] (2007) Accommodation information by disability. Morgantown, VW. Electronic version retrieved from http://www.jan.wvu.edu/media/atoz.htm

Kirsch B, Gewurtz R (2010) Organizational culture and work issues for individuals with mental health disabilities. In: Schultz IZ, Rogers ES (eds) Handbook of work accommodation and retention in mental health. Springer, New York

Kusznir A (2001, October) Understanding return to work challenges with depression. Paper presented at the second annual national symposium of the National Institute of Disability Management and Research. Ottawa, ON, Canada

Loisel P, Durand M, Berthelette D, Vézina N, Baril R, Gagnon D et al (2001) Disability prevention: New paradigm for the management of occupational back pain. Dis Manag Health Outcomes 9(7):351–360

MacDonald-Wilson KL (1994) Reasonable workplace accommodations for people with psychiatric disabilities. Reprinted from the Community Support Network News 12(1). Retrieved from Reasonable Accommodations: An on-line resource for employers and educators, at http://www.bu.edu/cpr/reasaccom/employ-read-macdon.html

MacDonald-Wilson KL, Rogers ES, Massaro JM, Lyass A, Crean T (2002) An investigation of reasonable workplace accommodations for people with psychiatric disabilities: Quantitative findings from a multi-site study. Community Ment Health J 38(1):35–50

MacDonald-Wilson KL, Rogers ES, Massaro J (2003) Identifying relationships between functional limitations, job accommodations, and demographic characteristics of persons with psychiatric disabilities. J Vocat Rehabil 18:15–24

Mac-Donald-Wilson KL, Russinova Z, rogers ES, Lin CH, Fergusin T, Dong S, et al. (2010). Disclosure of mental health disabilities in the workplace. In IZ Schultz and ES Rogers (eds) Handbook of work accommodation and rentention in mental health. New York: Springer

Mancuso LL (1993) Achieving reasonable accommodation for workers with psychiatric disabilities: Understanding the employer's perspective. Am Rehabil 21:2–8

Marrone J, Gandolfo C, Gold M, Hoff D (1998) Just doing it: Helping people with mental illness get good jobs. J Appl Rehabil Couns 29(1):37–48

McFarlin DB, Song J, Sonntag M (1991) Integrating the disabled into the work force: A survey of Fortune 500 company attitudes and practices. Employee Responsib Rights J 4(2):107–123

Means CD, Stewart SL, Dowler DL (1997) Job accommodations that work: A follow-up study of adults with attention deficit disorder. J Appl Rehabil Couns 28(4):13–16

Mpofu E, Watson E, Chan S (1999) Learning disabilities in adults: Implications for rehabilitation interventions in work settings. J Rehabil 65(3):33–41

Parente R, Stapleton MC, Wheatley CJ (1991) Practical strategies for vocational reentry after traumatic brain injury. J Head Trauma Rehabil 6(3):35–45

Provencher HL, Gregg R, Mead S, Mueser KT (2002) The role of work in the recovery of persons with psychiatric disabilities. Psychiatr Rehabil J 26(2):132–144

Sahi S, Kleiner BH (2001) Reasonable accommodation of psychiatric disability under the ADA. Equal Opport Int 20(5/6/7):133–137

Schultz IZ (2005) Impairment and occupational disability in research and practice. In: Schultz IZ, Gatchel RJ (eds) Handbook of complex occupational disability claims. Early risk identification, intervention, and prevention, 1st edn. Springer, New York, pp 25–42

Schultz IZ (2008) Disentangling disability quagmire in psychological injury and law. Part I. Disability and return to work: theories, methods and applications. Psychol Inj Law 1(2):94–102

Schultz IZ, Stewart AM (2008) Disentangling disability quagmire in psychological injury and law Part II Evolution of disability models: conceptual, methodological and forensic issues. Psychol Inj Law 1(2):103–121

Schultz IZ, Stowell AW, Feuerstein M, Gatchel RJ (2007) Models of return to work for musculo-skeletal disorders. J Occup Rehabil 17:327–352

Schultz IZ, Winter A, Wald J (2010) Evidentiary support for best practices in job accommodation in mental health: Macrosystem, employer-level and employee-level interventions. In: Schultz IZ, Rogers ES (eds) Handbook of work accommodation and retention in mental health. Springer, New York

Scott J, Dickey B (2003) Global burden of depression: The intersection of culture and medicine. Br J Psychiatry 183:92–94

Smith TE, Bellack AS, Liberman RP (1996) Social skills training for schizophrenia: Review and future directions. Clin Psychol Rev 16(7):599–617

Spirito-Dalgin R, Gilbride D (2003) Perspective of people with psychiatric disabilities on employment disclosure. Psychiatr Rehabil J 26(3):306–310

Statistics Canada (2003) Canadian Community Health Survey [CCHS] – Mental health and well-being (Rep. no. catalogue no. 82-617-XIE)

Stephens T, Joubert N (2001) The economic burden of mental health problems. Chron Dis Can 22(1):18–23

Storey K (2002) Strategies for increasing interactions in supported employment settings: An updated review. J Vocat Rehabil 17:231–237

Storey K, Certo NJ (1996) Natural supports for increasing integration in the workplace for people with disabilities: A review of the literature and guidelines for implementation. Rehabil Couns Bull 40(1):62–67

Wallace CJ, Tauber R, Wilde J (1999) Teaching fundamental workplace skills to persons with serious mental illness. Psychiatr Serv 50:1147–1153

Watson Wyatt Worldwide (2000) Staying at work 2000/2001 – The Dollars and sense of effective disability management. Watson Wyatt Worldwide, Vancouver

Weed RO, Brown TT (1993) Reasonable accommodation, job analysis, and essential functions of the job: Issues in ADA implementation for the private sector rehabilitation professional. NARPPS J 8(3):126–130

Weiss M, Murray C (2003) Assessment and management of attention-deficit hyperactivity disorder in adults. Can Med Assoc J 168:715–722

World Health Organization [WHO] (2001a) International classification of functioning, disability and health. World Health Organization, Geneva

World Health Organization [WHO] (2001b) The World Health Report 2001 – Mental health: new understanding, new hope. World Health Organization, Geneva

Yasuda S, Wehman P, Targett P, Cifu D, West M (2001) Return to work for persons with traumatic brain injury. Am J Phys Med Rehabil 80(11):852–864

Part V
Evidence-Informed Practice in Job Accommodation

Chapter 20
Inclusion of People with Mental Health Disabilities into the Workplace: Accommodation as a Social Process

Lauren B. Gates and Sheila H. Akabas

Introduction

A review of the literature reporting on the employment status for people with serious mental health conditions typically begins with a litany of statistics demonstrating their continued poor employment outcomes. The myriad of factors expected to open the doors to the world of work for people with mental health conditions, such as the passage of the Americans with Disabilities Act, improvement in treatment efficacy, emergence of strong advocacy by individuals with mental health conditions, and the desire and ability of individuals to work productively, have not had the anticipated impact (Dixon et al. 2003; Harris and Associates 1998; National Organization on Disability 2003; Schur et al. 2005; U.S. Census Bureau 2002a, b).

In the continued search for an understanding of why people with mental health conditions are not included in the workplace, one factor that comes to the fore is the misfit between the worker and the workplace (Akabas 1994; Bond 2004; Spataro 2005; Stone and Colella 1996). Workplace policies and practices do not match up with the needs of qualified individuals who may experience difficulty meeting the requirements of a specific job. Inclusion into the workplace is not just a function of reducing the extent to which individual characteristics and symptoms interfere with workplace policies and practices, but also a function of organizational responsiveness to these individuals as expressed through the attitudes and behaviors of management, supervisors, and coworkers (Akabas 1994; Gates 2000). It is dependent on the interaction of individuals with mental health conditions and their work groups. As such, inclusion is a social process and an interactive effort attempting to achieve mutual understanding and support for workplace actions and behaviors.

If inclusion is a social process, then assistance to support individuals in employment must also be social in nature. It must account for the ways in which an

L.B. Gates (✉)
Center for Social Policy and Practice in the Workplace, Columbia University School of Social Work, 1255 Amsterdam Avenue, New York, NY, 10027, USA
e-mail: lbg13@columbia.edu

I.Z. Schultz and E.S. Rogers (eds.), *Work Accommodation and Retention in Mental Health*, 375
DOI 10.1007/978-1-4419-0428-7_20, © Springer Science+Business Media, LLC 2011

individual's ability to meet work expectations affects and is affected by others. Effective support to individuals, therefore, must improve the fit between worker and workplace by meeting the needs of both the individuals with mental health conditions, as well as all others with whom these individuals interact in the course of getting their work done (Gates 2000; Ward and Baker 2005). Social service providers have begun to turn to accommodation as the strategy for helping individuals with mental health conditions improve the fit between worker and workplace, and thus their inclusion into the workplace. The purpose of this chapter is to present an understanding of accommodation as a social process to achieve inclusion of people with mental health conditions into the world of work.

Understanding Accommodation as a Strategy for Inclusion

The recent past has witnessed major changes in treatment standards for people with mental health conditions and in the ways in which the workplace is organized. The application of accommodation as an effective strategy to promote inclusion of people with mental health conditions into the workplace has emerged in response. The 1970s saw the rise of the disability rights movement, advances in psychopharmacology that resulted in new medications allowing people with mental health conditions to live independently, and shifts in political and economic policies that moved people with mental health conditions out of hospitals and into the community. Supported by, and emerging from, these changes was a shift in treatment approaches from a medical model to a recovery-oriented perspective. The medical model defined disability as an individual health problem that needed to be corrected so that the persons with disability could be more like those without disability (Hahn 1985; Harlan and Robert 1998). Alternatively, the recovery perspective was based on the understanding that people with mental health conditions were not condemned to a life of chronic illness that can be corrected only through medical intervention. It recognized that individuals with mental health conditions can lead productive, satisfying lives that include living and working in the community, side by side with people without disabilities (Deegan 2003; O'Day and Killeen 2002).

Rather than defined internally by a medical condition, the recovery perspective applies a social definition to disability, based on attitudes and social structures (Deegan 1985). From this perspective, the challenges people face integrating into the mainstream of everyday life are as much the result of how individuals clash with socially defined roles and environmental structures as they are the symptoms of the mental health conditions (Anthony 2000; Cook and Burke 2002). Impairment in one context may not be limiting in another. Consider someone who is bilingual, with a visual impairment. The individual may be limited in a job as a proofreader but may have no problem meeting the requirements of providing translation over the phone. Or consider an individual with a mental health condition who talks to herself in full voice. She may have trouble meeting the requirement of a library job where silence is valued, but this symptom of her condition may not pose a barrier to a job working as an assistant zookeeper where it is normal to talk to the animals.

Parallel to the shift in treatment approaches were changes in the way workplaces were organized and the way workplaces responded to people with disabilities. The medical definition of disability emerged during a time in which the system of production had shifted from the skilled craftsmen who performed complex and diverse tasks and had control over the entire production process, to the mass production of large-scale manufacturing guided by the principles of scientific management. Developed by Fredrick Taylor in the 1880s and 1890s, scientific management used scientific principles to determine the most efficient ways to perform the work so that production was maximized and costs were minimized (Taylor 1967). It did so by separating owners from management and management from labor, and by standardizing and simplifying jobs. As a result, job control was taken away from those who performed the tasks, and jobs were filled by unskilled workers. Work structures became inflexible, ruling out alternative ways to meet job demands. Those who could not complete job tasks as assigned were considered incompetent and unemployable. Because those with disabilities were often among those who were unable to complete the tasks as assigned, they were equated with inability to work. In this context, and consistent with the medical model, individuals with disabilities became employable when they were "fixed" and were able to meet the requirements of work in the same way as their able-bodied coworkers. Assistive technologies were developed to help people function like those without disabilities (Ward and Baker 2005). If, however, the individuals with disabilities could not change to fit the workplace, they remained excluded from it.

The shift to a recovery perspective ran parallel to a shift in workplace organization from the task specialization of the assembly line to job enlargement and enrichment of the high technology industries, or the service orientation of a growing number of jobs. In the 1960s and 1970s scientific management came under fire. Foreign competition, particularly from Japan, demonstrated that alternative approaches to production could be more efficient and profitable (Radnor and Barnes 2007). At the same time, low unemployment rates in the United States, record numbers of women, immigrants, and other people of difference entering the workplace, and a social climate in which a more humanistic approach to work was advocated, pressured employers to become more concerned about the quality of the work environment and meeting the needs of a diverse labor force in order to attract qualified workers (Spataro 2005). Employer response to these factors was to move away from the job structure of scientific management to an alternative system of production that placed greater emphasis on the quality of work life. In part, this was achieved through job redesign that included job enlargement and job enrichment. Job enlargement responded to the monotony of standardized, de-skilled jobs by increasing the variety or breadth of tasks that individuals performed. Job enrichment increased individual autonomy, control of job tasks, and responsibilities.

Although job redesign added richness and interest to work, it also added complexity, such that workers could no longer be used interchangeably. Further, the redesign coupled with the increased labor force diversity resulted in a wide variation in how jobs were performed. Policy and practices emerged as employers' response to meeting the challenge of maintaining a cohesive, productive work force with higher-skilled workers from diverse backgrounds and with a range of different needs (Akabas and Gates 1993). These policies and practices were designed to break down the physical and social barriers that denied access to any worker. As such, they created a culture in which, and offered the mechanisms through which, all employees, including people with disabilities, could seek and secure accommodation (Akabas 1994; Ledman and Brown 1993; Schur et al. 2005; Ward and Baker 2005). From this perspective, accommodation for people with mental health conditions became a response to just one among many groups of difference who were qualified and capable, but who may have had unique needs that had to be met to enhance their ability to fulfill job requirements.

Fueled by the recovery perspective and the changing workplace structures, legislation solidified the recognition of the similar status of people with disabilities and other groups of difference in employment. The passage of section 504 of the 1973 Rehabilitation Act acknowledged for the first time that people with disabilities deserved civil rights protections similar to other groups of difference (Mayerson 1992). The Act banned discrimination on the basis of disability for organizations that were recipients of federal funding. The passage of the Americans with Disabilities Act (ADA) in 1990, like the Civil Rights Act of 1964, extended protections against discrimination to public spaces and employment. Title I of the Act provides protection against discrimination in all aspects of employment for qualified individuals who have impairment and/or individuals *perceived* as having impairment or having a *history* of impairment; and employers are mandated to provide reasonable accommodation to remedy inequalities (Harlan and Robert 1998). Thus, the ADA recognized that disability had a social component, because it could be defined by the perceptions of others as well as by the actual impairment. It further recognized that accommodation was not the sole responsibility of the individual to change to fit the workplace, but also the obligation of the employer to change the way it supports workers. Legal challenges supported by Supreme Court rulings began to narrow the interpretation of the ADA, such that those with a history or societal perception of a mental health condition that was controlled by medications or treatment did not come under the protection of the Act. In 2008, however, Congress passed an amendment that set aside the Supreme Court's ruling, indicating that in the passage of the ADA it was the intent of Congress to spread a broad net of protection even when conditions are well controlled (US EEOC 2008).

In sum, the recovery perspective transformed the service world to support employment as a goal for people with mental health conditions. This is evidenced by the development of vocational services to support individuals in integrated competitive employment. Concurrent with this transformation, increasing diversity in the labor force and the change in the nature of work prompted employers to develop policies and practices to include groups of difference while maintaining production standards. As such, the strategies employers adopted to include all groups of difference had the potential to also include people with disabilities (Gates 2000; Hall and

Hall 1994; Schur et al. 2005). The social definition of disability that is derived from the recovery perspective aligned people with disabilities with other groups of difference facing barriers to employment. That is, barriers to employment are a function of workplace responsiveness, as well as of individual characteristics. It also recast accommodation as a process that cannot occur in isolation, but one that must take into account the mutual and interactive impact of any workplace change. These multiple social forces created a climate in which accommodation for qualified candidates could take place and promised a reciprocal relationship between accommodation and the social context, thus emphasizing accommodation as a social process.

Accommodation as a Social Process

Figure 20.1 summarizes the key components of accommodation as a social process.

Assess need for accommodation: The process begins with identification of the need for accommodation. Many individuals with serious and persistent mental health

Accommodation as a Social Process

Assess Need for Accommodation to Job Tasks, Job Routines, Work Environment and Work Relationships

Assess Organizational Readiness to Accommodate Individuals with Mental Health Conditions

Match Need for Accommodation with Job Requirements And Relationships at Work

Disclose Mental Health Condition

Request/Negotiate Accommodation

Set Accommodation In Place: Assess mutual impact of accommodation on individual and work group and support effective communication

Monitor Effectiveness of Accommodation on Barriers Caused by the Mental the Health Condition and Relationships at Work

Fig. 20.1 Accommodation as a social process

conditions will need support to best meet their job requirements (Akabas and Gates 2000; Baldridge and Veiga 2001; Florey and Harrison 2000; Harris and Associates 1998; MacDonald-Wilson et al. 2002). Gaps in functional capacity or side effects of treatment often interfere with completing job tasks, maintaining job routines, building productive workplace relationships, and navigating and/or managing the physical environment (Akabas et al. 2006; Neff 1968). Common accommodations to respond to these gaps include task substitution, scheduling breaks or flextime, or office redesign (Baldridge and Veiga 2001). Frequently, however, people with mental health conditions find that symptoms or side effects of treatment interfere with work relationships (Akabas et al. 2006; Gates 2000; Ward and Baker 2005). For example, people with mental health conditions may have difficulty communicating, violate social norms of the workplace, or require additional emotional support. Their coworkers may harbor stereotypes of people with mental health conditions, avoid interaction with them, overscrutinize them, feel uncomfortable interacting with them, resent them, or hold unrealistic expectations of them. Accommodations that respond to these interdependencies are fundamental to a productive work group.

Assessment of the need for accommodation requires a structured instrument that helps individuals with mental health conditions and their service providers identify strengths that support employment goals, such as work-related skills and abilities, or other life experiences. It also needs to identify the vocational and nonvocational challenges to employment. Vocational challenges might be functional limitations that interfere with meeting job requirements and maintaining relationships, or gaps in education or work history needed for the person's career goals. Nonvocational challenges are problems presented in other domains of an individual's life that might undermine employment, such as unstable housing, substance abuse, criminal justice history, domestic violence, or physical health problems such as inadequate control of diabetes or high blood pressure.

Assess organizational readiness to accommodate: The willingness of an employer to hire and support people with disabilities in general, and people with mental health conditions in particular, is reflected in the organizational culture (Akabas 1994; Gates 2000; Spataro 2005). Organizational culture can be conceived as the shared values, norms, and assumptions that guide the way people think, feel, and behave in order to solve problems in specific workplaces (Schein 1992, 1999). Organizational culture can be discriminatory. In such workplaces, although policies and practices are compliant with legal mandates, discrimination is expressed through promotion of certain characteristics or abilities that are evaluated more highly than others, or denial of participation in decision-making and access to information for those who are different (Spataro 2005). Organizational culture can be assimilative. Newcomers are welcomed as long as they are like everyone else, or change themselves to be like everyone else. Commonality among workers is emphasized and valued over their differences and unique qualities. Finally, organizations can be inclusive. In these workplaces, difference is prized. Diversity is viewed as a source of strength that enables the work organization to learn, adapt, and thrive (Akabas and Gates 1993; Gummer 1994; Mor Barak 2000; Wooten and

James 2005). See further discussion of organizational culture and mental health in Chap. 21 (Kirsh and Gewurtz 2010) of this book.

Culture that supports and values groups of difference finds expression through management policies and practices that support diversity outcomes (Akabas 1994). These might include:

- Leadership that shows commitment to diversity. For example, rewarding supervisors for hiring that supports diversity goals demonstrates leadership commitment.
- Recruitment and hiring strategies that include people of difference and are responsive to their unique qualifications. For example, linkages to vocational rehabilitation programs would be an expression of employer interest in people of difference.
- Employment benefits that provide access to the supports enabling employment to be sustained. Examples of these benefits include workplace accommodations, child care, options for paid and unpaid leave, disability management, return-to-work programs, work/family benefits, or comprehensive health benefits.
- Employee training that ensures understanding of policies and practices. Training available to management, supervisors, and line staff that offers skills, combats stereotypes, and increases understanding, awareness, and acceptance of groups of difference.
- Job structure and performance appraisal that communicate the importance of employees' unique contributions. For example, jobs might allow for flexible work schedules (i.e., flex-time), telecommuting or working from alternative locations such as from home (i.e., flexiplace), job sharing, or other work options.

Although diversity policies and practices offer new potential for including qualified workers with mental health conditions into the workplace, their application is limited by employers who have been slow to define people with disabilities among those covered by their diversity policies and practices. For example, fewer than half of Fortune 100 companies have written polices that expressly include people with disabilities in their definition of diversity (Ball et al. 2005). Employers continue to make false, negative assumptions that people with disabilities perform less well than their nondisabled coworkers, have higher rates of turnover and absenteeism, poorer safety records, and require too many special privileges and accommodations (Akabas et al. 2003; Dixon et al. 2003; McFarlin et al. 1991; Stone and Colella 1996). It is important, therefore, for individuals with mental health conditions and their social service providers to gauge employers' openness to groups of difference and their flexibility regarding alternative ways to meet work expectations in general and, to the extent possible, in relation to individuals with mental health conditions specifically.

Match need for accommodation with job requirements: The need for accommodation can only be assessed with respect to a *specific* job. As noted above, limitations in one context may not be problematic in others. The matching process requires not only a thorough assessment of the individual, but also detailed knowledge of job requirements and relationships at work. The job description might be the best source of information about job requirements in anticipation of the need for accommodation when starting a new job. Once on the job, however, individuals with mental health conditions and their supervisors are the best source of information about what a job entails.

Social service providers who offer support to individuals can structure discussions to best understand job requirements by asking about the tasks of the job, the routines of the job, the physical environment, and relationships at work (Akabas et al. 2006; Neff 1968). In particular, it is important to develop a systematic way to assess relationships at work because of the likelihood that the symptoms of the conditions or side effects of treatment will influence relationships with supervisors, coworkers, or others, and because openness to the request for accommodation is affected by these relationships.

One approach to understanding relationships is through the application of social network mapping (Gates et al. 1998; Tracy and Whittaker 1990). This technique identifies who is in the work group. It also evaluates the quality of the relationship with each work group member as perceived by the individual with a mental health condition along a range of dimensions, including closeness; social support as defined by House (emotional, instrumental, informational, and feedback; 1981); the extent to which others are critical of the individual; and the intensity of the contact as measured by frequency and duration.

Disclose the mental health condition: Disclosure is a necessary condition for accommodation to take place. Disclosure begins the dialogue between the individual (who must be qualified for the position, other than needing the accommodation) and the employer about the mental health condition and the need for accommodation. Disclosure is part of the process of building effective lines of communication, understanding and trust. By definition, disclosure is interactive.

There are practical reasons why disclosure is important. Employers may not be aware of the need for accommodation unless it is revealed by the worker. Individuals' performances may be appraised based on disability rather than ability. Without disclosure, individuals may not be able to access the supports available to them at the workplace. The individual with a mental health condition is not covered by the protections of the Americans with Disabilities Act without disclosure (for a fuller discussion of disclosure, see Chaps. 1 (Center 2010) and 10 (MacDonald-Wilson et al. 2010) in this book).

Disclosure, however, is also essential to building relationships that are supportive of accommodation by enhancing the understanding of others. As such, disclosure is an element of recovery to the extent that it is an expression of feeling comfortable with oneself. It may provide a sense of empowerment by giving individuals control over what information they choose to share. These possible benefits of disclosure are often overlooked because of the potential risks disclosure carries. The negative fallout of disclosure may undermine work status, such as by denial of promotion, rejection for hire, termination, or being passed over for challenging assignments. There can be a personal cost as well. Individuals may be avoided, isolated, overscrutinized, and misunderstood by coworkers.

Social service providers, as well as vocational, rehabilitation, and mental health counselors, can all play an essential role in helping individuals to make the decision whether to disclose by helping them weigh its benefits and risks. Disclosure is a choice. Often, however, individuals do not make an informed choice because they are familiar only with the risks and do not understand the benefits or the extent to which benefits may outweigh risks. The fear of resentment, embarrassment, and the potential loss of self-esteem that might be caused by revealing the mental health condition mask the benefits of an accommodation that disclosure might provide. Supervisors' and coworkers' reactions affect the attitudes of people with mental health conditions toward the workplace and their willingness to disclose (Baldridge and Veiga 2001, 2006; Balser and Harris 2008). For example, individuals with disabilities will not seek accommodation and, in fact, would rather quit than ask for support, when they believe that their requests will be perceived as a burden on others.

Providers can help individuals separate fear from reality, and educate the workplace to build understanding. One place to begin is to help individuals develop a disclosure plan. To ensure that the disclosure is managed and under the control of the individual with a mental health condition, it is important to prepare for when to disclose, to whom to disclose, how to disclose, and what information to share.

As part of a request for accommodation, disclosure can be integrated into a strengths-based approach (Saleebey 2009). Thus, the disclosure becomes embedded in the presentation of an individual's abilities. The individual begins by stating his or her qualifications and interest in the job. Next, the individual offers the disclosure and the suggestion for an accommodation that might help overcome the limitation posed by the disability. Last, the individual restates his or her interest and qualifications. For example, an individual with a mental health condition seeking a job as an aide in a nursing home might say:

> This job seems perfect for me. I enjoy being with older people and have had a lot of experience caring for older people at my volunteer job at the community program run by my church. Working the afternoon shift and weekends is no problem for me. I live nearby and can get to work easily at any time. My only concern is that you mentioned that there are a lot of tasks to learn. I have a mental health condition that makes it so that sometimes I become confused when I need to learn many new things at once. If you could let me learn each task one at a time I am sure that I would have no problem doing this job. In fact, I think I would be a great fit for this place. I am patient, reliable, and really enjoy working with older people.

Requesting the accommodation: Asking for an accommodation needs to take account of the social nature of the request. The accommodation must meet the needs of the individual with a mental health condition. It must help the individual overcome the barriers caused by his or her condition that interfere with meeting job requirements. The accommodation must also, however, meet the needs of the employer (Gates et al. 2001). Employers are most concerned about their bottom line and how the accommodation impacts productivity (Baldridge and Veiga 2006). Also of concern is the possible impact of the accommodation on work group morale (Balser and Harris 2008). In this light, the request for accommodation becomes a negotiation that responds to employer and work group concerns while meeting the needs of the individual. It acknowledges that there is no one perfect accommodation, but rather many ways to reach the shared goal of a productive worker. The accommodation is limited only by the lack of flexibility and creativity of the individual and the employer.

Usually, the request for an accommodation is made to the supervisor. Supervisors are of key importance in the process because they are most likely to know best what a job involves. They also play an essential role in helping the work group to understand, accept, and implement the accommodation (Gates 2000; Gates et al. 1996, 1998; Mor Barak 2000; Schur et al. 2005; Stone and Colella 1996). Supervisors establish performance expectations, conduct performance appraisals, and set the culture of the work group. The way supervisors respond to people with disabilities can contribute to the quality of the accommodation as well as individuals' sense of belonging and acceptance or alienation (Anderson and Williams 1996; Blanck et al. 2003; Florey and Harrison 2000; Klimoski and Donahue 1997; McLaughlin et al. 2004).

Others at the workplace, however, might also need to be included. For example, the individual with authority to sanction the accommodation might also need to be part of the negotiation. This individual might be from the Human Resource Department or the Legal Department. In unionized settings, a union representative might need to be present to ensure that the accommodation does not violate a collective bargaining agreement. Someone from risk management may need to be part of the process if there are safety issues (e.g., a need for assistance in evacuating the site in the event of an emergency).

Setting the accommodation in place: Because accommodation is a social process, setting the accommodation in place begins with an assessment of the potential impact of the accommodation on the work group. Accommodation might affect the tasks, routines, or environment in which coworkers do their jobs. Accommodation might also affect the way coworkers relate to each other (Gates et al. 1998; Klimoski and Donahue 1997; MacDonald-Wilson et al. 2002). For example, coworkers often have unfavorable attitudes toward people with disabilities, based on the preconceived idea that these individuals are less competent, when the cause of disability is believed to be under the control of the individual (e.g., paraplegia due to a motorcycle accident) and/or when the disability is disruptive, visible, and unattractive (Colella 2001; Colella and Stone 2005; McLaughlin et al. 2004). Further, coworkers are unsupportive to the extent that they are unfamiliar with people with disabilities (e.g., hold beliefs that a disability is contagious); or perceive that the individual with a disability will negatively affect either their ability to get their own jobs done or their status in the workplace, or that the accommodation received by the individual with disability constitutes special treatment that is undeserved (e.g., stigmatize those with disabilities). Based on these factors, people with mental health conditions are often viewed least favorably by coworkers, and this, in turn, might affect the coworkers' acceptance of the need for accommodation by workers with mental health conditions (Colella 2001).

Thus, setting accommodations in place should rely on the social process of effective communication among the supervisor, the individual with a mental health condition, and the work group, in order to dispel myths and misperceptions and clarify roles and responsibilities. Communication may be formal and achieved through training or supervision. It may be informal, through opportunities for interaction created by the supervisor. Central to effective communication, however, is ensuring that each staff person has access and is included in the communication network. For example, if the work group communicates primarily through email, then it is essential that all staff have access to a computer. If the mental health condition interferes with using a computer, then training is needed to remove this limitation, even if using a computer is not a stated requirement of the job. To have a truly inclusive workplace, people with mental health conditions must be "plugged into" the informal as well as the formal requirements of the job.

In sum, setting the accommodation in place is a key step that determines accommodation effectiveness by ensuring that all understand and accept how the accommodation affects their work, thereby creating a culture that is welcoming and participatory rather than resentful and stigmatizing. As a result, the outcome is one in which the work group is enriched, and each individual finds that there are unforeseen benefits from the accommodation for them, as well the assurance that they, too, will be accommodated if conditions require such action.

Monitoring accommodation effectiveness: Accommodation is not a one-shot deal. The need for accommodation can change when the symptoms of the condition or side effects of treatment change, or when the nature of the job or relationships at work change. When accommodation is recognized as a social process, there is the ongoing need to evaluate whether or not the accommodation best meets the needs of the worker with a mental health condition and his or her supervisor and work group. Effective accommodation occurs when the accommodation appropriately matches the gap in the worker's capacity with the specific component of the job. Some individuals are mismatched. For example, they receive accommodations to job tasks or routines when the gap in capacity is their inability to maintain work relationships. Some individuals are under-accommodated. In this instance, not all gaps in capacity are addressed by the accommodations. Finally, some individuals are over-accommodated. For these individuals, employers modify more aspects of their jobs than are needed. As a consequence, the individuals' sense of self-esteem and self-efficacy may be undermined.

Evaluation of accommodation effectiveness may lead the individual with a mental health condition to the conclusion that it is time to leave the current job. Because there is a tendency for providers to try to prevent this outcome, it is important to assess why an individual may decide to leave a job. In their effort to promote job retention, providers can lose sight of the normal job growth and transition that is a natural part of developing and following a career path. For some, the best "accommodation" may mean moving to another job.

The Significant Role of Communication: A Case Example

One way to foster effective communication is through training and consultation facilitated by a social service, or a vocational, rehabilitation, or mental health provider (Gates and Akabas 2007; Gates et al. 1998). Such providers are well suited to serve as facilitators of work group communication, because of their expertise related to mental health and their ability to understand the needs of both the workplace and the individual with a mental health condition. Training and consultation provides education to the work group about what it is like to be a worker with a mental health condition, and gives an opportunity for the group to identify challenges to inclusion and set goals to respond to the challenges that are specific to their work setting. It also offers a structured way to present the accommodation that has been proposed to respond to a specific limitation caused by the mental health condition, helps the group to clarify how the accommodation might affect their roles and responsibilities, and provides an opportunity for work group members to ask questions and share how they feel.

The authors of this chapter are investigating strategies for effective communication in order to promote the inclusion of peer providers on staff. One employment option for people with mental health conditions is to work in the position of peer provider. Peer providers are people with mental health conditions who may not have professional credentials, but who have direct experience with the mental health system as recipients of care. They are hired by community social service agencies as role models, counselors, educators, providers of assistance to meet the needs of daily living (such as housing and work), and as advocates to empower people with mental health conditions in the mental health system. Six community-based social service agencies in New York City volunteered to participate in the study, based on their interest in improving peer employment outcomes. Each had at least one work group with a peer on staff and that work group(s) was offered training and consultation to better include the peer.[1]

The training normalized the experience of people with mental health conditions at work by presenting the concept of an inclusive workplace and the role of diversity policies and practices in meeting the needs of all groups of difference in the workplace. It helped participants become aware of the challenges to employment faced by individuals with mental health conditions, and the misperceptions about workers with mental health conditions frequently held by their coworkers. Finally, it offered an approach for responding to challenges through work group strategies that clarified roles and built relationships.

The training was followed by regular consultation with the entire work group to establish effective lines of communication among staff, provide a forum for clarifying job roles and responsibilities, and compare the application of policies to peer providers and nonpeer staff. For three months, work groups met with the

---

[1]This study was funded in part by a grant from the Jacob and Valerie Langeloth Foundation, entitled *Peer Providers in Social Service Agencies: Creating Work Settings for Mutual Support*.

researchers every 2 weeks, followed by monthly meetings for 9 months. The first sessions were devoted to thinking through the application of the training material to their specific work groups. Based on the challenges they perceived for their groups, together they set goals to resolve them and developed specific indicators which signaled to the work group that the goals had been met. Examples of those issues included identifying: unrealistic expectations by nonpeer staff of what peer providers should do; inconsistent application of human resource and treatment policies to peer staff; poor communication to agency clients about the role of peer staff; lack of formal opportunities to communicate (e.g., staff meetings or formal supervision); and lack of opportunities for peer staff to participate in decisions related to their jobs. In short, these issues represented a breakdown of the social processes at work. Goals to respond to these challenges included: (1) review and discussion of job descriptions by the group in staff meetings and in individual supervision; (2) review of the human resource manual of workplace policy and procedures, to identify policies that were inconsistently applied and to revise practices to ensure equal treatment; (3) formalizing meeting and supervision times; (4) developing educational materials for clients about the role of each staff person; and (5) providing more control over disclosure among peer staff by allowing them to decide when to use personal experience as part of their support to clients.

Conclusions and Recommendations

Recovery is currently the leading perspective guiding service programs and treatment approaches for people with mental health conditions (Ralph 2000). From this perspective, the goal of services is to instill hope and to support individuals in making their own decisions in the process of achieving full integration into the mainstream of everyday life (Anthony 1993, 2000; Fisher 2003). Recovery goals are personal, varying for each individual. For many, however, recovery includes working competitively in the community. Work carries the same meaning for people with mental health conditions as it does for the general population. It is not only a source of financial independence and self-sufficiency, but an important element to physical health, a source of social interaction, the provider of a sense of purpose and meaning to one's life (Akabas and Kurzman 2005; House et al. 1988; O'Day and Killeen 2002).

For the goal of full inclusion into employment to be realized, many qualified individuals with serious persistent mental health conditions need accommodation. The effectiveness of the accommodation is dependent upon the extent to which it improves the fit between the individual with the mental health condition and the workplace. Thus, viewing accommodation as a social process activates organization-level policy and practices (e.g., performance appraisal, staff training and development opportunities, recruitment and hiring systems), as well as interaction and interdependencies among work group members. An inclusive workplace culture and the strategies employers adopt through their diversity policies and practices can support people with mental health conditions in reaching their employment goals.

Workplace characteristics are related to the likelihood that employers adopt inclusive policies and practices (Akabas et al. 2003; Bruyère et al. 2006; McMahon et al. 2008). Size is important. According to 2004 figures, small businesses in the United States (fewer than 500 employees) represent 99.7% of all employers and employ over half of the private workforce. Thus it is very likely that people with mental health conditions will find jobs with smaller employers (Bruyère et al. 2006). These companies tend not to have separate Human Resource (HR) departments, and HR functions are more loosely organized (Akabas et al. 2003; Bruyère et al. 2000). For example, HR functions are allocated to several employees who also have other roles in the company. As a consequence, specific diversity policies and practices may not be in place, and the workplace culture may not include openness to groups of difference. It is not uncommon to find small business owners who prefer to hire family members or others with whom they share the same race or ethnicity. Different employers can further affect willingness to accommodate. Surprisingly, not-for-profit employers can be more inflexible in their willingness than for-profit employers, because of the increased pressure for accountability in productivity that is the result of their operational transparency (Akabas et al. 2003).

Size and industry sector, however, do not free employers from the need for strategies to respond to disability. Mental health conditions can affect anyone at any time. Small employers as well as large employers, for-profit firms as well as not-for-profit organizations, need to ensure a productive workforce by understanding how to respond. Thus social service and other providers can play an essential role in facilitating accommodation as a social process, regardless of employer characteristics. At each step in the process, providers can help individuals with mental health conditions reach their employment goals by creating an understanding of the mutual impact of the accommodation on the individuals and the work group, helping employers apply their existing policies and practices to people with mental health conditions fairly, and providing people with mental health conditions with the skills to empower them in the workplace. Among smaller employers, the lack of formal policies can lead to greater flexibility. Among larger employers, providers and individuals with mental health conditions may be more likely to find structures and practices in place to facilitate the accommodation process. Table 20.1 summarizes services that providers can offer to support the accommodation of individuals with mental health conditions across employment settings.

Accommodation as a social process is fluid. The employment experience of people with serious mental health conditions continues to be a search for acceptance by and inclusion into the world of work. Social services, vocational, rehabilitation, and mental health providers can help individuals reach their employment goals, and assist employers to successfully include people with mental health conditions, when they understand accommodation as a social process. From this perspective, accommodation for people with mental health conditions is no different from the path others take to inclusion into the world of work. The outcome is a workplace enriched by the diversity of experience and the mutual understanding it brings.

Table 20.1 Services that support accommodation as a social process

Comprehensive individual assessment of employment and nonemployment factors that have the potential to interfere with work expectations

Workplace analysis of the readiness to support groups of difference generally, and people with mental health conditions specifically

General consultation/training for employers about the employment of people with mental health conditions and the accommodation process

Specific job analysis to identify gaps between individual capacity and required job tasks, job routines and environmental demands

Social network mapping to determine the mutual impact of accommodation on the work group and the individual with a mental health condition

Disclosure planning that helps individuals think through when to disclose, how to disclose, to whom to disclose, and what information to share

Workplace intervention, including accommodation negotiation and specific training and consultation for the work group of the individual with a mental health condition

Evaluation and monitoring of accommodation effectiveness

References

Akabas SH (1994) Workplace responsiveness: key employer characteristics in support of job maintenance for people with mental illness. Psychosoc Rehabil J 17(3):91–102

Akabas SH, Gates LB (1993) Managing workplace diversity. In: Klein J, Miller J (eds) The American edge. McGraw Hill, New York, pp 113–135

Akabas SH, Gates LB (2000) A social work role: promoting employment equity for people with serious and persistent mental illness. Adm Soc Work 23(3/4):163–184

Akabas SH, Gates LB, Koball G, Imperiali B (2003) Promoting effective connections between mental health care providers and employers. http://www.coalitionny.org/the_center/resources/. Accessed 24 Oct 2008

Akabas SH, Gates LB, Oran-Sabia V (2006) Work opportunities for rewarding careers: insights from implementation of a best practice approach toward vocational services for mental health consumers. J Rehabil 72(1):19–26

Akabas SH, Kurzman P (eds) (2005) Work and the workplace. Columbia University Press, New York

Anderson SE, Williams LJ (1996) Interpersonal, job and individual factors related to helping processes at work. J Appl Psychol 81(3):282–296

Anthony WA (2000) A recovery-oriented service system: setting some system level standards. Psychiatr Rehabil J 24(2):159–168

Anthony WA (1993) Recovery from mental illness: the guiding vision of the mental health service system in the 1990s. Psychiatr Rehabil J 16(4):11–23

Baldridge DC, Veiga JF (2001) Toward a greater understanding of the willingness to request an accommodation: can requesters' beliefs disable the Americans with Disabilities Act? Acad Manage Rev 26(1):85–99

Baldridge DC, Veiga JF (2006) The impact of anticipated social consequences on recurring disability accommodation requests. J Manag 32:158–179

Ball P, Monaco G, Schmeling J, Schartz H, Blanck P (2005) Disability as diversity in Fortune 100 companies. Behav Sci Law 23:97–121

Balser DB, Harris MM (2008) Factors affecting employee satisfaction with disability accommodation: a field study. Employee Responsib Rights J 20:13–28

Blanck P, Schur L, Kruse D, Schwochau S, Song C (2003) Calibrating the impact of the ADA's employment provisions. Stanford Law Pol Rev 14:267–290

Bond GR (2004) Supported employment: evidence for an evidence-based practice. Psychiatr Rehabil J 27(4):345–359

Bruyère S, Erickson WA, VanLooy S (2000) HR's role in managing disability in the workplace. Employ Relat Today 27(3):47–66

Bruyère S, Erickson WA, Van Looy SA (2006) The impact of business size on employer ADA response. Rehabil Couns Bull 49(4):194–206

Bruyère S (2000) Disability employment policies and practices in private and federal sector organizations. Cornell University, Program on Employment and Disability, Ithaca, NY

Center C (2010) Law and job accommodation in mental health disability. In: Schultz I, Sally Rogers E (eds) Handbook of work accommodation and retention in mental health. Springer, New York

Colella A (2001) Coworker distributive fairness judgments of the workplace accommodation of employees with disabilities. Acad Manage Rev 26:100–116

Colella A, Stone D (2005) Workplace discrimination toward persons with disabilities: a call for some new research directions. In: Dipboye RL, Colella A (eds) Discrimination at work. The psychological and organizational bases. New Jersey, Lawrence Erlbaum Associates, pp 227–253

Cook JA, Burke J (2002) Public policy and employment of people with disabilities: exploring new paradigms. Behav Sci Law 20:541–557

Deegan G (2003) Discovering recovery. Psychiatr Rehabil J 26(4):368–376

Deegan MJ (1985) Multiple minority groups: a case study of physically disabled women. In: Deegan MJ, Brooks NA (eds) Women and disability: the double handicap. Transaction, New Brunswick, NJ, pp 36–55

Dixon KA, Kruse D, Van Horn E (2003) Work trends: restricted access: a survey of employers about people with disabilities lowering barriers to work. Rutgers, the State University of New Jersey, John J. Heldrich Center for Workforce Development

Fisher D (2003) People are more important than pills in recovery from mental disorder. J Humanist Psychol 43(2):65–68

Florey AT, Harrison DA (2000) Responses to informal accommodation requests from employees with disabilities: multistudy evidence on willingness to comply. Acad Manage J 43:224–233

Gates LB (2000) Workplace accommodation as a social process. J Occup Rehabil 10(1):85–98

Gates LB, Akabas SH (2007) Developing strategies to integrate peer providers into the staff of mental health agencies. Adm Policy Ment Health Serv Res 34:293–306

Gates LB, Akabas SH, Kantrowitz W (1996) Supervisors' role in successful job maintenance: a target for rehabilitation counselor efforts. J Appl Rehabil Counsel 27(3):60–66

Gates LB, Akabas SH, Oran-Sabia V (1998) Relationship accommodations involving the work group: improving work prognosis for persons with mental health conditions. Psychiatr Rehabil J 21(3):264–272

Gates LB, Akabas SH, Zwelling E (2001) Have I got a worker for you: creating employment opportunities for people with psychiatric disability. Adm Policy Ment Health 28(4): 319–325

Gummer B (1994) Managing diversity in the work force. Adm Soc Work 18(3):123

Hahn H (1985) Introduction: disability policy and the problem of discrimination. Am Behav Sci 28(3):293–318

Hall FS, Hall EL (1994) The ADA: going beyond the law. Acad Manag Exec 8(1):17–32

Harlan SL, Robert PM (1998) The social construction of disability in organizations. Why employers resist reasonable accommodation. Work Occup 25(4):397–435

Harris L & Associates (1998) NOD/Harris survey of Americans with disabilities. Louis Harris & Associates, Washington, DC

House JS (1981) Work stress and social support. Addison-Wesley, Reading, MA

House JS, Landis KR, Umberson D (1988) Social relationships and health. Science 241(4865):540–545

Kirsh B, Gewurtz R (2010) Organizational culture and work issues for individuals with mental health disabilities. In: Schultz I, Sally Rogers E (eds) Handbook of work accommodation and retention in mental health. Springer, New York

Klimoski R, Donahue L (1997) HR strategies for integrating individuals with disabilities into the work place. Hum Resour Manage Rev 7(1):109–138

Ledman R, Brown D (1993) The Americans with Disabilities Act: the cutting edge of managing diversity. SAM Adv Manag J 58(2):17–20

MacDonald-Wilson KL, Rogers ES, Massaro JM, Lyass A, Crean T (2002) An investigation of reasonable workplace accommodations for people with psychiatric disabilities: quantitative findings from multi-site study. Community Ment Health J 38(1):35–50

MacDonald-Wilson KL, Russinova Z, Rogers ES, Lin CH, Ferguson T, Dong S, Kash-MacDonald M (2010) Disclosure of mental health disabilities in the workplace. In: Schultz I, Sally Rogers E (eds) Handbook of work accommodation and retention in mental health. Springer, New York

Mayerson A (1992) The history of the ADA: a movement perspective. http://www.dredf.org/publications/ada_history.shtml. Accessed 3 June 2008

McFarlin DB, Song J, Sonntag M (1991) Integration of the disabled in the work force: a survey of Fortune 500 company attitudes and practices. Employee Responsibil Rights J 4(2):107–123

McLaughlin ME, Bell MP, Stringer DY (2004) Stigma and acceptance of persons with disabilities: understudied aspects of workforce diversity. Group Organ Manag 29(3):302–333

McMahon BT, Rumrill PlD, Roessler R, Hurley JE, West SL, Chan F, Carlson L (2008) Hiring discrimination against people with disabilities under the ADA: characteristics of employers. J Occup Rehabil 18(2):112–121

Mor Barak M (2000) Beyond affirmative action: toward a model of diversity and organizational inclusion. Adm Soc Work 23(3/4):47–68

National Organization on Disability (2003) Employment facts about people with disabilities in the United States. http://nod.org. Accessed 8 July 2008

Neff WS (1968) Work and human behavior. Atherton, New York

O'Day B, Killeen M (2002) Does U. S. federal policy support employment and recovery for people with disabilities? Behav Sci Law 20:559–583

Radnor ZJ, Barnes D (2007) Historical analysis of performance measurement and management in operations management. Int J Prod Perform Manag 56(5/6):384

Ralph RO (2000) Review of recovery literature: a synthesis of a sample of recovery literature 2000. Portland, Maine: University of Southern Maine, Edmund S. Muskie School of Public Service. National Technical Assistance Center for State Mental Health Planning, National Association for State Mental Health Program Directors

Saleebey D (2009) The strengths perspective in social work practice, 5th edn. Pearson Education, Inc, New York, NY

Schein EH (1992) Organizational culture and leadership, 2nd edn. Jossey-Bass, San Francisco

Schein EH (1999) The corporate culture survival guide. Jossey-Bass, San Francisco

Schur L, Kruse D, Blank P (2005) Corporate culture and the employment of persons with disabilities. Behav Sci Law 23:3–20

Spataro SE (2005) Diversity in context: how organizational culture shapes reactions to workers with disabilities and others who are demographically different. Behav Sci Law 23:21–38

Stone D, Colella A (1996) A model of factors affecting the treatment of disabled individuals in organizations. Acad Manage Rev 21:352–401

Taylor FW (1967) The principles of scientific management. The Norton Library, New York

Tracy EM, Whittaker JK (1990) The social network map: assessing social support in clinical practice. Fam Soc 71(8):461–470

U.S. Census Bureau (2002a) Disability: Selected characteristics of persons 16 to 74. http://www.census.gov/hhes/www/disable/cps/cps102.html. Accessed 8 July 2008

U.S. Census Bureau (2002b) Disability labor force status-work disability status of civilians 16 to 74 years old, by educational attainment and sex. http://www.census.gov/hhes/www/disable/cps/cps202.html. Accessed 8 July 2008

U.S. Equal Employment Opportunity Commission. Notice Concerning the Americans with Disabilities Act (ADA) Amendments Act of 2008. http://www.eeoc.gov/ada/amendments_notice.html. Accessed 24 Oct 2008 from the U.S. Equal Employment Opportunity Commission

Ward AC, Baker PMA (2005) Disabilities and impairments: strategies for workplace integration. Behav Sci Law 23:143–160

Wooten LP, James EH (2005) Challenges of organizational learning: perpetuation of discrimination against employees with disabilities. Behav Sci Law 23:123–141

Chapter 21
Organizational Culture and Work Issues for Individuals with Mental Health Disabilities

Bonnie Kirsh and Rebecca Gewurtz

What Is Organizational Culture?

Organizational culture is an important construct in understanding the experiences of individuals in the workplace. Defined as the shared values, beliefs, and expectations among members of an organization (Moran and Volkwein 1992; Spataro 2005), the culture of an organization shapes much of what occurs within it and acts as an informal system of control. Specifically, organizational culture dictates norms, rules of behavior, expectations of members, how problems and challenges are addressed, and the integration of new members and members who are different (Spataro 2005). The culture of an organization emerges through the complex and continuous communication and interaction among members both within and across organizational structures (Keyton 2005). Schein (1992) refers to culture as "a basic set of assumptions that defines for us what we pay attention to, what things mean, how to react emotionally to what is going on, and what actions to take in various kinds of situations" (p. 22). In short, organizational culture has been identified as the central factor that holds the modern organization together (Goffee and Jones 1996).

Most scholars agree that organizational culture is a multidimensional, multilevel construct. Values, norms, and customs are usually considered the fundamental level of organizational culture and are often described as "taken-for-granted" or entrenched beliefs, assumptions, and unspoken rules that guide an organization (Schein 1992; Schur et al. 2005). These basic values are shared among members and taught to new members, but are not tangible or visible within the organization. A second level of components embodied by organizational culture includes qualities such as preferences, goals, and strategies that are more explicitly stated within organizations (Keyton 2005; Schein 1992; Schur et al. 2005). This level includes, for example, mission statements and philosophical principles that guide

B. Kirsh (✉)
Department of Occupational Science and Occupational Therapy, University of Toronto,
500 University Avenue, Toronto, ON, Canada, M5G 1V7
e-mail: bonnie.kirsh@utoronto.ca

I.Z. Schultz and E.S. Rogers (eds.), *Work Accommodation and Retention in Mental Health*,   393
DOI 10.1007/978-1-4419-0428-7_21, © Springer Science+Business Media, LLC 2011

an organization. A third level is often described as the artifacts of organizational culture, which consists of the visible and tangible things that an outsider would notice (Keyton, 2005; Schein, 1992; Schur et al. 2005). In addition to the physical attributes of the organization, these artifacts include social conventions, patterns of communication, routines, and procedures.

Despite its prominence in the operation and structure of an organization and the enormous influence it can have on how problems are solved and the experiences of individual members, much of the culture of an organization remains undocumented and difficult to discern (Keyton 2005). Most members would be unable to describe the culture of their organization despite being intimately influenced by it in their day-to-day work. Observers would have difficulty identifying important aspects of the culture of an organization except for overt behavioral manifestations such as the treatment of members, and tangible artifacts such as customs and social conventions. Even the features of an organization that are visible and tangible might be difficult to interpret and understand, because the underlying values and assumptions are not always clear (Keyton 2005). Although the different levels of organizational culture usually strengthen and reinforce each other, they can also conflict. For example, the unspoken values and rules may not be consistent with the stated goals or the physical attributes of an organization (Schur et al. 2005). An expressed commitment to hire individuals with disabilities in an organization might be part of the mission and philosophy of an organization. However, this goal might not be consistent with the norms of the organization or the physical environment, which remain intolerant and inaccessible to persons with disabilities.

Why Does Organizational Culture Matter?

Organizational culture can offer much insight into the way different members are perceived and treated and how differences among members are tolerated. In organizations that operate within a culture of differentiation, individuals who possess valued characteristics (they may well be white, able-bodied males) will enjoy more status and power than others (Robert 2003; Spataro 2005). Other organizations may strive to unite their diverse members under a common identity of belonging to the organization. In such organizations, differences are suppressed and the focus is on unity and collectivism. Although this focus may result in less conflict and improved cooperation, it can also reduce creativity and innovation, because different perspectives are downplayed (Chatman et al. 1998; Spataro 2005). Another type of organizational culture, one of integration, is focused on valuing differences and incorporating different perspectives into task completion. Although this last type of organizational culture may lead to increased conflict, it can also offer opportunities for innovation and discovery (Spataro 2005; Stone and Colella 1996). In organizations grounded within a culture of integration, differences among members are valued for the contribution they can make to overall performance and the new insights that are gained through collaboration among individuals with different ideas and backgrounds.

The culture of an organization determines which characteristics or qualities of workers are salient and which are likely to be overlooked or seen as insignificant (Spataro 2005). For example, gender differences or differences in physical functioning might be highly relevant in one organization but not in another. Such variability might depend on the values and norms, the mission statement and goals, and the physical attributes and social conventions of the organization. The ways in which differences and diversity are defined and perceived, and their impact on the day-to-day operations of an organization or the experiences of individual members, are embedded in the culture of the organization and can vary between organizations.

Organizational cultures have been shown to affect workers' commitment to and identification with the group and organization, as well as their sense of involvement with their work assignments (Etzioni 1961; Wiener and Vardi 1990). Schein (1984) pointed out that ultimately what makes it possible for people to function comfortably with each other and to concentrate on their primary task is consensus on the cultural assumptions. As stated by Hagner and DiLeo (1993), "the better we understand the dynamics of workplace cultures, the better we will be able to assist employees, including those with disabilities, to become socially included at work. The clues to inclusion are found in understanding the culture" (p. 30). The significance of organizational culture – of values, practices, beliefs, and worker perceptions – are critical to full inclusion and integration of any person entering a work situation.

Person-Culture Fit

Researchers have found significant relationships between the "fit" of employees with the prevailing organizational culture, and important work outcomes (O'Reilly 1989). Wilkins and Ouchi (1983) proposed that organizations have cultures that hold varying degrees of attractiveness for different types of individuals. The notions that particular kinds of individuals are attracted to particular organizations, and that those who do not fit an organization soon leave, have been explained by Schneider (1990) in his attraction-selection-attrition model. It may be the case that individuals are attracted to organizations they perceive to have values consistent with their own, and that employees adapt and adjust better to their work environment when the organization's characteristics match their personal orientations (Bretz and Judge 1994). By the same token, organizations may recruit individuals who share their values (Schneider 1987).

Many researchers (Chatman 1989, 1998; O'Reilly et al. 1991; Schneider 1987) have argued that the fit between personal and organizational values is an important determinant of whether employees stay on the job. A study by

O'Reilly et al. (1991) suggested that the level of congruence between an organization's culture and the value preferences of new employees is a predictor of turnover; when newcomers exhibited a profile of values close to that of their organization, their subsequent organizational commitment, job satisfaction, and likelihood to stay with the company tended to be higher. Vandenberghe (1999) replicated this research in the healthcare sector in Belgium and found that the congruence between the values of the hospital and the nursing recruits' preferred values was predictive of nurses staying with their organization one year after congruence was measured. The authors advise healthcare administrators to "incorporate value congruence between nursing applicants and the whole hospital as a criterion in hiring decisions" (p. 183). Kopelman and colleagues (1990) agreed that organizational culture values may be related to employee retention, explaining that an organization's cultural values influence its human resource strategies, including selection and placement policies, promotion and development procedures, and reward systems. In turn, these strategies result in psychological climates that foster varying levels of commitment and retention among employees. A study by Sheridan (1992) builds on this premise; it examined the relationship of organizational culture and retention of employees by following the retention of 904 college graduates hired into six public accounting firms over a 6-year period. Three firms were characterized as having a culture that emphasized the interpersonal relationship values of team orientation and respect for people, while two other companies were characterized as having a culture that emphasized the work task values of detail and stability. Professionals hired in the companies that emphasized the interpersonal relationship values stayed 14 months longer (45 months) than those hired in the firms emphasizing the work task values (31 months). Indeed, both strong and weak performers stayed much longer in the organizational culture emphasizing interpersonal relationships than in the work task culture. The author concluded that "the most parsimonious explanation of employee retention may simply be that an organizational culture emphasizing interpersonal relationship values is uniformly more attractive to professionals than a culture emphasizing work task values" (Sheridan 1992, p. 1052). He thus advises managers to foster cultural values that are attractive to most new employees rather than be concerned with the selection and socialization of individuals who fit a specific profile of cultural values.

Kerr and Slocum (1987), on the other hand, maintain that organizational culture values may moderate differences in the retention rates of strong and weak performers. They suggest that organizations with cultures that emphasize values of teamwork, security, and respect for individual members will foster loyalty and long-term commitment to the organization among all employees, regardless of their job performance. In contrast, organizations with cultures that emphasize personal initiative and individual rewards foster an entrepreneurial norm, whereby the organization does not offer long-term security and the employees do not promise loyalty. They suggested that weaker performers would soon leave such a culture, and stronger performers would stay in order to "exploit the organization until better rewards could be gotten elsewhere" (Kerr and Slocum 1987, p. 103). Therefore, these authors claim that some organizational cultures may yield high commitment for both strong and weak performers, while retention in other cultures may vary greatly according to the employees' job performance.

There is also evidence suggesting that the fit of the organizational culture with individual values can impact job satisfaction (Shore and Martin 1989). Silverthorne's (2004) study of person-organizational fit, conducted in Taiwan, indicates that fit is a key element in employee job satisfaction and organizational commitment. The researcher assessed different types of organizational culture and demonstrated that bureaucratic organizational cultures resulted in the lowest levels of job satisfaction and organizational commitment; innovative cultures were next highest and supportive cultures had the highest level of employee job satisfaction and organizational commitment.

Sosa and Sagas (2006) focused on organizational culture and its effects on job satisfaction in the context of intercollegiate athletics. Specifically, an organizational subculture was analyzed to determine if it influenced the job satisfaction of athletic academic administrators. The findings of this study suggest that distinct organizational cultures are distinguished by variations in competitiveness, social responsibility, supportiveness, innovativeness, emphasis on rewards, performance orientations, and stability. The authors suggest that a stronger organizational culture allows employees to realize greater job satisfaction, which should ultimately lead to increased productivity, commitment, and success of the organization.

Organizational Culture and Special Populations

Research on components of organizational culture as they impact the employment of special populations is useful in generating applications to the mental health population. Studies linking organizational culture with gendered experience have uncovered how behavioral expectations for women can be rooted in assumptions of the work culture (Cassell and Walsh 1997), and have illustrated ways in which cultural environments prevent women from "flourishing" (Cockburn 1991). Research into organizational culture and persons with disabilities further emphasizes the importance of culture in work integration. In their study of positive workplaces for persons with disabilities, Ochocka et al. (1994) found that successful employment was largely determined by the quality of fit between the individual and the workplace culture, and maintained that integration for people with disabilities is enhanced by an understanding of culture. Similarly, Hagner and DiLeo (1993) emphasized the importance of culture in "helping people with severe disabilities to achieve quality employment outcomes, as well as for employers wishing to better utilize diverse and traditionally untapped sources of labour" (p. 29).

Attention to organizational culture and its impact on employment for persons with disabilities is growing as the focus on inclusion and human rights mounts. Stone and Colella (1996) explored the factors that affect the treatment of individuals with disabilities in organizations, including the type of work they are assigned and the opportunities they are given for mentoring, advancement, and promotion. They developed a model that highlights the importance of the interaction between personal characteristics (i.e., attributes of individuals in the workplace), societal characteristics (i.e., relevant legislation), and organizational characteristics (i.e., components of the organizational culture) in determining how individuals with disabilities are treated at work and the opportunities they are afforded. They specifically highlight how false assumptions and perceptions about the abilities of persons with disabilities can influence the type of people who are recruited and hired in an organization, the nature of the interactions that occur between coworkers and supervisors, and the experiences of persons with disabilities at work.

Increasingly, research points to the impact organizational culture can have on attitudinal, behavioral, and physical barriers for workers with disabilities. There is general agreement that an emphasis on individualism, self-reliance, and competitive achievement prevents persons with disabilities from demonstrating their work potential, while an emphasis on cooperation, helpfulness, and social justice paves the way for realization of potential and contribution to the organization (Schur et al. 2005; Stone and Colella 1996). Schur et al. (2005) reviewed literature that addresses the relationship between social factors within organizations and the employment experiences of people with disabilities. These social factors affect the treatment of people with disabilities in the workplace, including the socialization of new employees, performance evaluations, return-to-work policies, and the provision of accommodations. The authors point to ways in which organizational values are expressed in policies and procedures that restrict employees with disabilities; as an example, a company's reliance on job analyses that identify ideal job characteristics rather than essential ones may exclude workers with disabilities. As stated by Boyle (1997), "the biggest problem is not the unsuitability of jobs but rather finding an organization that is willing to break the mold and allow individuals with disabilities a chance to prove their capabilities" (p. 264–265).

Research with employers and persons with disabilities has pointed to a number of workplace variables affecting employment. A study of conditions which enable or impede the performance of work by persons with multiple sclerosis points to human support (assistance with tasks, emotional support), health promotion (intermittent rest periods, absence of stress), and person-environment interaction (organization of the physical environment, self-pacing), as elements of the workplace which contribute to success (Gulick et al. 1989). Findings regarding the role of the supervisor in adjustment to work with a disabling condition have highlighted the importance of treating workers fairly, including them in decision-making and utilizing workers' skills (Gates 1993).

Organizational Culture and Mental Health Consumers

The culture and climate of the workplace have been found to be important determinants of employment outcomes for consumers of mental heath services. Kirsh (2000a) used the Organizational Culture Profile with mental health consumers to assess desired characteristics of the work environment and found a significant relationship between employment status and workplace climate. The level of "fit" between the values of persons and work environments, as reflected by person-environment fit scores, were important factors associated with continued employment for consumers.

Not unlike workplaces for other persons, desirable work sites for persons with mental health problems have been described as those that attend to employees' needs and that celebrate diversity (Akabas 1994). Influential components of the workplace have included both structural aspects, such as proximity to home and potential for flexible work schedules, as well as psychosocial characteristics, including open patterns of communication and trust (Kirsh 1996). In an article chronicling 17 accounts of consumers in employment projects, Secker and Membrey (2003) identified a number of workplace factors associated with job retention, such as flexibility, supportive interpersonal relationships, and approaches to staff management. Specifically, helpful approaches to staff management were described as including: (1) expressing a genuine interest in employees' welfare, (2) setting clear boundaries about getting the job done, and (3) providing positive feedback and constructive criticism. A study by Banks et al. (2001), which examined supported employment outcomes, indicated that social interaction and natural workplace supports improve employment outcomes for individuals with mental health disabilities. Wilgosh (1990) identified a supportive coworker climate as a factor most critical to developing "survival" skills, in her exploration of organizational climate and workers with mental health disabilities; and Nieuwenhuijsen (2004) showed that frequent communication between supervisors and employees on mental health leave was associated with a shorter duration of the sick leave until full return-to-work (with the exception of those with high levels of depression). Krupa's (2007) review of workplace interventions supports these findings about the importance of a supportive workplace, and includes additional features associated with mental health and mental illness, such as opportunities for control and decision-making, the full use of worker capacities and skills, employee involvement in the workplace, reasonable and well-integrated job demands, clear and predictable work expectations and conditions, interpersonal contacts and valued social positions in the workplace, and productivity that is connected to gains and rewards.

Type of disability has been found to be a potential variable intersecting with components of organizational culture, as attitudinal barriers have been found to be deeply ingrained in the culture and climate of organizations. In their study of workplaces and persons with disabilities, Ochocka et al. (1994) found that people with psychiatric disabilities tended to be particularly isolated from the culture, and they further noted that accommodations required for people with developmental or psychiatric disabilities involved attitudinal change within the culture. Saint-Arnaud et al. (2006) also found that attitudes varied according to perceptions of illness and disability. In their study on factors involved in the work reintegration process, they noted that the perceived reason for being off work was important in terms of the response from coworkers and supervisors. For example, missing work following a personal tragedy such as the death of a loved one was seen as socially acceptable and elicited support and compassion from colleagues in the workplace. However, requiring accommodations because of a serious mental illness was not met with the same degree of acceptance, understanding, or support. These perceptions influenced the return-to-work process and individuals' treatment at work upon their return.

How Does Organizational Culture Intersect with Stigma and Discrimination in the Workplace?

It is well documented that persons with mental health problems often experience stereotyping, exclusion, and varying degrees of stigma and discrimination at work. Studies show that the effects of stigma are far-reaching, and include diminished employability, lack of career advancement, and poor quality of working life (Stuart 2006). These negative experiences may be the result of the culture of an organization, which determines how certain characteristics are perceived, how individuals who are different are treated, as well as the nature and frequency of work and career opportunities. Negative attitudes from supervisors and coworkers affect the socialization of new employees with mental health challenges and restrict their full inclusion into the workplace. The marginalization that results limits job performance and opportunities for training and advancement (Schur et al. 2005). Employees with mental health problems who have returned to work following illness have found themselves in positions of reduced responsibility, lowered expectations, and increased supervision, and have experienced disheartening comments from coworkers (Schulze 2003; Wahl 1999a, b). Indeed, interpersonal difficulties have been cited as the root of the high rate of job loss (50%) for persons with serious mental illnesses (Becker et al. 1998). Hiring practices have also been shown to be affected by negative reactions to mental health problems amongst employers. For example, Drehmer (1985) found that people with mental illness were significantly less likely to be placed in a job than individuals with paraplegia or nondisabled clients, despite identical job qualifications and work histories. Wahl's (1999a) survey of over 1,300 mental health consumers revealed that approximately one in three consumers (32%) reported that they had been turned down for a job for which they were qualified after their mental health consumer status was revealed. From an organizational culture perspective, it is important to question how it is that conditions enabling these occurrences of discrimination, conflict, and marginalization have developed.

One study that sheds some light on the failures of organizations in prohibiting discrimination against workers with disabilities was conducted by Wooten and James (2005), who collected and analyzed archival case study data of discrimination in workplace lawsuits brought by employees with disabilities. They concluded that failures in eliminating disability discrimination reflect difficulties in organizational learning. Some organizations lacked routines within their operations to manage the challenges of discrimination; the authors speculated that this may be a consequence of the social construction of the work environment that emphasizes "ableness" and neglects persons with disabilities. In other workplaces, unspoken norms became routine, resulting in a dangerous behavior pattern within the culture; these routines perpetuated stigma and discrimination. Still other organizations employed defensive routines in the form of actions, policies, and norms of behavior that prevented management from taking responsibility for their decisions, blaming others instead. Furthermore, organizations that stigmatized or discriminated against persons with disabilities did not engage in vicarious learning, and therefore did not adopt others' successful management practices. Another organizational downfall identified by the authors is referred to as "window dressing," a superficial commitment to the concern of individuals with disabilities. The authors cite Boyle's (1997) study in which one individual with a disability expressed that "window dressing organizations" are more concerned about projecting a positive image than in making a commitment to helping persons with disabilities. The belief that the appearance of diversity within an organization is helpful to the reputation and public image of the organization (Robert 2003; Spataro 2005) might help the status of those with disabling conditions that are visible or accompanied by tangible markers, but it further marginalizes persons with mental health problems and other invisible conditions. Wooten and James (2005) conclude their study by recommending the eradication of dysfunctional organizational routines, suggesting that "if organizations want to make jobs accessible, they must move beyond the physical architect and consider the social architect, that is, the organizational culture that values and encourages fair treatment of disabled employees" (p. 137).

The need to modify organizational cultures so that they are more receptive to and tolerant of persons with mental health problems in particular is increasing as the rate of mental disability in the workplace increases. Currently, employers are at a loss as to how to approach and understand mental illness in the workplace, and few organizations have formal plans to address mental health and employment equity (Harnois and Bagriel 2000). Stuart (2006) asserts that "organizations will need to be proactive in identifying and managing mental health problems among their workers and in fostering an organizational culture that is supportive of mental health and psychosocial recovery" (p. 524). Wilgosh (1990) promotes the examination of such aspects as managerial support and concern for new employees, job challenge and variety, and working group cooperation and friendliness as potential entry points to changing work environments in order to increase receptivity to persons with disabilities, and to those with mental health problems in particular.

Disclosure and Organizational Culture

Research has demonstrated that persons with mental health problems draw on their
assessments and perceptions of organizational culture when making decisions
about disclosing their illness, as well as about the extent to which they ask for
accommodations and support at work (Kirsh 2000b). Such decisions are influenced
by elements of the organizational culture, including psychological and social
aspects of the work environment, the nature of relationships with coworkers and
supervisors, and the perceived level of acceptance and understanding (Gewurtz and
Kirsh 2009). Individuals must balance their need for accommodations and support
with their fear of being treated differently once their condition becomes known by
others in the workplace. Those who harbor such fears often resort to implementing
strategies to conceal their disability and cover up any limitations (Dyck and
Jongbloed 2000; Harlan and Robert 1998; Kirsh 2000b; Pinder 1995). For example,
individuals might work longer hours to compensate for their slower pace. Although
seemingly benign, studies have demonstrated that these concealment strategies can
cut into personal time and can have implications on health and recovery (Dyck and
Jongbloed 2000; Harlan and Robert 1998; Pinder 1995).

Furthermore, if individuals with mental health problems do decide to disclose
their condition, they may be met with a lack of understanding and a sense of disbelief about the legitimacy of their condition by their coworkers and supervisors.
Conditions that lack tangible and visible markers are poorly understood, and those
that involve intermittent and unpredictable periods of illness and wellness are
particularly susceptible to this disbelief (Gewurtz and Kirsh 2009). This is especially true when the accommodations that are requested are perceived as being
desirable to a large number of people without disabilities. Harlan and Robert
(1998) note: "When an accommodation might be perceived by able-bodied
employees as being equally useful to them as to employees with disabilities,
employers resist that accommodation" (p. 424). Coworkers and supervisors might
assume that the individual is being lazy and begin to question the legitimacy of
their request. Such accommodations can threaten the status quo and reveal the
subjective nature of rules and norms within the organization. Thus, if granted,
accommodations such as changes in scheduling, work processes and procedures,
and communication strategies, might lead to broad social changes that could have
significant implications for the organization. Despite their seemingly low financial
costs, such requests for accommodations are often viewed as disruptive and met
with resistance (Gewurtz and Kirsh 2009), even though this resistance can be in
direct violation of contemporary human rights and disability legislation (Harlan
and Robert 1998; Stuart 2006).

Interestingly, such disbelief and resistance is not a matter of monetary costs, and
research has shown that most employers spend very little on accommodations for
persons with disabilities (Lengnick-Hall et al. 2008). Rather, it seems to be about
preserving the status quo and resisting challenges to organizational culture
(Ochocka et al. 1994). Physical modifications such as special furniture, parking,

and adaptive equipment, which tend to require a one-time expense with little sustained effort, are often readily approved in the workplace (Harlan and Robert 1998). These accommodations are understood to be needed for work and daily living, and are likely not coveted by others in the work environment. However, requests that require a sustained effort (but little or no monetary cost), or that have implications for work structures, such as modifications to the work schedule, workload, and processes for communication between other employees, are seen as more problematic and are most often denied (Harlan and Robert 1998; Pinder 1995; Saint-Arnaud et al. 2006). It has been speculated that requests for such long-term, process-oriented accommodations threaten established workplace norms and could lead to broader employee demands (Harlan and Robert 1998). Unfortunately, this latter set of accommodations are those most often needed by persons with mental health problems. Although such persons might contribute to the diversity of an organization, their disabilities are largely invisible, episodic, poorly understood, and often require the very accommodations that are most often denied (Behney et al. 1997; Gewurtz and Kirsh 2009).

How Can Organizations Create Cultures that Foster Inclusion, Diversity, and Successful Work Outcomes for Persons with Mental Health Problems?

Existing evidence suggests that investments and commitments to diversity must pervade the culture of an organization; otherwise, the cost associated with diversity could outweigh the benefits and could cause havoc for the organization (Chatman et al. 1998; Kochan et al. 2003; Slater et al. 2008). In addition to the disability and human rights laws that already exist, there is a need for the culture of organizations to shift towards acceptance and valuing of diversity, so that the integration of those who are different is seen as important to the success of the organization. Organizations that adopt organizational practices to manage diversity and take advantage of the opportunities associated with diversity are likely to thrive. Specifically, Kochan et al. (2003) argue that:

> success is facilitated by a perspective that considers diversity to be an opportunity for everyone in an organization to learn from each other how better to accomplish their work and an occasion that requires a supportive and cooperative organizational culture as well as group leadership and process skills that can facilitate effective group functioning. Organizations that invest their resources in taking advantage of the opportunities that diversity offers should outperform those that fail to make such investments (p. 18).

Some researchers have examined ways in which organizational culture can be compatible with, and indeed reap the benefits of, diversity. Chatman et al. (1998), for example, used organizational simulation to explore how differences in organizational cultures moderate how demographically diverse people approach and solve problems. They found that an organization's focus on individualism or collectivism may affect the salience of organizational membership and influences social interaction, conflict, productivity, and perceptions of creativity. They concluded that organizations may get the most benefit from creating a culture that makes organizational membership salient, rather than one that emphasizes individualism and distinctiveness between members.

There is limited research to demonstrate specific features of organizational cultures which support diversity. Individual organizations may be hesitant to be part of any research endeavors out of fear of damaging their reputation or possible litigation (Kochan et al. 2003). However, cultural features that seem to be supportive of diversity are often suggested, based on anecdotal descriptions of actions taken by various organizations; and there does appear to be much agreement in the organizational and business literature regarding aspects of organizational culture that seem to foster diversity in the workplace.

A commonly cited approach to fostering diversity is training and education focused on diversity issues. Several authors have noted that education can increase awareness of the benefits of diversity on organizations, as well as why efforts should be put forth to ensure that different perspectives are present and utilized in the workplace (Kochan et al. 2003; Zintz 1997). Within the mental health literature, the value of education and training is increasingly stressed. The use of psychoeducational strategies aimed at creating environments which compensate for a functional disability has been suggested (Akabas 1994); and some authors have gone so far as suggesting that both employers and coworkers should be recipients of vocational services, along with their coworker with disability (Cook and Pickett 1995). Krupa (2007) explains that training and education can develop employer knowledge about work disabilities associated with mental illness and also about the rights and expectations of affected individuals regarding accommodation. Such training can help employers and coworkers increase their awareness of how to recognize signs of mental illness that may otherwise be confused with poor productivity or stress.

One of the most widely-cited approaches to fostering diversity is endorsement of the value of diversity by upper management and their overt actions to increase diversity at all levels of the organization. Such commitment to diversity from upper management may set a tone of inclusion for the entire organization. Many examples of integrating diversity within business strategies, visions, targets, and action plans exist (Lengnick-Hall et al. 2008; Schur et al. 2005; Slater et al. 2008; Wong 2008; Zintz 1997). Practices such as actively recruiting diverse candidates (Schur et al. 2005; Spataro 2005; Wong 2008; Zintz 1997), and promoting organizational values and norms that encourage effective use of diverse perspectives on teams (Chatman et al. 1998; Kochan et al. 2003) have been successful in increasing creativity, innovation, and productivity.

Currently, there is growing pressure on upper management to embrace the importance of mental health in the workplace and to implement policies and procedures that create corporate cultures that promote mental health. As an example, in Canada, the board of directors' guidelines on mental health and safety, issued by the Global Business and Economic Roundtable on Addiction and Mental Health 2004 "advises the corporate directors of public companies in Canada to place the topic of environment, health, and safety generally, and mental health specifically, on all future agendas of their boards and relevant committees." It also advises boards of directors "to express their support of senior management in its efforts to conduct and report on the costs and benefits of a strategic action plan to promote mental health as a business asset and prevent mental disability as an eventful form of risk management" (2004, p. 2).

Cultures that promote mental health and the inclusion of persons with mental health problems are gaining the attention of employers. For example, Barling (2007) has synthesized his research to enable its application to workplaces; he delineates a set of ten elements of a psychologically healthy workplace, which are: transformational leadership, workload and pace, work schedule, role clarity, job future, autonomy, workplace justice, reduced status distinctions, a social environment that emphasizes trust, and physical factors such as safety and ergonomics. Of these ten elements of a psychologically healthy workplace, leadership style has been stressed as the most important. Similarly, the U.S. Department of Health and Human Services has issued a guide to the "Mental Health Friendly Workplace" that advises employers how to influence their culture, including taking such steps as: welcoming all qualified job applicants; promoting work-life balance; supporting employees who seek treatment or hospitalization, and planning for return-to-work for those on disability leave; providing training for managers on workplace mental health, including identification of performance problems that may indicate worker distress; and providing communications to employees about accommodations policies and other topics that promote an antistigmatizing environment. Employers are gradually recognizing that changes at the organizational level can improve the mental health of the workplace and lead to improved employee work performance outcomes and overall company success.

In light of research that links increased person-culture fit to improved job tenure, satisfaction, and other work outcomes, there is a need for more focused attention on the work environment, as well as the intersection of individuals' values and behaviors with organizational culture. Job matching, as it is commonly referred to, suggests a process that considers the unique experiences, skills, and talents that an individual may contribute to a job. Identification of the values and norms that characterize the culture of the organization must be aligned with assessment of individual traits that will be valued in the workplace. To date, within mental health research and practice, there has been an emphasis on matching required skills to job requirements. Results of previously cited research indicate that matching individuals to the workplace culture is equally important, including potential sources of support and conflict in the work culture.

Conclusion

Organizational culture represents an important construct in understanding work and employment among persons with mental illnesses in the workplace. Existing evidence suggests that the culture of an organization can greatly influence the experiences of persons with mental illnesses in the workplace, the opportunities they are afforded, and issues associated with workplace accommodations, disability leave, and processes surrounding return-to-work. Future research in the area of work among persons with mental health disorders should pay attention to how organizational culture impacts employment outcomes, and how it can be used to improve outcomes. Although it may take time for the culture of an organization to evolve and change, the positive impact of cultures that are inclusive and that value diversity is evidenced in important outcomes such as employee satisfaction, commitment, job tenure, and loyalty; improvements in these and other work dimensions suggest that attention to diversity is worth the investment. Furthermore, employment specialists and vocational rehabilitation personnel involved in helping individuals with mental illnesses enter or return to the workforce should consider organizational culture when making recommendations and devising return-to-work plans. Organizations that pay attention to the mental health of their employees and foster a culture that values the contribution of those who are different will have a competitive advantage.

References

Akabas S (1994) Workplace responsiveness: key employer characteristics in support of job maintenance for people with mental illness. Psychosoc Rehabil J 17:91–101

Banks B, Charleston S, Grossi T, & Mank D (2001) Workplace supports, job performance and integration outcomes for people with psychiatric disabilities. Psychiatr Rehabil J 24:289–296

Barling J (2007) Ten key factors in building a psychologically healthy workplace. Keynote address presented at Canadian Institute for Health Research (CIHR) 2nd Canadian Congress on Mental Health and Addiction in the Workplace, Vancouver, British Colombia. http://www.comh.ca/conferences/2007/workplace/en/program/documents/JulianBarling.pdf

Behney C, Hall LL, Keller JT (1997) Psychiatric disabilities, employment, and the Americans with Disabilities Act background paper: U.S. Office of Technology and Assessment. http://earthops.org/ada_ota.html. Accessed 3 July 2008

Becker DR, Drake RE, Bond GR et al (1998) Job terminations among persons with severe mental illness participating in supported employment. Community Ment Health J 34:71–82

Boyle MA (1997) Social barriers to successful reentry into mainstream organizational culture: perceptions of people with disabilities. Hum Res Dev Q 8:259–268

Bretz RD Jr, Judge TA (1994) Person-organization fit and the theory of work adjustment: implications for satisfaction, tenure, and career success. J Vocat Behav 44:32–54

Cassell C, Walsh S (1997) Organizational cultures, gender management strategies and women's experience of work. Fem Psychol 7:224–230

Chatman J (1989) Improving interactional organizational research: a model of person-organization fit. Acad Manage Rev 14:333–349

Cockburn C (1991) In the way of women: men's resistance to sex equality in organizations. ILR Press, Ithaca, NY

Cook JA, Pickett SA (1995) Recent trends in vocational rehabilitation for people with psychiatric disability. Am Rehabil 20:2–12

Chatman JA, Polzer JT, Barsade SG, Neale MA (1998) Being different yet feeling similar: the influence of demographic composition and organizational culture on work processes and outcomes. Adm Sci Q 43(3):749–780

Drehmer DE (1985) Hiring decisions for disabled workers: the hidden bias. Rehabil Psychol 30:157–164

Dyck I, & Jongbloed L (2000) Women with multiple sclerosis and employment issues: a focus on social and institutional environments. Can J Occup Ther 67(5):337–346

Etzioni A (1961) A comparative analysis of complex organizations. Free Press, New York

Gates L (1993) The role of the supervisor in successful adjustment to work with a disabling condition: issues for disability policy and practice. J Occup Rehabil 3:179–189

Gewurtz R, Kirsh B (2009) Disruption, disbelief and resistance: a meta-synthesis of disability in the workplace. Work 34(1):33–44

The Global Business and Economic Roundtable on Addiction and Mental Health (2004) The Board of Directors Guideline on Mental Health and Safety. http://www.mentalhealthroundtable.ca/feb_2004/BOD_guideline_Feb_04.pdf

Goffee R, Jones G (1996) What holds the modern company together? Harv Bus Rev 74(6):133–149

Gulick E, Yam M, Touw M (1989) Work performance by persons with multiple sclerosis: conditions that impede or enable the performance of work. Int J Nurs Stud 26:301–311

Hagner D, DiLeo D (1993) Working together: workplace culture, supported employment and persons with disabilities. Brookline Books, MA

Harlan S, Robert P (1998) The social construction of disability in organizations: why employers resist reasonable accommodation. Work Occup 25(4):397–435

Harnois G, Bagriel P (2000) Mental health and work: impact, issues and good practices. World Health Organization and the International Labour Organization, Geneva

Kerr JL, Slocum JW Jr (1987) Managing corporate culture through reward systems. Acad Manag Exec 1:99–107

Keyton J (2005) Communication and organizational culture. Sage Publications, Inc, Thousand Oaks, CA

Kirsh B (1996) Influences on the process of work integration: the consumer perspective. Can J Community Ment Health 15:21–37

Kirsh B (2000a) Organizational culture, climate and person-environment fit: relationships with employment outcomes for mental health consumers. Work 14:109–122

Kirsh B (2000b) Work, workers, and workplaces: a qualitative analysis of narratives of mental health consumers. J Rehabil 66(4):24–30

Kochan T, Bezrukova K, Ely R, Jackson S, Joshi A, Jehn K et al (2003) The effects of diversity on business performance: report of the diversity research network. Hum Resour Manage 42(1):3–21

Kopelman RE, Brief AP, Guzzo RA (1990) The role of climate and culture in productivity. In: Schneider B (ed) Organizational climate and culture. Jossey-Bass, San Francisco, pp 282–318

Krupa T (2007) Interventions to improve employment outcomes for workers who experience mental illness. Can J Psychiatry 52(6):339–345

Lengnick-Hall ML, Gaunt PM, Kulkarni M (2008) Overlooked and underutilized: people with disabilities are an untapped human resource. Hum Resour Manage 47(2):255–273

Moran TE, Volkwein FJ (1992) The cultural approach to the formation of organizational climate. Hum Relat 45(1):19–47

Nieuwenhuijsen K, Verbeek J, de Boer AGEM, Blonk R, van Dijk FJH (2004) Supervisory behaviour as a predictor of return to work in employees absent from work due to mental health problems. Occup Environ Med 61:817–823

Ochocka J, Roth D, Lord J, Mac Gillivary H (1994) Workplaces that work, vol 34. Centre for Research and Education in Human Services, Kitchener, ON, pp 487–516

O'Reilly C (1989) Corporations, culture, and commitment: motivation and social control in organizations. Calif Manage Rev 31:9–25

O'Reilly CA III, Chatman JA, Caldwell D (1991) People and organizational culture: a profile comparison approach to assessing person-organization fit. Acad Manage J 3:487–516

Pinder R (1995) Bringing back the body without blame?: the experience of ill and disabled people at work. Sociol Health Illn 17(5):605–631

Robert P (2003) Disability oppression in the contemporary U.S. capitalist workplace. Sci Soc 67(2):136–159

Saint-Arnaud L, Saint-Jean M, Damasse J (2006) Towards an enhanced understanding of factors involved in the return-to-work process of employees absent due to mental health problems. Can J Community Ment Health 25(2):303–315

Schein E (1984) Culture as an environmental context for careers. J Occup Behav 5:71–81

Schein E (1992) Organizational culture and leadership, 2nd edn. Jossey-Bass, San Francisco

Schneider B (1987) The people make the place. Person Psychol 40:437–453

Schneider B (1990) Organizational climate and culture. Jossey-Bass, San Francisco

Schulze B, Angermeyer MC (2003) Subjective experiences of stigma. A focus group study of schizophrenia patients, their relatives and mental health professionals. Soc Sci Med 56:299–312

Schur L, Kruse D, Blanck P (2005) Corporate culture and the employment of persons with disabilities. Behav Sci Law 23(1):3–20

Secker J, Membrey H (2003) Promoting mental health through employment and developing healthy workplaces. The potential of natural supports at work. Health Educ Res 18:207–215

Sheridan JE (1992) Organizational culture and employee retention. Acad Manag J 35(5):1036, Retrieved September 12, 2008, from ABI/INFORM Global database. (Document ID: 135827)

Shore LM, Martin HJ (1989) Job satisfaction and organizational commitment in relation to work performance and turnover intentions. Hum Relat 42:625–638

Silverthorne C (2004) The impact of organizational culture and person-organization fit on organizational commitment and job satisfaction in Taiwan. Leader Organ Dev J 25(7):592–599

Slater SF, Weigand RA, Zwirlein TJ (2008) The business case for commitment to diversity. Bus Horiz 51:201–209

Sosa J, Sagas M (2006) Assessment of Organizational Culture and Job Satisfaction in National Collegiate Athletic Association Academic Administrators. AHPERD National Convention and Exposition, Salt Lake City, http://aahperd.confex.com/aahperd/2006/preliminaryprogram/abstract_8238.htm

Spataro SE (2005) Diversity in context: how organizational culture shapes reactions to workers with disabilities and others who are demographically different. Behav Sci Law 23(1):21–38

Stone DI, Colella A (1996) A model of factors affecting the treatment of disabled individuals in organizations. Acad Manage J 21(2):352–401

Stuart H (2006) Mental illness and employment discrimination. Curr Opin Psychiatry 19:522–526

U.S. Department of Health and Human Services. A mental health friendly workplace. Substance Abuse and Mental Health Services Administration http://www.allmentalhealth.samhsa.gov/documents/CEOBookletFINAL.pdf

Vandenberghe C (1999) Organizational culture, person-culture fit, and turnover: a replication in the health care industry. J Organ Behav 20(2):175–184

Wahl OF (1999a) Mental health consumers' experiences of stigma. Schizophr Bull 25:467–478

Wahl OF (1999b) Telling is risky business. Rutgers University Press, Piscataway, NJ

Wiener Y, Vardi Y (1990) Relationship between organizational culture and individual motivation: a conceptual integration. Psychol Rep 67:295–306

Wilgosh L (1990) Organizational climate and workers with mental disabilities. Can J Rehabil 4:9–16

Wilkins A, Ouchi WG (1983) Efficient cultures: exploring the relationship between culture and organizational performance. Adm Sci Q 19:357–365

Wong S (2008) Diversity: making space for everyone at NASA/Goodard space flight center using dialogue to break through barriers. Hum Resour Manage 47(2):389–399

Wooten L, James E (2005) Challenges of organizational learning: perpetuation of discrimination against employees with disabilities. Behav Sci Law 23:123–141

Zintz AC (1997) Championing and managing diversity at Ortho Biotech Inc. Natl Prod Rev Autumn:21–28

Chapter 22
Evidentiary Support for Best Practices in Job Accommodation in Mental Health: Employer-Level Interventions

Izabela Z. Schultz, Alanna Winter, and Jaye Wald

Introduction

Mental health disabilities are increasingly becoming a concern both internationally and within North America. The estimate that 450 million people worldwide were affected by at least one mental health disorder, as captured in the 2001 World Health Report released by the World Health Organization (2001), prompted many national and international initiatives to explore the personal and economic impact of mental health disorders. By the year 2020, mental health disorders are anticipated to account for 15% of the global disease burden (Murray et al. 1994). In the United States, the 1992 National Comorbidity Survey found the 12-month prevalence of major depressive disorder in respondents aged 15–54 years was 10.3%, and the prevalence of any anxiety disorder was 17.2% (Kessler et al. 1994). In Canada, it has been estimated that mental health disorders comprise close to 25% of all diseases and injuries, with 13% of that number attributable to depression alone (Dewa et al. 2004).

The estimates regarding the economic impact of mental health disorders are no less dire. A significant proportion of the economic costs of mental health disorders are due to their deleterious effects upon employment. There are "two main factors that make mental illness specifically a workplace issue. First, mental illness usually strikes younger workers. Second, many mental illnesses are both chronic and cyclical in nature, requiring treatment on and off for many years" (Standing Senate Committee 2004, p. 107). The members of the Global Business and Economic Roundtable on Addiction and Mental Health believe that the direct and indirect costs of mental health in the workplace equate to 14% of the net annual profits of all companies, or 16 billion dollars; roughly 3% of Canada's Gross Domestic Product (2004). Furthermore, according to Dewa et al. (2004), short- and

J. Wald (✉)
Department of Psychiatry, University of British Columbia,
2255 Wesbrook Mall, Vancouver, BC, Canada, V6T 2A1
e-mail: jwald@interchange.ubc.ca

I.Z. Schultz and E.S. Rogers (eds.), *Work Accommodation and Retention in Mental Health*,   409
DOI 10.1007/978-1-4419-0428-7_22, © Springer Science+Business Media, LLC 2011

long-term disability related to mental illness account for a third of all claims and approximately 70% of total disability costs. (Also see chapter by Dewa and McDaid (2010, this volume), on the epidemiological and economic impact of mental disorders.)

Consequently, it behooves employers, government, insurance and compensation policy makers, and researchers to attempt to improve the employment situation for workers with mental health disorders. This chapter will attempt to review the literature concerning the effective practices utilized to facilitate the retention and accommodation of employees with mental health disabilities in the workplace.

The absence of an evidence-informed approach to the issue of job accommodations for persons with mental health disabilities leaves policy makers, employers, rehabilitation and disability management professionals, researchers, and persons with disabilities themselves lacking the critical knowledge for the facilitation of employment and the enhancement of job retention for these workers. Although research has been undertaken in this area, it can be characterized as somewhat fragmented and methodologically inconsistent. More importantly, most of the research investigating the employment of individuals with mental health disabilities is concentrated upon the attainment of work and on serious mental illness. Indeed, the bulk of this research has concentrated on supported employment programs, job clubs, and other initiatives aimed at helping individuals with mental health disorders find employment. Once employed, however, it is often the case that individuals with mental health disabilities may need further support and accommodation to retain their employment. This systematic review of the accommodation literature is an effort to address the gap in knowledge and provide a direction for future research.

Barriers to Employment and Job Accommodation in Mental Health

People with disabilities, particularly mental health disabilities, face numerous barriers in obtaining equal opportunities – environmental, access, legal, institutional, and attitudinal barriers lead to social exclusion (Center 2010, this volume; Baldwin and Marcus 2010, this volume; Harnois and Gabriel 2000). For persons with mental illness, social exclusion is often the most challenging barrier to overcome and is usually associated with feelings of shame, fear, and rejection (Harnois and Gabriel 2000). When seeking employment, a very real concern for those with mental health disabilities is the question of whether to disclose the disability. To illustrate, some investigators found 81.5% of patients with a depressive episode and 82.0% of patients with Obsessive-Compulsive Disorder (OCD) expected to be discriminated against because of their mental illness when applying for a job. Moreover, 67.3% of patients with a depressive episode and 80.0% of those with OCD were advised to keep the mental illness a secret when applying for a job (Angermeyer et al. 2004; Stengler-Wenzke et al. 2004). Review the chapter by MacDonald-Wilson et al. (2010, this volume) on complex challenges related to disclosure of mental health disability.

A recent survey conducted by Schultz et al. (2010, this volume) investigated the attitudes of Canadian employers towards workers with mental health disabilities. The results indicated that over 50% of the Canadian employers surveyed (mid-size to large companies) had significant concerns about workers with mental health disability. The most significant concerns included: mental instability (83.1%), bizarre behaviors (80.7%), inability to tolerate work pressure and stress (79.5%), becoming violent (75.9%), and being unable to tolerate working conditions (74.7%). Clearly, the improved societal attitudes and increased acceptance of workers with physical disability are not paralleled by workplace inclusion of individuals with mental health disabilities. Social stigmatization continues to be a more salient barrier to employment than actual disability (Baldwin and Marcus 2010, this volume; Corrigan et al. 2007; Goldner et al. 2004; Greenwood and Johnson 1987; Haslan et al. 2005; McFarlin et al. 1991).

Because individuals with mental health disabilities often avoid the stigma and try to conceal their disabilities, there are consequences for the workplace. Harnois and Gabriel (2000) list the following workplace consequences: absenteeism (increase in overall sickness absence, particularly frequent short periods of absence); poor health, depression, stress, and burnout; physical conditions, including high blood pressure, heart disease, ulcers, sleeping disorders, skin rashes, headache, neck- and backache, and low resistance to infections; changes in work performance, also known as presenteeism (reduction in productivity and output, increase in error rates, increased amount of accidents, poor decision-making, deterioration in planning and control of work); staff attitude and behavior problems (loss of motivation and commitment, burnout, staff working increasingly long hours but for diminishing returns, poor timekeeping, labor turnover); and difficulties with relationships at work (tension and conflicts between colleagues, poor relationships with clients, increase in disciplinary problems).

Given the high estimates of unemployment among individuals with mental health disabilities (figures for the United States show between 75 and 85% of people with severe mental illness are unemployed, whereas in the United Kingdom estimates range from 61 to 73%) (Crowther et al. 2001); and given the statistical data reporting that millions of persons with mental illness are seeking entry to work (National Organization on Disability 2004); it is obvious that strategies to address the barriers and limitations these workers face once employed are important. In addition, there are estimates that many millions more are struggling to maintain employment (Akabas 1994). Although underdiagnosed, depression alone has been estimated to have a prevalence of approximately 10% among employees (Berndt et al. 1998). Research into the effectiveness of job accommodations and employment retention initiatives has begun, and this growing body of research literature attests to the need for job accommodations.

Job accommodations, as a workplace-based intervention, are a promising yet relatively understudied approach to improving employment outcomes for persons with mental health disabilities. The term "reasonable accommodation," as found in the Americans with Disabilities Act, refers to a modification to the job or the workplace that enables the disabled worker to successfully perform the essential functions of the job, or to enjoy equal benefits and privileges of employment (ADA 1990). Such initiatives have the potential to improve work reentry, reduce presenteeism, and facilitate work retention. However, the provision of job accommodations is a complex and multifaceted workplace-based intervention, and must consider the needs of the employee, the employer, and other stakeholders involved (e.g., third-party insurance companies and healthcare services). Furthermore, it is only one facet of the work entry and work retention process, and so more collaborative multidisciplinary research is needed to bridge job accommodations with individual-level and other employer-level interventions. As an intervention, it must also be viewed within a broader context that considers the social, cultural, and political factors that contribute to barriers to employment among persons with mental health disabilities.

Searching the Literature for Evidentiary Support for Job Accommodations

We initiated a comprehensive search of the existing literature into three aspects of mental health disabilities, searching for articles published within the last 10 years, with the intention of assessing the current state of the literature. The criteria for these searches are as follows:

(a) Complete review of the literature on employment rates, barriers, and factors facilitating work among persons with mental health disabilities, with particular emphasis on mood and anxiety disorders; persons with neuropsychological disabilities, including brain injuries, seizure disorders, brain tumors, neurovascular disorders, severe learning disabilities, and other neurological disorders affecting the brain; and persons with serious mental illness, such as schizophrenia and bipolar disorder.
(b) Complete literature review on the relationship between mental health/neuropsychological impairments and work capacity, focusing on identification of functional limitations and required work accommodations.
(c) Complete review of the literature on best practices in employment of persons with mental health/neuropsychological disabilities, with particular emphasis on job accommodations and other strategies to enhance employment and maintenance of employment.

The following databases were searched for the above-mentioned areas: PsycINFO, MEDLINE, PAL-MED, REHABDATA/NARIC, Academic Search Premier, EBSCO, and EMBASE. In addition, seminal journals and reference lists of relevant studies were hand-searched. The results yielded a master list of 688 articles. These articles then underwent evaluation for methodological rigor. An evaluation form was developed for this purpose and four research assistants (RAs) were trained in its use. Inter-rater reliability between the RAs was established. The literature was appraised on methodological design (controlled trial, cohort study, before/after study, case control study, case series, qualitative research or survey); design direction (prospective, retrospective, or same time); participant selection criteria (random selection, participants chosen by investigator, sample of convenience or other); clarity of definition of eligible participants (clear or implied); and the use of a control group. Minimum criteria were set that included having at least a survey design, identifiable design direction, identifiable selection criteria, and at least an implied definition of eligible participants.

From the master list, 275 articles met the methodological criteria for further evaluation. These 275 articles were assessed for topic relevance, and it was then determined that the scope of the study would be narrowed down to an emphasis on job maintenance and retention (omitting the securing of employment and all job search/supported employment/club house programs). Once this was determined, a final list of 58 articles met the criteria for inclusion in the systematic review. These final 58 were further evaluated for conclusions, and the findings were then integrated to produce a literature-informed best practices guideline.

Functional Limitations in Mental Health

There is a large body of literature showing that individuals with mental health disabilities often face numerous functional limitations in meeting basic work requirements, which in turn contribute to presenteeism, absenteeism, and unemployment. These findings identified worker-related limitations in terms of symptoms and their functional impact, as well as recognizing other mental health factors that contribute to their functional difficulties (e.g., low self-esteem, poor communication skills, adopting a passive role towards treatment seeking).

To begin, a review of the types of functional difficulties that persons with mental health disabilities typically experience is presented. These are then linked to the types of job accommodations they require. The literature commonly categorizes these difficulties under the general headings of cognitive, social, emotional, and physical limitations (MacDonald-Wilson et al. 2003), which are summarized below. However, it is important to note that these limitations rarely exist in isolation; for example, it would be likely that a person with major depression would have multiple limitations across these spheres, and would thus require several different types of job accommodations. Review the chapter by Rogers and MacDonald-Wilson (2010, this volume) on prediction of vocational capacity in mental health for further discussion of this topic.

Cognitive Limitations: The findings from the literature review suggest that for people with severe mental illness (e.g., schizophrenia and bipolar disorder) and neuropsychiatric disabilities, cognitive limitations (due to impaired attention, concentration, and memory) are highly prevalent and tend to be more common than interpersonal or social limitations (MacDonald-Wilson et al. 2003). There is a growing recognition that persons with other mental health disorders, including other mood and anxiety disorders (more "invisible" disorders), also experience various cognitive limitations, including concentration and memory problems.

The most common workplace difficulties for individuals with mental health disabilities that result in cognitive limitations include: learning the job, concentrating for extended periods, following schedules, attending work, assessing work performance, solving problems/organizing work, using basic language/literacy skills, making work-related decisions, initiating and carrying out job instructions and tasks, maintaining work stamina, and performing at a consistent pace (MacDonald-Wilson et al. 2003). To assist with these problems, several job accommodations are often required, with the most commonly used strategies being modified work duties, modified work environments, and assistive technologies.

Social Limitations: Impaired social functioning among persons with mental health disabilities is well recognized in the research literature (MacDonald-Wilson et al. 2003). The most common functional difficulties in this sphere include problems with: communicating and interacting with employers, coworkers, customers, or the general public; accepting instructions and responding to feedback and criticism; maintaining appropriate social behavior (i.e., "getting along"); and asking questions or requesting assistance.

MacDonald-Wilson et al. (2003) provide a list of the types of accommodations used to address social limitations for individuals with mental health disabilities experiencing problems in this sphere. These accommodations include assistance interacting with others and interpreting work/social cues. Using natural supports within the workplace (supervisors and coworkers) to train and model appropriate work behavior and skills can be helpful in reducing the impact of reduced social functioning. However, such strategies, although recommended throughout the literature, do not appear to be commonly used in practice.

Emotional Limitations: In contrast to other functional domains, relatively little research has examined limitations in emotional functioning among people with mental health disabilities. Studies have primarily described these limitations in terms of problems adapting to, and coping with, stressful circumstances at work (e.g., completing tasks under stress, communicating and interacting with others under stress, taking appropriate precautions); adjusting to changes in the workplace; and managing symptoms (MacDonald-Wilson et al. 2003).

The research indicates that accommodations for emotional limitations have primarily involved individual-level-based interventions such as stress management interventions, communication skills training, and Employee Assistance Programs. Little is known about how the workplace can best accommodate these types of limitations. The common workplace practices for this realm are work modifications (e.g., taking breaks), environment modifications (e.g., providing a quiet work area), and strategies for managing stressful events in the workplace, such as utilizing Employee Assistance Programs (EAPs).

Physical Limitations: It is not uncommon for people with mental health disabilities to have various physical limitations which can impact their ability to work (e.g., a person may develop posttraumatic stress disorder as well as chronic pain from injuries sustained in a car accident). Likewise, persons with physical or neurological disabilities may develop secondary mental disorders (e.g., a person with multiple sclerosis may develop depression) (Dewa and Lin 2000).

Key Job Accommodation Practices for Persons with Mental Health Disabilities

This section synthesizes the evidentiary support for job accommodation practices that emerged from the systematic review of the literature in the field. Because of the fragmentation of the literature and the absence of an overarching conceptual framework, common methodologies, and common outcome measures, the evidentiary consensus can be integrated into a relatively few best practices, as follows.

Modify work duties and requirements. A number of studies have shown that the modification of job duties is a useful accommodation for individuals with mental health disabilities (Akabas 1994; Coles 1996; Granger et al. 1997; Hirsh et al. 1996; Holzberg 2001; MacDonald-Wilson 1994; MacDonald-Wilson et al. 2002; Mancuso 1993; Means et al. 1997; Mpofu et al. 1999). Modifications to job descriptions are any arrangements made by an employer to change a job description in order to eliminate certain tasks from the job or change the way that tasks are scheduled. For example, Granger et al. (1997) cited gradual task introduction, adapting an existing job description, minimizing changes to job descriptions over time, and exchanging tasks with others as examples of this type of accommodation. Another study (Fabian et al. 1993) found that modifying job tasks, modifying hours or schedules, modifying work rules and procedures, modifying job performance expectations, and modifying the nonphysical work environment were among the reasonable accommodations cited for 30 supported employees.

> Best Practice #1: Modifications to a job description, including changing the way that tasks are scheduled and exchanging tasks with others, is a useful accommodation for employees with mental health disabilities.

Flexible scheduling. One of the most commonly cited accommodations for individuals with mental health disabilities is flexible scheduling (Akabas 1994; Coles 1996; Granger et al. 1997; Hirsh et al. 1996; MacDonald-Wilson 1994; MacDonald-Wilson et al. 2002, 2003; Mancuso 1993; Means et al. 1997; Weed and Brown 1993). Mancuso (1993) documented the employment experiences of ten workers with psychiatric disabilities, including their use of workplace accommodations.

This study found that the accommodation most often cited by workers was a flexible or part-time schedule. The employers most often cited modifications to work assignments or other supervisory interventions first, followed by flexible or part-time scheduling.

Best Practice #2: Flexible or part-time scheduling is an effective and commonly used means of accommodating individuals with mental health disabilities.

Best Practice # 3: Flexible scheduling is effective for all functional limitations resulting from mental health disabilities.

Modify work environment. Environmental modifications entail any changes to the area where the person will be performing the functions of the job (Coles 1996; Granger et al. 1997; Holzberg 2001; Mancuso 1993; MacDonald-Wilson 1994; Means et al. 1997; Weed and Brown 1993). Changes to physical space may serve to reduce auditory and visual distractions, as well as cue workers for specific job tasks. Mancuso (1993) lists several modifications, including the following: provision of an enclosed office; provision of partitions, room dividers, or otherwise to enhance soundproofing, and visual barriers between workspaces; offering reserved parking; and blocking noise through volume control of telephones, allowing for personal space, and access to a quiet rest area.

Best Practice #4: Changes to the workplace environment may reduce auditory and visual distractions and thus allow individuals with mental health disabilities to optimize concentration and work performance.

Job sharing. Job sharing can be an effective way to modify job tasks and schedules and is useful in addressing cognitive limitations that result in concentration and stamina difficulties (Akabas 1994; Granger et al. 1997; Hirsh et al. 1996; Weed and Brown 1993).

Best Practice #5: Job sharing is an effective accommodation for aiding employees with mental health disabilities. It addresses limitations in cognition and stamina, while ensuring that job tasks are accomplished.

Assistive technologies. An assistive device is any object or item that helps a person perform the job (Coles 1996; Granger et al. 1997; Hirsh et al. 1996; Holzberg 2001; Mancuso 1993; Means et al. 1997; Mpofu et al. 1999; Warren 2000; Weed and Brown 1993; Wehman et al. 1993). Some examples of assistive technologies include: computer hardware and software modifications, tape recorders, color coding, note-taking systems, organizers/PDAs, telephone adaptations, white noise machines, timers and alarms, spell checkers, glare guards, and data bank watches (Means et al. 1997). These tools can be particularly helpful for persons with neuropsychological disabilities (e.g., traumatic brain injury) (Warren 2000; Wehman et al. 1993).

Best Practice #6: Assistive devices (any object or item that helps a person perform the job) are useful accommodations for improving work performance and job retention among employees with mental health disabilities. Examples include: computer software, organizers, recorders, and timers.

Best Practice #7: Assistive devices can aid individuals with deficits in attention, memory, motor skills, and concentration. These are common functional limitations with mental health disabilities.

Natural supports/mentoring. Natural supports (e.g., mentoring, coworker, buddy, supervision) refers to the use of coworkers, supervisors, and other supports intrinsic to the job setting to facilitate job skill acquisition, maintenance, and integration (Akabas 1994; Akabas and Gates 2004; Atkins 2002; Banks et al. 2001; Bell et al. 2003; Beyer et al. 1999; Bond et al. 1997, 2001; Coles 1996; Crowther et al. 2001; Curl et al. 1996; Gates et al. 1998; Gerber and Price 2003; Granger et al. 1997; Haffey and Abrams 1991; Hirsh et al. 1996; MacDonald-Wilson 1994; MacDonald-Wilson et al. 2002, 2003; McHugh et al. 2002; Means et al. 1997; Rogan et al. 2003; Storey and Certo 1996; Weed and Brown 1993). Twenty-three articles that were included in this systematic review mentioned the effectiveness of utilizing existing workplace supports to: educate others, provide modeling of appropriate behavior, and help with job tasks in an ongoing fashion that is both cost-effective and beneficial to workers with mental health disabilities. Common natural supports are also useful in addressing cognitive and emotional limitations. Curl et al. (1996) developed a curriculum for training coworkers for this role. Using a "tell, show, watch, and coach" instructional sequence, coworkers simultaneously teach the trade, model setting-specific language, and facilitate typical interactions between new and old employees.

Best Practice #8: The utilization of coworkers and supervisors as mentors/ trainers may improve job retention and job satisfaction for employees with mental health disabilities.

Behavioral Feedback Interventions. Several studies indicated the usefulness of implementing behavioral feedback interventions for workers with mental health disabilities. Some examples included reinforcing appropriate work behavior, both social and task-oriented, teaching of alternate/acceptable behaviors, and having regular meetings for feedback and problem-solving (Atkins 2002; Bell et al. 2003; Jacobs et al. 1992; Weed and Brown 1993). Research results involving persons with mental illness supported the importance of specific work feedback and meetings with the worker as a behavioral intervention to enhance work performance and retention and "amplify the participant's sense of purpose and motivation" (Bell et al. 2003).

In a study by Jacobs et al. (1984), a weekly job maintenance group was available to workers with mental health disabilities. Training focused on getting along with others, stress management, enhanced daily living, and management of psychiatric symptoms. The training utilized a problem-solving approach, role-playing, and feedback-giving. The findings were limited to severe mental illness, which constitutes a minority of workers with mental health disabilities. Nevertheless, regular individual supervisory feedback and problem-solving around performance goals and issues constitute important job accommodations.

Best Practice #9: Behavioral interventions, such as regular meetings to provide feedback and reinforcement of work performance as well as to problem-solve around work and performance issues, are useful accommodations for improving job retention for persons with mental health disabilities.

Towards Evidence-Informed Practices: What Have We Learned and Where Are We Headed?

There is a significant discrepancy between the lengthy list of job accommodations designed for persons with mental disorders (Job Accommodation Network, 2010) in the aftermath of the implementation of the Americans with Disabilities Act (1990), and the quality of evidence supporting the effectiveness of these accommodations. Only eight job accommodation best practices identified in our systematic analysis can be considered "evidence-informed" because of the supporting research. Among them, the use of natural supports in the workplace has the best established evidentiary base, with 23 studies attesting to its effectiveness.

Considering the limitations of the utilized research designs, very few studies could be considered "evidence-validated." The research producing level-A evidence, i.e., randomized controlled studies, is almost nonexistent; and studies producing level B evidence are very limited. Furthermore, longitudinal studies that involve repeated assessments with the same individuals over time are rare (Gnam 2005).

Overall, the quantitative research on the efficacy and effectiveness of job accommodation interventions is in its early stages, despite much being published in the field. Research on the relationship between mental disorders and related functional limitations has been progressing more rapidly and provides a helpful basis for the development of job accommodations that match specific limitations. At this time, however, the chasm between the understanding of the limitations and the effectiveness of job accommodation interventions has not been bridged.

Of concern is the large variety of outcome measures used in studies on job accommodation, ranging from a single return-to-work episode, through the duration of disability or work absence, to self-report or employer report. The self-report of the respondent may or may not include the subjective role impairment, lost productivity, days spent in bed, cut-down days, and the perception of disability. Gnam (2005) cautions that "these measures could be biased if persons with mental disorders provide overly pessimistic appraisal of their functioning (p. 374)." Moreover, commonly utilized outcome measures do not capture the recurrent or cyclical nature of mental health disabilities (Baldwin and Marcus 2010, this volume), and there is a paucity of validated measures of reduced productivity ("presenteeism"). Last but not least, most of the studies on accommodations were completed primarily on individuals with serious mental illness, to the detriment of investigating the most prevalent conditions, such as anxiety and mood disorders and neuropsychological disorders including mild brain injury and other cognitive disorders. There is also a conspicuous paucity of studies on accommodating persons with addictions and those with autism spectrum disorders, attention deficit disorder, and learning disabilities. At the same time, because there is no established taxonomy of functional limitations in mental disorders and associated standardized functional assessments (see Rogers and MacDonald-Wilson (2010) chapter, this volume, for the review of current methods in this field), research in the field that attempts to take a transdiagnostic approach is seriously hampered.

With these caveats in mind, existing studies do support the utility of job accommodations involving the following categories: (1) flexibility in job design and execution: job description modification, flexibility in scheduling, part-time work, and job sharing; (2) changes to work environment; (3) utilization of compensatory technical aids; and (4) interpersonal supports, including utilization of natural peer supports and feedback-giving by the supervisor. Strikingly, many different types of accommodations identified by the Job Accommodation Network are missing from this list, especially those requiring major system-based changes through the employer's policies and practices, those involving interpersonal relations with management and peers, and those which are likely to remain low-frequency (e.g., the use of pets or music).

In addition, studies attempting to bridge employee-level intervention (treatment, rehabilitation, and skill training) with employer-level interventions (i.e., job accommodation interventions) are absent. On a positive side, the needs and benefits of employee-level interventions for people with mental health disabilities are well established in the research literature. These interventions have been shown to be associated with improved employment outcomes. They can be helpful in the management and reduction of symptoms and functional difficulties, enhance medication compliance, and improve stress management and communication skills. All these individual factors are important prerequisites for effective job accommodations.

Further, the literature is still in the early stages with respect to informing practices through the translation of functional strengths and limitations among persons with mental health disorders into the accommodating job and work environment design, including physical, organizational, and social components, and the interactions among them. However, relatively recently, promising advances in this field have been made (MacDonald-Wilson et al. 2003). Likewise, multisite studies helped address the issue of small sample sizes inherent in studies on job accommodation (MacDonald-Wilson et al. 2002).

Finally, Franche et al. (2005) point out correctly that the literature on job accommodations has not, as of yet, been focused on the most effective process by which an offer of work accommodation could be developed and provided to facilitate its acceptance and thus contribute to sustained employment. This is particularly true in the mental health literature. Qualitative research may offer much understanding of the social or interpersonal aspects of the accommodation process, including interactions among various stakeholders involved in the process (employer, insurer, and healthcare provider). Although it has been postulated that job accommodation is a social process (Gates and Akabas 2010, this volume; Gates et al. 1998), the improved understanding of the interactional dynamics among various players in the process will remain the long-term goal of research in the field. Notably, there is a paucity of studies on social accommodations in the workplace, especially those involving interactions with managers, coworkers, and supervisors, and a lack of a much needed taxonomy of social work functions that could facilitate research in this field.

The use of new methodologies, including mixed qualitative-quantitative designs, may help answer some of the key questions about the effectiveness of a wider range of accommodations, the processes involved, and matching individual strengths, needs, and limitations to the right accommodations.

Acknowledgement This chapter is based on the research completed for a project funded by Social Development Canada, "Towards Evidence-Informed Best Practice Guidelines for Job Accommodations for Persons with Mental Health Disabilities."

References

Akabas SH (1994) Workplace responsiveness: key employer characteristics in support of job maintenance for people with mental illness. Psychosoc Rehabil J 17(3):91–101
Akabas SH, Gates LB (2004) Mental health disorders with workplace consequences [on-line]. http://www.bu.edu/cpr/reasaccom/employ-read-akabas.html

Americans with Disabilities Act of 1990, 42 U.S.C.A. § 12101 et seq. (West 1993)

Angermeyer MC, Beck M, Dietrich S, Holzinger A (2004) The stigma of mental illness: patients' anticipations and experiences. Int J Soc Psychiatry 50:153–162

Atkins R (2002) Supported employment for people with severe learning disabilities. Br J Ther Rehabil 9(3):92–95

Baldwin M, Marcus S (2010) Effects of stigma and discrimination on the economic employment outcomes of persons with mental health disorders. In: Schultz IZ, Rogers ES (eds) Handbook of job accommodations in mental health. Springer, New York

Banks B, Charleston S, Grossi T, Mank D (2001) Workplace supports, job performance, and integration outcomes for people with psychiatric disabilities. Psychiatr Rehabil J 24(4):389–396

Bell H, Lysaker P, Bryson G (2003) A behavioral intervention to improve work performance in schizophrenia: work behavior inventory feedback. J Vocat Rehabil 18:43–50

Berndt ER, Finkelstein SN, Greenberg PE, Howland RH, Keith A, Rush AJ, Russell J, Keller MB (1998) Workplace performance effects from chronic depression and its treatment. J Econ 17:511–535

Beyer S, Kilsby M, Shearn J (1999) The organization and outcomes of supported employment in Britain. J Vocat Rehabil 12:137–146

Bond GR, Drake RE, Mueser KT, Becker DR (1997) An update on supported employment for people with severe mental illness. Psychiatr Serv 48(3):335–346

Bond GR, Becker DR, Drake RE, Rapp CA, Meisler N, Lehman AF, Bell MD, Blyler CR (2001) Implementing supported employment as an evidence-based practice. Psychiatr Serv 52(3):313–322

Center C (2010) Law and job accommodation in mental health disability. In: Schultz I, Sally Rogers E (eds) Handbook of work accommodation and retention in mental health. Springer, New York

Coles A (1996) Reasonable accommodations in the workplace for persons with affective mood disorders: a case study. J Appl Rehabil Couns 27(4):40–44

Corrigan PW, Larson JE, Kuwabara S (2007) Mental illness stigma and the fundamental components of supported employment. Rehabil Psychol 52:451–457

Crowther RE, Marshall M, Bond GR, Huxley P (2001) Helping people with severe mental illness to obtain work: systematic review. Br Med J 322:204–208

Curl RM, Fraser RT, Cook RG, Clemmons D (1996) Traumatic brain injury vocational rehabilitation: preliminary findings for the coworker as trainer project. J Head Trauma Rehabil 11(1):75–85

Dewa CS, Lin E (2000) Chronic physical illness, psychiatric disorder and disability in the workplace. Soc Sci Med 51(1):41–50

Dewa C, McDaid D (2010) Investing in the mental health of the labor force: epidemiological and economic impact of mental health disabilities in the workplace. In: Schultz I, Sally Rogers E (eds) Handbook of work accommodation and retention in mental health. Springer, New York

Dewa CS, Lesage A, Goering P, Caveen M (2004) Nature and prevalence of mental illness in the workplace. Healthc Pap 5(2):12–25

Fabian ES, Waterworth A, Ripke B (1993) Reasonable accommodations for workers with serious mental illness: type, frequency, and associated outcomes. Psychosoc Rehabil J 17(2):163–172

Franche RL, Frank J, Krause N (2005) Prediction of occupational disability. In: Schultz IZ, Gatchel RJ (eds) Handbook of occupational disability claims: early risk identification, intervention, and prevention. Springer, New York, pp 93–116

Gates L, Akabas S (2010) Inclusion of people with mental health disabilities into the workplace: accommodation as a social process. In: Schultz IZ, Rogers ES (eds) Handbook of work accommodation and retention in mental health. Springer, New York

Gates LB, Akabas SH, Oran-Sabia V (1998) Relationship accommodations involving the work group: improving work prognosis for persons with mental health conditions. Psychiatr Rehabil J 21(3):264–272

Gerber PJ, Price LA (2003) Persons with learning disabilities in the workplace: what we know so far in the Americans with Disabilities Act era. Learn Disabil Res Pract 18(2):132–136

Gnam WH (2005) The prediction of occupational disability related to depressive and anxiety disorders. In: Schultz IZ, Gatchel RJ (eds) Handbook of complex occupational disability claims early risk identification, intervention, and prevention. Springer, New York

Goldner E, Bilsker D, Gilbert M et al (2004) Disability management, return to work and treatment. Healthc Pap 5(2):76–90

Granger B, Baron R, Robinson S (1997) Findings from a national survey of job coaches and job developers about job accommodations arranged between employers and people with psychiatric disabilities. J Vocat Rehabil 9:235–251

Greenwood R, Johnson VA (1987) Employer perspectives on workers with disabilities. J Rehabil 53:37–45

Haffey WJ, Abrams DL (1991) Employment outcomes for participants in a brain injury work re-entry program: preliminary findings. J Head Trauma Rehabil 6(3):24–34

Harnois G, Gabriel P (2000) Mental health work: impact, issues and good practices. Joint publication of the World Health Organization [WHO] and the International Labour Organization [ILO]. WHO/ILO, Geneva, Switzerland, pp 1–66

Haslan C, Atkinson S, Brown SS, Haslan RA (2005) Anxiety and depression in the workplace: effects on the individual and organization (a focus group investigation). J Affect Disord 88:209–215

Hirsh A, Duckworth K, Hendricks D, Dowler D (1996) Accommodating workers with traumatic brain injury: issues related to TBI and ADA. J Vocat Rehabil 7:217–226

Holzberg E (2001) The best practice for gaining and maintaining employment for individuals with traumatic brain injury. Work 16:245–258

Jacobs HE, Kardashian S, Kreinbring RK, Ponder R, Simpson AR (1984) A skills-oriented model for facilitation employment among psychiatrically disabled persons. Rehabil Couns Bull 24(2):87–96

Jacobs HE, Wissusik D, Collier R, Stackman D, Burkeman D (1992) Correlations between psychiatric disabilities and vocational outcome. Hosp Community Psychiatry 43:365–369

Job Accommodation Network [JAN] (2010) Accommodation search. Mental health impairments. [JAN is a contracted service funded by the US Government, Department of Labour, Office of Disability Employment Policy, under agreement J-(-M-2-0022)]. http://www.jan.wvu.edu/soar/disabilities.html. Accessed 11 Mar 2010

Kessler RC, McGonagle KA, Zhao S, Nelson CB, Hughes M, Eshleman S et al (1994) Lifetime and 12-month prevalence of DSM-III-R psychiatric disorders in the United States: results from the national comorbidity survey. Arch Gen Psychiatry 51(1):8–19

MacDonald-Wilson K (1994) Reasonable workplace accommodations for people with psychiatric disabilities [on-line]. http://www.bu.edu/cpr/reasaccom/employ-read-macdon.html

MacDonald-Wilson K, Rogers ES, Massaro J, Lyass A, Crean T (2002) An investigation of reasonable workplace accommodations for people with psychiatric disabilities: quantitative findings from a multi-site study. Community Ment Health J 38(1):35–50

MacDonald-Wilson K, Rogers ES, Massaro J (2003) Identifying relationships between functional limitations, job accommodations, and demographic characteristics of persons with psychiatric disabilities. J Vocat Rehabil 18:15–24

MacDonald-Wilson KL, Russinova Z, Rogers ES, Lin CH, Ferguson T, Dong S et al (2010) Disclosure of mental health disabilities in the workplace. In: Schultz IZ, Rogers ES (eds) Handbook of work accommodation and retention in mental health. Springer, New York

Mancuso LL (1993) Case studies on reasonable accommodations for workers with disabilities [on-line]. http://www.mentalhealth.samhsa.gov/_scripts/printpage.aspx

McFarlin DB, Song J, Sonntag M (1991) Integrating the disabled into the work force: a survey of Fortune 500 company attitudes and practices. Employee Responsib Rights J 4:107–123

McHugh SA, Storey K, Certo NJ (2002) Training job coaches to use natural support strategies. J Vocat Rehabil 17:155–163

Means CD, Stewart SL, Dowler DL (1997) Job accommodations that work: a follow-up study of adults with attention deficit disorder. J Appl Rehabil Couns 28(4):13–16

Mpofu E, Watson E, Chan S (1999) Learning disabilities in adults: implications for rehabilitation interventions in work settings. J Rehabil 65(3):33–41

Murray CJ, Lopez AD, Jamison DT (1994) The global burden of disease in 1990: summary results, sensitivity analysis and future directions. Bull World Health Organ 72:495–509

National Organization on Disability [NOD] (2004) Landmark disability survey finds pervasive disadvantages: 2004 NOD/Harris Survey documents trends impacting 54 million Americans [on-line]. http://www.nod.org/index.cfm?fuseaction=Feature.showFeature&FeatureID=1422. Accessed 11 Mar 2010

Rogan P, Banks B, Herbein MH (2003) Supported employment and workplace supports: a qualitative study. J Vocat Rehabil 19:5–18

Rogers ES, MacDonald-Wilson KL (2010) Vocational capacity among individuals with mental health disabilities. In: Schultz IZ, Rogers ES (eds) Handbook of work accommodation and retention in mental health. Springer, New York

Schultz IZ, Milner RA, Hanson D, Winter AT (2010) Employer attitudes and job accommodations in mental health. In: Schultz IZ, Rogers ES (eds) Handbook of job accommodations in mental health. Springer, New York

Standing Senate Committee on Social Affairs, Science and Technology (2004, November) Mental health, mental disorder and addiction: overview of policies and programs in Canada (Report 1). Government of Canada Printing Office, Ottawa, ON, pp 1–247

Stengler-Wenzke K, Beck M, Holzinger A, Angermeyer MC (2004) Stigma experiences of patients with obsessive compulsive disorders. Fortschr Neurol Psychiatr 72:7–13

Storey K, Certo NJ (1996) Natural supports for increasing integration in the workplace for people with disabilities: a review of the literature and guidelines for implementation. Rehabil Couns Bull 40(1):62–67

Warren CG (2000) Use of assistive technology in vocational rehabilitation of persons with traumatic brain injury. In: Fraser RT, Clemmons DC (eds) Traumatic brain injury rehabilitation: practical vocational, neuropsychological and psychotherapy interventions. CRC, Boca Raton, FL, pp 129–160

Weed RO, Brown TT (1993) Reasonable accommodation, job analysis, and essential functions of the job: issues in ADA implementation for the private sector rehabilitation professional. NARPPS J 8(3):126–130

Wehman P, Sherron P, Kergel J, Kreutzer J, Tran S, Cifu D (1993) Return to work for persons following severe traumatic brain injury: supported employment outcomes after five years. Am J Phys Med Rehabil 72(06):355–363

World Health Organization [WHO] (2001) World Health Report 2001. Mental health: new understanding, new hope. WHO, Geneva, Switzerland

Chapter 23
Disability Management Approach to Job Accommodation for Mental Health Disability

Henry G. Harder, Jodi Hawley, and Alison Stewart

Mental Health Disability in the Workplace: The Need for Disability Management

A body of research exists that outlines the personal, social, and economic costs of mental health disability (e.g., Dewa and McDaid 2010, this book; Dewa et al. 2007; WHO 2005). More recently, employers are becoming aware of the financial and human costs of mental health disability in their work sites. In fact, mental health disability is currently one of the leading causes of workplace absenteeism, resulting in significant wage loss and insurance costs (Stephens and Joubert 2001). Other costs associated with mental health disability include presenteeism, or working with impaired function; and spillover, or the effect of one worker's activities on another. Spillover effects include workplace conflict which negatively affects coworkers and productivity, because of exposure to difficult work conditions (Dewa and Lin 2000; Sanderson and Andrews 2006).

Despite advances in our general understanding of mental health, mental health disability presents some unique features which continue to provide challenges to the workplace. For example, persons with mental health disabilities frequently experience recurring or repeated bouts of illness and time off from work. In addition, workers experiencing mental health disability often go unrecognized and untreated (Dewa et al. 2004). Perhaps most vexingly, there remains a pronounced stigmatization associated with mental health disability which contributes to poor disclosure, subsequently limiting access to available accommodations that could assist workers in successfully maintaining their work role (Baldwin and Marcus 2010, this book; Ellison et al. 2003; Hatchard 2008; Kirsh 2000a).

To further complicate matters, symptoms of mental health disability may include reduced interpersonal skills, poor judgment, emotionality, withdrawal, and other behaviors which restrict the social capital of workers. In other words, there is a

H.G. Harder (✉)
Disability Management & Psychology, University of Northern British Columbia, 3333 University Way, Prince George, BC, V2N 42N, Canada
e-mail: harderh@unbc.ca

I.Z. Schultz and E.S. Rogers (eds.), *Work Accommodation and Retention in Mental Health*, 425
DOI 10.1007/978-1-4419-0428-7_23, © Springer Science+Business Media, LLC 2011

reduction in both the number of people who can be expected to provide support, and the resources these individuals have at their disposal (Podolny and Baron 1997). This restricted or absent social capital often results in compromised relations among coworkers and supervisors, which negatively affects return-to-work planning and outcomes (St.-Arnaud et al. 2006).

There are compelling social and economic reasons to promote the inclusion, productivity, and retention of persons with mental health disabilities in the work force, including legislation, nondiscrimination laws, and policies that encourage or require inclusive workplace practices (e.g., Center 2010, this book). Managing mental health disability at the work site makes sense on multiple levels, and initiatives for inclusivity are increasing in frequency as both employers and employees recognize the benefits.

Unfortunately, there are few studies examining the return-to-work process for workers with mental health disabilities (Goldner et al. 2004). Krupa (2007) described psychosocial interventions that focus on the individual, the employer, and workplace organizational levels as the most promising for improving employment outcomes among individuals with mental health disabilities. Briand and colleagues (2007) identified the need to adopt a common language to describe the work disability situation, taking into account the complexity and interaction of the dimensions of the person as well as the various systems involved in the rehabilitation process. Neufeldt (2004) suggested using a systematic approach to transfer existing knowledge on mental health issues into the workplace. While research results support an interdisciplinary, collaborative approach to manage mental health disability in the workplace, there is little guidance on how to operationalize such a collaborative process. Disability management requires a coordinated effort that addresses individual needs, workplace conditions, and legal responsibilities (NIDMAR 1993). In order for disability management programs to extend their scope to include persons with mental health disabilities, various psychosocial rehabilitation interventions need to be integrated with existing practices and policies. Disability management then has the potential to represent a viable investment in the mental health of workers.

This chapter discusses the utilization of disability management programs as an approach to accommodate persons with mental health disabilities. The concepts of disability management are reviewed, including an examination of effective psychosocial interventions that can be integrated into a disability management program, in order to facilitate successful return-to-work outcomes for workers with mental health disabilities.

Disability Management Defined

The main emphasis of disability management is to prevent or minimize the impact of the disability and to assist in job retention for workers who are disabled or injured (Akabas et al. 1992; Shrey and Lacerte 1995). Disability management has been defined as "a workplace prevention and remediation strategy that seeks to prevent disability from occurring, or lacking that, to intervene early following the onset of disability, using coordinated, cost-conscious, quality rehabilitation service that reflects an organizational commitment to continued employment of those experiencing functional work limitations" (Akabas et al. 1992, p. 2). Disability management developed as a cooperative work site-based strategy to manage the human and financial costs of disability in the workplace. The Certification of Disability Management Specialist Commission (CDMSC) states that disability management is a "collaborative process that utilizes assessment, planning, implementation, communication, coordination, and evaluation in the provision of preventative and remedial services to minimize the impact and cost of disability, to enhance productivity, and to promote maximum recovery and function" (Goodwin et al. 2000, p. 231). In other words, an effective disability management program utilizes a company's financial and human resources to promote a healthy workplace, while ensuring a company's profitability (Akabas et al. 1992; Dyck 2009).

There is growing evidence that disability management is an effective mechanism to facilitate cost reductions and successful return-to-work for workers with injuries or disabilities. Employers report substantial savings and better return-to-work outcomes as a direct result of using a disability management approach (Briand et al. 2007; Dyck, 2009; Habeck 1996; Watson Wyatt Worldwide 2000/2001). Although the mental health of workers is increasingly becoming a major concern, one of the barriers to assisting workers with mental health issues in the workplace is the limited information and guidance available on how to effectively deliver programs and services specific to the needs of workers and work sites (Briand et al. 2007; Rosenthal and Olsheski 1999). Current research supports disability management strategies that include employer participation, cooperation between management and labor, coordination between the workplace and healthcare/rehabilitation providers, workplace accommodations, and supernumerary replacements, all of which contribute to reducing the duration of work disability and associated costs (Franche et al. 2005; Williams and Westmorland 2002). Although there is empirically based evidence to support the disability management process for physical injuries, there has been little research conducted regarding what works for persons with mental health disabilities. Thus employer-based disability management programs are in a unique position to facilitate return-to-work planning and the implementation of accommodations for these workers, while evaluating and monitoring their effectiveness in mental health.

Disability Management: Key Concepts, Interventions, and Techniques

Key concepts of a workplace disability management program include a proactive process to minimize the impact of injury or disability in order to protect the employability of the worker (Shrey and Lacerte 1995). Disability management promotes prevention strategies, rehabilitation treatment concepts, and return-to-work programs designed to control the costs of workplace injury and disability. Disability management is interdisciplinary, and involves physical, psychological, social, vocational, medical, legal, and organizational factors that impact on employment (Harder and Scott 2005). The interdisciplinary nature of disability management is critical to the accommodation of workers with mental health disabilities.

Employment is a fundamental right of citizenship as noted in the United Nations Declaration of Human Rights (General Assembly of the United Nations 1948). Employment is also a known social determinant of health. Substantial research has demonstrated the benefits of working and the financial hardship, increased health risks, and social isolation associated with unemployment or underemployment (Raphael 2006). Disability management recognizes the social value of work. A comprehensive disability management program places work at the center of its model, based on the premise that work is central to life (Harder and Scott 2005). Given the significant unemployment issues affecting persons with mental health disabilities, this particular disability management concept is fundamental to the accommodation of mental health disability.

Disability management does not perceive disability as resulting only from an impairment; rather, it is the result of interactions among the worker and the workplace, the healthcare system, and the financial compensation system (Harder and Scott 2005). This interaction between the worker, workplace, and the environment is the foundation for the success of supported employment programs and psychosocial interventions for mental health disability (Briand et al. 2007; Kirsh and Gewurtz 2010, this book; Kirsh et al. 2009; Krupa 2007, 2010, this book; Olsheski et al. 2002; Peer and Tenhula 2010, this book). Organizational strategies utilized in disability management programs are consistent with the principles, practices, and objectives of psychosocial rehabilitation. Primary strategies emphasized in the rehabilitation of individuals with mental health disabilities include strengthening workers' skills, competencies, and environmental supports, in order to enhance their functioning so they can participate independently in all spheres of their lives, including work (Garske and McReynolds 2001). Disability management and psychosocial rehabilitation both recognize the value and effectiveness of services which are delivered in the environment in which the individual needs to function, i.e., the workplace.

Disability management provides the establishment of structure, policies, and procedures, the commitment of appropriate resources, and the development of the common values required for an inclusive workplace. Given the increasing financial and human resource costs associated with mental health disability, it is practical that employers accommodate persons with mental health disabilities in their disability management efforts. Integration of disability management and psychosocial models can provide employers and professionals with tools to accommodate and maintain employees whose work performance is compromised by mental health disability. Integration of psychosocial interventions within the workplace is one of the most influential interventions for persons with mental health disabilities, yet is considered to be one of the most challenging principles to achieve (Loveland et al. 2007).

As more and more disability management programs are implemented and evaluated, evidence is emerging for factors that constitute an effective workplace response to disability. Relevant literature on this subject, largely drawn from research on musculoskeletal injuries, suggests the following determinants of successful workplace-based programs:

1. A supportive workplace culture, including management and labor commitment with supportive policies. Many workplaces have less than ideal labor–management relationships, and this problem can have a negative impact on effective disability management. Having the two groups come together and support disability management efforts is key to its success.
2. Early intervention and ongoing monitoring of disability. Early, but timely, intervention has been shown to be a very important element in successful disability management programs (Harder 2003). It is equally critical to put appropriate outcome measures in place so that the disability management program can show its effectiveness over time.
3. A collaborative and coordinated team approach. Having rehabilitation professionals, including disability management practitioners, as members of multidisciplinary treatment teams avoids maintaining too narrow a focus on the recovery of the person and broadens the clinicians' perspective to include the world of work.
4. Psychoeducational opportunities for managers and workers. Lack of information in the workplace can defeat everyone's best efforts. Providing information on medical and disability benefits and other workplace support structures is very important. Equally critical is helping everyone understand the factors that are behind prolonged absences and lack of cooperation.
5. Systematic case management procedures. Ensuring that the required paperwork gets done on time and that everyone who needs to be informed and involved in disability management has been contacted is a simple but critical matter that must not be overlooked.
6. An organized return-to-work program with supportive policies, modified work options, and workplace accommodations. Having agreed upon policies and procedures ensures that everyone is treated equally and the process is well understood. Development of trust in the process and players is a pivotal factor in the maintenance of the collaborative alliance between all parties.
7. Use of incentives in benefit design, cost accounting, and performance evaluation to encourage participation. It is critical to the longevity of a disability management program that it is provided on a cost-effective basis. Ensuring that this cost-effectiveness is verified right from the start of the process is often an overlooked yet critical factor. Providing for this information requires:
8. An integrated management system to monitor and evaluate outcomes. Being able to show the efficacy and effectiveness of a disability management program

is essential to all parties concerned (Dyck 2009; Goldner et al. 2004; Harder and Scott 2005; Shrey and Lacerte 1995; Williams and Westmorland 2002).

These components of a successful workplace program are consistent with the principles and practices of successful employment and intervention strategies for mental health disability in general (Briand et al. 2007; Kirsh et al. 2009; Krupa, 2007, 2010, this book; Marrone et al. 1998; Olsheski et al. 2002).

Evolution of Disability Management

Disability management programs were developed in response to what Habeck (1996) referred to as the "broken paradigm" of vocational rehabilitation services. Habeck described traditional vocational rehabilitation as "reactive, provider-based, and clinical." Traditional vocational rehabilitation services were generally provided after the onset of the disability, rather than addressing prevention of disability or injury, and services were generally not connected to the workplace. Environmental factors associated with the workplace were largely ignored. Over time, evidence started to emerge demonstrating that successful employment was not just a function of individual disability symptoms and characteristics; rather, it involved the responsiveness of the work environment to the needs of employees with disabilities (Havranek 1995; McFarlin et al. 1991; Roessler and Gottcent 1994). Such organizational responsiveness is demonstrated through management policies and practices that support diversity and influence employment outcomes for workers. In concert with nondiscrimination employment legislation, employer-based disability management programs developed from the recognition that clinical interventions needed to be connected to, and engaging of, the workplace. Employers could reduce costs and get employees back to work earlier by actively participating in the rehabilitation and return-to-work process (Harder and Scott 2005).

Importantly, the disability management paradigm considers not only the impairment, but also the psychosocial and socioeconomic factors affecting the individual, as well as the provision of workplace interventions as early as appropriate (Loisel et al. 2001). Moreover, application of disability management programs has expanded techniques and integrated various disability-related benefit programs, including workers' compensation, group health, and short- and long-term disability. This application of disability management practices across the spectrum of employer disability programs is widely termed "integrated disability management" (Ahrens and Mulholland 2000; Calkins et al. 2000; Dyck 2009). Integrated disability management means that regardless of etiology, or whether the injury was occupational or nonoccupational, the intention of the disability management program is to provide rehabilitation and return-to-work services in a consistent and coordinated manner. Only recently have disability management programs expanded services to include mental health disability.

A comprehensive model of disability management demands expanded competencies for professionals engaged in these programs. Disability management professionals are responsible for providing leadership in the development and delivery of effective programs. This requires: (a) relationship skills, including advanced communication, interpersonal, and leadership skills; (b) understanding and mediating the return-to-work program; (c) coordinating multiple services and service providers; (d) facilitating organizational change; (e) knowledge of nondiscrimination employment legislation; and (f) assisting organizations to appreciate the unique contributions of people with disabilities (Ahrens and Mulholland 2000; Rosenthal et al. 2007). Providing disability management services for individuals with mental health disabilities requires the additional skills and knowledge specific to the diagnosis and treatment of mental health disabilities (Olsheski et al. 2002).

The Working Alliance

If successful outcomes are to be realized, a critical component of disability management is the facilitation of effective working relationships among the individuals with a disability and their family members, employers, unions, disability compensation payers, health care providers, and other stakeholders (Dyck 2009; Harder and Scott 2005; Koch and Rumrill 2005; Pransky et al. 2004; Shrey and Lacerte 1995). The need for effective working relationships is of critical importance in the context of the challenging issues associated with mental health disability in the workplace. Establishing these relationships represents a formidable task, given the divergent needs, values, and interests of the various stakeholders in the disability management process. Although research exists around teamwork, team operations, and team roles, information is lacking about the processes teams must go through as they create, develop, and maintain collaborative team practices (Orchard et al. 2005).

A potential framework for facilitating cooperation and building collaborative relationships among various stakeholders in a disability management program is the concept of "working alliance" (Koch and Rumrill 1998, 2005). This concept was developed in response to the evidence of the strong link between the client–therapist alliance and outcomes. The term "therapeutic alliance" was originally coined to describe a patient's attachment to the therapist. Bordin (1979) reconceptualized the therapeutic alliance construct, noting that the alliance was a collaborative effort in which all stakeholders contribute equally to the process. This view is different from both the client-centered and psychodynamic approaches. Bordin emphasized the negotiation of goals and tasks as an important step for alliance building and for overcoming the typical challenges of a collaborative process. Research in counseling and vocational rehabilitation literature provides evidence that the working alliance is a critical factor in the development of client satisfaction and successful rehabilitation outcomes (Lustig et al. 2002; Martin et al. 2000). There is also evidence establishing the working alliance as a contributory factor towards successful rehabilitation outcomes for persons with mental health disabilities (Donnell et al. 2004; Spaulding et al. 2003).

More recently, the working alliance construct has been applied as a model for interdisciplinary collaboration (Koch and Rumrill 1998; Koch et al. 2005). This collaborative process requires the development of strong emotional bonds, the establishment of mutually agreed upon goals, and a shared commitment to carrying out the tasks that are required for goal achievement (Bordin 1979). These three components – bonds, goals, and tasks – provide the building blocks of the working alliance (Bordin 1994).

Bonding

Bonds provide a common purpose and shared understanding for the partners of the working alliance. In collaborative relationships, bonds are developed through the communication of empathy, warmth, and genuineness (Koch and Rumrill 2005). This requires understanding the personal, political, practical, legal, economic, and ethical aspects of each participant involved in the disability management of an individual with a mental health disability.

The benefits of developing a strong working alliance with persons with mental health disabilities are integral to enhancing successful return-to-work outcomes (Donnell et al. 2004). Research has demonstrated that worker involvement in the disability management process improves outcomes and worker satisfaction, and contributes to a stronger working alliance (Chan et al. 1997). The working alliance is achieved through professional disclosure, which provides an explanation of the collaborative interdisciplinary process, the goals and objectives of disability man-agement (returning to work) as well as the role of the disability management pro-fessional, and the benefits and limitations of participating in the process (Shaw and Tarvydas 2004). Clarification of roles and expectations should be considered as an ongoing process, since discrepancies may arise along the way, often as a result of misperceptions of another person's responsibilities, roles, actions, or goals (Holmes et al. 1989).

Collaboration based on a relationship of interdependence and built on respect, trust, and understanding of the unique and complementary perspectives of each stakeholder in the disability management process can be accomplished with role socialization of the team members and respect for differences in values, beliefs, and responsibilities. In an interdisciplinary, participatory, collaborative process, each member of the team needs to discuss and share knowledge about: (a) their understanding of their own roles and expertise; (b) confidence in their abilities; (c) commitment to the values and ethics of their profession and/or the agency or institution they represent; (d) willingness to understand disciplinary practices; and (e) social accountability (Orchard et al. 2005). Once an interdisciplinary, participatory, collaborative process has been established, team members need to reach agreement on where disciplinary boundaries are unique and where they can be shared.

Goals

The next step of the working alliance involves the establishment of goals. The purpose of any working alliance involved with the return-to-work of an individual with a mental health disability is safe and timely return-to-work in a sustainable and cost-effective manner. In disability management programs, goals may involve short-term, intermediate, and long-term objectives. The immediate or short-term goal is to return the worker to sustainable employment, thus providing the personal, social and economic benefits that employment brings to the worker. The intermediate goal of disability management is to reduce the cost of disability for the employer and to improve the workplace culture, including the development and implementation of supportive policies and practices to embrace inclusion and diversity. The intermediate goal may also include preventative measures and the creation of a healthier work environment, specifically a work environment which is free of discrimination and understands and supports the inclusion of mental health disability in the workplace. A long-term goal of disability management, while not necessarily explicit, is to decrease the stigma and marginalization of persons with mental health disabilities in society as a whole.

Tasks

Once the goal of return-to-work has been collaboratively established, the members of the working alliance determine the specific activities that each team member will engage in to instigate or facilitate change (Bordin 1979). For the working alliance to be effective, all team members or stakeholders must perceive the tasks as relevant. There needs to be shared responsibility for the completion of the tasks. It is important for the working alliance to anticipate a successful outcome and to reinforce task performance, which can be accomplished through periodic appraisal and acknowledgment of tasks that the working alliance has completed (Koch and Rumrill 1998). However, tasks for achieving the goals may require significant negotiation and delegation among the working alliance members (Koch and Rumrill 2005). For example, effective psychosocial interventions for individuals with mental illness may suggest the presence of a workplace mentor or support person for the worker with a mental health disability. While the employer may recommend a supervisor to assume this role, the worker may identify a colleague with little training in behavioral competencies to support employment. Negotiation may be required to provide necessary accommodations that acknowledge the workers' understanding of their own unique employment challenges, strengths, and needs, while recognizing workplace issues from the employer's perspective.

Given the limited experience of employers, insurance payers, and other stakeholders with issues of mental health disability, significant psychoeducational efforts may be required regarding effective interventions for successful and sustainable employment outcomes for persons with mental health disabilities. An essential task for the working alliance will be development of the workplace psychoeducational process. This process needs to include the following components: (1) a disclosure plan which reviews the benefits and risks of the worker's decision to disclose their disability – should the decision be made to disclose, the process needs to include when and what should be disclosed, and to whom (see chapter by MacDonald-Wilson et al. 2010, this book); (2) a mechanism for determining who should be included in the psychoeducational process; and (3) the specific psychoeducational training content to be incorporated into the process, including appropriate nondiscrimination employment legislation policies (Gates 2000).

A key to the effective functioning of the working alliance is the manner in which disagreements and conflicts are handled; specifically, how to: (a) accept disagreement, (b) identify issues that are likely to cause dissent, and (c) develop a process to address dissent (Orchard et al. 2005). A conflict resolution approach for addressing discrepancies in the working alliance is a useful strategy for strengthening such alliances (Bordin 1979).

The first task in a disability management program is to promote investment in a healthy labor force. The culture of the workplace is a critical factor influencing the inclusion and retention of workers with mental health disabilities, potentially creating a workplace environment that is positive, flexible, collaborative, and inclusive (Akabas 1994; Kirsh 2000b; Kirsh and Gewurtz 2010, this book). To increase awareness of the relationship between workplace practices and positive mental health among employees, disability management addresses workplace development, changes in workplace policies and practices, and the promotion and creation of a culture of inclusion and accommodation. Studies have demonstrated that an increase in management supervisor awareness of accommodations contributed towards an organizational structure for accommodating workers, and raised the acceptance of employees with disabilities when a disability management program was in place (Bruyère 2000).

Case Example

Matilda is a married mother of three children. She was employed as a child care supervisor, which required her to make daily drives to and from various daycare locations. In the performance of her job duties, she was involved in a serious automobile accident that resulted in many physical injuries, for which she was treated. She recovered from these injuries, but developed psychological trauma symptomatology, including intrusive thoughts, avoiding activities and places associated with the accident, diminished interest in social activities, depressed mood, and an inability to work.

Eventually, Matilda was referred to her workplace-based disability management program. A disability management professional introduced Matilda to the collaborative disability management process that is described above. The objectives, benefits, limitations, and professional roles and responsibilities of the disability management process were outlined, and Matilda was provided with benefits counseling related to her workers' compensation, disability benefits, and other entitlements.

The disability management (DM) professional who was educated and experienced in mental health disabilities, recognized Matilda's emotional fragility and began by building rapport with her and aiming to understand the social and occupational demands, and the psychological and health issues that Matilda experienced, as well as the coping skills she possessed. The DM professional also worked to identify potential barriers to Matilda's returning to work. Development of this understanding is fundamental to planning and delivering appropriate interventions, as well as facilitating a successful return-to-work. Initially, Matilda was feeling pressured by her employer to return to work and did not know how to respond to the employer's inquiries. The DM professional provided critical assistance at the very beginning of the return-to-work process by addressing these issues.

The DM professional discussed with Matilda disclosure of her mental health symptoms as critical to engaging the collaborative process required for a successful return-to-work. Without her employer knowing about her trauma and resulting symptoms, planning a successful return-to-work for Matilda would have been difficult or impossible. The benefits of disclosure included gaining coworker and supervisor support to deal with potential workplace challenges. Once Matilda decided to disclose, a plan involving the what, how, when, and who of the disclosure process was developed. The disclosure plan was graduated to disclose to the employer, then to coworkers and friends, all in preparation for eliciting understanding and support in an eventually successful return-to-work.

With Matilda's permission and understanding of the collaborative process and the working alliance, the DM professional contacted the employer, union representative, healthcare providers, and other key participants in the process, in order to establish a working alliance and to outline the goals and tasks of Matilda's return to work. Matilda's employer expressed the need to have her back at work immediately to complete activities critical for the daycare's continued operation. The employer needed education with respect to the impact of the trauma that Matilda endured and the resulting decreased tolerance for stress and reduced energy level she experienced. At this point, the DM professional considered introduction of occupational accommodations to the employer as a way to increase the employee's functioning and capabilities once she returned to work. To achieve this goal, the DM professional, given knowledge of Matilda's job and her current symptoms, determined what would constitute a supportive workplace, and enlisted the employer's support to help create such a workplace environment. At the same time, the DM professional was sympathetic to the employer's concerns about the stress it posed for the organization.

The DM professional may need to arrange medical, psychological, and/or vocational assessments to assist with rehabilitation planning, with special attention to the needs expressed by the employee for his or her own return-to-work. In Matilda's case, psychological treatment was warranted to assist her to deal with the symptoms of her traumatic automobile accident. Treatment involving cognitive-behavioral therapy (CBT) was recommended for her symptoms. This included psychoeducation (i.e., providing Matilda with information and education about her current symptoms, their treatment, and prevention of relapse); skills training, including anxiety management (i.e., relaxation) and exposure therapy/trauma processing (i.e., helping individuals face feared situations like driving and trauma-related images and memories); and cognitive restructuring interventions, to assist Matilda to monitor and reduce maladaptive thoughts. Communication with Matilda's psychological treatment provider was critical to understanding her work capacity and to assist with a successful return-to-work.

Matilda's prior work capabilities and competencies were compromised by her psychological symptoms. Skills training was needed to assist Matilda to increase her self-confidence and self-efficacy, and to improve rehabilitation outcomes. It included participating in a defensive driving training program, an exercise and meditation program, stress management, self-care, pacing at work, communication and interpersonal skills training, and problem-solving skills training. The DM professional coordinated the appropriate mental health, medical, and ancillary services or interventions, including brief counseling for the family, with a focus on communication and adjustment to trauma. Addressing the major source of Matilda's posttrauma symptoms and the related avoidance of driving was central to her plan for returning to work. The DM professional negotiated with the insurance carrier to provide for Matilda's transportation while she participated in a driving training course. The psychologist and family members provided additional support while Matilda increased her confidence and her ability to drive safely.

Including family and friends in the return-to-work planning process is often useful; the involvement of social supports have been demonstrated to be effective in supporting work among individuals with mental health disabilities. The DM professional was able to ensure clear communication among all the stakeholders about expectations for Matilda's return to the workplace, clarify strategies and responsibilities among all the stakeholders, and establish time frames for her successful return to work.

As Matilda demonstrated increased resilience and motivation to return to work, her accommodation needs were addressed. The accommodations included not only physical changes to the work site, but also changes to job tasks, routines, and relationships among coworkers and supervisors. The most salient accommodations for mental health disabilities often involve training, orientation, and engagement of supervisors and coworkers, including sensitivity training regarding stress tolerance, pace of work, and interpersonal interactions. Accommodations may include decreased hours initially, allowing for a graduated return-to-work, an on-site mentor or support person, and reduced workload demands. Matilda's initial return-to-work was graduated, and she returned to a less demanding position; this had a number of benefits, including removing the extra burden on coworkers. The employer was relieved of the burden of reduced productivity and increased costs; and at the same time, there was reduced pressure on the returning worker to perform to a specified productivity level.

One of the challenges of the return-to-work process was assisting Matilda to maintain the skills she developed, such as pacing and stress management, while integrating these skills into her day-to-day life. Central to workplace retention at this juncture is the support and assistance of family, friends, and employers, along with continuing professional support. As Matilda resumed her work duties, it was important for all stakeholders in the return-to-work effort to assess the impact of their collaboration on Matilda's work outcome and make recommendations for future interventions. The DM professional can be central to leading this debriefing discussion.

Conclusion

Disability management is charged with preventing or minimizing the impact of disability on the employer and the employee, and assisting in job retention. In this context, disability management represents a comprehensive set of skills and services that are necessary to effectively address the case management, planning, and resource management needs of persons with mental health disabilities. Disability management professionals, particularly those with vocational rehabilitation and mental health backgrounds, can assist employers in assessing the needs of workers with mental health issues and in developing appropriate programs and interventions to accommodate and retain these workers. In order to develop successful working alliances with workers, employers, union representatives, family members, and healthcare professionals, disability management professionals need to be aware of appropriate psychosocial interventions, knowledgeable about employment nondiscrimination legislation, able to confront stigma and marginalization in the workplace, and understand the complex process of implementing accommodation and workplace modifications. Further, disability management professionals need to be continually cognizant of policy and practice changes which may positively (or negatively) impact the retention and productivity of persons with mental health disabilities. An effective working alliance has the potential to be created within a comprehensive disability management program based on an interdisciplinary and collaborative process.

Empirically based evidence supports workplace disability management as an effective approach for accommodating injured workers. With an increased awareness of mental health disability and its impact on the workforce, disability management has the potential to significantly reduce costs associated with persons with mental health disabilities. Significant new research and ongoing well-evaluated disability management applications and policies can contribute to the retention of workers with mental health disabilities who require appropriate workplace interventions and supports.

References

Ahrens A, Mulholland K (2000) Vocational rehabilitation and the evolution of disability management. J Vocat Rehabil 15(1):39–46

Akabas SH (1994) Workplace responsiveness: key employer characteristics in support of job maintenance for people with mental illness. Psychosoc Rehabil J 17(3):91–101

Akabas SH, Gates LB, Galvin DE (1992) Disability management: a complete system to reduce costs, increase productivity, meet employee needs, and ensure legal compliance. AMACOM, New York

Baldwin M, Marcus S (2010) Stigma, discrimination, and employment outcomes among persons with mental health disabilities. In: Schultz IZ, Sally Rogers E (eds) Handbook of work accommodation and retention in mental health. Springer, New York

Bordin ES (1979) The generalizability of the psychoanalytic concept of the working alliance. Psychother Theory Res Practice 16(3):252–260

Bordin ES (1994) Theory and research on the therapeutic working alliance: new direction. In: Horvath A, Greenberg L (eds) The working alliance: theory, research and practice. Wiley, New York, pp 13–37

Briand C, Durand MJ, St. Arnaud L, Corbière M (2007) Work and mental health: learning from return-to-work rehabilitation programs designed for workers with musculoskeletal disorders. Law Psychiatry 30:444–457

Bruyère SM (2000) Disability employment policies and practices in private and federal sector organizations. Cornell University School of Industrial and Labor Relations Extension Division, Program on Employment and Disability, Ithaca, NY

Calkins J, Lui JW, Wood C (2000) Recent developments in integrated disability management: implications for professional and organizational development. J Vocat Rehabil 15:31–37

Center C (2010) Law and job accommodation in mental health disability. In: Schultz IZ, Sally Rogers E (eds) Handbook of work accommodation and retention in mental health. Springer, New York

Chan F, Shaw LR, McMahon BT, Koch L, Strauser D (1997) A model for enhancing rehabilitation counselor–consumer working relationships. Rehabil Couns Bull 41:122–137

Dewa CS, Lin E (2000) Chronic physical illness, psychiatric disorder, and disability in the workplace. Soc Sci Med 51(1):41–50

Dewa C, McDaid D (2010) Investing in the mental health of the labor force: epidemiological and economic impact of mental health disabilities in the workplace. In: Schultz IZ, Sally Rogers E (eds) Handbook of work accommodation and retention in mental health. Springer, New York

Dewa CS, Lesage A, Goering P, Caveen M (2004) Nature and prevalence of mental illness in the workplace. Healthc Pap 5(2):12–25

Dewa CS, McDaid D, Ettner SL (2007) An international perspective on worker mental health problems: who bears the burden and how are costs addressed? Can J Psychiatry 52(6):346–356

Donnell CM, Lustig DC, Strauser DR (2004) The working alliance: rehabilitation outcomes for persons with severe mental illness. J Rehabil 70(2):12–18

Dyck DE (2009) Disability management: theory, strategy & industrial practice, 4th edn. LexisNexis, Markham, ON

Ellison ML, Russinova Z, MacDonald-Wilson KL, Lyass A (2003) Patterns & Correlates of Workplace Disclosure Among Professionals & Managers with psychiatric conditions. Journal of Vocational Rehabilitation 18(1):3–13

Franche RL, Cullen K, Clarke J, Irwin E, Sinclair S, Frank J, The IWH Workplace-Based Return-to-Work Intervention Literature Review Group (2005) Workplace-based return-to-work interventions: a systematic review of the quantitative literature. J Occup Rehabil 15(4):607–631

Garske GG, McReynolds C (2001) Psychiatric rehabilitation: current practices and professional training recommendations. Work 17:157–164

Gates LB (2000) Workplace accommodation as a social process. J Occup Rehabil 10(1):85–98

General Assembly of the United Nations (1948) Universal declaration of human rights. http://www.un.org/en/documents/udhr. Accessed 11 Nov 2009

Goldner E, Bilsker D, Gilbert N, Myette L, Corbière M, Dewa C (2004) Disability management, return to work and treatment. Healthc Pap 5(2):76–90

Goodwin BA, Taylor DW, Fong C, Currier K (2000) Perceived training needs of disability management specialist in five knowledge domains of disability management practices. Rehabil Educ 14:229–241

Habeck R (1996) Differentiating disability management and rehabilitation. NARPPS J 11(2):8–20

Harder HG (2003) Early intervention in disability management: factors that influence successful return to work. Int J Disabil Community Rehabil 2(2):1–8

Harder HG, Scott LR (2005) Comprehensive disability management. Elsevier, Toronto, ON

Hatchard K (2008) Disclosure of Mental Health. Work: A Journal of Prevention, Assessment & Rehabilitation 30(3):311–316

Havranek J (1995) Historical perspectives in rehabilitation counseling profession. In: Shrey DE, Lacerte MD (eds) Principles and practices of disability management in industry. GR Press, Winter Parks, FL

Holmes GE, Hull RW, Karst RH (1989) Litigation avoidance through conflict resolution: issues for state rehabilitation agencies. Am Rehabil 15(3):12–15

Holmes JG, Rempel JK (1989) Trust in Close Relationships. In Close Relationships, edited by Clyde Hendrick, Newbury Park, CA: Sage pp. 187–220

Kirsh B (2000a) Factors associated with employment for mental health consumers. Psychiatr Rehabil J 24(1):13–21

Kirsh B (2000b) Organizational culture, climate and person-environment fit: relationships with employment outcomes for mental health consumers. Work: J Prevent Assess Rehabil 14:109–122

Kirsh B, Gewurtz R (2010) Organizational culture and work issues for individuals with mental health disabilities. In: Schultz IZ, Sally Rogers E (eds) Handbook of work accommodation and retention in mental health. Springer, New York

Kirsh B, Stergiou-Kita M, Gerwurtz R, Dawson D, Krupa T, Lysaght R, Shaw L (2009) From margins to mainstream: what do we know about work integration for persons with brain injury, mental illness and intellectual disability? Work 32:391–405

Koch LC, Rumrill PD (1998) The working alliance: an interdisciplinary case management strategy for health professionals. Work 10:55–62

Koch LC, Rumrill PD (2005) Interpersonal communication skills for case managers. In: Chan F, Leahy ML, Saunders JL (eds) Case management for rehabilitation health professionals, 2nd edn. Aspen Professional Services, Linn Creeek, MO

Koch L, Egbert N, Coeling H (2005) The working alliance as a model for interdisciplinary collaboration. Work 25:369–373

Krupa T (2007) Interventions to improve employment outcomes for workers who experience mental illness. Can J Psychiatry 52(6):339–345

Krupa T (2010) Approaches to improving employment outcomes for people with serious mental illness. In: Schultz IZ, Sally Rogers E (eds) Handbook of work accommodation and retention in mental health. Springer, New York

Loisel P, Durand M, Vézina N, Baril R, Gagnon D et al (2001) Disability prevention: new paradigm for the management of occupational back pain. Dis Manag Health Outcomes 9(7):351–360

Loveland D, Driscoll H, Boyle M (2007) Enhancing supported employment services for individuals with a serious mental illness: a review of the literature. J Vocat Rehabil 27:177–189

Lustig DC, Strauser DR, Rice ND, Rucker TR (2002) The relationship between the working alliance and rehabilitation outcomes. Rehabil Couns Bull 46:25–42

MacDonald-Wilson KL, Russinova Z, Rogers ES, Lin CH, Ferguson T, Dong S, Kash MacDonald M (2010) Disclosure of mental health disabilities in the workplace. In: Schultz IZ, Sally Rogers E (eds) Handbook of work accommodation and retention in mental health. Springer, New York

Marrone J, Grandolfo C, Gold M (1998) Just doing it: helping people with mental illness get good jobs. J Appl Rehabil Counsel 29(1):37–48

Martin DJ, Garske JP, Davis MK (2000) Relationship of the therapeutic working alliance with outcomes and other variables: a mega-analytic review. J Counsult Clin Psychol 68:438–450

McFarlin DB, Song J, Sonntag M (1991) Integrating the disabled into the workforce: a survey of Fortune 500 company attitudes and practices. Employee Responsib Rights J 4(2):107–123

National Institute on Disability and Rehabilitation Research (NIDMAR) (1993) Rehab brief: strategies to secure and maintain employment for people with long term mental illness. Rehab Brief 15(10):1–4

Neufeldt AH (2004) What does it take to transform mental health knowledge into workplace practice? Towards a theory of action. Healthc Pap 5(2):118–132

Olsheski JA, Rosenthal DA, Hamilton M (2002) Disability management and psychosocial rehabilitation: considerations for integration. Work 19:63–70

Orchard CA, Curran V, Kabene S (2005) Creating a culture for interdisciplinary collaborative professional practice. Medical Education Online. http://www.med-ed-online.org. Accessed 11 Nov 2009

Peer J, Tenhula W (2010) Employment interventions for mood and anxiety disorders. In: Schultz IZ, Sally Rogers E (eds) Handbook of work accommodation and retention in mental health. Springer, New York

Podolny JM, Baron JN (1997) Resources and relationships: social networks and mobility in the workplace. Am Sociol Rev 62:673–693

Pransky G, Shaw WS, Franche R-L, Clark A (2004) Disability communication and prevention among workers, physicians, employers, and insurers: current models and opportunities for improvement. Disabil Rehabil 26(11):625–634

Raphael D (2006) Social determinants of health: an overview of the implications for policy and the role of the health sector. http://www.phac-aspc.gc/ph-sp/phdd/overview_implications/01~overview.html. Accessed 12 Nov 2009

Roessler RT, Gottcent J (1994) The work experience survey: a reasonable accommodation/career development strategy. J Appl Rehabil Counseling 25(3):119–126

Rosenthal DA, Olsheski JA (1999) Disability management and rehabilitation counseling. Present status and future opportunities. J Rehabil 65(1):31–39

Rosenthal DA, Hursh N, Lui J, Isom R, Sassion J (2007) A survey of current disability management practice: emerging trends and implications for certification. Rehabil Couns Bull 50(2):76–86

Sanderson K, Andrews G (2006) Common mental disorders in the workforce: recent finding from descriptive and social epidemiology. Can J Psychiatry 51(2):63–75

Shaw LR, Tarvydas V (2004) The use of professional disclosure in rehabilitation counseling. Rehabil Couns Bull 45(1):40–48

Shrey DE, Lacerte MD (1995) Principles and practices of disability management in industry. GR Press, Winter Parks, FL

Spaulding WD, Sullivan ME, Poland JS (2003) Treatment and Rehabilitation of Severe Mental Illness. Guilford Press: New York

St.-Arnaud L, St.-Jean M, Damassem J (2006) Towards an enhanced understanding of factors involved in the return-to-work process of employees absent due to mental health problems. J Community Mental Health 25(2):303–315

Stephens T, Joubert N (2001) The economic burden of mental health problems in Canada. Chron Dis Can 22(1):18–23

Watson Wyatt Worldwide (2000/2001) Staying@Work: the dollars and sense of effective disability management. Watson Wyatt Worldwide, Toronto, ON

Williams RM, Westmorland M (2002) Perspectives on workplace disability management: a review of the literature. Work 19:87–93

World Health Organization (WHO) (2005) Mental health policies and programmes in the workplace. World Health Organization, Geneva, Switzerland

Part VI
Future Directions

Chapter 24
Best Practices in Accommodating and Retaining Persons with Mental Health Disabilities at Work: Answered and Unanswered Questions

Izabela Z. Schultz, Terry Krupa, and E. Sally Rogers

Introduction

Although growing in the last decade, the literature on employment outcomes for persons with mental health disorders continues to be fragmented, with research evidence not yet sufficient to develop best practice guidelines for employment interventions or job accommodations based on the published literature alone. While there are evidence-based employment interventions for individuals with severe mental illness, there is less of an evidence base for individuals with anxiety and mood disorders and mild to moderate brain injury, including both those who are not job attached and those already in the workforce. The intention of this book is to provide readers with an understanding of current models and interventions, research, and practice regarding work among individuals with mental health disorders, and to address the unresolved issues and barriers to the integration and transfer of this knowledge to practice. Many questions were addressed by the contributors to this book, but many remain unanswered. This chapter will review and integrate the themes presented in this book, as well as highlight future research, policy, and practice directions.

Need for a New Framework

There is a gap between the predominance of medical and treatment-focused models for persons with mental health disabilities and the reality of the workplace environment. Medical models, while useful for optimizing treatment and controlling symptoms, are not useful as a framework for maximizing work capacity or job accommodations in the workplace (Schultz et al. 2007). A system-based, social

I.Z. Schultz (✉)
Vocational Rehabilitation Counselling Program, University of British Columbia,
Scarfe Library 297, 2125 Main Mall, Vancouver, BC, V6T 1Z4, USA
e-mail:ischultz@telus.net

I.Z. Schultz and E.S. Rogers (eds.), *Work Accommodation and Retention in Mental Health*, 445
DOI 10.1007/978-1-4419-0428-7_24, © Springer Science+Business Media, LLC 2011

model of mental health disability, with service and research emphases on function and functional accommodations, would seem more suitable (Gates and Akabas 2010) in combination with the biopsychosocial model integrating clinical and occupational interventions.

Currently, no guidelines for job accommodations and employment retention for persons with mental health disabilities have been published, despite the mounting research worldwide on similar interventions for individuals with physical and pain-related disabilities. Yet societal, professional, and economic concerns are escalating over the low rates of employment of persons with mental health disabilities, decreased productivity, "presenteeism," premature job terminations, and lower rates of job retention. These factors, together with what some refer to as a spiraling "mental health disability epidemic" in the workplace (Bond et al. 2001; Dewa and Lerner et al. 2010; McDaid 2010), suggest a problem of significant magnitude.

The rationale for development of guidelines is also embedded in epidemiological and health economics research. This research evidence shows a high and increasing prevalence of mental health disorders, accompanied by a strongly negative economic impact for society, the workplace, healthcare and compensation systems, and for persons with mental health disabilities themselves (Baldwin and Marcus 2010; Dewa and McDaid 2010; Gnam 2005). Moreover, healthcare and rehabilitation services for persons with mental health disabilities still primarily rely on the medical model that separates mind and body and focuses on psychopathology (Schultz et al. 2000, 2007; Schultz and Gatchel 2005). Such models are ill equipped to provide occupationally relevant and effective interventions that facilitate long-term employment.

The provision of appropriate work accommodations, with the goal of enhancing employment retention for individuals with mental health disabilities, is challenging for many reasons. The complex and cyclical nature of mental health disabilities can be exacerbated by a wide range of stressors, inherent in daily life and work environments. The specific nature and impact of these stressors still need to be better understood. Furthermore, societal stigma and employers' negative attitudes and fears often lead to a lack of awareness and preparation for employees with mental health disabilities (Schultz et al. 2010b, Baldwin and Marcus 2010; Wang 2010). These problems are compounded by the absence of an integrative conceptual model of employment interventions in mental health and solid research evidence informing workplace policies and practices for employees with mental health disabilities.

Multisystem Interventions in Occupational Mental Health

Work accommodations and interventions for persons with mental health disabilities show promise for improving employment outcomes, but to be maximally effective, they should be offered in an integrated and multisystemic manner. Interventions and accommodations for persons with mental health disabilities can be conceptualized at the following three levels:

(1) macrosystem interventions: focusing on society, culture, legislation, and policies;
(2) employer-level (mesosystem) interventions; and
(3) employee-level (microsystem) interventions.

These interventions will be discussed in this order.

Macrosystem Interventions

These large-scale interventions aim to change societal attitudes at large and generally involve legislative, policy, cultural, or systemic changes in both the public and private sectors. A review of the literature suggests that the field could benefit from a society-wide *public awareness* campaign using a social marketing approach, such as public advertising, one that targets attitudes towards mental health and associated societal-behavioral changes (Schultz et al. 2010c). Negative societal attitudes, stigma, fears, and lack of knowledge are critical barriers to employment of persons with mental health disabilities that can be best addressed at the level of the macrosystem. Such attitudes and concerns permeate society, and the expectations held by employers, coworkers, disability compensation and management systems, and society at large. Social stigma is also a major barrier in the workplace for persons with mental health disabilities (Baldwin and Marcus 2010; Haslan et al. 2005; Wang 2010) and is often more disabling than the primary mental health condition. The anticipated outcome of social marketing interventions would be the gradual dissipation over time of the stigma surrounding mental health problems and the integration of mental health into an expanding concept of health and wellness (Gallie et al. 2010; Wang 2010; Kirkwood and Stamm 2006).

Another important macrosystem intervention includes *legislative and public policy support* for mental health parity in healthcare, insurance, compensation, and rehabilitation systems, as well as fairness in employment practices for persons with mental health conditions (Baldwin and Marcus 2010; Center 2010; Gallie et al. 2010; Standing Senate Committee 2006).

Importantly, *multisystem interaction*, involving the integration of clinical and vocational rehabilitation services, is essential, as demonstrated by the strong and growing evidence for integrated clinical and occupational interventions using a biopsychosocial model in the field of musculoskeletal pain disability (Schultz et al. 2007; Schultz and Gatchel 2005). Integration of mental health and vocational services using a multidisciplinary team approach has been found effective in enhancing employment retention of persons with mental health disabilities (Burke-Miller et al. 2010; Bond et al. 1997; Drake et al. 2003; Macias et al. 2001; Arthur 2000; Berndt et al. 1998; Bishop 2004; Bond et al. 1997; Holzberg 2001; Johnstone et al. 2003; Means et al. 1997; Scheid 1998; Sherbourne et al. 2001; Smith et al. 2002). Furthermore, in Canada, the high profile Global Business and Economic Roundtable on Mental Health (2005) has highlighted the importance of addressing the psychological stress experienced by employees in contemporary workplaces and the need to support their intellectual and emotional capacities in order to meet growing business demands for innovation and global competition (Krupa 2007).

To facilitate multisystem interaction, employers need an organizational infrastructure that allows for communication and collaboration with other stakeholders, i.e., healthcare and insurance/compensation systems. Reciprocal multisystem interactions that involve the healthcare system, compensation system, employer, union, and the worker, with a common goal of promoting job retention, are essential to effect lasting paradigm changes in attitudes, organizational structure, policy, and practices.

Employer-Level (Mesosystem) Interventions

These workplace organization-based interventions focus on the extent to which the workplace itself can be constructed to facilitate job retention and health. We know from existing research that there are workplace features that are associated with psychological health. A workplace culture that values opportunities for control and decision-making; the full utilization of worker capacities and skills; opportunity for a variety of workplace activities; employee involvement; reasonable and well-integrated job demands; clear and predictable work expectations and conditions; interpersonal contacts; valued social positions in the workplace; and productivity connected to gains and rewards – all seem to be associated with greater psychological benefits for employees (Kirsh and Gewurtz 2010; Krupa 2007; Vézina et al. 2004; Warr 1987). Such a culture constitutes an optimal environment for work accommodations and retention interventions.

Employer-level interventions can be divided into Level I interventions and Level II interventions. Level I interventions focus on primary prevention of occupational disability and disease, and promotion of health and wellness in the workplace. Level II interventions focus on secondary prevention of occupational disability, which includes early detection, identification, and intervention with mental health problems in the workplace to prevent or reduce emerging disability.

Level I Intervention: Primary Prevention

Results of a recent series of studies (Schultz et al. 2010a, b; Gallie et al. 2010; Krupa 2010a, b) and a synthesis of the literature (Schultz et al. 2010c) converge to suggest that there are a number of employer-based interventions that are likely to facilitate employment outcomes for persons with mental health disabilities.

Organizational Restructuring: Integrative Approach

The literature suggests that a workplace organizational structure that integrates and synchronizes employer resources in occupational health and safety, health and wellness, disability management, return-to-work, corporate training, and Employee Assistance Programs (EAPs), in service of health promotion and disability prevention, and in the climate of inclusion of workers with mental health disabilities, would be beneficial (Gallie et al. 2010). Experts seem to concur that provision of EAPs is a hallmark of a caring employer, yet EAPs alone (as presently constructed), are insufficient to retain individuals with mental health disabilities in the workplace. EAPs are helpful in addressing life issues and emotional concerns, but often do not focus on employment problems and job accommodations. EAPs should be part of a multifaceted intervention that addresses *both* the clinical and occupational needs of employees.

Written Policies on Health and Wellness

Ultimately, workplace attitudes are translated into specific policies and practices that help attach an employer to his or her employees. These policies and practices, in turn, affect the attitudes of management and employees. A commitment to the well-being and productivity of employees with mental health disabilities should be expressed in written company policies along with policies on health, wellness, stigma, discrimination, and employee privacy/confidentiality protection. Developing policies on diversity and including mental health disability on that spectrum is one additional way of committing to greater workplace tolerance (Kirsh and Gewurtz 2010).

Budgeting for Costs of Job Accommodations

It is important to have specific policies regarding the payment and costs of providing job accommodations. Such costs may include the following: extending additional paid and unpaid leave during a hospitalization or other absence; allowing additional time for workers to reach performance milestones; extending probationary periods; allowing an employee to make phone calls during the day to personal and professional supports; providing a private space in which to make such phone calls; providing a private space to rest or talk to supportive coworkers; allowing an employee to work at home; and allowing workers to consume fluids at their workstation (for example, due to medication side effects) (as cited by Bruyère 2000). The scant research on costs of accommodations for individuals with mental health disabilities suggests that employers may not have large out-of-pocket expenses; rather, accommodations in mental health often involve creative and flexible solutions that are relatively low in cost (MacDonald-Wilson et al. 2002).

Training of Management and Supervisors

Training of management and supervisory staff in early detection and intervention for mental health-related performance difficulties, combined with awareness raising and stigma reduction, and developing the best organizational conditions for a mentally healthy work place, are of critical importance (Gallie et al. 2010). Training employers and supervisors in the nature and importance of job accommodations is also important (MacDonald-Wilson et al. 2001, 2002).

Level II Interventions: Reducing Occupational Disability and Enhancing Employment Outcomes

There is growing recognition in both research and practice that supporting the employment of individuals with mental health disabilities is a complex process, requiring an integrated organizational framework to ensure that interventions are provided in a comprehensive, coordinated, fair, and cost-effective manner. Two of the most well-developed and well-disseminated organizational frameworks for meeting the employment needs of people with mental health disabilities are disability management (for those already work attached) and supported employment approaches (generally for those with more severe mental health disabilities). Job-employee matching, postulated as a human resource strategy enhancing employment retention in mental health, could be incorporated into either of these frameworks. Recognition and systematic targeting of employer-based economic, social, and organizational barriers to job accommodation would be a prerequisite to the development and implementation of best evidence–informed practices in work accommodation and improved employment outcomes in mental health.

Disability Management

Disability management programs are employer-directed programs and practices that are focused on preventing work disability and facilitating rapid return-to-work (Currier et al. 2001; Habeck and Hunt 1999; Harder et al. 2010). Such programs typically involve a single management plan designed to ensure implementation of key disability management practices (Westmorland and Buys 2002). Such approaches can support labor and management relationships by balancing both economic and productivity needs with the sustained health and well-being of the workforce (Scully et al. 1999). This promising approach is restricted, however, to those already in the workforce, and thus does not apply to those who are not work attached: for example, those whose mental health disability is severe enough to present barriers to obtaining employment. Other limitations of the approach may include the lack of sensitivity to complex interactions between an individual's mental health and organizational practices, and the employer's lack of preparation for dealing with individuals with mental health disabilities.

Supported Employment

Supported employment refers to models of service delivery for individuals who have a disability and are unemployed and which focus on securing employment quickly and providing "wrap around supports," rather than long periods of training and preparation prior to job placement (Anthony and Blanch 1989). The goal of the evidence-based Individual Placement and Support model (IPS) of supported employment in particular is to help individuals with severe mental illness rapidly enter into competitive jobs in the community (Bond 1998). While the IPS model has repeatedly demonstrated its effectiveness in helping individuals achieve competitive employment (Bond et al. 1997; Crowther et al. 2001a, b; Drake et al. 1996), it has not yet been validated outside of the spectrum of severe mental illness for more prevalent mental health conditions such as mood and anxiety disorders. In addition, job retention, even among individuals receiving the IPS model, remains unacceptably low (Bond et al. 2008). Such problems could be partially addressed by job-employee matching and a systematic, integrated approach to job accommodations.

Job-Employee Matching

A number of studies emphasize the importance of matching a worker's interests and abilities with the appropriate job in order to improve job retention (Haffey and Abrams 1991; Mueser et al. 2001). The importance of determining functional limitations and capacity has been underscored in the literature and linked to more effective job accommodations (MacDonald-Wilson et al. 2003). The notion of a "person-environment fit" or an ecological approach to job matching (Szymanski and Vancollins 2003), with attention to the goals of the individual as well as to the workplace and the organizational culture, is likely to be even more critical for individuals with mental health disabilities.

Recognizing and Targeting Barriers to Job Accommodation

Barriers to implementation of job accommodations at the workplace, in addition to negatively-held attitudes and fears, often arise from a lack of education, knowledge, and awareness among employers, coworkers, and individuals with mental health disabilities. Employers lack understanding and knowledge about mental health symptoms and their impact on work performance. Many employers are minimally aware of the accommodation needs of employees with mental health disabilities or the types of accommodations that can be helpful (Schultz et al. 2010b). The expert interviews on job accommodations documented in this book also uncovered the unrealistic expectations about recovery and work functioning held by employers (Schultz et al. 2010a).

In one of the previous chapters, we reported a survey finding that while only 40% of employers reported having previously hired persons with mental health disabilities, those who had were more likely to use job accommodations (Schultz et al. 2010b). Similarly, results suggested that a past positive experience with individuals with mental health disabilities was associated with employers being more receptive to hiring such individuals in the future (Schultz et al. 2010a).

Organizational, policy, and resource-related challenges at the employer level also need to be underscored. Most barriers to job accommodations are due to systemic and contextual factors that limit employers' willingness and ability to implement such initiatives. Our research suggests that the size of the company and the type of industry are key factors in whether employers provide job accommodations. Larger companies (as compared to small and mid-size companies) typically have the financial and structural resources to offer such services, and as a result more often have established policies regarding hiring persons with mental health disabilities and offering job accommodations. Having a company policy about hiring individuals with disabilities tends to be predictive of the use of job accommodations. These two factors are generally stronger predictors of providing job accommodations than other factors, including employer attitudes. Our research findings indicate that the systemic limitations faced by employers are key barriers to the provision of such services even if they have more positive attitudes and a willingness to hire persons with mental health disabilities (Schultz et al. 2010a, b).

Also of concern is employers' sole reliance on a narrow understanding of the disability management model. This model implies a coordination of services and the maintenance of employment, often in the absence of direct intervention with the employee and without sufficient recognition of the clinical and psychosocial complexities of the workplace. Likewise, excessive reliance on traditional external Employee Assistance Programs for counseling, in the absence of occupational interventions with attention to system-related factors, has been noted to be problematic (Gallie et al. 2010). At the same time, relationship-based accommodations including natural supports in the workplace (e.g., using coworkers) are often underutilized by employers. Employers are least aware of job accommodations for persons with mental health disabilities that require flexibility in redesigning or reorganizing work processes: for example, working from home, job sharing, establishing a distraction-free work space, and the appropriate use of work scheduling and breaks (Schultz et al. 2010c).

Best Evidence–Informed Work Accommodations

The contributing chapters in this book identify a number of employment interventions and job accommodations for persons with mental health disabilities, together with research and expert evidence for their use. Notably, research to validate various interventions and accommodations is in its infancy and caution should be exercised when utilizing such approaches. Emphasis should be placed on those interventions and accommodations that have the best evidentiary support to date.

Nonetheless, a number of very useful best practices has emerged from a research literature synthesis (Schultz et al. 2010c), expert interviews (Schultz et al. 2010a), and an employer survey (Schultz et al. 2010b). The research evidence available suggests that the following interventions and accommodations could be useful to ameliorate the social, emotional, or cognitive limitations associated with mental health disabilities:

- modified work duties,
- flexible scheduling,
- modified work environment,
- job sharing, and
- assistive technologies.

However, research is still lacking on: (1) the characteristics of those who are most likely to benefit from these accommodations, (2) what organizational circumstances are needed to optimize the outcomes of accommodations, and (3) at what point in time, by whom, and how specific work accommodations are to be introduced in the workplace. Moreover, the effectiveness of work accommodations, especially those involving social interactions and processes, rather than technology and changes in organization, remains an open research question.

Employee-Level (Microsystem) Interventions

The employee-level interventions focus on enhancing workers' mental health functioning, productivity, and readiness for job accommodations.

Early Identification, Diagnosis, and Treatment

Early identification and diagnosis among employed workers can facilitate recognition of employment difficulties related to mental health symptoms that may be alleviated through appropriate pharmacological or psychological treatments. Early recognition and diagnosis have gained prominence, particularly for mental health disorders such as depression and anxiety that are highly prevalent in the workplace, that are disruptive to the individual's work performance, and that may be effectively treated, but that frequently go undetected (Bender and Kennedy 2004; Lerner et al. 2010; Wald 2010). Careful and comprehensive diagnosis increases the likelihood that people will receive the most appropriate interventions available (Lerner et al. 2010; Peer and Tenhula 2010; Wald 2010). In addition, it has been suggested that employers depend on the recommendations of qualified health professionals to validate requests for reasonable accommodations, particularly for mental health disabilities that are "invisible" and subsequently subject to suspicion (Kirsh and Gewurtz 2010; Krupa 2010a, b; Krupa et al. 2005).

Access to treatment is not a guarantee of effective treatment response or enhanced workplace productivity. Pharmacological treatments are not uniformly effective and are known to have adverse side effects that can interfere with work function. In addition, individuals may not receive treatments that are consistent with even minimal clinical practice guidelines in mental health (Allen 2004). Bender and Kennedy (2004), for example, report that despite practice guidelines, the medication prescription patterns of disability claimants often do not meet recommendations for dose and duration.

Client-Centered Assessment and Planning

Job-person matching is one of the key dimensions of client-centered assessment in the workplace (Kirsh and Gewurtz 2010). Effective employment interventions are grounded in assessment and planning processes that are characterized as client-centered, comprehensive, ecological, and dynamic. Client-centered practice is based on the assumption that individuals with mental health disabilities can and should be actively involved in both identifying their unique employment problems, strengths, and needs, and in the design, implementation, and ongoing evaluation of work-related interventions (e.g., Anthony et al. 2002; Rogers et al. 2006).

Coping Skills Training

Employment interventions have focused on the development of coping skills in those areas that have proven to be particularly problematic for individuals with mental health disabilities. These efforts are directed at enhancing the individual's response to difficult work situations through effective management of their emotions and the development of an effective repertoire of behavioral skills. They are also intended to encourage a personal attitude of hope and control, and consequently the individual's active involvement in their own employment performance and retention (Krupa 2007). Evaluations of these interventions have demonstrated their potential for positive change in specific competencies and the application of these competencies to the work setting (for example, Lerner et al. 2010; Peer and Tenhula 2010; Smith et al. 1996; Wallace et al. 1999).

Self-Awareness Counseling

Studies have suggested that there is a relationship between successful employment outcomes, personal self-awareness, and an attitude of control (Gerber et al. 1992; Goldberg et al. 2003). Self-awareness, as it relates to employment, refers to the individual's recognition of personal strengths and limitations, including an understanding of the specific nature of their disability and how it is experienced in the context of work. Control refers to the individual's attitude that personal efforts can be directed towards adjusting to and succeeding in employment. Self-awareness and control are considered essential to effective job accommodations (Krupa 2007). Ultimately, employers depend on workers to be self-advocates to define their needs in the workplace, to engage in planning for accommodations, and to evaluate the success of these efforts (Gerber and Price 2003).

Social Network Development

A social environment, constructed in a way that facilitates an individual's employment success, is a consistent and important theme in the literature (for example, Carling 1993; Gates 2000; Gerber et al. 1992; Goldberg et al. 2003; Tse 2002). Natural supports, acquaintances, or coworkers who are not paid service providers but are invested in an individual's sustained employment, have long been associated with positive work outcomes. They are believed to enhance the individual's own efforts by providing guidance and advice, practical and material supports, and by increasing the esteem that encourages belief in self and emotional stability (Goldberg et al. 2003; Krupa 2007). Efforts directed to identifying and recruiting appropriate natural supports, building awareness of the key support activities and their potential impact, developing consensus about reasonable employment expectations among the support network, as well as establishing formal contracts to define expected supports, are all hallmarks of this type of employment intervention (Krupa 2007).

Work Hardening

Although generally underreported in the literature, work hardening interventions have been applied in an effort to improve the employment outcomes of people with mental health disabilities (Finch and Wheaton 1999). The assumption behind work hardening is that cognitive, emotional, and psychological functions (for example, fatigue, concentration, and tolerance for stress) can be strengthened in a systematic way using graded programs (Krupa 2007). Case studies of work hardening appear in the literature and use progressively more complex simulated or real workplace tasks (Radley 2004; Wisenthal 2004) and focused attention training designed specifically with the goal of improving workplace function (Tryssenaar and Goldberg 1994).

The provision of on-the-job training can be an effective means for helping persons with mental health disabilities undergo work hardening. Such an approach may allow new employees to develop job skills with additional support and supervision. and to apply systematic strategies for managing their functional difficulties on the job.

Job Coaching

Job coaching refers to the process of having a service provider assist the employee to obtain a job, to perform work tasks once employed, or to meet the cultural and social demands of the workplace. In the literature, this intervention is typically discussed in relation to individuals with severe mental health disabilities who have ongoing employment support needs (e.g., Burke-Miller et al. 2010; Fraser et al. 2010; Targett and Wehman 2010) and whose engagement in the world of work or return-to-work is complicated by a greater functional incapacity, a significant time period away from the workforce. and/or a limited employment history. Job coaching is described as a flexible employment intervention, delivered in a manner that meets the needs of the individual and the employment situation (see Burke-Miller et al. 2010; Fraser et al. 2010; Targett and Wehman 2010). Job coaching may include a wide range of activities, including job development, assistance with learning important job tasks and expectations, assistance with understanding and managing social dynamics in the workplace, identification of needed workplace accommodations, and encouragement to persist at work (Cook and Pickett 1994; Marrone et al. 1998; Murakami 1999; Nimgade 2003; Yasuda et al. 2001).

Conclusions and Recommendations for Practice

The design, implementation, and evaluation of job accommodations and interventions for persons with mental health disabilities are more complex and multidimensional than the development of analogous interventions and job accommodations for persons with physical disabilities. Such interventions are evolving from legislative, policy, attitudinal, employer, and disability stakeholder perspectives as well as from a research perspective.

Development of evidence-based interventions to assist individuals to return to work, enter the workforce for the first time, or return to existing employment continues to be needed. Despite the compilation of a seemingly comprehensive list of job accommodations by the US Department of Labor Job Accommodation Network (2007), research and evidentiary support for most accommodations is just emerging. Unfortunately, it is sparse, fragmented, and difficult to integrate. In addition, employers are unaware of many important accommodations. Despite the considerable stigma surrounding mental health conditions, employers' awareness of job accommodations depends less on their attitudes and more on the size of the company, the existence of written policies in regards to accommodations, and prior experience in hiring a person with a mental health disability (Schultz et al. 2010b). Solutions to the problem of poor awareness and underutilization of job accommodations are multisystemic and multilayered, involving the need for a significant paradigm shift in conceptualizing and accommodating mental health disability in the workplace among key stakeholders.

Systemic solutions center around the following issues:

1. The need for macro-level interventions to change public attitudes towards persons with mental health disabilities, including overcoming stigma, social marketing, awareness raising, and changes in legislative health and employment policies;
2. Enhancement of multisystem interactions among employer, healthcare, and rehabilitation systems, and the insurance and compensation system, to improve joint clinical and occupational outcomes in mental health disability, bridging the chasm between medical model-based mental health services and employer-based vocational services; and
3. At the employer level, key interventions need to be performed from both primary and secondary prevention perspectives:

   (a) Primary prevention in the workplace should focus on promotion of the health and wellness of a diverse work force; awareness raising and anti-stigma initiatives, combined with the development of written policies on relevant areas including job accommodations reflective of management commitment; appropriate organizational restructuring to integrate various health, human resource, and occupational services within the company in the service of accommodating and retaining employees with mental health disabilities; budgeting for accommodations and training management in early detection and intervention with employees with mental health disabilities, and in job accommodations;

   (b) Secondary prevention should target both supported employment for employees with persistent and serious mental health disabilities and an expanded disability management approach combined with modern, proactive, and biopsychosocial vocational rehabilitation approaches for a wide range of disabling mental health conditions.

Workplace-based interventions should be integrated with clinical interventions at the individual level to enhance a person's work readiness and work retention. Such interventions include early diagnosis and intervention, client-centered assessment and planning, self-awareness and disclosure counseling, training in coping skills, job coaching, and social skill and social network development.

Optimal approaches to effective job accommodation for persons with disabilities should include the following steps and features:

1. Conduct assessments: following disclosure of a mental health disability, utilize qualified individuals to conduct workplace-based functional ("ecological") assessments collaboratively with workers, management, and unions to identify job accommodation needs using an individualized, worker-centered approach

that emphasizes the employee's capacities, strengths, and compensatory skills that promote job performance and retention.

2. Offer training: in addition to training of direct line supervisors and coworkers on how to best promote mental health through social relations in the workplace, orientation, training, and supervision for new employees with mental health disabilities including training in feedback giving and getting, problem-solving and realistic goal-setting. This training can be generic and targeted toward the specific needs of individuals with disabilities.

3. Enhance social support: increase workplace social support and utilize natural supports at work (coworkers) by engaging coworkers as trainers and mentors; recognize negative workplace attitudes; improve awareness and educational opportunities in mental health disabilities for coworkers and management; when appropriate, provide employees with sensitivity training on mental health disabilities and how to work with individuals who may be exhibiting overt symptoms or have cognitive difficulties, such as distractibility, short-term memory problems, and impairments in organization.

4. Promote flexibility: be flexible with job accommodations that involve modification of work task requirements; changes to the work environment and location; scheduling and hours of work; use, frequency, and duration of breaks; time to complete work; work organization; and the application of assistive technology.

5. Use an employment specialist: in the case of complex or unfamiliar work conditions, or challenging labor-relations situations, engage an internal or external employment/return-to-work/vocational rehabilitation/case management specialist familiar with accommodations and job retention issues among persons with mental health disabilities.

6. Use multidisciplinary resources: in complex cases, particularly in serious mental illness, it is important to ensure that the employee has access to multidisciplinary resources to help manage work performance and changes at work, monitor the status of their symptoms, assess the effectiveness of workplace accommodations, and identify other treatment and occupational needs including expanded EAP interventions. It is also important to ensure that workers are collaborating in the process of maintaining both their physical and their psychological "wellness" by following recommended treatments and using self-management strategies aimed at optimal health. However, organizational readiness, availability of resources, and logistical issues may present barriers to this complex undertaking.

7. Protecting privacy and confidentiality: be sensitive to the employee's possible guardedness around mental health issues and the need to protect their privacy and confidentiality (MacDonald-Wilson et al. 2010). The employer is not entitled to know the worker's diagnosis, but should understand the functional capacities and limitations that arise from the diagnosis and that affect work performance.

8. Provide performance evaluations and feedback: recognize functional problems arising from the employee's mental health disability and differentiate them from true underperformance that requires a disciplinary approach. Identify

environmental, management, and social factors in the workplace that may help the employee enhance performance and implement desired changes. Recognize the role of stress in reduced function among persons with mental health disabilities, and the types of stressors that are problematic. Provide regular supportive feedback to the employee, and use both supervisors and peers as coaches. Ensure access to special skills training, for example, organizational skills, and assertiveness or conflict resolution skills that the employee may require to augment the utility of job accommodations and their overall job performance (Schultz et al. 2010a).

Last but not least, there is an international need for advanced (both undergraduate and graduate) interdisciplinary training of vocational rehabilitation, disability management, and return-to-work professionals, using integrated biopsychosocial and evidence-informed models of service delivery to persons with mental health disabilities who desire to enter the workforce or who are already in the workplace. This critical educational gap may undermine future efforts to include persons with mental health disabilities in the workplace.

Future Research Directions

The literature and the chapters contributed to this book identified significant research gaps in the following areas:

1. Methodology: There is a lack of well-developed transdisciplinary methodologies for research on multisystem interactions affecting employment outcomes, job retention, and job accommodations among workers with mental health disabilities.
2. Research paradigms: Most of the existing research is based upon a psychopathology/medical model. Typically, funding agencies sponsor studies in either clinical or vocational domains, thus reinforcing the chasm between clinical and occupational research. There is no consistent research funding with a mandate for bridging these two paradigms.
3. Scope of diagnostic categories: Serious mental illness, anxiety and mood disorders, and brain injury are some of the most researched diagnoses; very little relevant research has been conducted in highly prevalent pervasive neuropsychological conditions affecting both cognitive and emotional functioning in adults, such as learning disorders and autism spectrum disorders, although some literature on attention deficit disorder is emerging.
4. Predictors of work disability: Research on predictors of occupational disability in mental health, except for emerging studies in brain injury, posttraumatic stress disorder, and depression, has been limited. Research does suggest that diagnoses along the schizophrenia spectrum are associated with poorer vocational outcomes; cognitive difficulties and low educational attainment also predict poorer vocational outcomes (Rogers and MacDonald-Wilson 2010).

5. Functional assessment: There is a lack of validated assessment instruments that measure functional work capacity; existing psychological and neuropsychological tests are diagnostically rather than functionally oriented. Yet standardized and validated functional "ecological" assessments are critical for the development of job accommodations and to facilitate future research in the field (Rogers and MacDonald-Wilson 2010).

6. Organizational interventions and job accommodations: There is a lack of research on which primary and secondary workplace prevention programs are associated with improved job retention and accommodations, and under what conditions (who, where, when, and what) they are effective. Moreover, there is no consensus on how to measure clinical, occupational, and employment outcomes in the current literature.

7. Effectiveness of job accommodations: Research on the effectiveness of job accommodations is in the emerging stage; methodological difficulties include small sample sizes, the use of samples of convenience, a lack of tools for pooling data from several studies, a lack of control groups and randomized designs, difficulties in standardizing protocols for job accommodations in workplace environments, and the multitude of extraneous ("environmental noise") factors affecting outcomes, together with difficult-to-capture intra- and interorganizational and clinical treatment factors. Moreover, multivariate and multilevel analysis models are underutilized in the area of job accommodations. In addition, effectiveness of accommodations is likely dependent on the type and stability of functional work limitations, timing of interventions, consistency of application, a variety of employer-related and coworker-related support factors, as well as task demands and worker control over the work tasks. Such conditions and factors demand difficult, labor-intensive and costly-to-execute research designs. A combination of qualitative and quantitative research studies, using mixed methods designs, undertaken by teams of researchers in different geographic locations yet using the same methodology, is likely to be a most promising approach.

8. Include the voice of the consumer: There is a need to develop more understanding of the lived experience of working with a mental health disability. Attention needs to be directed to how peer support and social interaction-related approaches and strategies can be effectively used in the workplace context.

In conclusion, despite major societal, clinical, employer, and research advances in the area of mental health in the workplace, a major gap continues to exist between healthcare services and the occupational needs of workers with mental health disabilities on the one hand; and research evidence on what works with whom, where, and when in the workplace on the other. A new research paradigm for mental health disabilities in the workplace, integrating multisystem research with combined clinical and occupational approaches to disability, is needed. Promising research paradigms and examples can be found in the more advanced field of return-to-work in the area of musculoskeletal disability (Feuerstein 2005; Franche et al. 2005; Linton et al. 2005; Loisel et al. 2005a, b; Pransky et al. 2005; Schultz and Gatchel 2005; Sullivan et al. 2005). Such new paradigms need to be supported, however, by research funding whose current mandates parallel the existing, yet removable, divide between clinical and workplace research.

Only by integrating the efforts of researchers, policy-makers, healthcare practitioners, employers, educators, disability compensation systems, and persons with mental health disabilities can the challenge of mental health disability in the workplace be addressed.

Acknowledgement Parts of this chapter were informed by research reports prepared for a Social Development Canada-funded project, "Towards Evidence-Informed Best Practice Guidelines for Job Accommodations for Persons with Mental Health Disabilities."

References

Allen P (2004) For the employer productivity is critical. Healthc Pap 5(2):95–99
Anthony WA, Blanch AK (1989) Research on community support services: what have we learned? Psychosoc Rehabil J 12(3):55–81
Anthony WA, Cohen MR, Farkas M, Gagne C (2002) Psychiatric rehabilitation, 2nd edn. Boston University, Center for Psychiatric Rehabilitation, Boston, MA
Arthur AR (2000) Employee assistance programs: the emperor's new clothes of stress management. Br J Guid Counsell 28(4):549–559
Baldwin MW, Marcus SC (2010) Stigma, discrimination, and employment outcomes among persons with mental health disabilities. In: Schultz IZ, Rogers ES (eds) Handbook of work accommodation and retention in mental health. Springer, New York
Bender A, Kennedy S (2004) Mental health and mental illness in the workplace: diagnostic and treatment issues. Healthc Pap 5(2):54–68
Berndt ER, Finkelstein SN, Greenberg PE, Howland RH, Keith A, Rush AJ et al (1998) Workplace performance effects from chronic depression and its treatment. J Health Econ 17:511–535
Bishop M (2004) Determinants of employment status among a community-based sample of people with epilepsy: implications for rehabilitation interventions. Rehabil Counsel Bull 47(2):112–120
Bond GR (1998) Principles of individual placement and support model: empirical support. Psychiatr Rehabil J 22:11–23
Bond GR, Becker DR, Drake RE, Rapp CA, Meisler N, Lehman AF et al (2001) Implementing supported employment as an evidence-based practice. Psychiatr Serv 52(3):313–322
Bond GR, Drake RE, Mueser KT, Becker DR (1997) An update on supported employment for people with severe mental illness. Psychiatr Serv 48(3):335–346
Bond GR, Drake RE, Becker DR (2008) An update on randomized controlled trials of evidence based supported employment. Psychiatr Rehabil J 31(4):280–290
Bruyère SM, (ed)(2000) Employing and accommodating workers with psychiatric disabilities. [Brochure]. Ithaca, NY: Cornell University. (Originally written in 1994, and updated in 2000 by LL Mancuso. http://digitalcommons.ilr.cornell.edu/edicollect/5
Burke-Miller J, Cook J, Razzano L (2010) Integration of mental health and vocational services. In: Schultz IZ, Rogers ES (eds) Handbook of work accommodation and retention in mental health. Springer, New York
Carling PJ (1993) Reasonable accommodations in the workplace for individuals with psychiatric disabilities. Consult Psychol J 45(2):46–67
Center C (2010) Law and job accommodation in mental health disability. In: Schultz IZ, Rogers ES (eds) Handbook of work accommodation and retention in mental health. Springer, New York
Cook JA, Pickett SA (1994) Recent trends in vocational rehabilitation for people with psychiatric disability. Am Rehabil 20(4):2–12

Crowther RE, Marshall M, Bond GR, Huxley P (2001a) Helping people with severe mental disorder obtain work: systematic review. Br Med J 322:204–208

Crowther RE, Marshall M, Bond GR, Huxley P (2001b) Vocational rehabilitation for people with severe mental illness (Cochrane Review). In: The Cochrane Library, 4, Update Software, Oxford

Currier KF, Chan F, Berven NL, Habeck RV, Taylor DW (2001) Functions and knowledge domains for disability management practice: a delphi study. Rehabil Counsel Bull 44(3):133–143

Dewa CS, McDaid D (2010) Investing in the mental health of the labour force: epidemiological and economic impact of mental health disabilities in the workplace. In: Schultz IZ, Rogers ES (eds) Handbook of work accommodation and retention in mental health. Springer, New York

Drake RE, Becker DR, Bond GR (2003) Recent research on vocational rehabilitation for persons with severe mental illness. Curr Opin Psychiatry 16(4):451–455

Drake RE, McHugo GJ, Becker DR, Anthony WA, Clark RE (1996) The New Hampshire study of supported employment for people with severe mental illness. J Consult Clin Psychol 64:391–399

Feuerstein M (2005) Introduction: the world challenge of work disability. J Occup Rehabil 15(4):451–452

Finch JR, Wheaton JE (1999) Patterns of services to vocational rehabilitation consumers with serious mental illness. Rehabil Counsel Bull 42(3):214–228

Fraser RT, Strand D, Johnson C (2010) Employment interventions for persons with mild cognitive disorders. In: Schultz IZ, Rogers ES (eds) Handbook of work accommodations and retention in mental health. Springer, New York

Franche R-L, Baril R, Shaw W, Nicholas M, Loisel P (2005) Workplace-based return-to-work interventions: optimizing the role of stakeholders in implementation and research. J Occup Rehabil 15(4):525–542

Gallie KA, Schultz IZ, Winter A (2010) Company-level interventions in mental health. In: Schultz IZ, Rogers ES (eds) Handbook of work accommodation and retention in mental health. Springer, New York

Gates LB (2000) Workplace accommodation as a social process. J Occup Rehabil 10(1):85–98

Gates LB, Akabas SH (2010) Inclusion of people with mental health disabilities into the workplace: accommodation as a social process. In: Schultz IZ, Rogers ES (eds) Handbook of work accommodation and retention in mental health. Springer, New York

Gerber PJ, Ginsberg R, Reiff HB (1992) Identifying alterable patterns in employment success for highly successful adults with learning disabilities. J Learn Disabil 25:475–487

Gerber PJ, Price LA (2003) Persons with learning disabilities in the workplace: what we know so far in the Americans with Disabilities Act era. Learn Disabil Res Pract 18(2):132–136

Global Business and Economic Roundtable on Addiction and Mental Health (2005) http://www.mentalhealthroundtable.ca/about_us.html. Accessed 12 Sept 2005

Gnam WH (2005) The prediction of occupational disability related to depressive and anxiety disorders. In: Schultz IZ, Gatchel RJ (eds) Handbook of complex occupational disability claims early risk identification, intervention, and prevention. Springer, New York, pp 371–384

Goldberg RJ, Higgins EL, Raskind MH, Herman KL (2003) Predictors of success in individuals with learning disabilities: a qualitative analysis of a 20-year longitudinal study. Learn Disabil Res Pract 18:222–236

Habeck RV, Hunt HA (1999) Disability management perspectives. Developing accommodating work environments through disability management. Am Rehabil 25:18–25

Haffey WJ, Abrams DL (1991) Employment outcomes for participants in a brain injury work re-entry program: preliminary findings. J Head Trauma Rehabil 6(3):24–34

Harder H, Hawley J, Stewart A (2010) Disability management approach to job accommodation for mental health disability. In: Schultz IZ, Rogers ES (eds) Handbook of work accommodation and retention in mental health. Springer, New York

Haslan C, Atkinson S, Brown SS, Haslan RA (2005) Anxiety and depression in the workplace: effects on the individual and organization (a focus group investigation). J Affect Disord 88:209–215

Holzberg E (2001) The best practice for gaining and maintaining employment for individuals with traumatic brain injury. Work 16:245–258

Johnstone B, Vessell R, Bounds T, Hoskins S, Sherman A (2003) Predictors of success for state vocational rehabilitation clients with traumatic brain injury. Arch Phys Med Rehabil 84:161–167

Kirkwood AD, Stamm BH (2006) A social marketing approach to challenging stigma. Prof Psychol Res Pract 37(5):472–476

Kirsh B, Gewurtz R (2010) Organizational culture and work issues for individuals with mental health disabilities. In: Schultz IZ, Rogers ES (eds) Handbook of work accommodation and retention in mental health. Springer, New York

Krupa T (2010a) Approaches to improving employment outcomes for people with serious mental illness. In: Schultz IZ, Rogers ES (eds) Handbook of work accommodation and retention in mental health. Springer, New York

Krupa T (2010b) Employment and serious mental health disabilities. In: Schultz IZ, Rogers ES (eds) Handbook of work accommodation and retention in mental health. Springer, New York

Krupa T (2007) Interventions to improve employment outcomes for workers who experience mental illness. Can J Psychiatry 52(6):339–345

Krupa T, Kirsh B, Gerwurtz R, Cockburn L (2005) Improving the employment prospects of people with serious mental illness: five challenges for a national mental health strategy. [Special electronic supplement on mental health reform for the 21st Century in partnership with the School of Policy Studies and with the Centre for Health Services and Policy Research of Queen's University, Kingston, ON]. Canadian Public Policy, 31(Suppl.), 60–63. http://economics.ca/cgi/jab?journal=cpp&view=v31s1/CPPv31s1p059.pdf

Lerner D, Adler D, Hermann RC, Rogers WH, Change H, Thomas P et al (2010) Depression and work performance: the work and health initiative study. In: Schultz IZ, Rogers ES (eds) Handbook of work accommodation and retention in mental health. Springer, New York

Linton SJ, Gross D, Schultz IZ, Main C, Côté P, Pransky G et al (2005) Prognosis and the identification of workers risking disability: research issues and directions for future research. J Occup Rehabil 15(4):459–474

Loisel P, Buchbinder R, Hazard R, Keller R, Scheel I, van Tulder M et al (2005a) Prevention of work disability due to musculoskeletal disorders: the challenge of implementing evidence. J Occup Rehabil 15(4):507–524

Loisel P, Durand M-J, Baril R, Gervais J, Falardeau M (2005b) Interorganizational collaboration in occupational rehabilitation: perceptions of an interdisciplinary rehabilitation team. J Occup Rehabil 15(4):581–590

MacDonald-Wilson KL, Rogers ES, Anthony WA (2001) Unique issues in assessing work function among individuals with psychiatric disabilities. J Occup Rehabil 11(3):217–232

MacDonald-Wilson KL, Rogers ES, Massaro J (2003) Identifying relationships between functional limitations, job accommodations, and demographic characteristics of persons with psychiatric disabilities. J Vocat Rehabil 18:15–24

MacDonald-Wilson KL, Rogers ES, Massaro J, Lyass A, Crean T (2002) An investigation of reasonable workplace accommodations for people with psychiatric disabilities: quantitative findings from a multi-site study. Community Ment Health J 38(1):35–50

MacDonald-Wilson KL, Russinova Z, Rogers ES, Lin CH, Ferguson T, Dong S et al (2010) Disclosure of mental health disabilities in the workplace. In: Schultz IZ, Rogers ES (eds) Handbook of work accommodation and retention in mental health. Springer, New York

Macias C, DeCarlo LT, Wang Q, Frey J, Barreira P (2001) Work interest as a predictor of competitive employment: policy implications for psychiatric rehabilitation. Adm Policy Ment Health 28(4):279–297

Marrone J, Gandolfo C, Gold M, Hoff D (1998) Just doing it: helping people with mental illness get good jobs. J Appl Rehabil Counsel 29(1):37–48

Means CD, Stewart SL, Dowler DL (1997) Job accommodations that work: a follow-up study of adults with attention deficit disorder. J Appl Rehabil Counsel 28(4):13–16

Mueser KT, Becker DR, Wolfe R (2001) Supported employment, job preferences, job tenure and satisfaction. J Ment Health 10(4):411–417

Murakami Y (1999) Job coaching for persons with mental disabilities. Work 122(2):181–188

Nimgade A, Costello MC (2003) Return to work for a company president with traumatic brain injury. J Head Trauma Rehabil 18(5):464–7

Peer J, Tenhula W (2010) Employment interventions for mood and anxiety disorders. In: Schultz IZ, Rogers ES (eds) Handbook of work accommodation and retention in mental health. Springer, New York

Pransky G, Gatchel R, Linton SJ, Loisel P (2005) Improving return to work research. J Occup Rehabil 15(4):453–458

Radley J (2004) Three sessions and back to work. Occupational Therapy Now, 6(5)

Rogers ES, Anthony WA, Farkas M (2006) The choose-get-keep approach to psychiatric rehabilitation. Rehabil Psychol 51(3):247–256

Rogers ES, MacDonald-Wilson KL (2010) Vocational capacity among individuals with mental health disabilities. In: Schultz IZ, Rogers ES (eds) Handbook of work accommodation and retention in mental health. Springer, New York

Scheid TL (1998) The ADA, mental disability, and employment practices. J Behav Health Serv Res 25(3):312–325

Schultz IZ, Crook J, Fraser K, Joy P (2000) Models of diagnosis and rehabilitation in pain-related occupational disability. J Occup Rehabil 10(4):271–293

Schultz IZ, Duplassie D, Hanson D, Winter A (2010a) Systemic barriers and facilitators to job accommodations in mental health: experts' consensus. In: Schultz IZ, Rogers ES (eds) Handbook of work accommodation and retention in mental health. Springer, New York

Schultz IZ, Gatchel RJ (2005) Research and practice directions in risk for disability prediction and early intervention. In: Schultz IZ, Gatchel RJ (eds) Handbook of complex occupational disability claims. Early risk identification, intervention, and prevention. Springer, New York

Schultz IZ, Milner R, Hanson D, Winter A (2010b) Employer attitudes towards accommodations in mental health disability. In: Schultz IZ, Rogers ES (eds) Handbook of work accommodation and retention in mental health. Springer, New York

Schultz IZ, Stowell AW, Feuerstein M, Gatchel RJ (2007) Models of return to work for musculoskeletal disorders. J Occup Rehabil 17:327–352

Schultz IZ, Winter A, Wald J (2010c) Evidentiary support for best practices in job accommodation in mental health: macrosystem, employer-level and employee-level interventions. In: Schultz IZ, Rogers ES (eds) Handbook of work accommodation and retention in mental health. Springer, New York

Scully SM, Habeck RV, Leahy MJ (1999) Knowledge and skill areas associated with disability management practice for rehabilitation counselors. Rehabil Counsel Bull 43(1):20–29

Sherbourne CD, Wells KB, Duan N, Miranda J, Unutzer J, Jaycox L et al (2001) Long-term effectiveness of disseminating quality improvement for depression in primary care. Arch Gen Psychiatry 58:696–703

Smith JL, Rost KM, Nutting PA, Libby AM, Elliott CE, Pyne JM (2002) Impact of primary care depression intervention on employment and workplace conflict outcomes: is value added? J Ment Health Policy Econ 5:43–49

Smith TE, Bellack AS, Liberman RP (1996) Social skills training for schizophrenia: review and future directions. Clin Psychol Rev 16(7):599–617

Standing Senate Committee on Social Affairs, Science and Technology (2006) Out of the shadows at last. Transforming mental health, mental illness and addiction services in Canada (Final report, pp. 1–484). Government of Canada Printing Office, Ottawa, ON

Sullivan MJL, Feuerstein M, Gatchel R, Linton SJ, Pransky G (2005) Integrating psychosocial and behavioral interventions to achieve optimal rehabilitation outcomes. J Occup Rehabil 15(4):475–490

Szymanski E, Vancollins J (2003) Career development of people with disabilities: some new and not-so-new challenges. Aust J Career Dev 12:9–16

Targett P, Wehman P (2010) Return to work after traumatic brain injury: a supported employment approach. In: Schultz IZ, Rogers ES (eds) Handbook of work accommodatins and retention in mental health. Springer, New York

Tryssenaar J, Goldberg J (1994) Improving attention in a person with schizophrenia. Can J Occup Ther 61(4):198–205

Tse S (2002) Practice guidelines: therapeutic interventions aimed at assisting people with bipolar affective disorder achieve their vocational goals. Work 29(2):167–179

US Department of Labor Job Accommodation Network (2007) Accommodation information by disability. Morgantown, WV. [JAN is a contracted service funded by the U.S. Government, Department of Labour, Office of Disability Employment Policy under agreement J-9-M-2-0022]. Electronic version retrieved from http://www.jan.wvu.edu/media/atoz.htm

Vézina M, Bourbonnais R, Brisson C, Trudel L (2004) Workplace prevention and promotion strategies. Healthc Pap 5(2):32–44

Wallace CJ, Tauber R, Wilde J (1999) Teaching fundamental workplace skills to persons with serious mental illness. Psychiatr Serv 50(9):1147–1153

Wald J, (2010) Anxiety disorders and work performance. In:Schultz IZ, Rogers, ES (eds) Handbook of work accommodations and retention in mental health. Springer, New York

Wang JL (2010) Mental health literacy and stigma associated with depression in the working population. In: Schultz IZ, Rogers ES (eds) Handbook of work accommodation and retention in mental health. Springer, New York

Warr PB (1987) Work, unemployment and mental health. Oxford University Press, Oxford, UK

Westmorland M, Buys N (2002) Disability management in a sample of Australian self-insured companies. Disabil Rehabil 24(14):746–754

Wisenthal A (2004) Occupational therapy provides the bridge back to work. Occupational Therapy Now 6(5)

Yasuda S, Wehman P, Targett P, Cifu D, West M (2001) Return to work for persons with traumatic brain injury. Am J Phys Med Rehabil 80(11):852–864

Index

LaVergne, TN USA
14 January 2011
212465LV00003B/30/P

9 781441 904270